By Ingrid Cranfield:
The Challengers: British and Commonwealth Adventure since 1945

Edited by Ingrid Cranfield and Richard Harrington:
Off the Beaten Track: A WEXAS Travel Handbook

The Independent Traveller's Handbook

edited by Ingrid Cranfield in association with Richard Harrington

More than a handbook, this is the traveller's 'Bible'. Some 45 experts, many internationally famous, have lent their expertise and experience to this work. Their contributions cover all aspects of travel, from riding a camel to hitchhiking, from cooking to survival, from buying a secondhand camera to renting a car overseas. The book's coverage encompasses many of the world's less accessible places—from Kathmandu to Tombouctou, from Bogotà to Buenos Aires.

Over 150 pages of reference material include maps, a summary of useful reading, a directory of services, words and phrases in eleven languages, information on weather, currencies, shipping companies, mapping agencies, motoring organizations—and many other lists and tables.

A book to help you with planning and preparation, to occupy a prominent place in your backpack, suitcase, or car, to cheer you on and cheer you up and give you ideas about saving, changing, spending, and even making that all-important travel commodity: money. A book even to save your life, maybe.

Read it for amusement, refer to it for facts and figures, rely on it for good advice. Whether first-timer or seasoned traveller, you cannot fail to learn something from it—or afford to leave home without it.

THE INDEPENDENT TRAVELLER'S HANDBOOK

Edited by
INGRID CRANFIELD

in association with
RICHARD HARRINGTON

with a foreword by Sir Edmund Hillary, K.B.E.

Published by
HEINEMANN : LONDON
and
WEXAS INTERNATIONAL LTD.

William Heinemann Ltd.
10 Upper Grosvenor Street, London W1X 9PA

LONDON MELBOURNE TORONTO
JOHANNESBURG AUCKLAND

First published 1980

434 14826 1

Printed in Great Britain by The Pitman Press, Bath

This Handbook is dedicated to the Marco Polo spirit that inhabits every real traveller.

A man must carry knowledge with him, if he would bring home knowledge.

Samuel Johnson

Contents

PART 5 GETTING IT TOGETHER: Fit and ready to travel

PART 6 TAKING IT SERIOUSLY: Expeditions and other travel with a purpose

PART 7 KEEPING OUT OF TROUBLE: Stumbling-blocks

PART 8 KEEPING TRACK: Travel spin-offs and records

SECTION II MAPS

SECTION III DIRECTORIES

SECTION IV REFERENCE

List of Maps

Foreward

by SIR EDMUND HILLARY, K.B.E.

Those of us fortunate enough to have been involved in major adventures are continually being asked the question 'What is there left to do?'. The poles have been reached; the highest mountains climbed; the ocean depths penetrated; and men have walked on the moon.

Much more careful examination of maps is needed before territories may be found that are unknown to adventurous man. But unexplored places do still exist and modern technology makes possible the overcoming of difficulties that defied the abilities of previous explorers.

In the last generation there has been something of an adventure boom, with the new freedoms of leisure and financial opportunity permeating every layer of our society. The motives that drove great explorers in the past are as strong as ever in their modern counterparts—but not only are the great adventures being pursued. Thousands of people, young and old, are finding satisfaction in more modest undertakings. Imagination and enthusiasm are still vital factors, but for the average adventurer they must be supplemented by practical information, a degree of business acumen, some specialised expertise and even a sensitive nose for publicity.

The adventurous spirit continues to exist in all its shades, from those seeking physical daring for its own sake to the comfortable citizen anxious for a temporary respite from the pressures of society or the specialist pursuing evidence for his theories in the world's remoter places.

Creativity and originality are important factors in adventure but foolhardiness is not to be recommended. Every modern explorer, of whatever complexion, is wise to minimise unnecessary physical risk, discomfort and expense. There is a wide fund of previous experience that can be called on to ensure sound and sensible procedures. Thanks to the emergence of information pools such as WEXAS there is a mass of help available to the would-be off the beaten track traveller. The most relevant currently available material is contained in this handbook.

Edmund Hillary

Preface

The Independent Traveller's Handbook started life as a new edition of *Off the Beaten Track,* but like a character in a novel, it soon displayed a fine direction of its own and exerted a personality to match. Readers of *Off the Beaten Track* 1977 may perhaps just barely detect a trace of the parent in the progeny. Of some eighteen articles contained in *Off the Beaten Track,* nine have found their way into the *Handbook.* All the other articles in the *Handbook,* with one exception, have been specially commissioned for it from contributors chosen for their expertise, their shared philosophy of independent travel and their ability to communicate. Collectively, these articles cover as wide a field as the distance between these two covers will allow.

The directory of services and book list have expanded considerably and become more international in flavour; and the number of lists, charts and tables has more than doubled.

Despite the change of name, the book is still aimed largely at the independent traveller off the beaten track This term is truly mercurial: try to put your finger on it and it slips away. One man's beaten track is, after all, another man's remote El Dorado; one man's wilderness is another's back yard. So instead of trying to pin the term down, we've put a cage around it.

We have assumed that our readers are mainly British, North American, European or Australasian and that to them going off the beaten track means going abroad. There are unspoilt and even untrodden areas on every continent, but in the context of this book Britain, Europe and North America are virtually excluded as destinations. Australia and New Zealand are a kind of halfway house, being both a source of numbers of independent travellers and an off the beaten track part of the world to others. For no very good reason, the USSR receives very little attention here.

Another somewhat arbitrary distinction has had to be made between adventure sports and adventure travel. In theory the division is obvious enough; in practice there are grey areas between the two. A traveller may use sporting means of getting from A to B; a sportsman may travel long distances, but the rewards of his travelling are mainly physical, not cultural or social. Broadly speaking, this book aims at the traveller and not at the sportsman. For this reason, we omit all but passing references to kayaking, ballooning, hang-gliding and climbing, but retain sailing, motorcycling and mountain exploration. Similarly, we pay scant attention to pure backpacking of the backwoods variety, but more detailed attention to long distance trekking in a foreign country.

Even the word 'independent' has its degrees of meaning. There are travellers so independent they won't even glance at a travel handbook; others who are independent-minded but happy to leave practical details of organisation to tour operators and such.

Our reader is most likely to be an independent thinker but willing to learn, enterprising and eager for experience and for real contact with people of other lifestyles, footloose and free but a thorough planner, and careful with his not too overwhelming funds. By using the information and advice in this book, the user will certainly quickly recoup the money he spent to buy it. We hope he may indeed make a profit.

The book is divided into two main sections. The first contains articles by experts that cover administrative, social and personal, financial, geographical and political aspects of the whole course of a trip, from its conception to the time when it has long since been wound up. The second section is a reference work in its own right, containing bibliography, lists, charts, directories and tables.

Since many hundreds of organisations have been named in the book, it is obviously impossible to vouch for the accuracy of the information throughout. The editor would gladly welcome corrections and suggestions from readers. By the same token—sheer weight of numbers—it is not possible to speak for the reliability of the organisations named. Inclusion of the name of a firm or association in this book should not be taken to signify the testimony or approval of the editor or publishers.

Ingrid Cranfield
Editor

Please address any correspondence to:
The Editor
The Independent Traveller's Handbook
WEXAS International
45 Brompton Road
Knightsbridge
London SW3 1DE

Contributors

Major Sir Crispin Agnew of Lochnaw Bt., Royal Highland Fusiliers, was born in 1944 and began his career in exploration 22 years later as a member of the Royal Navy Expedition to East Greenland. He has since led Services expeditions to East Greenland and Chile and been a member of expeditions to Elephant Island, Nuptse—where he reached about 26,000ft—and Everest. He has climbed in the UK and Alps since 1960. He is also Chief of the Clan.

Maurice and Maralyn Bailey were born in 1932 and 1941 respectively. They have been sailing for many years and in 1968 started to live aboard their boat. They are best remembered for their epic 118-day ordeal in the Pacific Ocean after they had been cast adrift in their liferaft when a wounded sperm whale had attacked and sunk their yacht *Auralyn.* Since their rescue they have built *Auralyn II* and completed a 20,000 mile circumnavigation of South America.

Hilary and George Bradt have forsaken their professions of occupational therapist and librarian to become full-time writers and publishers. They produce the Backpacking Guide Series for travellers interested in seeing the world on foot.

John Carlton is a keen walker and rucksack traveller. He has been a member of the YHA for about 25 years and has also worked in the travel trade for the same time. After a long period of service on voluntary committees of the Association, decided eight years ago to bring both interests together and may currently be found behind the counter at YHA. Travel in London. He has visited most countries in Western and Eastern Europe, Morocco, Canada and USA.

Jim Couper is a 35-year-old Scottish-born Canadian who has driven over 90,000 miles in 48 different countries on several trips during which he supported himself by working as an editor, photographer, teacher, landlord and bookstore owner.

Graham Cownie was born in 1956 and spent most of his first ten years in East Africa, where he first got a taste of the expedition and safari tradition. He began climbing while at school and has since been on climbing trips to Scotland, Wales, the Pyrenees and the Alps. He has also been on overland expeditions to the Middle East and Scandinavia. He is an officer in the RAF, currently reading for a Psychology degree.

Ingrid Cranfield was born in Australia where she obtained a degree in Geography. She was Senior Assistant in the Map Room of the Royal Geographical Society from 1968 to 1972 and has subsequently worked as a freelance writer, editor and translator. She is the author of *The Challengers: British and Commonwealth Adventure since 1945* (Weidenfeld & Nicolson 1976) and was co-editor with Richard Harrington of the WEXAS handbook *Off the Beaten Track* (WEXAS and Wilton House Gentry, 1977).

René Dee was born in Switzerland in 1946 and educated in Britain, where he became a regular soldier in the Intelligence Corps and served in Singapore and Malaysia. In 1967, after leaving the army, he travelled overland to India and Nepal and in 1969 led a series of three-week adventure holidays to Morocco, from which stemmed his love of desert travel. By 1972 he had formed his own company specialising in treks by camel and mule, which ceased to trade owing to economic hardships in 1975. His private expedition *Caravan 77,* intended to cross the western Sahara by camel caravan, had to be postponed owing to political unrest. He is now Marketing and Promotions Manager for Treasure Treks Ltd.

John Douglas, author and photographer, is a Director of Geoslides Photo Library. He is a travel consultant specialising in the sub-Arctic and in the Third World.

Gerald Dunn is a Fellow of the Royal Geographical Society, a member of the Royal Institute of Navigation and the inventor of a dial for solar navigation, commercially known as the Cruiserfix, which is a simple and effective alternative to the magnetic compass for navigating trackless terrain.

Dr. John Frankland is a General Practitioner who has been a caver for 26 years and a medical adviser to the Cave Rescue Organisation for 13 years. He has advised many British caving expeditions and explored caves in Europe and North and South America.

Jon Gardey grew up in California, lived in Alaska and England and now resides in the USA. He is the author of a book on Alaska and his writing and photography have appeared in most of the major magazines in the world. He travels regularly to the remote parts of the globe.

David Gillespie was born in 1947, obtained a degree in Business Studies and worked as a computer systems analyst until the travel bug bit him. He travelled extensively privately and then slowly became involved with the commercial travel and expedition world. Onetime bo'sun on oceangoing yachts, he is now Operations Manager for Exodus Travels Ltd., running overland expeditions.

Robin Hanbury-Tenison is a well-known explorer, author and broadcaster. He has taken part in over a dozen major expeditions in South America, Africa and the Far East, served on the Council of the Royal Geographical Society and, in 1977/8, led their largest scientific expedition ever to the Gunong Mulu National Park in Borneo. He is Chairman of Survival International, the organisation that seeks to prevent the extinction of the world's remaining tribal groups.

Jo Hanson worked with horses as a teenager until a broken leg gave her time off to write three pony books. She travelled round the world on the proceeds, getting a job in a mental hospital in Australia and as a journalist in India. Married with one child, she travels whenever possible, preferably to Africa.

Richard Harrington is a widely travelled freelance travel writer.

Philip Harrison is an ex-Army man who now works in the insurance business. He has raced boats for many years and is a member of the Royal Ocean Racing Club. Since 1975 he has been Honorary Secretary of the Scientific Exploration Society.

David Hodgson, AIIP, was senior staff photojournalist and later picture editor with Features International. He has travelled throughout the world on assignments and his work has appeared in many major magazines, including *Life, Match* and *Stern.* He is the author of eight books on photography.

Pearl Hoffman is President of Pearl's Freighter Trips, Inc., of Great Neck, USA, and an authority on freighter travel.

Jack Jackson is an experienced expedition leader and overland traveller. He is co-author, with Ellen Crampton, of *The Asian Highway* (London, Angus & Robertson, 1979), a detailed and authoritative manual for any overlander.

Diane Johnson joined the YHA at fourteen. Her travels have taken her overland to India, on the Trans-Siberian Railway to Japan, to Iceland, Morocco, Kenya, Tanzania, Southern Africa, Israel and North America, staying in hostels wherever possible. She works as YHA Education Liaison Officer in London.

Gene Kirkley is a 53-year-old native Texan with a degree in forestry. He is a freelance outdoor writer and photographer in hunting, fishing, boating, camping and travel, a Director of the Outdoor Writers Association of America, member of the Southeastern Outdoor Press Association and past President of the Texas Outdoor Writers Association. Contributor to *Outdoor Life, Southern Outdoors, All Outdoors, Texas Sportsman, Texas Fisherman* and other publications. Writes a weekly newspaper column 'East Texas Outdoors' and does four 15-minute radio shows each week by the same name. He is active in public relations in the outdoor field and is also a rodeo announcer.

John D. Lindstrom was born in Eugene, Oregon, in 1938 and has explored in the Cascade Range of Oregon and the North Cascades of Washington. He has a BA in Biology and an MA in Photojournalism and now teaches photography at Everett Community College, Washington State. He is most interested in geology, mountain exploration and 35mm photography.

Ian McQueen is a Canadian engineer who first travelled to Japan for Expo 70 in Osaka and remained in the country for fifteen months, travelling 12,000 miles by motorcycle. He returned in March 1978 to research a guide book on Japan for publication late in 1979 by Lonely Planet Publications of South Yarra, Australia.

John Michael is a 42-year-old Welshman, a cartographer for the Department of the Environment, based in Nottingham. He is a keen photographer and has been a four-wheel-drive 'fanatic' since 1972 when he purchased a Carawagon from new. He has done 140,000 miles since, much of it abroad. He has led expeditions to the Bilma Sands (1976-7) and Trans-Australia (1978) and gives audio-visual lectures on his trips.

Wendy Myers was born in Sheffield, England, in 1941 and left home at the age of 18 to see the world. Seven years later she returned home. The story of her hitchhiking, living and working experiences abroad is recounted in her book *Seven League Boots* (Hodder & Stoughton, 1969). She is a SRN, a State Certified Midwife, a Health Visitor and has studied at the Hospital for Tropical Diseases, London. She has done medical work in the Solomon Islands and Upper Volta and is currently employed by the Save the Children Fund in Nepal.

The late **Henry Myhill** was the author of *Motor Caravanning–A Complete Guide* (Ward Lock) and overseas correspondent of *Motor Caravan and Camping.*

Bob Nance is an Australian who took up sailing in 1963, when he was 24. He has since cruised nearly 100,000 miles, including three passages around Cape Horn. For five years he worked on tugs delivering fuel along the northwest coast of British Columbia.

Bez Newton has been travelling—usually by the least expensive means available to him—since he was sixteen. Since that time he has supported himself by a variety of occupations, the most satisfying of which, he says, have been editing a surfing magazine, writing paperback novels and freelance journalism.

Chris Parrott was born in 1946 in Aldershot, England, and has lived in France, Singapore, Spain and Brazil, as well as travelling extensively in Europe, the Middle East and the Americas. Until July 1977 he was Head of Geography and Physical Education in the British School, Rio de Janeiro. He is Editor of *Trailfinder Magazine* and occasionally leads overland trips to Peru and Bolivia. Engaged in writing a geography text book of South America.

Dr. John Payne has led three expeditions, to Norway, Spitsbergen and southern Greenland. By profession a psychiatrist, he has been particularly interested in the effects of expeditions upon people, and the inter-relationship of people on expeditions. A paper describing some of his work, entitled *Changes in Self-Concepts of School Leavers on an Arctic Expedition,* has been published in the Journal of Educational Psychology. He is Senior Health Service Officer in the Student Health Service of the London School of Economics and Political Science.

Christopher Portway has been a freelance travel writer for seven or eight years, is the travel editor of a women's magazine and is a member of the Guild of Travel Writers.

After graduating from Durham University with a degree in Geography, **Rowena Quantrill** married and has since lived in Belgium, Cuba and the Philippines. She and her husband have four children, with whom they have travelled widely, including an overland bus trip from Kathmandu to London.

Ray Reece is a tall (6ft 3in) electronics consultant who at the age of 41 set a world record for the fastest cycle circumnavigation of the world: 13,000 road miles in 143 days between June and November 1971. He was educated in Malta and the UK, speaks Arabic fluently and is also a water-polo international. He sings, plays guitar, drums and writes music.

Mary Schantz is a freelance writer who contributes to *Mother Earth News* and writes a regular column on cookery for *Wilderness Camping* magazine. She also teaches 'Trek Out', a YMCA-sponsored camping and wilderness survival programme. She has done most of her camping in Alaska and in the northern Appalachians.

Gilbert Schwartz, a teacher and veteran traveller, spent over a year researching and compiling his book *The Climate Advisor* (Climate Guide Publications, New York), which has met with considerable success and is now in its second printing.

Tom Sheppard, MBE, has covered more than 4,500 miles in desert overlanding and is an expert on four-wheel-drive vehicles. Recipient of the Royal Geographical Society's Ness Award for Exploration following his 1975 West-East Sahara expedition, he left the RAF in 1976 with the rank of Squadron Leader and now works for British Aerospace.

Ted Simon rode a Triumph 500cc motorcycle round the world, starting in 1973 and returning in 1977. He travelled extensively in Africa, Latin

America, Australia and Asia and his articles appeared in the *Sunday Times*. His book on the journey is called *Jupiter's Travels* (Hamish Hamilton, UK, and Doubleday, USA).

Anthony Smith is a zoologist by training and a writer, broadcaster and presenter of television programmes, including the *Wilderness* series on BBC Television. His first expedition was to Iran with an Oxford University team in 1950, since when he has ridden a motor-cycle the length of Africa, written an account of the Royal Geographical Society/Royal Society Mato Grosso Expedition of 1967 and built and flown hydrogen-filled balloons and airships. He was co-founder of the British Balloon and Airship Club and sits on the RGS Expeditions Committee.

Richard Snailham was born in 1930. He read Modern History at Oxford and was a schoolteacher until becoming in 1965 a Senior Lecturer at the Royal Military Academy, Sandhurst. He has been on expeditions to the Middle East, Ethiopia, Zaïre, Jamaica and Ecuador and recently led parties of tourists to Ethiopia and China. He was co-founder with John Blashford-Snell and first Honorary Secretary of the Scientific Exploration Society, is the author of three expedition books and lectures widely on his travel experiences.

Dr. Peter Steele was born in 1935 and qualified as a doctor in 1960. He has climbed in Britain, the Pyrenees, the Alps, North Africa, Mexico, Nepal, and the Cordillera Vilcabamba. He has worked at a mission hospital in Kathmandu and with the Grenfell Flying Doctor Service in northern Labrador, where he travelled the coast by dog team and boat. With his wife, Sarah, he has driven overland to India, and with his family, comprising two children under the age of four, travelled on muleback across the Himalayan kingdom of Bhutan. In 1972 he hitchhiked around South America with his ten-year-old son. He was physician to the International Expedition to Mount Everest in 1971. Author of several books on exploration and exploration medicine, he is now in practice in Whitehorse, Yukon Territory, Canada.

Rick Strange is a 37-year-old professional freelance travel photographer specialising in travel brochures, but he has also sold pictures throughout the world for encyclopaedias, advertisements, calendars, feature articles and text books. He attended Stowe School in Buckinghamshire and studied at the Germain School of Photography in New York, where he obtained eight diplomas and the only Bronze Medal Award for Outstanding Colour Photography to be awarded there in five years. He is now Executive Director of the J. Allan Cash Worldwide Photographic Library in London and has a personal photo collection of over 20,000 colour transparencies from nearly 50 countries.

Adrian Warren has a B.Sc. in Zoology. He first led an expedition, to Mount Roraima, in 1971, when he was only 22. He has been on nine other expeditions to South America and visited the Galapagos Islands and East Africa to study animals. He was a freelance stills and film cameraman before joining the BBC Natural History Unit where he now works.

Peter Fairney Williams is a physical education specialist serving in the Royal Air Force with the rank of Wing Commander. He is the author of seven books on outdoor activities which cover canoeing, hillwalking and all aspects of camping.

Tony Wheeler is a prime example of the dangers of off the beaten track travel. Seven years ago he set off on a little jaunt from England across Asia and still hasn't got back. In the interim he has set up travel guide publishers Lonely Planet Publications and currently has a band of keen travellers scouring the globe for off-beat travel adventures. Tony still reserves at least a couple of trips a year for himself.

Making it Happen

Off Beat Travel

Finding your way through the airfares jungle

by Richard Harrington

'Deregulation' was where it all started back in 1977, and fare-cutting was the name of the game. That year, the first Laker Skytrain took off, and the US Civil Aeronautics Board took the first steps towards making itself ultimately redundant. In 1978, the process went onc stage further, and IATA gave up its power to decide international airfares. From then on, airlines were to set their own fares, subject only to bilateral government approval. It looked as if cheap air travel, which started with the charter boom back in the late 60's, was here to stay.

Now at last the air seems to be clearing a little in the wake of so many new fare structures that even the airlines and travel agents are complaining that they cannot keep up with the innovations. So where does that leave the Average Traveller, keen to get the best value for money, and bewildered by a vast array of fares, especially on the crowded North Atlantic route between Europe and North America?

I won't attempt to hold your hand and lead you carefully through the current fare maze. The whole system is still changing so fast that any very specific examples would only be out of date by the time you read this. However certain broad principles—some new and some old—underlie the system, and these seem to be fairly constant.

Let's start with scheduled services. In 1978, the three tier system was introduced on certain routes and is already spreading to others. Whether you call it First Class, Coach and Economy or whether you employ terms like Budget and Discount for the low cost third class, the result is the same: three classes with services and restrictions (or freedoms) to match. The more you pay, the more you get, not just in quality of in-flight service and seating, but, more important, in terms of what you can do with your ticket—like getting a refund on it, changing the dates, switching to another airline and so on. The three tier system seems likely to stay and to spread to all the most frequently used international routes.

Talking of airline service, you may have noticed that airline *couchettes* are back (last seen in the 50's on Pan Am's Clipper service across the Atlantic). Japan Airlines has *couchettes* available in First Class on its Far East route, provided that you pay a supplement. With a premium First Class service already available on Concorde, it seems that variation in quality of airline service and in ticket type is going to spread in the wake of deregulation. It only remains to be seen whether governments will impose many controls on airline pricing, or will go as far as to allow competitive pricing on the same route. On quality of in-flight service, the free-for-all has already arrived.

The idea of advance purchase tickets started a decade ago with the British Airways Earlybird fares to the Caribbean, spread to the concept of the ABC (Advance Booking Charter) flight, and went from there to the scheduled APEX fare now spreading from one route to another as airlines compete to attract ethnic and tourist markets that are able to plan well in advance and could not afford air travel except at low attractive rates.

Airlines have various names for their advance purchase tickets, but don't be fooled by terms like Maple Leaf, Super Apex, Poundstretcher and Earlybird. The principle is the same for all, whether you have to book 90 or 21 days in advance. In general, dates cannot be changed once you have booked one of these tickets, and on cancellation there may be a hefty refund penalty to pay. In some cases, there may even be no refund at all, so make sure that you have cancellation insurance (as part of a general travel insurance package— never travel uninsured) if you decide to take one of these fares.

Excursion fares have been around a lot longer, and they're still the kind of fare that a lot of leisure (as opposed to business) travellers prefer. In price, these fares fall somewhere between advance purchase tickets and annual economy class tickets. There are fixed minimum and maximum stays allowed (for example 10 to 30 days from Europe to the Middle East), and on some routes there may be two or three fares available with different restrictions. These usually are a matter of different minimum and maximum stays, and availability (or not) of stopovers. On most excursion fares, return dates can be changed within the maximum stay limitation, and part refunds are normally available on cancellation, if only the difference between the round trip excursion fair and the full economy one-way fare. But watch out, for the difference between these figures may be very small indeed.

I've referred above to the new three tier system. In general, passengers travelling on so-called discount tickets (APEX and Standby) fly in the third class section where this system operates. In a two class system, they obviously travel in the second class. Some airlines, like Icelandic, operating between Europe and North America, have a one-class system—all economy class.

Those travelling on economy class tickets or excursion tickets on a three class system travel in the second class. However with certain types of excursion ticket, travel in the third class section may be required.

First class speaks for itself. Note however that when it comes to quality of First Class service, enormous disparities exist between airlines, and some of the gaps are widening.

I've left to last the Standby fare, common in the US in the 60's in the form of the Youth Fare, next seen in the Laker Skytrain concept, and now increasingly prevalent on a variety of routes where airlines seek to sell off at the last minute seats that might otherwise go empty. The idea of the Standby seat was based on the idea that there exists a creature called, by Freddie Laker, the 'Forgotten Man'. The 'Forgotten Student' might have been a better term, since this fare is one that appeals particularly to those with an urge to travel overseas and the time to wait around for the next flight or the one after if the first one or two are full up. Those who have temporarily dropped out of their jobs to see the world, as well as teachers with long vacation periods, are likely to be the main clients for the Standby fare. If you have the necessary patience (avoid peak periods, especially Christmas), if you don't mind checking in many hours in advance on the day of travel and returning for the flight later, then this might be the fare to save you a lot of precious money. Standby check-in systems are likely to show increasing diversity and refinement in the next few years; partly to make Standby more attractive (perhaps to make it appeal to the elderly low income traveller), and partly to prevent hordes of students at airports from slowing down the process of checking in passengers on other fares.

Before deregulation, civil aeronautics regulations aimed at protecting charter and scheduled carriers more or less equally. Governments reckoned that airlines should make profits, and that a profitable airline was a safe one. However politics intervened at times, and it seems that one of the main results of deregulation has been decreasing protection for charter airlines. The answer may have come in the form of the Public Charter, to use the American term. As I write, the concept is still a new one, aimed at replacing the now established ABC concept of charter travel. The idea of Public Charters is little more than a free-for-all, and is tantamount to eliminating altogether the distinction between charter and scheduled services. The idea is basically: no advance booking, tickets valid up to one year, one-ways available, refunds on unused return portions of tickets. In fact, only one thing separates such tickets and services from those of scheduled airlines, and that is a rule debarring charter airlines from selling their seats themselves to the public. The idea of a 'charter' is still in there somewhere. Another company has to charter the aircraft, thus buying the use of *all* the seats, and then sell these off to its clients.

If deregulation goes so far as to allow charter airlines to sell their seats directly to the public, then there is no charterer and these airlines become *de facto* scheduled carriers. Or, since scheduled airlines can and do charter out their aircraft, there would be a situation in which there were airlines either having aircraft operating scheduled services or chartering out planes to anyone willing to pay a price that would cover use of the whole aircraft. There would then simply be airlines, some operating charter services, some operating scheduled services and some both. But there would cease to be any distinction between the two traditional types from a regulation and licensing point of view. This seems to be the likely course of events and may even be a reality by the time you read this. After all, Laker Airways was solely a charter carrier until the day the Skytrain service began. The Skytrain Standby service from London to New York and Los Angeles marks the real point at which the traditional distinction begins to blur.

If price is a prime consideration, you may not have much choice as to what airline you fly. Certain organisations selling low cost travel can still offer rates that no regular travel agent can match—provided you don't mind flying an airline that wouldn't always figure in your Top Ten. (Many airlines, however, that can be booked in this way do rate amongst the best known and most reliable.)

If you are prepared to pay the airline's going rate, either direct from the airline or through a regular travel agent (the price should be the same either way—if it's not, you've been misquoted), then the choice of airline is yours, provided that space is still available. I have a few favourite airlines and I don't mind sharing them with you. Some of them I haven't even flown on, but I listen to what other people tell me, and gradually an impression begins to form in my mind about certain carriers.

If service on a charter airline is what you're after, (assuming 'charter' airlines are still around by the time you read this) then I reckon that the Canadian airline Wardair must take the prize. Bear in mind that airlines that offer good in-flight service are also generally good on the ground and have a good safety record. All seasoned travellers are only too aware that it's on the ground—not in the air—that things tend to go wrong with airlines. That's right, I'm talking about airline reservation departments, one of the crosses we regular travellers have to bear in life.

Amongst scheduled carriers, the Far Eastern ones have developed a remarkable reputation for in-flight service, especially Singapore Airlines, Thai International (not so good on the ground), Cathay Pacific and Japan Airlines. In Europe, most of the carriers offer poor in-flight service on European routes, being much better on intercontinental routes. When it comes to efficiency and general punctuality, there's more to differentiate the European airlines. Swissair, Finnair and Lufthansa rate highly, along with KLM and SAS, while British Airways and Alitalia rate bottom of my European League on just about any score you can imagine.

I won't go into details of how to pick which airline for which route. Remember, however, that most Third World airlines have a lot to learn, and while statistically air travel is many times safer than driving in a car, a small handful of airlines do have an appalling safety record. I won't tell you which they are, but if you want some good news, you may be interested to know that to the best of my knowledge Qantas and Air New Zealand have never had a passenger fatality.

There are four ways you can book an air ticket. You may book direct with an airline, but if you do so by telephone (don't even try by letter) be prepared for the run-around. It will probably happen in more than 50 per cent of cases, especially for complex itineraries. Be prepared for the reservations clerk (assuming you even get through on the 'phone in the first place) to tell you that the fare you want doesn't exist. The chances are that the clerk is about three months out of date in a field where fares are changing almost daily. Try ringing three airlines or even the same airline three days in a row. The chances are you'll get quoted a different fare each time. If so, take the cheapest. It may or may not be the right one, but the airline's mistake could save you a lot of money.

The next place you could turn to for a reservation is a regular travel agent, the kind that has a windowful of pretty pictures supplied by tour operators, airlines, hotel chains or tourist boards.

Only 'phone your travel agent if you know him well. Otherwise go and book personally. In general, you'll get a better service than by going direct to an airline. You may also stand more chance of getting an answer on the 'phone if you decide to call up.

Bear in mind that now airlines are paying different rates of commission (above and below the counter) to more productive travel agents, some (many) agents will steer you towards certain airlines with stories of exotic native in-flight cuisine and suchlike. Beware. The agent is probably thinking of the 15 per cent override commission he's likely to get. He may also try to steer you to a ticket that's more expensive than the one you need or want. The more you pay, the more commission he gets. Deregulation has made this kind of travel hazard all the greater, and after all, travel agents are only human, like any other businessmen.

The bucket shop is a British phenomenon that used to exist in the US too until about 1975. It's a travel agency of the no-windows-with-pretty-pictures variety. It still exists in the UK (mostly in London) as a conduit via which certain reputable and not so reputable airlines dump unsold seats on the market at discount rates. Some of these tickets may be genuinely discounted over the official filed tariff, others will be inclusive tour air tickets (group or individual) sold off without the ground package that is meant to accompany them. And many tickets will be those available through any ordinary travel agency—a whole plethora of new promotional fares dreamed up by airline marketing men. Usually, since the airlines regard such tickets as discount fares anyway, bucket shops will sell them at the same price as would an airline or a regular travel agency.

If you use a bucket shop, do so warily. Most are honest, but most work in volume on very low margins and as a result their standards of efficiency leave a lot to be desired. This shows up in a number of ways: late ticketing, difficulties in changing ticket dates and high penalty payments to do so, high cancellation charges, difficulties in obtaining refunds, problems in getting confirmation, problems in getting written receipts for payments, lack of clear, specific, written check-in information. So if you live in the UK and plan to use a bucket shop, visit it personally, pay only a small deposit on booking, get a receipt and a written confirmation of the booking, and get flight details that are as specific as possible. And don't hand over the balance (you'll probably be asked for it in cash) until you get your tickets in exchange. Again, get a written receipt. Don't accept vouchers in lieu of outward or return tickets. They'll only cost you time and headaches, and maybe money.

A book could be written about low cost air travel and it still wouldn't exhaust the subject. Like this article, it would also be increasingly out of date with each day that passes. There is no commercial field in the world that is changing as fast as that of civil aviation. With a little knowledge of the trends that are taking place in this field, you'll be better equipped to get the best value for money when you travel by air, and, looking at it another way, you'll be less likely to run into one hassle after another. However, I hate to wish you an entirely hassle-free experience when you travel by air, because whatever the airline, whoever you book through, there's always bound to be at least one hassle—usually while you're abroad. The fact is that air travel operates across national and cultural boundaries, and the result for many years to come is bound to be airline incidents and misunderstandings. As I've mentioned above, these are often in the reservations department. To be even more specific, they usually occur when you come to confirming your return flight. You can't be too careful about doing this thoroughly.

Don't ever expect international air travel to be a bed of roses. Something, big or small, is always bound to go wrong. Just make sure that you're insured, and try to reduce the chances of things going wrong by planning thoroughly. I hope that this article will have given you a little more understanding. That too will help reduce the hassles. Who knows, everything might even go without a hitch next time you fly.

Travel by Freighter

by Ingrid Cranfield
Compiled with the help of Pearl Hoffman

Very few of the readers of this book will ever allow the thought 'passenger cruise' to cross their minds, let alone the words to pass their lips. Not for our independent traveller the luxury accommodation, all-found entertainment and packaged miniature land excursions of the large passenger liners. Freighter travel, with its relaxed and unpretentious atmosphere, is a different scene altogether. And yet in a sense, the freighter passenger is more dependent on other people and contingencies than the cruise participant. Cargo on a freighter takes precedence and the passengers accordingly fall into second place.

The International Conventions and Conferences on Marine Safety have defined a freighter (or cargo liner, which is the same thing) as 'a vessel principally engaged in transporting goods which is licensed to carry a maximum of 12 passengers'. 'Vessels principally engaged' in goods traffic, yet carrying more than 12 passengers, are called combination cargo-passenger ships.

Perhaps the biggest attraction of freighter travel, for the person with few pressing time commitments who savours the unexpected, is the flexibility of departure dates and itinerary common to cargo ships. The initial date of departure may vary by a few days; a port may be added to the itinerary or deleted from it; arrival times may be brought forward or put back. And if the duration of the passage is longer than anticipated, the passenger pays no additional charge. Freighters visit offbeat ports not usually serviced by cruise liners and stay longer in each port than do such passenger ships. While no shore excursions are arranged for freighter passengers, plenty of advice on sights to see will be forthcoming from the crew or, in advance, from the shipping line or its agents.

Most freighters currently in operation are of postwar construction, with cruising speeds of 16-23 knots. They usually offer well appointed and spacious accommodation amidships (where there is little vibration from propellers.) Common rooms are frequently air conditioned and many vessels have not only a lounge but also a games room and a library.

Passengers mix freely with the crew and take meals with the officers, which affords ample opportunity for the passengers to quiz their hosts about life at sea and ports of call to be visited along the way. Casual wear is the order of the day—and usually of the evening as well.

There are other advantages. For comparable accommodation and meals, the freighter passenger normally pays considerably less than the passenger on a cruise liner, and small extras, such as between-meal snacks, are often included in the overall price.

A voyage on a freighter is generally smoother than on a cruise ship as most of the ship's weight is carried below the water level. The vessel therefore lies low in the water, much of it out of reach of the wind and surface swell. Additionally, the weight of the cargo acts as a stabiliser; and the wide beam of the vessel counteracts any tendency to roll.

Many freighter passengers are retired people over the age of 60 years. Older people particularly should note that a freighter carries no doctor on board. The shipping companies require would-be passengers over 65 years to submit a letter from a G.P. stating that they are fit to travel. Medical supplies are however always carried on board and all officers are trained in first aid. In the event of serious illness or emergency, a doctor on another ship or a shore station can be called upon for advice or summoned to the freighter, or the patient put ashore or transferred to another ship.

A few tips for freighter travellers:

Consult the shipping line concerned for information; often better still, ask your travel agent to advise on competitive sailings and other connections for onward travel. The cost of tickets bought from an agent is no more than from the shipping company itself. Ask the company for a pass to visit the freighter on which you are considering a voyage before you make a reservation.

2. Plan and make your reservation as far ahead as possible, as certain popular routes and sailing times fill up rapidly.

3. Ask about special permits that may be required. Passengers embarking in the USA who are not themselves US citizens must obtain an Income Tax Clearance Certificate (Sailing Permit). This is obtainable on application in person from an office of the US Department of Internal Revenue.

4. Insure your baggage fully, as the liability of the shipping company is severely limited. Valuables may be placed in the ship's safe, but the risks are assumed by the owner.

5. Consult the company or travel agent about the shipping of vehicles, furniture or household effects or if you wish pets to accompany you on your voyage. Mail can be sent to passengers care of the port agent; letters should be marked with the name and sailing date of the vessel and its destination.

6. The Purser, Ship's Steward and other officers are at your disposal on board. Discuss with them any problems you may encounter with shipboard life, especially such minor matters as tipping.

7. Pack a pair of low-heeled, rubber-soled shoes: they are a must for safety when your vessel hits rough seas.

The Two Wheeled Traveller

by Ray Reece

Perhaps a better title would be *So you want to get away from it all, but can you stand the pain?* You can be sure of one thing: that no matter where you travel on a bicycle you are going to have an experience that no money can buy. As the cycle is going to be your best friend, and sometimes your only friend, we'll start with it first.

Home touring or transcontinental touring? There is quite a difference, both in approach and equipment; but whichever you are doing you do so under your own steam and you had better believe it. Playing it tough or cool may look good at first, but one day you'll be saying a little prayer for anything you can do to make life a little easier on the way. In long distance touring, one just *must not* leave things to chance, for the friendly cycle dealer or drugstore you hope will put everything right in the next town usually doesn't exist.

Old or new? If it's a once in a lifetime big tour, there is a temptation, if you're loaded (not many bikies are), to treat yourself to a completely brand new machine, but the chances are your old one, especially if you are aiming at more than 3,000 miles, will be more trustworthy, particularly the frame. It is however worth renewing all the rotating parts. The frame size should be approximately 10 in. less than your inside leg measurement, and of the best quality tube, or else you'll be spending more time in the welding shops than out, if travelling in Asia! Frame angles should be between 71° and 73° only.

The wheels *must* receive special attention, for they are going to carry more load and get more punishment than you ever thought possible. The rims must be steel, for in the event of a disaster, they can still be hammered, welded, or even kicked back into a rideable shape. As for size some controversy exists, for 27in high pressure wheels are not only in many parts of the world as rare as gold dust but are also somewhat lacking in reliability and comfort. In my experience, however, they are still the best choice. In an emergency a 26in tyre, of the type English tourists traditionally prefer, can be fitted into a 27in frame, but not usually the other way round. You have to carry spares anyway, and the 27in tyres are lighter, less bulky, roll better, and provide an extra turn of speed which can often get you out of trouble. You can cover more ground when conditions are good and the smoother tread is much less likely to pick up grit and punctures on even the roughest dirt track, especially if you keep the pressure about 10 per cent below average. Beware, though, of ruts and potholes. Reckon on about 3,000 miles for a rear tyre and about 6,000 for the front. Wide flange hubs are the best choice, but not the quick release type, which are too easily stolen. The spokes must be top, repeat top, quality, using a gauge thicker than normal (e.g. 12 gauge double butted) at least on the rear wheel, and built to only a moderate tension, *not* tied and soldered.

The real hassle, if you weigh around 200lb, is spoke breakage on the gear side. Normally this means carrying a sprocket remover, but depending on the ratios and the type of freewheel, a spoke can be pushed through, sometimes by slightly enlarging the sprocket lightening holes beforehand, and fitted without any dismantling. All bearings are best packed with grease (Molygrease), which will keep the dirt and the water out. Spend some time getting each adjustment really 'neat' and you'll have no problems. Brakes and side-pulls are fine but use wet weather blocks. Pedals must be good quality. Toe clips and straps are strongly recommended, especially for struggling up mountains. Gear-change mechanisms must be metal, as they straighten more easily. Ten speed is fine, and can be repaired or patched up more easily than hub gears. Ratios from mid-thirties up to 80 inches. You'll be glad of mudguards which will keep the chain and gears clean and smooth, and you much drier, for even in the hottest country it always manages to bucket down sooner or later. Handlebars are a personal thing, but with drops you have a choice of position, and it can sure take a load off your butt!

Which brings us to the saddle. Get the best leather job you can afford and treat it with dressing or linseed oil until supple. It will be your only comfort at times.

Panniers are best for luggage. Don't carry too much on the front of the bike, as it makes steering sluggish and tiring, and don't carry anything at all on *your* back. Your tool-kit: keep it light. But you will need at least a combination spanner that will fit every nut on the cycle, a screwdriver, spoke key, chain rivet extractor, puncture kit, small flat file, scissors, two-tube epoxy, a small brush for chain and gear cleaning, PVC tape, spare brake blocks, and front and rear wheel nuts, spare spokes, and pump connection. Don't forget a penknife and a bit of string.

Your cycle is your life-line, but everywhere you go, even in the most civilised countries, somebody will think he needs it more than you. So use a good cable lock, always. Pass it through the frame and wheel, and then around something fixed or heavy. Typical advice when camping in Turkey: "Tie it to a sensitive part of your body" (and they don't mean your big toe!). Wherever possible keep it in sight or better still take it with you, meeting the raised eyebrow and protests with a smile; people will usually understand.

Which way to go? Don't just get a route map and play it by ear. Every bit of planning will pay off. On a cycle you've got to be more particular, but you will really get something out of it instead of giving everything you've got just to make it through. Check out contours, temperatures, rainfall, and unusual happenings like floods, monsoons, snow at altitude, sandstorm seasons, drought, etc.

Check the jabs you need, get them early in case you get a reaction, and when possible get visas before starting out, as there is often a delay in getting them along the way. Sometimes you can be turned back at a border on a technicality. One poor fellow I met had to backtrack 400 miles each way just to get a yellow fever jab. Don't ever try dodging the formalities at a frontier, no matter how easy it seems, for if you are stopped or when you come to leave the country again you will be in big trouble. Get the stamp on your passport. Some countries will also want to stamp the cycle/frame number in as well, so make sure it is visible. What you take in you must take out or you will be charged customs duty, on the assumption that the bike has been sold. Exchange only the minimum amount of currency in each country, as it is often very difficult or expensive to convert it back when leaving. Keep all receipts of goods purchased or currency exchanges on you, as this will avoid problems at frontiers or with police, especially if they are looking for stolen goods. When crossing a border, try and look as much like your passport photograph as possible, i.e., if you were clean-shaven in the photograph, don't grow a beard. Stay cool, for officials are not impressed by freaks or jokers; and don't carry anything that resembles drugs for they are very sharp and you could end up in the slammer for an indefinite period, no matter who your relations are!

One very good tip is to make a small cotton or waterproof zippered bag to hang around your neck. In it at *all* times keep your passport, bills, traveller's cheques and medical tickets. Money belts are in my experience too obvious, uncomfortable, and not safe. Remember that in some countries a passport is worth more on the black market than everything else you have put together!

Those who want to skip some of the distance by other transport, BEWARE. Supervise handling and stowage yourself. In many countries cycles are carried in separate trains for passengers and get lost. Cycles carried on roof racks of buses get swiped off at low bridges. On ships, ask a crewman for some rope and tie the bike down securely. Fortunately on aircraft there is seldom a problem, but don't dismantle your machine unless requested to do so.

Training is a tough problem. First decide on what weekly mileage you hope to achieve. 280 miles is comfortable, or if you are very fit 350 to 450. 700 miles a week is possible but this is strictly for the 'superman' category. Aim at doing at least half of your scheduled weekly mileage each week for about 4 to 5 weeks before you leave, until the week before you take off; then forget the bike, and recharge your batteries!

On the road: in Europe, particularly the east, watch out for the big trucks. In Italy everyone is a Grand Prix driver! In Turkey beware of the flying buses, which have pilots not drivers. In Afghanistan look out for everyone—they drive on *both* sides of the road. In Pakistan and India, mind the bullock carts, though the main hazard is what they leave behind. The most dangerous threat to a cyclist is a rabid dog. With the exception of the United Kingdom, this hazard is everywhere. Don't ever take a gamble: any contact with a suspect animal must mean going straight to the nearest hospital. Any other course of action may be fatal.

Accommodation is an area you can play by ear, often to great advantage. While it may be great curled up in a comfortable hotel occasionally, don't miss the ultimate experience of cycle-touring with its total freedom of choice and contact with the common man. Beggar or millionaire, you will learn something from everyone you meet. For the camper a good down sleeping bag is a must, even if he'll be sometimes using it on the floor of local inhabitants' homes.

When off the beaten track, always try to camp near another group or habitation. Ask permission if necessary. When completely alone, or in hostile regions, don't make camp until just on dusk. Be sure you haven't been watched and stay out of sight, particularly of the road and its car headlamps. Never ride at night. Women should not be without some muscle in the party if venturing beyond Western Europe.

Some advice once given to a Westerner travelling to New York was 'keep a smile on your face and your hands in your pockets.' Well, this goes for riders abroad too, but be sure to keep one hand on your bike as well. Treat others as you would have them treat you, even if you can't speak the lingo, and in any tricky situation just offer a handshake—it will usually be accepted. If not, make mileage. And always remember: watch out for those motorists!

Motorcycling

by Ted Simon

It seems pointless to argue the merits of motorcycles as against other kinds of vehicle. Everyone knows more or less what the motorcycle can do, and attitudes to it generally are quite sharply defined. The majority is against it, and so much the better for those of us who recognise its advantages. Who wants to be part of a herd? Let me just say that I am writing here for people who think of travelling through the broad open spaces of Africa and Latin America, or across the great Asian land mass. Riding in Europe or North America is straightforward, and even the problems posed in Australia are relatively clear cut. As for those fanatics whose notion of travelling is to set the fastest time between Berlin and Singapore, I am all for abandoning them where they fall, under the stones and knives of angry Moslem villagers.

Here, then, are some points in favour of the motorcycle, for the few who care to consider them. In my view it is the most versatile vehicle there is for moving through strange countries at a reasonable pace, experiencing changing conditions and meeting with people in remote places. It can cover immense distances, and will take you where cars can hardly go. It is easily and cheaply freighted across lakes and oceans, and it can usually be trucked out of trouble without much difficulty, where a car might anchor you to the spot for weeks. If you choose a good bike for your purpose it will be economical and easy to repair, and it can be made to carry quite astonishing amounts ot stuff if your systems are right.

In return the bike demands the highest level of awareness from its rider. You need not be an expert (I had not ridden bikes before myself) but you must be enthusiastic and keep all your wits about you. It is an unforgiving vehicle which does not suffer fools at all. As well as the more obvious hazards of potholes, maniacal truck drivers and stray animals, there are the less tangible perils like dehydration, hypothermia and plain mental fatigue to recognise and avoid.

The bike then, poses a real challenge to its rider, and it may seem to verge on masochism to accept it, but my argument is that by choosing to travel in a way that demands top physical and mental performance you equip yourself to benefit a thousand times more from what comes your way, enabling you quite soon to brush aside the discomforts that plague lazier travellers.

You absolutely must sit up and take

...her ...rs; ...le ... are ...adjust more ...the climate, both ... the eulogy upon the ... generalisations, it is ...cular. There is no ...ke forms, nor one for all ... s a splendid machine ... with ...id reputation for touring ... infallible, and it *is* expensive. ...bikes need a lot of maintenance ... they are ruggedly engineered ... sily repaired, given the parts ... unjabi workshop to make them ... apanese bikes have a shorter useful ...e, but they work very well ... and their dealer networks are incomparable. They are hard to beat as a practical proposition provided you go for models with a tried record of reliability.

On the whole I would aim for an engine capacity between 500cc and 750cc. Lightness is a great plus factor. Too much power is an embarrassment, but a small engine will do fine if you don't mean to hump a lot of stuff over the Andes, or carry another person as well.

I travelled alone almost all the way, but most people prefer to travel in company. As a machine the motorcycle is obviously at its best used by one person, and it is my opinion that you learn faster and get the maximum feedback on your own, but I know that for many such loneliness would be unthinkable. Even so, you need to be very clear about your reasons for choosing to travel in company. If it is only for security then my advice is to forget it. Groups of nervous travellers chattering to each other in some outlandish tongue spread waves of paranoia much faster than a single weary rider struggling to make contact in the local language. A motorcycle will attract attention in most places. The problem is to turn that interest to good account. In some countries (Brazil, for example) a motorcycle is a symbol of playboy wealth, and an invitation to thieves. In parts of Africa and the Andes it is still an unfamiliar and disturbing object. Whether the attention it attracts works for the rider or against him depends on his own awareness of others and the positive energy he can generate towards his environment.

It is very important in poor countries not to flaunt wealth and superiority. All machinery has this effect anyway, but it can be much reduced by a suitable layer of dirt and a muted exhaust system. I avoided having too much glittering chrome and electric paintwork, and I regarded most modern leathers and motorcycle gear as a real handicap. I wore an open face helmet for four years and never regretted it, and when I stopped among people I always took it off to make sure they saw me as a real person. My ideal was always to get as far away as possible from the advertised image of the smart motorcyclist, and to talk to people spontaneously in a relaxed manner. If one can teach oneself to drop shyness with strangers the rewards are dramatic. Silence is usually interpreted as stand-offishness, and is almost as much a barrier as a foreign language.

Obviously you should know your bike and be prepared to look after it.

Carry as many tools as you can use, and all the small spares you can afford. Fit a capacitor so that you don't need a battery to start. Weld a disc on the swing stand to hold the bike in soft dirt. Take two chains and use one to draw the other off its sprockets. This makes frequent chain-cleaning less painful, something that should be done in desert conditions. Take a tin of Swarfega or Palmit; it's very useful where water is at a premium and for easing tyres off rims. Buy good patches and take them (I like 'Tip-Top'). You won't get them there. The Schrader pump, which screws into a cylinder in place of the spark plug, is a fine gadget, and one of the best reasons for running on two cylinders. Aerosol repair canisters, unfortunately, do not always work. The quickly detachable wheel arrangement on the Triumph saved me a lot of irritation too.

Change oil every 1,500 miles, and don't buy it loose if you can avoid it. Make *certain* your air filter is good enough. Some production models will not keep out fine desert grit, and the consequences are not good. Equally important are low compression pistons to take the strain off and to accept lousy fuel.

I ran on Avon tyres and used a rear tread on the front wheel, which worked well. A set of tyres gave me 12,000 miles or more. The hardest country for tyres was India, because of the constant braking for ox carts on tarmac roads. It was the only place where the front tyre wore out before the rear one because, of course, it's the front brakes that do most of the stopping.

Insurance is a problem that worries many people. Get it as you go along. I was uninsured everywhere except when the authorities made it impossible for me... buying it (Costa R... Rhodesia and South A... most definitely illegal a... recommend it: if you get co... have only yourself to blame.

Other things I found ...ntial ... a stove, a good all-purp... knife, some primitive cooking eq... ...nt and a store of staples like ... beans. Naturally you need to ... up to a gallon if possi... ...er too, ability to feed myself w... was a great protection a... ...ker as well as an incentive ...n further off the beaten ... ...etrol ous stove to use is the ... stove; it did very we... square things pack easier than ...es. In the end I finished with a complex and sophisticated kitch... ...e of my boxes, but of course tha... a matter of taste.

Finally a few things I ...ned not to do:

Don't ride without arms, knees and eyes covered, and watch out for bee swarms, unless you use a screen which I did not.

Do not carry a gun or any offensive weapon unless you want to invite violence.

Do not allow yourself to be hustled into starting off anywhere until you're ready; something is bound to go wrong or get lost.

Do not let helpful people entice you into following their cars at ridiculous speeds over dirt roads and potholes. They have no idea what bikes can do. Always set your own pace and get used to the pleasures of easy riding.

Resist the habit of thinking that you must get to the next big city before nightfall. You miss everything that's good along the way and, in any case, cities are the least interesting places.

Don't expect things to go to plan, and don't worry when they don't. Perhaps the hardest truth to appreciate when starting a long journey is that the mishaps and unexpected problems always lead to the best discoveries and the most memorable experiences. And if things insist on going too smoothly, you can always try running out of petrol on purpose.

Travel by Camel

by René Dee

In this mechanised and industrialised epoch the camel does not seem to be an obvious choice of travelling companion when sophisticated cross-country vehicles exist for the toughest of terrains. Add to this the stockpile of derisory and mocking myths, truths and sayings about the camel and one is forced to ask the question: why use camels at all?

As purely an alternative means to the vehicle of getting from A to B when time is the most important factor, the camel should not even be considered. As a means of transport for scientific groups who wish to carry out useful research in the field the camel is limiting. It can be awkward and risky transporting delicate equipment and specimens. However, for the individual, small group and expedition wishing to see the desert as it should be seen, the camel is an unrivalled means of transport.

From my own personal point of view, the primary reason must be that, unlike any motorised vehicle, camels allow you to integrate completely with the desert and the people within it—something which is impossible to do at 50 mph enclosed in a 'tin can'. A vehicle in the desert can be like a prison cell and the constant noise of the engine tends to blur all sense of the solitude, vastness and deafening quiet which are an integral part of the desert.

Travel by camel allows the entire pace of life to slow down from a racy 50 mph to a steady 4 mph, enabling you to unwind, take in and visually appreciate the overall magnificence and individual details of the desert.

Secondly, camels do of course have the ability to reach certain areas inaccessible to vehicles, especially through rocky and narrow mountain passes, although camels are not always happy on this terrain and extreme care has to be taken to ensure that they do not slip or twist a leg. They are as sensitive as they appear insensitive.

Thirdly, in practical terms they cause far fewer problems where maintenance, breakdown and repairs are concerned. No bulky spares are needed nor expensive mechanical equipment to carry out repairs. Camels do not need a great deal of fuel and can exist adequately (given they are not burdened down with excessively heavy loads) for 5-10 days without water. Camels go on and on and on and on and on until they die; and then finally one has the option of eating them, altogether far better tasting than a Michelin tyre.

Lastly, camels *must* be far more cost effective if you compare directly with vehicles; although this depends on whether your intended expedition/journey already includes a motorised section. If you fly direct to your departure point, or as near as possible to it, you will incur none of the heavy costs related to transporting a vehicle, not to mention the cost of buying it. If the camel trek is to be an integral portion of a motorised expedition, then the cost saving will not apply as of course hire fees for camels and guides will be additional.

In many ways, combining these two forms of travel is ideal and a very good way of highlighting my primary point in favour of transport by camel. If you do decide on this combination, make sure that you schedule the camel journey for the very end of your expedition and that the return leg by vehicle is either minimal or purely functional, as I can guarantee that after a period of 10 days or more travelling slowly and gently through the desert by camel, your vehicle will take on the characteristics of a rocket ship and all sense of freedom, enquiry and interest will be dulled to the extreme, and an overwhelming sense of disillusion and disinterest will prevail. Previously exciting sights, desert towns and Arab civilisation, will pall after such intense involvement with the desert, its people and its lifestyle.

For the individual or group organiser wanting to get off the beaten track by camel his first real problem is to find them, and to gather every bit of information possible about who owns them, whether or not they are for hire, for how much, what equipment/stores/provisions are included, if at all, and, lastly, what the guides/owners are capable of and whether they are willing to accompany you. It is not much good arriving at Tamanrasset, Tombouctou or Tindouf without knowing some, if not all, of the answers to these questions. Good pre-departure research is vital but the problem is that 90 per cent of this information won't be found in any tourist office, embassy, library or travel agent. Particularly if you're considering a major journey exclusively by camel, you'll more than likely have to undertake a preliminary fact-finding recce to your proposed departure point to establish contacts among camel owners and guides. It may well be that camels and/or reliable guides do not exist in the area where you wish to carry out your expedition. I would suggest, therefore, that you start first with a reliable source of information such as the Royal Geographical Society, which has expedition reports and advice which can be used as a primary source of reference, including names and addresses to write to for up-to-date information about the area that interests you. Up-to-date information is without doubt the key to it all. Very often this can be gleaned from the commercial overland companies whose drivers are perhaps passing your area of interest regularly and may even have had personal experience of the journey you intend to make. Equally important is the fact that in the course of their travels they build up an impressive collection of contacts who could well help in the final goal of finding suitable guides, smoothing over formalities and getting introductions to local officials, etc. Most overland travel companies are very approachable so long as you appreciate that their time is restricted and that their business is selling travel and not running an advisory service.

In all the best Red Indian stories, the guide is the all-knowing, all-seeing person in whom all faith is put. However, as various people have discovered to their cost, this is not always so. Many so-called guides know very little of the desert and its ways. How then to find someone who really does know the route/area, has a sense of desert lore and who preferably owns his own camel? I can only reiterate that the best way to do so is through personal recommendation. Having found him, put your faith in him, let him choose your camels and make sure that your relationship remains as amicable as possible because you will be living together for many days in conditions which will be familiar to him but alien to you, and you need his support. Arrogance does not fit into desert travel, especially from a *nasrani*. Mutual respect and rapport are essential.

Once you've managed to establish all this and you're actually out there, what are the do's and don'ts and logistics of travel by camel? Most individuals and expeditions (scientifically oriented or not) will want, I imagine, to incorporate a camel trek within an existing vehicle-led expedition so I am really talking only of short-range treks of around 10-15 days' duration, up to 250 miles. If this is so, you will need relatively little equipment/stores and it is essential that this *is* kept to a minimum. Remember that the more equipment you take, the more camels you will need, which will require more guides, which means more cost, more pasture and water, longer delays in loading, unloading, cooking and setting up camp and a longer wait in the mornings while the camels are being rounded up after a night of pasturing.

Be prepared also for a very swift deterioration of equipment. In a vehicle you can at least keep possessions clean and safe to a degree, but packing kit onto a camel denies any form of protection—especially since it is not unknown for camels to stumble and fall or to roll you over suddenly and ignominiously if there is something not to their liking, such as a slipped load or uncomfortable saddle. My advice is to pack all your belongings in a seaman's kitbag which can be roped on to the camel's side easily, is pliable, hard-wearing and, because it is soft and not angular, doesn't threaten to rub a hole in the camel's side or backbone. (I have seen a badly placed baggage saddle wear a hole the size of a man's fist into the back of a camel). If rectangular aluminium boxes containing cameras or other delicate equipment are being carried, make sure that they are well roped on the top of the camel and that there is sufficient padding underneath so as not to cause friction. Moreover, you'll always have to take your shoes off whilst riding because over a period of hours, let alone days, you could severely wear out the protective hair on the camel's neck and eventually cause open sores.

Water should be carried in goatskin *guerbas* and 20 litre round metal *bidons* which can again be roped up easily and hung either side of the baggage camel under protective covers. Take plenty of rope for tying on equipment, saddles, etc., and keep one length of 50ft intact for using at wells where there may be no facilities for hauling up water. Don't take any cooking stoves—the open fire is adequate and far less likely to break down. Don't take any sophisticated tents either, which will probably be ruined within days and anyway just aren't necessary. I have always used a piece of cotton cloth approximately 20ft square which, with two poles for support front and rear and with sand or boulders at the sides and corner, makes a very good overnight shelter for half a dozen people. Night in the desert can be extremely cold, particularly of course in the winter, but the makeshift 'tent' has a more important role during the heat of the day when it provides shelter for the essential two-hour lunch stop and rest.

Your daily itinerary and schedule should be geared to the practical implications of travelling by camel. That is to say that each night's stop will, where possible, be in an area where pasture is to be found for the camels to graze. Although one can take along grain and dried dates for camels to eat, normal grazing is also vital. The camels are unloaded and hobbled (two front legs tied closely together), but you will find that they can wander as much as 3-4 kilometres overnight and there is only one way to fetch them—on foot. Binoculars are then very helpful as spotting camels over such a distance can be a nightmare. They may be hidden behind dunes and not come into view for some time.

Other useful equipment is goggles for protection in sand storms, a prescription-suited pair of sunglasses and of course sun cream. Above all, take comfortable and hardwearing footwear for it is almost certain that you will walk at least half the way once you have become fully acclimatised. I would suggest that you take Spanish fell boots or something similar, which are cheap, very light, give ankle support over uneven terrain, are durable and very comfortable. The one disadvantage of boots by day is that your feet will get very hot, but it's a far better choice than battered, blistered and lacerated feet when one has to keep up with the camels' steady 4 mph pace. Nomads wear sandals but if you take a close look at a nomad's foot you will see that it is not dissimilar to the sandal itself, i.e. as hard and tough as leather. Yours resembles a baby's bottom by comparison—so it is essential that you get some heavy walking practice in beforehand with the boots/shoes/sandals you intend to wear. (If your journey is likely to be a long one, then you could possibly try sandals, as there will be time for the inevitable wearing-in process of your feet involving blisters as well as stubbed toes and feet spiked by the lethal acacia thorn).

For clothing, I personally wear a local free-flowing robe like the *gandoura*, local pantaloons and a *chèche*—a 3-metre length of cotton cloth which can be tied round the head and/or face and neck for protection against the sun. You can also use it as a rope, fly whisk and face protector in sand storms. In the bitter cold nights and early mornings of winter desert travel, go to bed with it wrapped around your neck, face and head to keep warm. If local clothing embarrasses and inhibits you, stick to loose cotton shirts and trousers. Forget your tight jeans and bring loose-fitting cotton underwear. Anything nylon and tight fitting next to the skin will result in chafing and sores. Do, however, also take some warm clothing and blankets, including socks and jumpers. As soon as the sun sets in the desert the temperature drops dramatically. Catching cold in the desert is unbearable. Colds are extremely common and spread like wildfire. Take a good down sleeping bag and a groundsheet. Your sleeping bag and blankets can also serve as padding for certain types of camel saddle. In the Western Sahara you will find the Mauritanian butterfly variety, which envelops you on four sides. You're liable to slide back and forth uncomfortably and get blisters unless you pad the saddle. The Tuareg saddle commonly used in the Algerian Sahara is the pommel type. Here one sits astride a more traditional-looking saddle with a fierce looking forward pommel which threatens man's very manhood should you be thrown forward against it. In Saudi Arabia, female camels are ridden and seating positions are taken up behind the dromedary's single hump rather than on or forward of it.

Never travel alone in the desert, without even a guide. Ideal group size would be seven group members, one group leader, three guides, eleven riding camels and three baggage camels. The individual traveller should take at least one guide with him and three or four camels.

Be prepared for a mind-blowing sequence of mental experiences, especially if you are not accustomed to the alien environment, company and pace, which can lead to introspection, uncertainty and even paranoia. Travel by camel with nomad guides through any section of desert is the complete reversal of a lifestyle we normally live. Therefore it is as important to be mentally prepared for this culture shock as it is to be physically prepared. Make no mistake, travel by camel is hard, physically uncompromising and mentally torturing at times. But a *meharée* satisfactorily accomplished will alter your concept of life and its overall values, and the desert's hold over you will never loosen.

Hitchhiking

by Wendy Myers

For the person who likes conversing with strangers, relishes using newly acquired languages and yet enjoys being (often suddenly) quite alone, far off the beaten track, then hitchhiking must surely be hard to beat, as a principal mode of getting about. I say 'principal' because, in my opinion, nobody should set off on a journey intending entirely to be reliant upon other people's goodwill. Those who set out on foot, hoping for a lift but feeling happy to walk, will be pleasantly relieved when one comes, whereas the attitude that it is somebody's duty to transport one from A to B, free of charge whenever one wishes, could mean a long series of disappointments.

One April morning I left England to 'walk around the world'. Seven years and over one hundred countries later I returned home, with enough interesting material to fill the equivalent of eight books and much information gained about hitchhiking, which had been my principal method of travelling.

Let us assume that the reader proposes, as I did, to see as much of the world as possible. Therefore, what to take? 'As much as can be comfortably carried while walking' is a good criterion. My own rucksack weighed 36lb, which I carried with reasonable ease. As a guide to clothing, I suggest strong shirts, shorts and trousers for travelling, plus a sweater for cold weather. Tee-shirts, to be worn next to the skin, will both decrease pressure from rucksack straps and mop up perspiration. Strong, comfortable shoes are a must, while sandals and 'flip-flops' come in handy for hot climates or if one has blisters. Swimming gear plus toilet articles should find a place, and some light 'all over' garment, to relax in, or when other clothes are wet. Invitations will crop up, and anyway it would be an unusual traveller who did not hope to include formal visits and spectacles in his itinerary, thus necessitating a set of 'smart' clothes. A small first-aid kit is a sensible acquisition, and can include water-purifying tablets and anti-malarial pills. Finally, three 'musts': sun-hat, water-bottle and diary. . . .

Don't forget valid vaccination certificates. Money-wise, traveller's cheques of small denominations combine convenience with safety, and for those hoping to boost funds by working overseas, photostat copies of any personal certificates or diplomas could be useful. Small scale maps can frequently be purchased cheaply at petrol stations in the appropriate country, while membership of the International Youth Hostel Federation gives access to maps plus cheap accommodation through most of Europe and in several other countries as well.

To save time and trouble one can obtain visas before setting off, although as most have prefixed dates, preparations in this field tend to be somewhat limited. However a word of cheer: it is often far quicker and easier to procure a visa as one gets closer to the country concerned. For example, even for wartime South Vietnam, my visa was swiftly issued by the Japanese Embassy (their then representatives) in Cambodia. Also in Cambodia, visas for both North Vietnam and The People's Republic of Communist China were granted to me following a waiting period of only six weeks—during which letters and applications had to be sent to the respective capitals.

It must be understood that people give lifts for a number of acceptable reasons, and, if these can be discovered as soon as possible and complied with, a pleasant journey should ensue. It could be that, facing a long tedious road ahead, a driver wants to ensure against falling asleep, or maybe he wishes to practise his English, or just 'likes foreigners'. Perhaps help with driving is hoped for. . . . I had a wonderful tour around New Zealand's North Island with a holidaying resident who had been ill and was on the look-out for a possible co-driver. On another occasion I rode with an illiterate Australian cattle drover for whom I had to read the names on the signposts. Drivers also sometimes find hitchhikers who are foreigners like themselves useful sources of information on places to stay, the local territory and other potential passengers. In Third World countries, it is not only foreigners who hitchhike but the locals themselves, who are usually either young middle class students, poor people or military personnel going home on leave. The first group are the most fruitful contacts for a driver as they prove, on the whole, to be well-informed and hospitable.

Conduct will make or mar the future for hitchhikers: those who snore for hours in the back seat, eating food they find, are hardly a good advertisement.

In my experience it is very rare actually to have to 'thumb down' a vehicle—and waving an outstretched arm in front of moving traffic is downright dangerous. Upon seeing a hiker drivers will stop—if they want to. A girl with no luggage, thumbing a lift will often stop a car, but if a boyfriend suddenly emerges from hiding with two large rucksacks, the driver of a small vehicle will understandably be infuriated. The politest way of requesting a lift was one I discovered in Fairbanks, Alaska. 'If you want to get a ride from here, 'phone the radio people', I was told. 'Your message will be broadcast immediately. This is a remote town, so we all try to help each other'. I did just that and somebody 'phoned to offer me a lift on the same day.

Hikers should be prepared to give something in exchange for a ride such as interesting conversation, help with map reading or an English lesson. If they join the vehicle for a longish time, say more than a day, they should offer to pay their share of petrol and tolls and of course also pay for their own food and lodging. All the same, they should remember that they are guests on somebody else's property. That means no requests for adjustments to windows, heating etc., unless prompted by the driver. When wet and muddy, it is considerate to wait and see whether one's host wishes to protect the car seat or floor before one enters. The hiker should be prepared to leave a vehicle when asked to. Drivers may require time to themselves for the last lap of a journey; moreover they should certainly not be made to feel guilty if their destination falls several miles short of one's own. Personal hygiene and dress are important also. Of course one will look like 'a hiker', but verminous, unwashed hair, or an attire which is regarded as unsuitable for that part of the world can be an embarrassment. On the whole, people who give lifts are themselves kind and unselfish.

All over Europe, Australia and New Zealand, drivers of cars and trucks are 'geared' to hitchhikers, so let us look beyond. In many Far Eastern countries car travel is popular now and hikers will readily be offered lifts through Sri Lanka, Thailand, Malaysia, Singapore and Japan. In rural India cars are less common but there drivers often go out of their way to aid a traveller—which may not only involve transport but board and lodging as well. In Laos, Vietnam and the Philippines, hitchhiking on trucks is relatively easy, but not always advisable for girls. The same applies to South America, along whose roads *camions* with cheerful drivers (often ex-cowboys), rumble for days and nights on end.

Through most of East, West and South Africa, there is considerable traffic. In central African countries where long distances over inhospitable terrain must be covered, trucks are the commonest vehicles found, and this applies to the Sahara Desert region. To cross the Sahara the two most popular routes northwards run from Gao and Agadez. The latter is a longer, more varied route and that is the one I chose. An Algerian driver smuggling Nigerian sheep took me all the way across the Sahara to his home in El Golea. With this hospitable desert Arab and his companions I spent bitterly cold nights beneath a canvas spread on the sand, got chased by police between In Salah and Raggane, cheered when we escaped (with the sheep), and kept the pre-Ramadan fast.

In jungle-covered lands such as Zäire, the Sudan and the Gambia, roads are less frequented than rivers. The same applies to the Amazon region of Brazil and to parts of Laos. Thus one's transport may vary. . . . from the pillion of a motor-bike to enormous trucks with thirty-four wheels on the ground. When one gets right off the beaten track it is not uncommon to find people who own little boats or private 'planes. Two of the occasions on which such free lifts were offered to me involved a boat journey along the Mekong River from Vientiane to Savannakhet in Laos and a flight between St. Louis (Senegal) and

Nouakchott (Mauritania). At times one may cover vast distances in the same vehicle, such as when a new car is being delivered from one country to another.

Here I should mention one exception to walking on for miles between lifts, and that is through vast stretches of sparsely populated territory, where it is either very cold or very hot. It has not been unknown for hikers to perish in such conditions, for most are either unequipped with adequate warm gear for the former or enough water for the latter. Such a place is Australia's Nullarbor Plain. The traveller who enters it on foot must beware. Here, the temperature can soar to 45°C in the shade and in many places it has not rained for years. At a place called Karalee stands the only hotel between Coolgardie and Southern Cross. Right on the railway line, it is the only place within 116 miles that can supply one with a drink. Villages shown on the map become increasingly smaller on the ground as one penetrates the desert, diminishing to one house in the scrub with a sign up saying 'No petrol here'.

Rivers too can foil the hitchhiker. In the same country, the road between Wyndham and Darwin may suddenly become impassable from mid-November, when the rivers begin to run.

The traveller off the beaten track will encounter a great variety of types of lodging, from free sleeping places such as railway stations, barns and beaches to paid accommodation. Youth Hostels give excellent value for money; so do YWCAs and YMCAs. Invitations to stay with people should never be expected, but it would be a rare hiker who returned home without having enjoyed the thrill of living as 'one of the family' with hosts of other lands.

One soon discovers that there are situations where both hiking and hitchhiking would be either unwise or impossible. Then one must investigate the possibility of paid transport, which ranges in price from the hundreds of dollars charged for visiting the Pacific islands on a liner to the mere 150 cents (17½p) one pays per 24 hours to see Sarawak on a bicycle.

When funds for food, transport and accommodation run low these can be boosted by temporary jobs. During 'the season' New Zealand welcomes all willing apple pickers and tobacco stringers. If one has exciting tales to tell, radio and newspaper people are often interested, while teaching English is always a reliable stand-by. Giving blood is a way of making money in some countries.

'But there must be endless difficulties for the hitchhiker to face', people mutter. Well, let us look at a few potential ones, starting with language. Most people who offer lifts speak some English or another 'international' tongue such as French, Spanish or German. And if not, a friendly smile and the use of sign language are universally accepted methods of communication. Hitchhiking, on the whole can be considered dangerous only in western countries.

Then there are the obvious hazards faced by a girl travelling alone. As one who knows, I can confidently assure readers that if a girl is modestly attired and makes clear her reason for hitch hiking, problems should be minimal. My only really frightening experience resulted from travelling with drivers who, unpredictably, were to get them selves intoxicated with drugs (such as from chewing *coca* leaves) or alcohol. A direct 'Will you come to bed with me?' type of approach from a normal male can invariably be rebuffed by laughing it off, discussing his family, expressing great anxiety at such a proposal, or by pleading sudden sickness and leaving the vehicle rapidly. . . .

Finally, it is sad but true that many a journey through otherwise unspoiled lands is hindered by political disputes and sometimes violence. Such was my experience in Ethiopia, Sudan and war-smitten North and South Vietnam.

My advice here would simply be: 'obey the rules' and 'never interfere'. A politely worded, sensible sounding request to continue one's journey is generally granted, as long as one remains courteous and carries travel documents which are in impeccable order.

Here then are some tips gleaned from personal experience. Prices will no doubt have altered since my own journey in the nineteen-sixties; so may place names, such as Kampuchea instead of Cambodia. But people themselves won't have changed, and he who travels as a worthy ambassador of his own nation will win many hearts overseas.

So, good luck to you on your travels off the beaten track, and, should lifts at times be few and far between, remember these words of a Chinese philosopher:

'The journey of a thousand miles begins with one step'.

Sailing: The Rough and the Smooth

by Bob Nance

For most of us, voyaging by yacht has romantic connotations. The reality is that it is a time-consuming and costly business. A good ground rule to apply is to estimate the time and cost involved, double both and then add a little more.

In fitting out a yacht it is wise to use the best quality equipment. Regard each piece of equipment as a long term investment and it will pay off in maintenance costs and safety. Cut costs, if necessary, in some other area so as to be able to purchase some valuable item of equipment.

There are always differences of opinion amongst sailors about the quantity and type of electronic equipment suitable for a yacht. The larger yachts may have radar, Loran or Omega systems, which are of great value for navigation, but they are expensive and require a power source. Away from Europe and North America, maintenance and replacement parts could be the cause of delays. Many smaller yachts, including my own, *Tzu Hang,* carry no electronic equipment other than a small radio direction finder and radio receiver. There have been many instances where lives have been saved by the use of a radio transmitter. By choice, I do not carry one: I prefer to rely on our own resources should we strike trouble and always make crew aware of this fact.

The trickiest part of using a radio transmitter is to know which crystals for which frequencies should be used in different parts of the world.

A radio direction finder can be of great value when making landfalls or cruising along coastal areas in most parts of the world; and it is desirable also to have a good depth sounder.

When provisioning for a voyage to remote parts, it is usually better and cheaper to stock up on bulk foods such as rice and flour and on canned foods before leaving Europe or North America. Though bulk foods vary little in price throughout the world, their quality in Africa or South or Central America tends to be inferior. Canned foods last for a long time. I have carried cans aboard *Tzu Hang* for up to four years and found the contents still in good condition. Meat needs to be bought bottled or canned, but fruits and vegetables and some other provisions can be bought fresh at local markets. Buying fresh foods where possible is preferable and generally cheaper. Fresh food keeps well at sea, even without refrigeration. Recently we cruised for over two months in the Patagonian Canals of Chile without being able to purchase fresh supplies and without refrigeration. Remember that high islands will normally have some fresh fruit available, while low islands and atolls are unreliable for all fresh provisions, including water.

Fresh supplies can be supplemented by fish, for those skilled enough to catch them, but it is worth noting that some reef fish in the tropics are poisonous. Shellfish are usually abundant in colder areas, but it often takes a diver equipped with a wet suit to catch them.

A last word on provisions: freeze-dried foods are excellent, but the large quantities of water required to make them up may prohibit their use on a yacht.

In selecting crew, beware of overcrowding, which is one of the major sources of trouble. A skipper should avoid being dependent on a large crew: at worst, he may be held to ransom by a mutinous crew, aware that he cannot proceed without them.

We all tend to think we are free of officialdom once on a yacht, but there are still plenty of formalitites to be observed. In planning a cruise, apply for visas and permits ahead of time. The consulate or embassy of the country concerned will advise you what papers are required for entry and exit. Knowing what is in store is the best way to avoid irritations, fines and restrictions.

It is proper, and avoids creating friction with officials, to fly the quarantine flag and the flag of the country you are entering as a courtesy. Whether you arrive at night, at a remote spot or in a harbour, always seek out officials as soon as possible, if necessary making your way to an official entry point or town. Most countries have quite a lot of paperwork to enter a yacht, which can entail seeing several officials and, in some cases, checking in and out of each port. In many countries, a patrol boat coming across your yacht anchored and yourself perhaps ashore without papers for entry will have you promptly arrested and your boat towed under guard to the nearest police or military station. In Australia, the United States, Canada and New Zealand, to name but a few countries, your boat will most likely be confiscated if you enter without going through customs. In these countries, and in the Pacific and the Caribbean Islands, officials generally process incoming yachts

quickly and efficiently. But it is always as well to allow plenty of time for this processing and essential to conform to the regulations.

It is acceptable to anchor in a remote bay or anchorage in many countries provided that you fly the quarantine flag and do not go ashore. In case of emergency, the Captain can go ashore alone to make contact with officials.

Occasionally you may encounter corrupt officialdom and be asked for illegal payments. Start by refusing to pay and request to see the head of the department concerned. If the official does not retract and the amount is small, it is probably better to show discretion and pay up. Carrying firearms is usually very troublesome: declare them and in many ports they will be taken from you, to be returned only when you depart.

Don't set out on a voyage without leaving behind you a reasonable bank balance that can be drawn upon in emergencies. As everywhere else, so too at sea money is a necessary part of living. Your reserve fund need not, however, be enormous. Nobody can plan for catastrophic expenses. It is still possible to work your way around the world to meet your travel expenses, but to rely on this means of support nowadays is risky, as each year there are more and more restrictions and greater numbers of people travelling.

Maintenance of the yacht will be of prime importance and as much spare equipment should be carried as possible. Paints vary in quality, but can be obtained everywhere, though bottom paint is frequently not available and usually exorbitantly expensive. Again it is best to carry a good supply with you. When hauling the yacht or getting work done at yards and yacht clubs, remember always to make sure of the costs and have them written down. Verbal arrangements sometimes lead to misunderstandings.

Plan ahead, take reasonable precautions, maintain the courtesies and you will find that the benefits of sailing off the beaten track are great and far outweigh the inconveniences.

Overland by Public Transport

by Chris Parrott

It's not everyone who has the resources to plan, equip and insure a full-scale Range Rover expedition across one of the less developed continents, though it's the sort of thing we all dream about. Ready to answer these dreams are several overland companies who offer three weeks to three months of adventure, be it in a three ton truck camping all the way from London to Johannesburg, a minibus and cheap (usually called 'budget') hotels for two weeks in Morocco, or six weeks by luxury coach and luxury hotels from Istanbul to New Delhi. The trouble with this kind of trip—apart from the obvious drawback of not being able to choose your fellow-travellers and possibly having to live for three months with someone who has got up your nose since Day 1—is that these overland companies are *companies.* They're in business to make money, and after the transport costs and overheads have been met, there has to be some profit to make it all worth their while.

The better the company, the more efficient the organisation, the higher are the overheads, and so the price to the customer.

You can do it all more cheaply on your own, by public transport. Generally speaking, wherever overland companies take their trucks, there public transport goes too. And often public transport goes where overland companies cannot—up (or down) the Nile, over the snowbound Andes to Ushuaia in Tierra del Fuego, across Siberia to the Pacific.

Of course, Damascus to Aleppo is not quite the same as getting on a coach to Washington D.C. at the New York Greyhound Terminal, nor does 'First Class' imply in Bolivia quite what it does on the 18.43 from Paddington to Reading.

The Damascus to Aleppo bus is an ancient Mercedes welded together with the remains of past generations of Damascus/Aleppo buses, and propelled in equal proportions by a fuming diesel engine, the Will of Allah, and the passengers (from behind). It makes unscheduled stops while the driver visits his grandmother in Homs, when the driver's friend visits the Post Office in the middle of nowhere, and when the whole bus answers the call of nature—the women squatting on the left, and the men standing on the right (the French usually display more cool at moments like this.'

First Class in Bolivia means hard, upright seats already full of people and chickens spilling over from second class, whimpering children, no heating even in high passes at night in winter, passageways blocked by shapeless bundles and festering cheeses, impromptu customs searches at 4 a.m., and toilets negotiable only by those equipped with Wellingtons and a farmyard upbringing. Trains rarely arrive or depart on time, and the author has experienced a delay of 26 hours on a journey (ostensibly) of eight hours. But these trains are nothing if not interesting.

The secret of the cheapness of this means of travelling lies in the fact that it is *public,* because the very fact that it is public transport indicates that it is the principal means by which the public of a country moves from place to place. It follows that if the standard of living of the majority of people is low, so will the cost of public transport be low. A 20-hour bus ride from Lima to Arequipa in Southern Peru costs only $10; a 20-hour bus ride in Brazil from Rio de Janeiro to the Paraguayan border costs about $20; whilst a 20-hour bus ride through France or Germany would cost $40 or more. It all depends on the ability of the local population to pay.

An English company, Trail Finders of Earls Court, London, has taken advantage of this and has introduced a series of low-cost group tours using public transport to Asia, up the Nile, around the Central Andes, and across Russia. These trips all have the advantage that you can travel fairly cheaply, but someone else does the worrying about whether the next bus will be full.

Of course, there are certain disadvantages to travel by public transport:

a) Photography is difficult at 70 mph, and though most drivers will stop occasionally, they have their schedules to keep to.

b) You may find out that all transport over a certain route is fully booked for the week ahead, or there is a transport strike.

c) You may find that your seat has been sold twice. In circumstances like this, tempers fray and people begin to speak too quickly for your few words of the local language to be of much use.

Efficiency of reservation arrangements varies from one part of the world to the next. The following may serve as a general guide to travelling in the underdeveloped parts of the world.

Booking. Whenever you arrive in a place, try and find out about transport and how far ahead it is booked up. It may be, for example, that you want to stay in Ankara for three days, and that it's usually necessary to book a passage four days in advance to get to Iskenderun. If you book on the day you arrive, you have only one day to kill; if you book on the day you intended leaving, you have four days to kill. This is a basic rule and applies to all methods of transport.

Routing. Try to be as flexible as possible about your routing and about your means of transport. There are at least six ways to get from La Paz in Bolivia to Rio de Janeiro in Brazil. Check all possible routes before making a final decision.

Timing. Don't try and plan your itinerary down to the nearest day—nothing is ever that reliable in the less developed world (or the developed world for that matter.) You should allow a 10-20 per cent delay factor if, for example, you have to be at a certain point at a certain time to catch your 'plane home.

Possessions. Baggage is often snatched at terminals. Be sure, if you are not travelling within sight of your bags, that they have the correct destination clearly marked, and that they do actually get loaded. Breakfast in New York, dinner in London, baggage in Tokyo happens all too often. Arriving or leaving late at night or early in the morning you are particularly vulnerable to thieves; this is the time when you must be most on your guard. Never leave anything valuable on a bus while you have a quick drink or dispose of a quick drink, not even if the driver says the bus door will be locked.

Borders. Prices rise dramatically whenever a journey is made during the course of which a national frontier is crossed. Usually it's cheaper to take a bus as far as the frontier, walk across, and then to continue your journey by the local transport in the new country. 'International Services' are always more expensive, whether airlines, buses, trains or boats. (The author recalls that a donkey ride to the Mexican frontier was 20 pesos, but to have crossed the international bridge as far as the Belize Immigration office, an extra 40 yards, would have increased the cost to 40 pesos.)

Each particular medium of transport has its own special features.

Trains are generally slower than buses, and the seats may be of wood. There is often no restriction on the number of seats sold, and delays are long and frequent. However, slow trains make photography easier, and the journeys are usually more pleasant than on buses if not too crowded. It's often worth going to the station a couple of days before you're due to leave and watching to see what happens. It will tell you whether you need to turn up two hours early to be sure of a seat.

Trains in Peru, India, Pakistan and Zaïre (Elizabethville-Port Francqui) offer reasonable services. The most famous train service of all must surely be that which chugs between Peking and Moscow: six inexpensive days and nights on the Trans-Siberian Express.

Buses reflect the sort of terrain they cross; if the roads are paved and well maintained, the buses are usually modern and in fair condition. If the journey involves unmade mountain roads, your bus and journey are not going to be very comfortable.

Buses sometimes go where other vehicles do not, e.g. in parts of the Middle East, between Panama and Mexico and over the Andes mountains. Here passengers are liable to have to push the vehicle for much of the way. In some countries, e.g. Colombia, where there are bandit scares, a traveller is safer on a bus than, for example, hitchhiking or travelling independently on the road; moreover a 300 mile bus journey here can be had for less than the price of a cheap meal at home.

Boats. This, if you're lucky, could mean an ocean-going yacht that takes passengers as crew between, say, St. Lucia and Barbados, or a cement boat from Rhodes to Turkey or an Amazon river steamer. With a little help from your wallet, most captains can be persuaded to accept passengers. A good rule is to take your own food supply for the duration of the trip and a hammock if there is no official accommodation.

Cargo boats ply the rivers Amazon, the Congo and Ubangi in Zaïre, the Niger in Mali, the White Nile in the Sudan, the river Gambia and Ecuador's River Guaya, where an all-night crossing costs about 13 *sucres* (30p).

Planes, in areas where planes are the only means of communication, are often very cheap or even free. Flying across the Gulf of Aden to Djibouti, for example, costs as little as sailing. A good trick is to enquire about privately owned planes at mission schools (in Africa) or at aeroclubs. Someone who is going 'up country' may be only too pleased to have your company. Similarly, in parts of South America, the Air Forces of several countries have cheap scheduled flights to less accessible areas, though of course, one must be prepared for canvas seats and grass runways.

Hitch-hiking. Where a new road has been opened there is often a delay of a few years before regular public transport is introduced. In this case, hitchhiking is a recognised means of travelling and one usually waits at some appropriate picking-up point—say a ferry or petrol station— and enquires what room a driver has and what remuneration he is prepared to accept. Have everything agreed before you accept and make sure he is going where you are going and is not likely to leave you in the middle of the Sahara or the Amazon with no food or transport. Half the payment now, half on arrival, is a good rule.

Passenger-cargo trucks, a kind of halfway house between public transport and hitchhiking, provide cheap, if uncomfortable, ways of getting about in Africa, India and South America.

One final thing: if you do go on your own, or with a friend, don't worry about being lonely. The locals will attempt to communicate in sign language and there'll always be someone else like you trying to do the trip, wherever you are. . .

The Great Overland Routes

by Ingrid Cranfield

To claim to have driven overland across Asia, through Africa or around South America is nowadays likely to cause no raised eyebrows, though the experience itself is, as it always was, well worth having. Twenty years ago such journeys were major adventures; today they seem almost commonplace.

In 1922-23 a new kind of vehicle, the halftrack, invented in Russia and developed by André Citroën, was tested on the first Trans-Sahara crossing. The journey of 2,000 miles took 22 days and finished in Tombouctou.

In 1924, another Trans-African journey took place in eight vehicles. This was La Croisière Noire, under the direction of George-Marie Haardt and Louis Audoin-Dubreuil, which left Béchar in Algeria and reached Kampala, Uganda, before splitting into four groups. Three of these headed for the east coast and reached Madagascar; the fourth, led by Haardt, met great obstacles on the way to Mozambique. First there were swamps, which bogged them down. Travelling at five miles per day, they navigated for a time entirely by compass. Then came vast plains with soft going, firmer ground and a bush fire, when the vehicles barely escaped as their tyres burst and their caterpillar treads burned. From Mozambique the team went to rejoin the other groups at Madagascar and a small section of the party then continued to Cape Town. The journey carried out a full scientific programme and returned with film, photos, drawings and thousands of specimens of mammals, birds and insects.

By the late 1920s, journeys overland through Africa were becoming fashionable. It was not, however, until 1954 that the first overland Trans-African trip, using modern expedition vehicles, took place. This was the Oxford and Cambridge Trans-African Expedition, which bought their own Land-Rovers and drove them from London to Capetown and back in the four months of the summer vacation. The 'home rep.' was Adrian Cowell, who has since become known for his filming and writing on South American Indians. Cowell, too, conceived the idea for an overland expedition the following year, to drive from London to Singapore and back. No-one at that time had ever got further than Calcutta, though a few had tried. Roads leading east from the Patkai and Naga Ranges into Burma had been cut and bulldozed in 1944, held open for a time and then allowed to fall into disrepair.

So was born the Oxford and Cambridge Far Eastern Expedition. Under the leadership of Tim Slessor the team succeed in travelling to Singapore in six months and six days, and then back again (not entirely by land), a journey of 32,000 miles through 21 countries.

They proved long-distance overlanding was possible. And they discovered a truth revealed again and again to countless overlanders since, that 'places seen are no longer just names but *places*—because one has been to them'. They experienced conditions far worse than any the roads have to offer today: on one stretch (Ramadi to Baghdad) 'appalling potholed earth' for 65 miles; on another (Takow to Tachilek) for 218 miles, a road 'mostly bad, very hilly and narrow'. They put their washing with water and detergent in a pressure-cooker and let the vehicles' vibration do the rest. They were criticised for their military approach but knew that 'any fool can be uncomfortable and, as a result, ill-tempered—and a Land-Rover is a very small place.'*

And so they laid the basis for the last two decades of overlanding. Thousands of people, mostly young and mostly western, have driven in their wheelmarks since. But that does not mean that the principal overland routes have been stripped of their excitement or their challenge. To be 'first' is a special achievement; but every traveller claims his own different kind of achievement when he sees something for his 'own' first time.

None of the three routes described below should ever be undertaken lightly. They are great routes, scenically, culturally and historically. They should be embarked upon with due mental and physical preparation, caution—and even with some sense of awe.

Trans-Asia
London—Kathmandu

The Trans-Asia overlander starts his journey in London, takes to a Channel ferry and scuttles across Europe, via Belgium, the Rhine valley, Austria, Yugoslavia and Greece, until he arrives, with relief and anticipation, in Istanbul, the crossroads of two continents. Turkey has much to offer the traveller with a sense of history for it is strewn with the remains of ancient civilisations — Roman, Greek, Persian and Christian. There are Roman baths, amphitheatres, temples, crusader castles and whole ruined cities to be found along the road through Asia Minor and along Turkey's Mediterranean coast.

*Tim Slessor: *First Overland. The story of the Oxford and Cambridge Far Eastern Expedition.* London, Harrap, 1957.

One variant of the route now turns inland and crosses the high pass known as the Cilician Gate. In this part of central Turkey, villages hewn out of rock and troglodyte dwellings and churches are to be found. From here there are two routes into Iran. One passes Mt. Ararat and enters northern Iran, heading for Tehran. The other passes Lake Van and heads by way of the Zagros Mountains for Isfahan, famed chiefly for its buildings and secondarily for its adherence to traditional crafts. Skimming the edges of the Great Salt Desert, the traveller heads for Persepolis, the ancient royal city from which King Darius sallied forth to extend the bounds of his vast Persian Empire 2,500 years ago. An alternative is to turn south from Turkey into Syria and Jordan, cross Iraq, enter Iran from the west and head for Persepolis.

Eastern Irán holds rough conditions in store for the motorist; first there is the Dasht-e-Lut, the huge desert region of white salt pans, then there is the dusty, arid landscape of southern Iran. Herat is the first town reached in Afghanistan.

There is one good road through Afghanistan, the southern route, which is a highway and much frequented by commercial vehicles. Central Afghanistan boasts the mountains, lakes and high passes of the Hindu Kush, as well as the two colossal 2,000-year-old Buddhas carved in the face of sheer cliffs at Bamiyan, but the route is inhospitable and even impassable in winter and both this and the northern route are virtually beyond the capabilities of tourist vehicles. From Kabul it is a short day's drive to the Pakistan border.

The road then winds up past the forts of the Khyber Pass, scene of invasions into India by Aryans, Persians, Greeks, Tartars, Moghuls and Afghans. On the other side is Peshawar, the Indus Valley, Rawalpindi and Lahore, on the Indian subcontinent's vast flat expanse of plains. The route passes through Amritsar, heart of the Indian Punjab and the holy city of the Sikhs, with its Golden Temple set in a carp-filled artificial lake. Travellers wishing to visit Jammu and Kashmir, Ladakh (Little Tibet) or the Karakoram mountains beyond, turn northwards here. Others travel east through the Indian plains, the route encompassing Delhi, Agra with its famous 17th century mausoleum, the Taj Mahal, the deserted one-time Moghul capital Fatehpur Sikri, and Benares on the Ganges with its many temples dedicated to Siva, the 'great destroyer'.

Crossing the Ganges near Patna, the traveller enters southern Nepal and a region of low-lying tropical jungle. This gives way to foothills and finally, beyond mountain passes some 2,000 metres high, the peaks and valleys of the Greater Himalaya itself. A new road leads to the capital of Nepal, the once isolated and inaccessible town of Kathmandu. Since Burma does not currently admit tourists by land—there being no linking road!—most overlanders end their journey here.

Trans-Africa
London to Johannesburg

The Trans-Africa route, north to south, begins either in northern Tunisia or in Morocco. The former passes through the salt-pans of Nefta, the souk of Kairouan and into Algeria; the latter through Fez with its Medina and across the High Atlas mountains to the Algerian desert. The two route variants meet at Ain Salah in central Algeria. From here, it is a long journey across the rock plateau of the central Sahara (a detour takes in the weird desolation of the Hoggar mountains) to Tamanrasset, Algeria's southern administrative centre, home of the famous Tuareg cameleers, the 'blue men of the desert'.

The route proceeds via Assamaka, the isolated border check-point into Niger. In the soft sand of this country, camel trains may appear from the middle of nowhere, and the winds are capricious, making for treacherous going. At the edges of the desert, tribal herdsmen are to be seen gathered around each well with their cattle, which graze on the brittle savannah. Agadez marks the southern limit of the desert; from here monotonous scrub characterises the three day journey to Zinder and the good road beyond that leads to the Nigerian border. Kano, the thousand-year-old, bustling city, marks a change in direction and the beginning of the journey through the lush equatorial forest of central Africa. The route turns due east and leads to the farmland of northern Cameroon, with its red dust tracks that turn to mud after tropical downpours. Further to the south and east is Bangui, capital of the Central African Empire and 450 miles to the east again, along an extremely bad dirt road, Bangassou, centre of a coffee-growing region.

The roads into Zäire are badly maintained, the bridges across the Ubangi River decrepit and dangerous, the ferries ancient and rundown. Several roads lead through the forests of north-eastern Zaïre to the western slopes of the Ruwenzori mountains, with their eucalyptus woods. A descent of the escarpment brings the traveller out onto flat country populated by zebra and wildebeest. From Lake Kivu the road crosses from Zaïre into Rwanda and thence winds its way towards Tanzania. Graded dirt roads cross dry, sparsely settled scrubland to Mwanza on Lake Victoria, from where it is a day's drive to the entrance to Serengeti National Park. Other game parks in East Africa are Masai Mara, Amboseli, Tsavo (Kenya), Manyara and Ngorongoro (Tanzania). Tanzania's central region, the Masai Steppes, is dominated by Kilimanjaro, 5,895 metres, an extinct snow-capped volcano.

Travellers usually take the coast road beside the Indian Ocean passing through Bagamoya and Dar es Salaam. From Tunduma on the Tanzania/Zambia border, the tarmac road (once known as the 'Hell Run') heads south-west towards Zambia's copperbelt, its capital Lusaka and the spectacular 120 metre high, mile-wide Victoria Falls in the south. 2,000 miles from Dar es Salaam is the great Zambesi river, crossed by ferry at Kasungula. The river forms the boundary between Zambia and Botswana, where dirt roads lead south into the Republic of South Africa. Johannesburg is commonly the finishing point for the Trans-Africa journey.

A four-wheel-drive vehicle is highly recommended on the African route, especially in the rainy season. High ground clearance, such as a Volkswagen bus gives, is absolutely essential. The Trans-Africa route is extremely difficult—quite unlike Trans-Asia—and should be attempted only in a reliable vehicle by people who know what they are about.

Around South America

There is no real 'standard' route through South America, though most travellers, by necessity or design, pass through the same key cities and locations.

It is customary to begin in Venezuela (Caracas) or Colombia and head southwards along South America's Pacific edge. The coast here boasts beautiful golden beaches, clear water and crystal streams cascading down from the 5,800 metre summits of the Sierra Nevada. To the south is the big industrial port of Barranquilla and then Cartagena, an impressively fortified town dating from 1533, through which for nearly 300 years gold and treasures were channelled from throughout the Spanish colonies. Passing through hot swampland and then inland up the attractive forested slopes of the Cordillera Occidental, the traveller emerges on a high plateau where Bogotà is sited, at 2,620 metres. The Gold Museum has over ten thousand examples of pre-Colombian artefacts. An hour away are the salt mines of Zipaquira, inside which the workers carved an amazing 23 metre high cathedral.

South from Bogotà are the Tequendama Falls, the splendid valley of the Magdalena River and, high up on the Magdalena Gorge, the village of San Agustin. Here hundreds of primitive stone statues, representing gods of a little-known ancient Indian culture, guard the entrances to tombs. The road then loops back over high moorland to Popayan, a fine city with monasteries and cloisters in the Spanish style. The tortured landscape near here has been said to resemble 'violently crumpled bedclothes'; tilled fields on the opposite mountain faces 'look nearly vertical'.

So the road crosses into Ecuador. Just north of Quito the equator, *La Mitad del Mundo,* cuts the road, a few hundred metres from the grand stone monument built, unfortunately in the wrong place, to mark the meridian. Quito itself is at 2,700 metres, ringed by peaks, amongst them the volcanoes of Pinchincha. It has much fine colonial architecture including, according to *The South American Handbook* eighty-six churches, many of them gleaming with gold.

Travellers then cross the Andes, passing from near-Arctic semi-tundra, through temperate forest, equatorial jungles and down to the hot total desert of the Peruvian coast, punctuated by oases of agricultural land where irrigation has distributed the meltwaters from the Andes over the littoral. Here too the ancient empires of the Chavin, Mochica, Nazca and Chimu people flourished. Ruined Chan-Chan, near Trujillo, was the Chimu capital; nearby Sechin has a large square temple, 3,500 years old, etched with carvings of victorious leaders and dismembered foes.

The coast near Lima is picturesque and rich in fish and birdlife, owing to the Humboldt Current. Lima itself has both shanty towns *(barrios)* and affluent suburbs, parks and fine beaches. Well worth seeing are the National Museum of Anthropology and Archaeology and a number of other museums. There are alternative routes here. One can branch off the Lima road at La Oroya and pass through the Huancayo plains and mountainous countryside before reaching Cuzco, or follow the fast coast road, through the desert, past the famous, vast and little understood Nazca lines, before cutting back inland into the mountains, either via Abancay or along the better road via Arequipa. All the routes cross the Andean heartland of Peru and are correspondingly tortuous.

Cuzco sits in a sheltered hollow at 3,500 metres. This was the capital of the Inca empire. Inca stonework forms the base of many of the Spanish buildings and the ancient city layout survives to this day. Overlooking Cuzco's red roofs is the ruined fortress of Sacsahuaman. Nearby too, are the ruins of Pisac and Ollantaitambo and, reached by train down the valley of the Urubamba, (further upstream called the Vilcanota) the 'Lost City of the Incas', Machu Picchu. This magnificent ruined city sited nearly 500 metres above the river was overgrown with jungle until its discovery in 1911. After the sacking of Cuzco, the Virgins of the Sun are said to have fled to this city whose existence was unknown to the Spanish.

From Cuzco the road crosses the watershed of the Andes to the dry and dusty Altiplano, a high treeless plateau stretching from here to and across much of the Bolivian upland. Here lies Lake Titicaca, at 3,810 metres the world's highest navigable lake, blazing a deep blue because of ultraviolet rays. On the floating reed islands of the lake live the Uru-Aymara Indians. Across the border in Bolivia are the ruins of Tiahuanaco, relic of an ancient race; the main feature is the carved 'Gate of the Sun'.

La Paz lies in a valley just below the rim of the Altiplano, at 3,350 metres. It is the world's highest capital, famous for its bizarre Indian markets, street

corner pharmacies and rich store of folk music.

South of the city, the roads gradually peter out over the quicksands and salt-pans that stretch between Oruro and the borders of Chile and Argentina. This area should be traversed circumspectly and only in the dry season. Northern Chile is desert country, known as the High Atacama. For those continuing south towards the tip of the continent, the most often preferred route is that which runs the length of Chile. Central Chile has farmland and vineyards, while southern Chile is a mountainous region that has been called 'the Switzerland of the Americas'. It is usual to cross the border into Argentina at a point south of Santiago and visit Bariloche, situated in a national park and now a fashionable ski resort.

The part of the continent south of 40°S is known as Patagonia and it is through this region of tundra scrub that roads run straight for mile upon mile. Worth visiting in Argentinian Patagonia are Lago Argentino with the spectacular Moreno Glacier, the oil town of Comodoro Rivadavia and the Valdés Peninsula, a remarkable spot for viewing wildlife in its natural habitat.

At the southernmost tip of South America is the island of Tierra del Fuego, reached by ferry across the Magellan Straits. The scenery is wild and desolate, the weather often vicious.

Heading north again, one crosses the Rio Negro and enters the *pampas* (large treeless plains) where cattle graze on ranches that stretch over the horizon and beyond. Buenos Aires is on the road to Asunción in Paraguay.

Travellers in South America often, because of politics or personal choice, dispense with the southernmost regions and choose an alternative route from Oruro. It is possible to go through Chile to Santiago, on to Bariloche and then turn north; or through the southern Bolivian towns of Cochabamba, Sucre and Potosi, over the rocky Sierras into Argentina, as far as Salta and then eastwards; or south-eastwards through the fertile Paraguayan *chaco* (a warmer, more humid kind of *pampa*) to Asunción.

A good road runs from the plateau country of eastern Paraguay to the Brazilian border, crossed usually via the International Friendship Bridge over the Rio Parana. An hour away are the Iguaçu Falls, where between two and three hundred cataracts fall over a gigantic curved rim of cliffs at two main levels. Catwalks built over the river permit fine, if wet, appreciation of the exceptional views.

Dense tropical jungle lies between here and the cities of Curitiba and São Paulo. A beautiful coast road from Santos leads to Rio, one of the world's most exciting and beautiful cities—chic, vibrant and varied.

At this point, overlanders head north-west through the hilly jungles of Minas Gerais. Brasilia, that ultimate in planned cities, is an oasis of modernity in the forest. From here two main alternatives present themselves. One is to go north along the Belém Highway to Estreito, then take the Trans-Amazon Highway to Altamira on the lower Xingu river, head west to Santarém on the Amazon and go on upriver to Manaus—a not too comfortable journey by local riverboat of three days. The other way is to follow a new track west to Cuiabá in the heart of the Mato Grosso and from there turn north for Santarém.

The Amazon jungle roads are almost all earth tracks and therefore offer rough going, because they are either dusty or muddy. The jungle is rich in animal life, with armadillos, giant anteaters, jaguars and anacondas; but the indigenous tribes have mostly long since melted away from the immediate vicinity of main roads and it is rare to meet native people who have been untouched by the process of westernisation.

Riverboats ply the Rio Negro and the Rio Branco, tributaries of the Amazon and they provide a break from overlanding and a convenient, if primitive, way of visiting remote villages. From Manaus, the route continues north across the border into Venezuela. The road between Boa Vista (Brazil) and the gold-mining town of El Dorado (Venezuela) winds through spectacularly beautiful country, passing sheer-sided Mount Roraima at the junction of three countries. A sidetrip here is recommended (or a 'plane trip) to the 1,000 metre Angel Falls. At Perto Ordaz the traveller crosses the Orinoco and soon is back at his starting point of Caracas.

Travelling Off the Beaten Track with Children

by Rowena Quantrill

Many people believe that once they have children, travel must be restricted to the safe and conventional, at least until the children have reached their teens. This need not be so. Children are remarkably adaptable and with forethought and planning can be taken almost anywhere. We have travelled extensively and happily, even when our children were very small, and recently completed an overland journey from Kathmandu to London which our four children, aged between six and twelve, look back on as one of the best experiences of their lives.

There are, however, two important facts to remember when travelling with children. First, it will be harder work than travelling alone, and second, children can usually only put up with a limited amount of sightseeing or shopping. If you want to visit the bazaar or the nearest temple in the morning, find somewhere to swim or scramble on rocks in the afternoon, or vice versa. There are also other advantages and disadvantages. On the plus side children learn responsibility and practical information; the unfamiliarity of the outside world tends to knit a family more closely together; and discounts in travel and accommodation are often such that it is cheaper to bring children with you than to leave them behind, where you will have to pay to have them cared for or send them on an alternative holiday e.g. to camp, and worry about them into the bargain. Most hotels allow children up to the age of 12, or even older, to share a hotel room with a parent free. Drawbacks include the need, especially for a woman travelling alone with children, to find sitters from time to time along the way; and the fact that children often miss their friends and complain bitterly about it. Parents looking for babysitters abroad will find that local student organisations often offer such a service; hotels too may provide supervised nurseries and playgrounds and an evening sitting service.

One of the things that worries people

most about taking children off the beaten track is the possibility of serious illness. Provided however that you take sensible precautions there is no need to worry more about children than adults. Obviously the whole family should have any recommended inoculations for the area you will be visiting and you should ask the advice of your doctor about what medicines to take with you. Insect repellent is useful for most places. Take great care with drinking water in any underdeveloped area. Try to find out from expatriates in the country you are visiting which local ice-cream, if any, is safe to eat, as the children are bound to want some.

Most people going to more remote areas have, of necessity, to travel light. The easiest thing to cut down on with children is clothes. So long as what you take is easy to wash and dries quickly (polyester/cotton mixes are good) you can get away with two outfits. Remember to take along a sewing kit and some spare buttons for on the spot repairs. Do make sure also that you have something waterproof for them and a warm sweater, as even in the tropics it can turn cold. Children generally tolerate heat better than adults but feel the cold more and a cold child is a miserable one. It is usually worth carrying a small supply of emergency food e.g. peanut butter, cheese, cans of baked beans, etc. If your child will eat fairly adventurously this will probably be enough but if he's fussy it may be worth taking along a lightweight stove, packets of soup, etc., so you can cook the occasional meal. (This is supposing you aren't camping and cooking all your meals anyway.) If you give the children vitamin pills or drops every day you will worry less about their getting a balanced diet. When travelling in a hot country it is vital to carry sufficient water containers with you as thirsty children complain endlessly and an adequate fluid intake is important for good health.

What games and amusements you take will depend very much on your children, but don't overdo it. I would recommend: as many favourite books as you can manage, plenty of paper, colouring books, pens and pencils, a pack of cards and, for older children, a game such as chess or Scrabble which can be played while travelling. Word games and games built around the moving scenery outside the car (or bus or train) window—counting things, identifying things—have the added advantage of requiring no equipment. If children want to bring a soft toy let them, as this will provide security in the various strange and different places they have to sleep. Do, however, make sure it is a reasonable choice—my daughter is very attached to a stuffed hippopotamus! If you are travelling in a group it is very important that your children behave in a reasonably polite and considerate manner and if they do you may well find other people will do a good deal of the amusing for you. With younger children it is important to keep up routines like reading a bedtime story and do be prepared to read to them at other times too, even if all you really want to do is look at the view.

It pays to find out as much as you can beforehand about places you are visiting and pass on the more striking facts to the children: they will be far more interested if they know that something is the largest in the world, or was once the crater of a giant volcano, or is the home of a living goddess the same age as themselves or whatever. Older children will appreciate being given clearly written guidebooks to read beforehand or route maps to study. They also usually respond well to being given responsibility for a particular aspect of a journey, e.g. learning Arabic numerals so that you will know what the coins are worth.

What I have said so far applies mainly to children of about three years and upwards. The problems of younger children are somewhat different. If you want to travel with a small baby, do try to breastfeed; it cuts out about half the difficulties in one go. If travelling in a vehicle with a bottlefed baby, get one of those food warming devices that plugs into the cigarette lighter on the dashboard. Obviously you will need to carry more with you—disposable nappies are bulky but better than the endless washing and drying otherwise involved. Prickly heat powder may prove useful in a hot climate. The easiest way to carry a baby around is in a papoose on your back and he will probably remain contented for hours travelling in this manner. Flying with a baby can be a trial. Hold the baby on your lap during takeoff and landing and whenever the 'Fasten Seat Belts' sign comes on. The change in air pressure may come as a shock to a baby and cause him to cry but his discomfort will be quickly relieved by a few sips of water. Make sure that you keep the nappy-changing operation quite separate from the eating and serving of food (yours and other people's). The toddler stage is possibly the most difficult of all and it is probably best to try to avoid long journeys with children this age.

In one way the children themselves can be a positive travel asset to their parents. In the Catholic countries especially, where small children are venerated, your child may be the point of introduction to other children and hence their parents. Between small children of different nationalities and colour there are no barriers. Where real contact with local people might otherwise be impossible, it often happens easily through the spontaneity and natural friendliness of children. From your children you will gain a new perspective on travel for you will be seeing things through their eyes.

I hope I have said enough to convince you that with proper care and organisation it is not difficult to take children on your travels even to remote areas. I'm sure that you will all be glad they came and any schooling they might miss will be more than made up for by the experiences of different environments, cultures and landscapes which they will have seen and by your pleasure in sharing your love of travel with them.

The Lone Woman Traveller

by Jo Hanson

'Nix amore nix good'. These words by a hopeful multilingual suitor have followed me round the world in one form or another for the last twenty years, and they will ring a bell for every travelling 'gel'. Being worried like a bone by men is the first extra burden the woman adventurer has to carry, and it doesn't matter how advanced in years or poundage she might be. It is a bore too. 'If only one of them would think up something new', you find yourself groaning, exhausted by the repetitious glances, remarks and advances.

(Sometimes they do. Turks have been known to throw themselves on the ground to look up a passing skirt. Pakistanis assume soulful looks and quote Urdu couplets.)

Many countries south and east of Italy, plus much of South America, breed men who look upon the woman-on-the-loose as loose and a target for frustrations. You have to learn to live with this. Judge the climate of each place. In some, like Bangkok, a sweet half-smile might be sufficient brush-off. In others, Algiers for instance, the same would rate as a raving invitation and draw the males from miles around as a wounded fish draws sharks. In this situation it is easiest to avoid eye contact altogether, dreary though it might be to stumble along looking everywhere but at your fellow men.

Read, if you haven't already, *With Love, Siri and Ebba* (£1 by surface and £2 by air from 2 Neals Yard, London WC2). Their account of two nights in Libya is hair-raising, with suitors ten-deep being beaten off by 'saviours', who later have to be beaten off in their turn by the girls, while a hotel owner meanwhile pops into their room via the window to try his luck. This is an extreme predicament, and single women under 33 have since been banned from Libya(!) but how would you handle this if you were alone? Ideally you would not get into the mess in the first place, but as a friend of mine pointed out: 'If you go around saying no to everything and everybody because of the possible dangers, you miss out on an awful lot. We travel for new experiences, so we might as well stay at home if we want to put chain mail on.' In short, as in most sexual matters, you end up trusting to luck with as much good judgement as possible thrown in.

The chances of being assaulted abroad are probably the same as in Britain today. If you are actively afraid, here is the advice of an American criminologist, Frederic Storaska. There are three 'don't's' and five 'do's'.

1. Don't antagonise the would-be rapist by making him angry or by further upsetting his fragile emotional balance. Do not scream or struggle.
2. Don't commit your behaviour, that is, don't do anything you cannot 'take back'. Do not initiate violence, which will very likely precipitate him to greater violence.
3. Do nothing that can hurt you. Avoid any form of defence that could be turned against you or that could misfire and harm you without deterring your attacker.

Complete surrender is not advised, nor is any kind of violent attack unless it is certain to incapacitate the rapist —and very few women have sufficient knowledge or training to achieve this.

Do:

1. Retain—or regain—your emotional stability. Fear may put your assailant off, or it may flatter him: either way, it does no harm. But don't panic to the point where you are incapable of observing and reasoning.
2. Treat the rapist as a human being. Don't insult or belittle him. Communicate with him if you can.
3. Gain his confidence. Let him know that you represent no threat to him. Don't heighten his own fear by threatening to call others to your aid or to report him.
4. Cooperate until you can safely react. Stall, do what he says, wait for your moment to gain control.
5. Use your imagination and your good judgement. Improvise. Invent a

reason, medical or otherwise, but convincing, why you 'reluctantly' cannot cooperate. Turn the attacker off. Use talk to defuse the situation and as a kind of therapy. Remember that the rapist is a man in trouble, who needs help and human empathy.

Many girls of course find foreign men far more attractive than the home-grown variety, and want to have affairs with or even marry them. The novel *English Woman, Arab Man*, by Alison Leigh-Jones (published by Paul Elek, London), gives an honest picture of the joys and disappointments of a liaison of this kind.

Not wishing to moan about all manhood from Milan onwards, I should add that genuine types do sometimes turn up amongst the wolves in their sheepskin coats. They are far more likely to be found in country areas, like the black Catholic catechist I spent a calm night with in a volcanic village in the Cameroons. Siri and Ebba describe in their book how they lived for a week with a nomad in his thorn hut in the Sudan. He carved hair sticks for them and tried to make Siri give up smoking.

Staying with a saviour-type Assistant Stationmaster in Quetta while I waited for his weekly train, I found what he really wanted was my help in getting him to Britain. This is another wilting predicament, which can involve you in endless red tape, trips to Heathrow, and possible fraudulent practices. Best to make up a wild story of non-availability, such as a proposed ten-year stay amongst the aborigines of Irian Jaya.

This ruse can also be tried when you are asked to post presents back to people who have done you a good turn. No one would mind sending a book or a child's sweater, but occasionally you are expected to crate tape recorders and other luxury goods halfway across the world. In fact to be safe all round, do not give any address for yourself to anybody if you can gracefully avoid it; after all, you're of no fixed abode, aren't you? However, don't fail to get the address of your benefactors and send a postcard or chatty letter.

Any normal woman of course feels an obligation to a person or family who puts her up or helps her on her way, and probably thinks 'Should I offer some money?'. Alleviate this by carrying with you as many small presents as you can. They need not be expensive: in fact white elephants from your own home will often suffice. The tie your Dad was going to chuck out because he hated the zigzag pattern would probably look great on an African; the toy cars your brother got tired of would light up the face of a Nepalese boy, and the sickly perfume passed on to you by Aunt Edna might be the only luxury ever given to an Untouchable woman. Wrap each gift firmly, write on the outside what it is, and give away the biggest or heaviest first.

A lot of time in underdeveloped countries is spent in sitting around waiting for things to happen. A good book takes the boredom out of this and helps you keep cool—an important asset.

The second extra burden peculiar to the woman traveller is also to do with her biology: her periods. In many places it is even difficult to find a place to pee, let alone tackle tampons. ('Madam!', cried a Sikh with an expansive wave of his arm when I asked for the toilet in his Punjabi village, 'The whole of India is a toilet!') If you are on the pill, do what the women doctors do and take it continuously to stop menstruation altogether—ask your GP the most suitable brand for this. Otherwise, carry as much sanitary gear as you can, as it is scarce or expensive outside Europe and the USA. You can of course always tear up an old T-shirt in an emergency. You will be in good company: poor women around the world cope like that all the time. If you would like to know more about their existences, try *The Private Life of Islam* by Ian Young (Allen Lane, London).

A last word for lone travellers of either sex. You may or may not meet suitable companions, whether temporary or lifelong, on your travels, but don't count on it: be prepared to be self-sufficient and enjoy your own company. If you live alone or have travelled alone in your own country, you'll know that it's possible. A little extra caution is necessary, and a little extra expense sometimes inevitable (single room supplements in hotels, no discounts such as large groups can claim), but your freedom of choice will more than compensate for this. You will be glad that you are you, on the move and pleasing yourself most of the time.

Motoring

Choosing and using four wheels

The Expedition Vehicle

by Tom Sheppard

Many expeditions will evolve on a 'Where can we go with the vehicle we have?' basis, but it is instructive, even in these cases, to consider the criteria for selection of an expedition vehicle where choice exists. The route, how much of it will be off tarmac, how much of it will be off tracks, the total payload, the number of people and vehicles in the party, and the distance between fuel and replenishment points will all have an important influence on choice of vehicle.

To sample both ends of the scale, a reliable hatchback/station wagon with good ground clearance (say not less than 14in rims) will be quite adequate for a climbing or camping trip in Morocco or Turkey where not too much gravel track is encountered, but for Sudan, say, where distances are large, tracks very poor and logistics very demanding, a capacious middleweight four-wheel-drive (4x4) vehicle would be almost essential.

For long distances without replenishment, a mix of vehicles will often be the answer—a medium 4x4 truck such as the Bedford 4-tonner (an outstanding cross-country vehicle) can act as main load carrier and mobile fuel depot mothering a clutch of three or four Land-Rovers which can also come to its aid with tow ropes when its inevitably lower power-weight ratio leaves it floundering on a deep sand incline. Two or three vehicles are often a prerequisite anyway from a safety point of view in remote areas; the demise of a single vehicle in such regions could otherwise have dire consequences. Taking care of the vehicle en route includes *never* overloading it beyond the makers' recommended maximum GVW (Gross Vehicle Weight); which in turn means picking a vehicle with the right capacity in the first place.

Some idea of the required load can be obtained if one knows the proposed route and petrol/provisioning points along it. What remains of a vehicle's allowable maximum after earmarking the fuel load can be allocated to crew and remaining equipment. With this basic information a few vehicle specifications can be examined to see how their payload and general suitability match up to the requirement. Although the major decision of two-wheel-drive (4x2) or four-wheel-drive (4x4) will be dictated almost invariably by the terrain to be covered, it is sometimes possible to get by on 4x2 where short ranges are involved and the cargo comprises mainly people. In these circumstances—for example, working a short distance 'up-country' from a base camp on, say, a scientific project—a muscular cargo can work wonders in pushing a bogged vehicle free. Peugeot pick-ups and the like cover preposterous terrain in Africa on this basis. Use of a 4x2 here can be cheaper both in outlay and in fuel costs. Normally, however, a 4x4 should be used where any significant amount of poorly maintained unsurfaced track or off-tracks driving is to be done.

The petrol vs. diesel choice may come up. Briefly the pros and cons of diesels: expensive to buy, heavier than petrol equivalent, less power for given number of cc's, tricky to service if you get injector/fuel problems, *but* cheaper to run, good low speed slogger, generally good durability and reliability. Regarding petrol engines, remember that high octane petrol (essential with a high compression ratio) is not always available overseas. Never use low octane petrol where high octane is needed.

Like the leopard-skin hat band and bush jacket of the old white hunter movies, roof racks and external jerry can holders seem often to be *de rigueur* for expedition vehicles but the maximum GVW must still be observed. It is better to do without them if possible since both items invariably increase a vehicle's tendency to roll or pitch on undulating tracks, which in turn makes greater demands on suspension and shock dampers. In this context the centre of gravity of load and of the whole vehicle is worth some thought. The ideal vehicle would have nearly all its mass low down and centrally within its wheelbase and track—no great rear overhang, no top-heavy loads, no side panniers. It would also have a 50/50 front/rear weight distribution. The more a vehicle is to be used away from surfaced roads the more important these criteria are. The classic how-not-to-do-it picture of an overloaded, tail-heavy Land-Rover shouldering a colossal roof-rack and bogged to the rear hubs in some Saharan soft spot shimmers familiarly into focus here and we can certainly learn from it.

From that and from some of the points mentioned so far the shape of the ultimate long range, off-road expedition vehicle begins to emerge—and therefrom the characteristics of a vehicle faced with less demanding tasks can also be established. This ideal vehicle would have a payload of about a ton, large wheels and tyres (a radial tyre around 9.00 x 16) to give good ground clearance and flotation, a shortish wheelbase of 90-105in to give good clearance over underbelly bumps, a forward-control layout to give the best driver visibility on bad terrain and get the load evenly distributed between front and rear axles, and a power-weight ratio of better than around 30 bhp per ton of gross weight. Into this category fit the Volvo C303, the 101in wheelbase 1-tonne military Land-Rover, some of the Fiat PC65 and PC75 range and, albeit not making the power-weight optima, certain models of Unimog. (This, the Fiats and the Volvo also have axle-end reduction gears that effectively increase the under-axle ground clearance even further.)

The pros and cons of other vehicles now fall quickly into place and, although they may do the job, their weaknesses can more objectively be seen. The long wheelbase Land-Rover, virtually the standard expedition vehicle, very effective and reliable, is a bit lacking in under-axle and underbelly clearance, carries most of its payload over the rear axle (a case where external jerry cans at the front *can* help), has a lot of tail overhang and is a bit down on power-weight ratio. (NB Watch for new versions of the newly re-engined V8 3½ litre Land-Rover.) The Range-Rover, probably the best vehicle in its class in the world, is outstandingly agile and comfortable but again, at maximum weight, a little tail heavy; it is rather a short range vehicle in terms of logistic payload and, in the most extreme desert conditions, needs the larger 7.50x16 radials that can just be squeezed into the wheel arches.

By the same criteria the American 'recreation vehicle'—Blazer, Power Wagon, Bronco etc—is peculiarly suited to the US market (cheap petrol, easy logistics) but unsuited to serious expedition work—too much overall weight, small payload, preposterous power-weight ratio and fuel consumption (both too high), primitive suspension in most cases; tyres, like the whole vehicle, frequently optimised for *machismo* cosmetics rather than for truly efficient performance. The high GVW versions of the larger US 4x4 pickups can make a bit more sense but the wheelbase becomes long in these options, inviting underbelly clearance problems, and again care must be taken to select an engine that will give adequate power without necessitating allocation of the whole payload to fuel drums.

Payload (in tonnes) per metre of wheelbase is an academic but quantifiable parameter of effectiveness—roughly how much metal and engineering you have to buy to move a given load. It is also indicative of the vehicle's overall efficiency and ability to support itself on an expedition—a kind of 'toughness coefficient' or 'porterage parameter' (PP). No-nonsense machinery like the Volvo C303 or 1-tonne Land-Rover rate PP figures of .5 and .48, while the Fiat PC75 is as high as 1.28, as befits its beefy configuration. That war horse of many African expeditions, the Bedford 4-tonner, rates 1.0; US vehicles optimised for low density cargoes rate lower figures—Jeep CJ7 is .14, Plymouth-Trailduster .16, Blazer .2. The long wheelbase Land-Rover is .39 (see table pp 56, 57). Beware of going for too high a PP as high cargo density usually also brings high tyre pressures—bad news in sandy desert or soft mud.

All the philosophy in the foregoing can be applied to choice of a 4x2. As with 4x4s the actual make of vehicle selected will be influenced by the availability of en route spares. Keep the vehicle specification simple—avoid power steering, automatic transmission, or air conditioning unless you are sure about reliability and servicing availability. Many of these extras will also cost you payload. The likely weather will affect thoughts on vans, hard-tops, soft-tops and pickups, but with care removable weather-proofing can be made effective and thief-proof too. This makes some of the light 4x2 pickups such as the Peugeot, Datsun and Toyota well worth considering. Most of this class of machine are simple and robust, have big wheels, reasonable fuel consumption and excellent payload; in Africa and the Middle East, Peugeot spares will be relatively easy to get and the Japanese makes have an almost equal foothold.

Tyres have an important contribution to make—and despite the advertisements there is more to them than whether they look good or have white lettering on the sidewalls. Radials are best on all surfaces for both 4x2s and 4x4s, and appropriate tread types should be used—for example, the Michelin XS is unsurpassed for desert work and has a very clever tread pattern for getting the best traction in soft sand. Attention to tyre pressures is essential and even though it could mean frequent inflation and deflation as the going changes, the manufacturer's figures according to axle load and terrain must be adhered to. Basically the rule is low pressure for best flotation and traction is soft going—low speeds only—and high pressures for rock and roads. Radials can take much lower pressures without heat build-up and danger of de-lamination due to sidewall flexing. A guide to tyre pressures with radials might be 75 per cent of road pressures for on-off-road use not above 30 mph, and 40 per cent of road pressures for 15 mph-maximum soft sand going; but check the maker's figures and beware trying this with cross-plies.

New tyres and tubes should be used for long overland expeditions (tubeless are impractical) and it is essential that they are put on in scrupulously clean conditions with the correct rubber lubricant on the rims—this will help later removal for puncture repairs. Even a tiny particle of grit between cover and tube can cause an apparently mysterious puncture due to the constant flexing. Only one spare wheel is necessary though taking two or more tubes is a prudent precaution. It is wise to practise before departure the seemingly dying art of tyre removal with tyre levers and the subsequent puncture repair; in tropical conditions this can otherwise be daunting on large tyres which, to the novice, can give every indication of being welded to the rims. Most important of all is to drive with consideration for the tyres and never scuff the sidewalls on kerbs or rocks.

Trailers are frequently suggested as a solution to the payload problem. Despite a deservedly bad name for lateral instability on poor tracks and difficulty in reversing out of bogged down situations, a trailer can, with these limitations in mind and with high flotation tyres, be a valuable aid. It spreads a given payload over six wheels instead of four and, with a crew of three or four people, can fairly easily be manhandled. For a long wheelbase Land-Rover or a Range Rover, the 15cwt ex-military type (with shock absorbers) is preferable to the lightweight ½-tonners; the massive NATO-type tow-hook and local strengthening for it are worthwhile investments. Trailer nose-load on the car tow-hook between 0 and 100lb seems to work well but may vary from trailer to trailer when optimum is sought; remember to allow for this when calculating vehicle load.

The last word must be on careful packing and lashing down of equipment in the vehicle. Carpet underfelt is ideal for lining the floor before fitting the boxes and jerry cans and the aim should be to eliminate all rattles. If a vehicle is as quiet as a Rolls Royce it tends to get driven that way—a good thing for vehicle and occupants alike on a long expedition.

Editor's note: Good news for 4x4 overlanders is the recent introduction of two long wheelbase versions of the Land-Rover, powered by the V8 engine of the 3½ litre Rover Saloon, which has been modified to give maxi-

mum torque at lower engine speeds—166lb/ft at 2000rpm. There is a four door station wagon with a tail-gate, and a two door pickup with hard or soft top and an optional truck cab. The new models have restyled fronts flush with the wings to house the larger engine.

The V8 Land-Rover was designed both for off-road economy and for speed and comfort on the improved roads of Africa and Asia. It has continuous four-wheel-drive and differential lock, which on tests in Africa have proved to reduce petrol consumption in sand and to save wear and tear on suspension over corrugations, owing to improved pace. Sand tyres to match the increased engine size can make easier work of difficult traction conditions.

The vehicle has many of the optional features of the standard Land-Rover—*de luxe* front seats, interior sun visors, door and interior mirrors, trip speedometer, lifting/towing rims front and rear, bonnet lock, rear mud flaps and hazard warning lights. For expedition work there is only one major drawback: there is no room to fit jerry can brackets in the front, and rear brackets tend to put undue stress on the back panels.

Initially the V8 is for export only and will not be sold in Britain. It comes in four colours and markets for about £7,500.

Recently tested on a gruelling 15,000 mile trip across the Sahara, two *Stonefield* trucks have proven their value for expeditions and off road use, especially where carrying capacity is of some importance. Stonefields are somewhat larger than Land-Rovers, incorporating a forward control cab in the 10ft (pickup) or 11ft 6ins x 6ft bodywork and having a payload of 1.6 tons (4x4) or 3 tons (6x4). The length can be extended rearwards or into the back of the cab. Both types of truck are capable of towing a further 1½ tons. The turning circle is the same as that of a Land-Rover. The vehicles use 138hp Ford V6 or 150hp Chrysler V8 petrol engines. No diesel version is yet available.

The Stonefield has full-time four-wheel-drive and automatic transmission, which is intended to protect the vehicle from the ravages of poor driving. A spare gearbox is economical to carry, being about one-third the price and weight of the Land-Rover unit. The vehicle has an automatic differential lock on the central differential. As a result of the Sahara trials, the factory is convinced of the need to fit heavier duty suspension. Overall, however, and despite gross overloading, the vehicles proved comfortable, reliable and economical, giving nine miles to the gallon in the desert.

The pickup comes in one basic version, to which specialist bodywork can very readily be added. The price for the basic model of Stonefield chasis is currently around £13,500.

Vehicle Maintenance on Expeditions

by Tom Sheppard

The whole philosophy behind keeping the expedition vehicle fit can be summed up in two words—preventive maintenance. A really thorough, no compromises service and overhaul *before* the expedition will pay off far better than trying to cope with problem as they arise in the field. Vitally allied to reasonable and sensitive driving (see next chapter), maintenance on the expedition will probably then be limited to the 20 minute nightly inspection and rectification of minor defects found *en route*.

A very great deal depends upon the team member taking on the task of mechanic—more even than would be expected. An eagle eye for detail, a mature ability to face facts, a total devotion to his task, an ability to influence his fellow team members, and as much mechanical knowledge as possible are some of the qualities he must possess. And he will have to display them during the pre-departure overhaul, for here no detail of the vehicle's condition must escape his notice, nothing can be glossed over, no hope-for-the-best attitudes adopted. The hard facts must be faced and acted upon when oil leaks, mechanical wear and the implications of remedial work necessary are being considered. Faults will not go away or cure themselves just because it will be expensive or difficult to fix them—any more than they will in the field just because there happens to be a sand storm blowing.

One of the first facts to face is the expedition mechanic's own experience and knowledge. He should know his own limitations, know and admit what he does not know and work with more experienced people in preparing the vehicle. But even here there is cause for caution for many 'experts' in vehicle maintenance are jaundiced, ever open to expedient compromise and will rarely have the keen expeditioner's enthusiasm for thorough preparation. The crossing of this tightrope will be good training for those to follow.

A new, run-in vehicle is the best way to start an expedition but budgets do not always stretch to this and a critical and thorough examination of the expedition machine must start as soon as it is acquired. A workshop manual is essential. Start at the ground and work up. Are the tyres right for the job and in top condition? (It is best to start with new ones.) Wheel bearings—free, with correct end float? Brake drums off to check linings and operating cylinders, also hub oil seals. Are the axle cases bent? (A ripple on the top surface will normally tell.) Axle breathers must be serviceable, also differential housing, oil seals and gaskets. Spring-eye and shock absorber rubbers and mountings, spring-to-axle U-bolt fixings; check no spring leaves broken. Prop shaft sliding and universal joints should not have excessive rotational or vertical play. Check underside of engine and gearbox for oil leaks—especially from around the clutch bell-housing drain hole since a slight leak there might indicate an oil seal failure in the rear main bearing and the possibility of oil getting to the clutch plate—disastrous if it happens in the field. Engine- and gearbox-to-chassis rubber mountings should be checked, also condition of all brake pipes and hoses and the security and condition of the whole exhaust system.

Similar thorough visual inspection in the engine compartment should be carried out, again being on the lookout for oil leaks, condition of hoses, clips, wiring, fuel pipes and unions, exhaust manifolds, radiator. A compression test on each cylinder, plug and carburetter condition and a check of engine tune on an electronic timer will be essential and leads to a full functional check of all aspects of the vehicle—steering, brakes, clutch performance and adjustment, engine power, gearboxes and transfer box controls. If in any doubt on shock absorbers, renew them. It is worth fitting contactless electronic ignition if this is not already standard; the contact breaker type is susceptible to dust erosion and needs frequent re-tuning to maintain best power and economy in the field. The Lumenition type has been used in the Sahara at under-bonnet temperature over 82°C—a very severe test.

Careful reference should be made to the workshop manual for limits and settings—beware the know-all mechanic who uses figures that are 'OK for most engines'. The whole examination is little more than a methodical and commonsense check of the complete vehicle and should naturally be combined with a major schedule service (such as a 12,000 mile check) and oil and filter change. The inspection will yield also the basis of the nightly inspection check list. Vehicles work hard when high daily mileages over bad roads are covered and it is essential to carry out a methodical inspection each evening—preferably stopping to camp early enough to allow it and any remedial work to be done in good daylight. A groundsheet, pair of overalls, old beret and pair of working gloves are invaluable for working on the vehicle; washing water and laundry are always a problem on expeditions and the Ragged Look for expedition folk went out of fashion years ago. Check tyre pressures and coolant level first thing in the morning when cold; don't fill the radiator to the brim as the coolant will only expand out through the overflow.

The toolkit must be assembled with an eye to the particular vehicle and the pre-departure work on it will have brought to light most of the 'specials' required such as extra long socket extensions, wry-neck spanners, thin-rimmed ring spanners, special sized sockets and whether the vehicle nuts and bolts are metric, AF or a mixture of thread types. Vibration and bad tracks can often cause fatigue failure of vehicle parts in the field and blacksmithing or sheet metal repairs may be called for. Hacksaw, files, metal shears, a large assortment of nuts and bolts and pop rivets, a rivet gun, large hand drill and set of drills should be taken with this in mind. A small clamp-on vice that can be mounted on a bumper or tailgate is well worth the weight. Two jacks are better than one, and if one is a high-lift bumper jack it can often be used in recovery situations enabling sand ladders to be placed under the wheels. Tyre inflator—spark plug type—will be needed frequently, as well as a wide ranging repair kit for tyre tubes and for the vulnerable sidewalls of radial covers. Use bags for the tools rather than steel chests; bags do not rattle.

The main spares kit needs disciplined and realistic thought. A thorough pre-expedition overhaul is better than a large spares pack; careful and infallible driving can lessen the *en route* requirement even more—springs and half-shafts, for example, are broken by drivers and do not fail of their own volition. However, hoses, fan belts, clutch and brake cylinder rubbers, electrics such as fuel pumps, coils, condensers, HT leads and alternators are less predictable and are worth taking. Carburetter diaphragms, oil and fuel filters, air filter elements, electrical repair materials such as wire, terminals, 12v soldering iron, tape, fuses, bulbs should be taken, as well as a length of plastic tubing to use as a gravity fuel pipe with a can if the fuel system fails.

Keep a notebook in which you record mileage, price of petrol bought, accidents and any parts failures or defects. Note especially those items that will need attention at the next full servicing.

Finally cleanliness. Beating sand and dust on a hot-climate expedition may sound impossible but with care it need not be. Remember that every level-plug and filler cap removed carelessly can shake dust into the oil it is designed to keep in. A one-inch paint brush in a jam-tin half full of petrol should be used religiously to clean down these or any sealing surfaces or parts before removing or working on them—or even to clean a suspected oil leak so that its condition can be noted next evening. It is again just a question of facing facts—a few grains of sand can destroy a bearing.

Driver Training

by Tom Sheppard

Restrained, sympathetic and alert driving is probably the most important characteristic to inculcate into any expedition team going on a long and arduous expedition, especially if it is likely to cover many miles on tracks or across open country. We are unhelpfully surrounded by four-wheel-drive magazines that feature brutish driving, airborne vehicles, flying dust, chrome spotlights and dummy exhaust pipes as though these were the ingredients of professionalism. In a sense, since it is normal for the media to dramatise trivia and lionise the extrovert, we have at least an inverted guide on what to avoid. The vehicle is the very core of the expedition; certainly considerable cost, if not life itself, will be at stake if it is not looked after well and driven considerately. Whilst the best 4x4's do convey a feeling of being unstoppable, anything will break given enough abuse, so the urge to prove one's manhood (or liberated status) by driving with excessive panache must be resisted.

But points of technique are important as well as just care, since there will be times, especially in soft conditions, where all the vehicle's power output will have to be used.

Gearbox and Transmission. Know your gearbox and, if it is manual, especially know which gears have synchromesh; those that do not you will have to double-de-clutch into, changing up or down. Practise this until a smooth silent change can be made into all gears—including first—even over rough going. Avoid slipping the clutch. Keep your foot off it unless starting, stopping or changing gear. This may require willpower but it will save the life of your clutch. Most four-wheel-drive vehicles (4x4's) have a low ratio selector (sometimes called a range change) that will gear the vehicle down by a factor of about 2:1 whatever gear is selected in the main gearbox; it will normally also take the transmission from two-wheel-drive (4x2) into four-wheel-drive (4x4) as it does so (the Red Knob moved aft in the Land-Rover). This kind of vehicle will also usually have a means of selecting 4x4 in high ratio (Yellow Knob pushed down in the Land-Rover); the control can be either separate or an intermediate position on the range change lever.

Follow the manufacturer's recommendations regarding when range changes can be made; in many older vehicles this must be done while stationary. Whatever the technique recommended, the overriding consideration is to make these changes without clunking, crashing gears or shock loading the transmission. Even though it may not seem necessary from a traction point of view, use of high ratio 4x4 when on rough or corrugated tracks will halve the fatigue loads on the rear propeller and half-shafts by spreading the transmission loads across both axles. On a 'part-time 4x4 vehicle (one which is normally 4x2 with selectable 4x4), 4x4 should not, of course, be used on paved roads as it can cause transmission 'wind-up', extra tyre wear and heavy steering loads. There are an increasing number of 'full-time' 4x4's around—for example the Range Rover and many US vehicle options—and here there is no high range 4x4 selector since the vehicle is in four-wheel-drive the whole time (high *or* low range) and has a differential between front and rear prop shafts to equalise axle speed differences and eliminate 'wind-up' stresses. This differential will often be lockable to avoid front or rear wheel-spin, for example in the extreme case of rear wheels being in mud and front wheels on rock.

A more common problem associated with transmission and which is probably responsible for more than half of all 'failed traction' situations—bogging in soft-going included—is the 'diagonal suspension' case. Imagine a driver's eye view approaching a shallow V-shaped ditch going diagonally, 30° left of dead ahead direction, from distant left to foreground right. In crossing it the vehicle will reach a condition where the front right wheel is on the far side of the ditch, the rear left is on the near side of the ditch (the diagonal suspension situation) and the other two remaining wheels are hanging down into the bottom of the V. When the ditch is deeper than the extent of the wheel movement allowed by the vehicle's springs, these wheels (left front and right rear) will spin in thin air and, owing to the axle differential, all the power will go to them and traction will be lost. Different versions of this classic basic situation will manifest themselves in a hundred ways—scrambling up bumpy slopes, crossing ruts, on rough ground or in soft going. Almost invariably it will be 'diagonal' wheelspin that stops a 4x4 in severe going. A lockable or limited slip differential in the axle(s) is the answer but if this is not available it helps to be on the lookout for, and quickly recognise, this condition and stop before wheelspin makes things worse by scooping sand or earth beneath the afflicted wheels. The torque of the propeller shaft power tends to tilt front and rear axles in opposite directions relative to the chassis in a soft-sand-wheelspin bogging and this is one of the main reasons for the cardinal rules in any bogging: *avoid wheelspin, admit defeat early, reverse out*.

By quitting early, excessive underwheel scooping is avoided and reversing out conveniently tends to tilt the axles in the reverse direction thus enhancing your traction for getting out backwards anyway. As in line one, it comes down to restrained, sympathetic and alert driving.

A final point on transmission: where a transmission brake is fitted (a brake drum on the prop shaft behind the gearbox as in the Land-Rover, Range

Rover or Bedford MK) it should *never* be used except when the vehicle has been brought to rest with the wheel brakes and is stationary. If used on the move it can cause half-shaft failure and severe stress in the transmission.

Slippery Conditions, Poor Traction, Slopes. Jumping off a meringue or piecrust is the best analogy to poor traction conditions—mud, sand, bog, snow or slippery hills. Low surface strength and low shear strength are common to all and the demands of both must be reduced. Fitting big, high flotation tyres and lowering tyre pressures reduces vertical load per square inch and going from two- to four-wheel-drive halves the shear strength required of a given piece of ground. Both these remedies must be allied to *smooth* driving, use of the highest reasonable gear and very gentle use of the accelerator. This implies, too, getting the required gear and speed before the obstacle—a gearchange at the wrong time can sometimes cause a bogging or a failed ascent. This driving technique is especially important on slippery inclines where too low a gear (say first, low range) will make wheelspin all too easy to induce; a climb can more reliably be made by using second gear and a gentle right foot. Avoidance of wheel skidding must be borne in mind also when descending steep slopes—the same gear must be used to go down as would have been appropriate for going up; most usually second gear low range, occasionally even third, *never* first. (First gear low range is most useful for slow, controlled traversing of really rough savage going such as rock, boulders or 'doorsteps'; there is generally too much torque for poor traction conditions and wheel spin results.) Too low a gear in a descent enhances the possibility of a sliding glissade where the engine cannot 'catch up' with the rate at which the vehicle is sliding down the slope.

Obstacle Clearance, Recce and Marshalling. If in doubt, recce on foot-hills, gullies, bumps in the track. On the umpteenth hot, sticky day of a rough expedition in the tropics the temptation to press on and hope for the best will be high, but the cost of getting it wrong—just once—will be higher. The passenger should be used as a marshaller where clearances (especially over rocks under the vehicle) are tight. There is a tendency for everyone to shout at once in these situations but the driver should take his directions from just one marshaller (about 40ft ahead and facing the driver with a good view of all four wheels). These directions should be visual rather than by shouting and mishearing over engine noise. An agreed system of marshalling signals should be practised during pre-expedition driver training; go right, go left, advance, stop and reverse are all that is necessary. The driver must obey and trust the marshaller completely and watch him, not the track, when being guided. In difficult going of the kind where marshalling is needed, shovels should be regarded as an emergency low gear and must regularly be used to safeguard the vehicle; if the obstacle is likely to hazard the vehicle then 'landscaping' can remove the offending rock or bump or provide a feasible patch for a wheel. Do not skimp the digging; invest another five minutes in shovelling and be sure of getting out first try.

Special Rules for Desert Driving. All sand offers better flotation and traction when cool, dewy or damp; a bad bogging in the heat of the afternoon will be easier to get out of around dawn (and will also cost less water—see *Desert Survival*). Soft sand on tracks, churned by previous vehicles, is different from sand in open desert where there is a definite top crust structure. This crust will be broken by the passage of one vehicle and following vehicles should not follow in the same wheel marks. On tracks the opposite can *sometimes* be true; a previous vehicle in some types of sand can favourably compact it for the one behind.

In desert, savannah or bush areas, unsurfaced tracks which carry heavy traffic and are not regularly graded or scraped will often develop a surface of transverse corrugations. These parallel ridges, at right angles to the vehicle's path, can be as high as 10-15cm and spaced 30-60cm apart. Taken at the wrong speed they can nearly shake a vehicle to pieces. A 'best' speed, dictated by your vehicle's suspension rate and tyres, will result in at least the vehicle body and occupants having a relatively smooth ride though the springs and shock absorbers will be going through hell. This speed is generally 50-70 kph (30-40 mph). Do not go any faster; remember you are virtually skipping from crest to crest and adhesion for braking or rapid steering response will be very much reduced.

Salt flat—sometimes shown on maps as *Sebkha* or *Chott*—consists of a crust of unpredictable strength frequently disguising soft salty bottomless mud beneath. A bogging in this can often be forever. Never drive at random over this. If there is a well used track it should be safe, but even a metre or so away from this can be disastrously soft.

In dusty conditions seal doors and windows with masking tape—in spite of the heat.

Driving through Water. There may be times when you have to drive through water. People are easier to un-bog than vehicles so follow the recce-on-foot rule and wade through first to test the condition of the bottom (rocky? level? soft?) and also the depth. If the depth exceeds the height above ground of the fan, the ensuing under-bonnet spray-bath could cause ignition failure; remove the fan belt, use a WD40 spray on the ignition harness. Sometimes in marginal cases an old coat or sack over the engine (well clear of the fan) will keep the ignition dry. On Land-Rover and Range Rover where one is provided, insert the wading plug into the clutch bell-housing and, on any vehicle, try if you can to block the axle breathers before wading. If you do not, the sudden cooling of the axle casings can cause water to be sucked in through the breathers and contaminate the oil. Finally always drive slowly in water to minimise spray and bow-waves.

Recovery. Shovels, aluminium sand ladders (5½ft long, 10in wide, rungs at 6in pitch), a *long* tow rope, a high lift jack and a winch are, in that order, the best recovery aids to free a stuck vehicle in the event that it has passed the stage of being able to reverse out. The 'sand' ladders can be used in any conditions to put beneath or ahead of the wheels to give support and traction. Where it is necessary to make do planks, gravel or branches (e.g. palm fronds) can be laid down to support the vehicle. Often diagonal suspension hang-ups will occur where it will be necessary to dig away beneath one or both of the supporting wheels and lower the vehicle to a parallel-axle condition. A long tow rope with a tow (forwards or back) from an accompanying vehicle is the most effective recovery method but the effort must be coordinated—the clutches on the towed and towing vehicles must be engaged at the same time. Again an external marshaller can best give the signal. A tandem or multi-vehicle tow is even more effective in the worst conditions and again coordination by signal is vital. So-called 'snatch' towing is brutal, dangerous and in nearly every case results in damage to one or both vehicles or the two rope. Winches are slow, heavy, expensive and hard to match to road-wheel power and there is nothing to winch onto in real desert; however they have their uses in recovering a thoroughly 'dead' vehicle or in emergencies such as righting a vehicle that has rolled over. The most surprisingly effective recovery aid, in conclusion, is the human shoulder. Despite their small horse-power output, a group of pushing people can work wonders, even making sure that a towed recovery will work first time.

Training. For any expedition involving track or off-road driving, special sessions of driver training on rough country are really worthwhile—as well as enormous fun. Special permission can nearly always be obtained for use of military training areas. It is far better to make your mistakes there first than with a heavily laden machine overseas. Finish off the session with an egg-and-spoon autocross (spoon held out of the passenger's window); it makes for just the kind of smooth but effective driving required. Driving is tiring; on an expedition with a two-man crew, have Driver A hand over to Driver B at midday and vice versa next day thus ensuring a mid-session night's sleep.

Off tarmac, never drive at night; it is too easy to hit unseen pot-holes and to get lost. On hard roads night-drive only if absolutely necessary; in inhabited regions of Africa, India and the Middle East unlit vehicles, bicycles and bullock-carts abound and compete with sleeping livestock for your road-space.

International Guide to Four-Wheel-Drive Vehicles

Compiled by Tom Sheppard

This is a selective list of the world's 4x4 vehicles. Omitted are some of the pick-ups—of which there are endless varieties—that are unsuitable for long distance travel; specialised conversions, one-off or small volume production items; and the many different engine and body options which are merely variants on a basic model.

The vehicles—pickups, station wagons and soft-tops—are listed in ascending order of wheelbase. The fourth column shows the payload (in tonnes) per metre of wheelbase. This is an indication of the machine's ability to carry a worthwhile expedition load, a 'Porterage Parameter' or PP (see text). (N.B. Owing to yearly specification changes, the many options available on certain vehicles and the sheer lack of specific information supplied by manufacturers, these figures should be taken as a guide only.) Very broadly, a vehicle for a long-range expedition with 400 miles or more between re-supply points demands a PP around .35 or above. With knowledge of the terrain to be crossed, this must be balanced against a reasonable power-weight ratio and off-road fuel consumption.

Manufacturer	*Model*	*Country of Origin*	*'PP' (See pages 45 and 55)*
Fiat	6615 (Military)	Italy	.45
Suzuki	SJ20/SJ10	Japan	
Suzuki	LJ80	Japan	.10
Volkswagen	Iltis (Military)	Germany, Fed. Rep	.32
Daihatsu	F20	Japan	.20
Jeep Ebro	Bravo	Spain	.25
Volvo	C202	Sweden	.48
AMC Jeep	CJ5	USA	.19
Lada	Niva	USSR	.21
Steyr-Puch	Pinzgauer 4x4	Austria	.46
Steyr-Puch	Pinzgauer 6x6	Austria	.68
Daimler-Benz	Unimog U600	Germany, Fed.Rep.	.56
Land-Rover	Land-Rover 88in wheelbase	UK	.20-.30
Nissan	Patrol	Japan	.20 (Est)
Toyota	FJ40	Japan	.22 (Est)
UMO	UAZ 452D	USSR	.41
Volvo	C303	Sweden	.52
Fiat	1107 (Military)	Italy	.26
Fiat	Campagnola	Italy	.30 (Est)
ARO	Land Car 244	Rumania	.31
Portaro	Pampas	Portugal	.30
AMC Jeep	CJ7	USA	.14
Daimler-Benz/ Steyr-Puch	Geländewagen (short wheelbase)	Germany, Fed.Rep/ Austria	.28 approx.
Fiat	PC65	Italy	1.26
	Tundra	USSR	.32
ARO	TV12	Rumania	.52
Subaru	Brat	Japan	.17 (Est)
Land-Rover	Range Rover	UK	.31
International Harvester	Scout II	USA	.35
Land-Rover	1-tonne (Military)	UK	.48
Jeep Ebro	Bravo L	Spain	.27
Jeep Ebro	Commando S	Spain	.21
Jeep Ebro	Campeador (pickup)	Spain	.47
Daimler-Benz	Unimog 421	Germany, Fed.Rep.	.51
Chevrolet/Isuzu	LUV 4x4/KB40	Japan/USA	.36
Saviem	TP3	France	.69
AMC Jeep	CJ6 (out of production)	USA	.17
Ford	Bronco (post 1978)	USA	.17
Plymouth	Trailduster	USA	.16

Manufacturer	*Model*	*Country of Origin*	*'PP' (See pages 45 and 55)*
Toyota	Land Cruiser FJ55	Japan	.26
Lancia	209.400 (Military)	Italy	.79
Chevrolet	Blazer	USA	.20
Ford/Toyota	Courier Brahma 4x4	Japan/USA	.20 (Est)
Fiat	PC75	Italy	1.28
AMC Jeep	Wagoneer	USA	.34
AMC Jeep	Cherokee Chief	USA	.24
Land-Rover	Land-Rover 109in wheelbase	UK	.32-.39
Stonefield	P3000/P5000 4x4	UK	.44
Stonefield	P3000/P5000 6x4	UK	64
Daimler-Benz/ Steyr-Puch	Geländewagen (long wheelbase)	Germany,Fed.Rep/ Austria	30 approx.
Daimler-Benz	Unimog 'S' (U110/404)	Germany, Fed.Rep.	.67
Toyota	Land Cruiser pickup	Japan	.41
Dodge	Power Wagons 100/ 150/200	USA	.25 (Est)-.40
Ford	F150/250/350	USA	.25-.40
International Harvester	150/250 pickups	USA	.25 (Est)-.36
Chevrolet	K10/20/30 pickups	USA	.18/.45/.50
AMC Jeep	J10/20 pickups	USA	.30-.46 (Est)
International Harvester	Terra pickup/Traveler	USA	.36/.30 (Est)

Cournil, a limited-production French manufacturer, produces Land-Rover-type vehicles in petrol/diesel and short/long wheelbase forms.

Preparing Your Four-Wheel-Drive Vehicle for the Big Trip

by Richard Harrington

Editor's Note. For addresses of most manufacturers and suppliers mentioned in this article, please see the list of 4WD outfitters in the directory section of this book.

If you're American, you probably call it *four wheeling* or *off roading.* If you're English, you're more likely to talk about four-wheel-drive *overland travel.* The differences don't end with the terminology. The 4WD explosion in North America in the last ten years (and it's still exploding) has produced a widening gap in the way that non-commercial four-wheel-drive vehicles are used on either side of the Atlantic.

In the US today, most Chevy Blazers sold (around 150,000 a year at the last count), to name only the most popular 4WD truck, do most of their miles on the highway, rather than off it. And off road use is mainly of the weekend recreational variety. New Mexico, Idaho, Colorado, Arizona, Southern California and Baja California have become the four-wheeling playground of thousands of Americans, making the most of open country while it's still free and still legal.

Let's look at Europe on the other hand. There too, ever since England's Rover Company brought out the fabled Range Rover in 1970, four-wheel-drive vehicles have acquired as much status appeal as utilitarian value. There are probably as many Range Rovers around the streets of London as around the rest of England. But there is one big difference between the continents. In Europe, there is little 4WD recreational use at weekends. Suitable off road country is usually too far away. So where does a European get full off road value out of his rig? For the lucky few, the answer lies to the south in the Sahara or to the east on the legendary road overland to Kathmandu in Nepal or Madras in India.

So what better project than to bridge the Atlantic divide by taking two Chevy Blazers across the Sahara? But there's a problem. First of all, the Sahara has seen plenty of English Land-Rovers over the last thirty years, so some Land-Rover spares are available and more than a little Arab expertise at tying broken axles together with a piece of string. But no-one ever heard of a Chevy Blazer in the Sahara before. Hence the plan to take two identically equipped vehicles (to back each other up in emergencies), plus an extensive range of spares (with an eye on the weight problem however) and a lot of mechanical expertise among the crew.

That's not all, though. Most US 4WD vehicles, including the Blazer, are designed for recreational use over a few days, rather than sustained rugged driving thousands of miles over a period of months. There aren't that many English Land-Rovers in the US, but when it comes to long expeditions south of the border, a lot of Americans still prefer to drive the old reliable Land-Rover, developed over the years for just this kind of trip.

So can you take a Chevy Blazer across the Sahara? The answer lies in good planning and preparation of your truck. The intense competition between the 4WD manufacturers in the US and between aftermarket manufacturers has produced a lot of research and a lot of accessory equipment, as has the whole 4WD competition scene. A lot of the new gear is decorative—of the chrome fenders and running boards variety. California rigs sport a lot of fancy lights that aren't legal on the highway and probably never get a chance to be used off it. But out of all the fancy paraphernalia comes a residue of really useful equipment that is what this article is all about.

A lot of this article is going to be lists. Where a comment is needed I've provided it. You may not agree with everything. You may consider something essential that I've left out. Some of it may seem so elementary that you're going to wonder why I mention it. Remember that this is just one *aficionado*'s list, for one particular make of vehicle, for one particular trip. Some overlanders stick to the rule 'If in doubt, leave it out', on the grounds that most things needed can be picked up somewhere along the way and that this way they save time, space and money.

But the rule applies mostly to travellers in populated areas (not the Sahara)—and you need always to consider the size of the doubt.

I believe a Chevy Blazer could make it where only Land-Rovers are used to going.

My choice of vehicle is the 1980 Chevy Blazer with Hydramatic (automatic) transmission and the big V-8 engine for maximum power under a heavy load, and with Cheyenne interior for maximum comfort. Some of the accessories on my list are supplied by Chevrolet as optional extras and others are supplied by aftermarket accessory manufacturers. For maximum low end torque I choose the rear axle with a 3.73:1 Final Drive Ratio. (The 4.11:1 axle is only available with the 6 cylinder engine). With a fully laden vehicle expected to weigh in at as much as 8,000 pounds, the fully protected underside of the rig will need to be beefed up with springs, shocks and axle trusses to support easily the full GVW (gross vehicle weight), not just at standstill but under stress on a rough potholed road at reasonable speed.

Underside Accessories

Front differential skid plate with drain hole. (1in. min. between plate and diff.) (From: Hickey).

Rear differential cover with drain hole. (1in. min. between cover and diff.) (From: Hickey).

Transfer case skid plate with drain hole. (1in. min. between plate and transfer case.) A combined engine, transmission and transfer case skid plate is available from Con-Ferr.

43 gallon replacement fuel tank, well baffled, with dashboard gauge and on/off switch, or 31 gallon ex-factory replacement tank. Ensure that all fuel lines are of steel, are properly installed and wrapped in asbestos sleeve where they pass close to exhaust, to prevent vapour lock. Secure properly to frame with clips. (From: Hickey).

Skid plate for replacement fuel tank. (Chevrolet offers a skid plate for the standard 25 gallon tank only).

Front axle truss. (From: Hickey).

Rear axle truss. (From: Hickey).

Heavy duty coil springs ex-factory (Or: RPO F60 springs; Hellwig Products' TF 116C for front and OL 214C light duty overload springs for rear).

Heavy duty shock absorbers, one each wheel. (I recommend Gabriel Hi-Jackers, the right rear shock angled forward from the axle, and the left rear shock angled rearward to counter torque reaction. Available from Fairway Ford.) For extra protection, unmount the rear shock absorbers and remount with a collar made from tough rubber hosing, which has a bore at least ½in. larger than the diameter of the shock absorber and is at least 2in. shorter.

Reliable underseal coating against rust.

Vent hoses leading from axle breather orifices. Put a baffled vent at end of hose plus foam air cleaner boot. Likewise vent hoses for crankcase, transmission and transfer case.

5 x Armstrong Radial Tru-Trac 11R-15 tubeless tyres with loading range B. Check you have a leaflet showing the maximum weight the tyres can support at various pressures. The more you deflate for flotation, the less weight the tyres can support without rupture. (From: Dick Cepek). As back-up, you will be taking tubes (see list of spares).

Radial tyres have the disadvantage that

they cannot be mixed with nonradials and outside western Europe and North America may be hard to replace—especially as one blowout may mean replacing all four tyres.

Waterproofing for differentials, hub assemblies, crankcase, transmission and transfer case should be considered if a lot of driving through water is likely.

Protection plate for inter-axle locking differential.

You may consider front and rear torsion stabiliser bars, but in my view these are unnecessary on the latest Blazer. Likewise a hydraulic steering stabiliser and/or a direction stability and shimmy damper kit.

We've all come to accept the idea of a locking rear axle differential and a locking inter-axle differential for extra traction in tricky sticky situations. However there's still a lot of doubt about whether the extra traction you'd get from a limited slip diff. in the front axle (e.g. the Detroit Locker by Detroit Automotive Products) would be worth the risk of front-end drift. (Detroit A P claim there is no front-end drift. with their Locker, as do the manufacturers of the Dual Drive Differential available from Western Off-Road Wholesalers).

Rubber gravel flaps behind all four wheels.

15 x 8 wheels (highest quality only—ensure they can support the weight you envisage carrying per wheel, front and back). (From: Dick Cepek).

Front power disc brakes (ex-factory).

Non-Underside Accessories

Brush guard/push bar. (From: Hickey).

Solid tow hook at front and rear. (From: Hickey).

Transmission oil cooler, preferably ex-factory, otherwise one from: Hayden; Flex-O-Cool; Air-Cool). Hayden manufacture a combined engine oil and transmission cooler. Otherwise a heavy duty engine oil cooler is available ex-factory.

Power steering cooler. (From: Hayden Mfg. Co., Corona, CA, 91720).

Seat belts for two front seats. Fixed not recoiling type.

Air conditioning, ex-factory. (This is likely to be a point of considerable dispute for a trans-Sahara expedition. The arguments against are: one more thing to go wrong; fuel is precious and with one more belt to drive, the engine consumes more fuel; contrast between interior and outside temperatures results in poor acclimatisation and ability to withstand illness. My preference is for air conditioning. If you think you can get the extra fuel to see you through several hundred miles, if you reckon you can afford the extra fuel and if you think you're fit, the air conditioner used sparingly from time to time will do wonders for morale when it's 120°F outside in the sun. And the extra fuel is only consumed when the air conditioner is actually switched on.)

Power steering. This is another thing that can go wrong. The advantage is that in the desert, it reduces driving effort, and effort reduced equals body liquid retained in an area where every gallon of water is precious. (The same is true of air conditioning. It can do wonders for your water supply.)

Boomerang side mirrors giving ample view to the rear.

Console box between the front seats to

provide an armrest. It opens revealing an icebox and cool drinks ready to hand. (From: Hickey.)

Dashboard thermometer showing temperatures inside the vehicle. This is especially useful if you have air conditioning.

105 amp heavy duty alternator to handle two upgraded 70 amp batteries mounted on a split system. (From: Leece-Neville Co.).

One upgraded 4000 watt 70 A/H battery, preferably ex-factory, or a Delco Freedom battery. Second battery should be more powerful. Main battery should be 115 A/H with a performance of 550 cranking amps and a 27F rating. The 70 A/H battery should have a performance of at least 400 cranking amps.

Split charge automatic battery isolator, preferably ex-factory. (Or from: Sure Power Products, 1426 21st Street, Oregon 97222; or try the BMC Protector 111, available from Viking.)

In-line filters between all fuel tanks and petrol pump(s).

AC Dual-Stage (polyurethane and paper) air filter and spare. Or cleanable foam filter. You may prefer an oil bath filter like the K & N E150 and K & N air filter oil, both available from Fairway Ford.

Wood steering wheel (a luxury item that's a personal preference of mine for hot climates). (From: Hickey.)

Locking utility and valuables box bolted under the hood. (From: Symmetra-Box Mfg., 1592 Old Bayshore Hwy., San Jose, CA 95112.) Combination lock best if you keep it well oiled and keep a copy of the combination hidden somewhere you can find it.

Interior hood lock.

Tinted glass all round—the darkest tint you can get if it's for the Sahara. You'll attract a lot of curious attention, more than you want, so one way glass would be even better, in addition to a deep tint. However I don't know of any manufacturer making one way glass windows for vehicles.

Hilites long range/flood combination light with installation kit and double rocker switch for flood or long range beam. A lock is available for these lights which are available direct from Hilites.

Scheel orthopaedic seat to replace stock front passenger seats. (Fixed not swivel mountings.) (From: Scheel Auto Seats, 1007 South Fremont Avenue, Alhambra, CA. 91801.) This seat is a luxury well worth the cost on a long hard back-breaking trip. It has armrests and a headrest.

Microwave burglar alarm. (From: Mountain Man Industries, P.O. Box 1147, Grand Junction, CO. 81501.) The *only* kind of vehicle alarm worth getting. Forget the rest for travel in North Africa—the other kinds don't work if someone breaks the glass out of a vehicle and leans in to steal things without opening the door or rocking the vehicle hard.

Good no-nonsense horn with loud 'honk', mounted inside the vehicle against theft.

Rear screen washer and wiper (unless you have a camper top to the vehicle.)

One accessory you'll want to take off if you're driving outside the US is the vehicle's catalytic converter, if you've got one already. You certainly won't find unleaded fuel in the Sahara, and anyway you'll get more miles to the gallon on leaded fuel without causing any traceable pollution in the Sahara, as long as it remains relatively free of vehicles. Talking of fuel, a word is probably called for here on diesel engines. More and more diesel 4WD vehicles are being developed in the US. The Land-Rover has been around for years in petrol and diesel versions. However while diesel is cheaper and gives more miles to the gallon, it's also harder to get than petrol, even in the Sahara. That gas station you were counting on reaching 300 miles across the desert just before the tank read empty will probably (if you're lucky and they've been restocked recently!) supply petrol, but there's no guarantee of diesel. Also diesel is very noisy—bad for morale on a long tiring trip. Experience shows that most Land-Rovers making the Sahara trip have petrol engines.

Additional Useful Accessories

Two lightweight sand ladders 5ft. long.

Aluminium sand anchor. (From: Western Off-Road Wholesalers).

Folding narrow-bladed trenching shovel. (From: Viking).

10,000lb. nylon snatch strap. (From: Hickey).

Automatic cab and engine compartment fire extinguisher with safety lock. (E.g. the Flame-Out. Another is the Fire-Eater from International Equipment Corp., 1110 North Hudson Avenue, Suite B, Los Angeles, CA 90038.) You should also have a 5lb dry-type extinguisher conveniently located (e.g. OMC). (From: Hickey).

An accessory that plugs into the cigarette lighter and allows you to run a 110 volt or 220 volt electrical appliance (like an electric shaver) off the vehicle's 12 volt battery system. (From: Good Time Recreational Supply).

1in. thick plywood jack pad one foot square.

Bumper Jack (also known as sheepman jack) in addition to the vehicle's standard manufacturer's jack. 'I recommend the Hijacker).

Bull bag Mk II (will raise 3500lb weight) to raise a wheel off the ground with minimum effort, thus conserving precious body fluid in the event of a flat in hot conditions. (From: Hickey).

Two steel 5 gallon GI fuel jerry cans and 6in diameter funnel/pourer with filter. Must be properly secured to the vehicle, but never in front, and locked on if mounted on vehicle exterior. (From: Western Tire Center). Incidentally, never use jerry cans for carrying water, as they rust and give an unpleasant taste to the water.

Cigarette lighter socket mini air compressor. (From: W. R. Brown Corp., 2701 N. Normandy Avenue, Chicago, Illinois, or from Hickey).

Electric front bumper winch. Possibly a debatable item for a Sahara expedition. A good one will probably weigh around 130lb, so is all that extra weight really necessary? The only thing you'll have to pull against in the desert is your sand anchor, possibly supplemented by a spare tyre. However in a last resort, the winch may be your only chance, when pushing, unloading, deflating tyres, digging, bumper jack, sand ladders, reversing and anything else you can think of has failed. However in the desert, with proper use of sand ladders, you certainly won't need it as much as you might in muddy terrain. (From: Hickey; Sidewinder model).

Foot air pump with gauge (in case your

mini air compressor fails). (From: Good Time Recreational Supply).

Key-locking wheel lug nuts, 6 per wheel plus key. (From: Hickey).

Double suction cup for windscreen to reduce chance of shattering. Take a set of tinted ski goggles in case it does shatter and you have to drive with the wind blowing in your face. Consider taking an emergency folding windscreen or a length of wire screening to mount on the windscreen when driving on stony ground.

Wheel weighing scale. You needn't take this with you, but get a good idea of how much weight you're carrying at each of the four corners of the vehicle before you depart and how evenly the weight is distributed. Remember not to exceed the allowable maximum for tyres, wheels, axle and suspension. Keep weight balanced and the front/rear ratio the same throughout the trip. Your local truck weigh station should be able to test the weight of your rig, but in the Sahara such services don't exist, so you'd better be good at calculated guesses or you could be in trouble.

Two additional 12 gallon saddle fuel tanks (one each side behind rear wheel wells) with electric switchover valve, dash gauge and on/off switch. (From: Hickey and Western Off-Road Wholesalers). With the large main tank and two filled jerry cans, this should bring fuel capacity up to 60-70 gallons, giving a range of 420-490 miles at least, allowing for a pessimistic seven miles per gallon. This should be sufficient range for all but a few of the more commonly driven stretches of *piste* (rough road) in the Sahara, but won't allow for much real off-roading unless miles per gallon can be improved (in spite of the big V-8 engine) or more fuel carried in extra jerry cans. On an imperial gallon, you should get at worst about 9-10 miles per gallon, except on very poor sand surface. On a good *piste,* you might even get 11-12 miles per gallon. One tip on fuel. Try to fill up early in the morning before it gets too hot. When petrol warms up, it expands, so by buying early you can get up to 5 per cent more gas into your tank at no extra cost.

Heavy duty radiator and 6 or 7 blade radiator fan for extremely hot driving conditions. (Flex-a-lite 6 blades heavy duty fan and spacer kit available from Simpson's).

Voltmeter mounted on dashboard. (Note that if 16 volts or more shows on a 12 volt system, the voltage regulator is probably broken). I suggest you take the ex-factory gauge package which comprises voltmeter, temperature gauge, oil pressure gauge.

AM/FM radio/cassette player and good cowl aerial and under-hood static suppressor. (Available from Western Auto). This is not really a luxury. Good for weather reports when you're not too far from inhabited areas receiving radio broadcasts (if you understand the language). And your favourite music on tape works wonders for morale when the going gets tough. However forget about CB radio. Not only is it illegal in most countries, but you'd find no one else to talk to except the police or the army, who would probably pick you up for spying in certain countries I won't name.

Dashboard clock (ex-factory). Some of us head for the wilderness to forget about time. Nevertheless a clock or watch is a necessity on such a trip for making various logistical calculations.

Floor mats (ex-factory).

Front compartment headliner (ex-factory). (A little luxury to make life sweeter.)

Map light which plugs into the cigarette lighter (From: Hilites.)

Magnet.

Stone guards homemade from heavy-duty cage wire, screening, etc., to protect front and rear lights. (Only Land-Rovers come equipped with stone guards.)

Tropical roof, a thin skin of sheet metal fitted ½in-1in above the vehicle's roof to keep the interior cooler. There is no sense in having a tropical roof if you have a full-length roof rack with a solid floor.

Spare Parts

A thought to begin with is that some overlanders take not one but two sets of spares. Consider your route, your vehicle and your own resourcefulness and then decide for yourself.

Set of fuses.

Engine oil (1½ times engine capacity) (transferred to an all-metal can to avoid leakage).

Power steering, automatic transmission top-up fluid (2 quarts), transfer case fluid.

Gear oil (3 quarts) plus hose to get it into gear case.

Brake fluid (2 quarts). Tape the can shut to prevent leaks.

Bottle of distilled battery water.

Differential lubricant as recommended by manufacturer. (For Blazer this is GM part 105022—Positraction lube).

One can of bearing grease.

Door-Ease lubricant.

Lubriplate white grease.

Heatproof plumber's grease.

WD-40 spray.

Permatex Form-a-Gasket Number One.

Permatex Form-a-Gasket Number Two.

Permatex Silicone Form-a-Gasket.

Spare set of spark plugs.

Two spare inner tubes (correct size) and valve core.

Spare nuts, bolts, washers (rubber and metal), wood screws, sheetmetal screws, lock washers, cotter pins, nails, retainer clips.

Spare fanbelt and accessory engine belts. Also a second spare fan belt only long enough to go round the crankshaft pulley and water pump, thus bypassing alternator pulley.

Set of gaskets for: side and top valve covers, head, thermostat, water pump, fuel pump, manifold, timing cover, all gearbox pans, differentials, axles, hubs.

Seals for: transmission, front and rear hubs, pinion.

Can of 3 in 1 oil.

Set of spare 12 volt bulbs for all exterior lights, interior roof light and main dash lights.

Complete set of spare hoses and clamps (3 feet of heater hose).

Spare fuel pump. (Have an electrical pump as the basic pump. The spare pump should be a manual pump mounted in parallel. There should be a spare diaphragm for this manual pump. There should be a valve in line

to shut off fuel from either pump if it fails, otherwise fuel could flood the crankcase.)

Four tyre valve cores for tubeless tyres. (From: Hickey).

Spare distributor cap.

One spare rotor.

One spare tyre and wheel. (Some argue for two spares for a trip like this. That means a lot of extra weight and space taken up. I prefer to take my chances with one spare and tubeless and tube repair kits, in addition to the backup of the second identical vehicle, which would also have one spare. The spare parts and tools carried on the trip should be shared equally between the two vehicles.)

One spare complete carburetter.

One set wheel bearings and races (front and rear, inner and outer).

One U-joint.

Complete set of spare spark plug leads.

Three feet steel petrol line.

One foot vacuum line.

Blackseal engine sealer.

Roll of Scotch thread sealant and lubricant.

Two foot strip bicycle inner tube.

Loctite Lock'n'Seal.

Eight inch square piece of cork gasket material.

Eight inch square piece of asbestos gasket material.

One by two foot sheet of standard gasket paper.

10in of ¼in diameter copper tubing.

Assorted springs.

Solderless electrical connectors (two each of open-ended, close-ended, male pin).

Spare thermostat.

3ft rubber hose ⅛in inside diameter plus clamps.

Spare transmission fluid filter.

One roll electrical tape (type that will not melt in hot humid conditions).

One tube RTV silastic rubber.

10ft 14-gauge insulated electrical wire.

One small roll of mechanic's wire.

One can radiator stop leak.

Tyre boot for flats (Baja Boot from Dick Cepek).

One gas tank repair kit (From: Dick Cepek).

Pressurised goop (one quart) for tubeless tyre flats.

Inflator/sealant for tubeless tyres. (This is not the same thing as goop). (Try Seal'n'Air from Dick Cepek).

Windscreen washer additive.

Adequate supply of anti-freeze for cooling system (works as coolant as well as anti-freeze).

Centre bolts for leaf springs and nuts (4 each).

One pair of U-bolts and nuts for leaf springs.

One spare alternator.

One spare starter motor if the vehicle cannot be crank-started.

One spare manifold and gaskets.

Can of CRC556 spray-on liquid to waterproof and dustproof the distributor and other vulnerable parts.

Can of spray type petroleum distillate such as WD40, Gly, P-38, as lubricant and for drying up water in ignition.

Wiper blades.

Finally never leave home without a spare set of all your vehicle keys carefully suspended on a piece of strong wire somewhere out of sight (and where they won't rattle) in the engine compartment, accessible from underneath the vehicle, but not visible from underneath the vehicle. Or have one member of the expedition carry the spare set of keys permanently on his person.

Tool Kit

Steel tool box with tote tray. (From: Western Auto).

Standard tool kit and jack purchased with vehicle.

Mixed screwdriver set including long and medium standard (slot) and Phillips screwdrivers. Always have one very small slot screwdriver for electrical work. Preferably the whole set should be magnetic tipped.

One large and one small adjustable crescent wrench.

One large pair and one small pair of visegrips.

One cross wrench for wheel lug nuts.

One 2lb sledge hammer.

One brake bleeding kit.

Three shop rags.

One tyre pressure gauge with dial. (From: Western Off-Road Wholesalers).

One set of tyre irons (bead breaker kit). (From: Dick Cepek).

One tubeless tyre repair kit. (From: Dick Cepek).

One patch kit for tube tyres. (From: Dick Cepek).

One hacksaw (folding) and spare blade.

One spark plug wrench.

One points gauge.

One tyre valve core wrench.

Small file (rough and smooth).

Set of open end/box end wrenches (¾in through ⅜in). (From: Western Auto).

Petrol syphon with hand pump (e.g. Liquid Hand Pump from: Howard Products Co., P.O. Box 57246, Dallas, TX. 75207, or Free-Flow syphon from Western Auto).

One wheel block. (From: Dick Cepek).

Medium size mechanic's hammer.

One pair of diagonal cutters.

One pair of needlenose pliers.

One pair of channel-lock pliers.

One awl.

Sheetrock knife and blades.

One hand drill with bits.

One timing light.

One hydrometer.

One compression gauge with flexible stem.

One pair of jumper cables. (From: Dick Cepek).

Engine compartment inspection light which plugs into cigarette lighter. (From: Western Auto).

Last but far from least are the following: owner's manual, manufacturer's workshop manual, *Chevrolet and GMC 4-wheel Drive Maintenance* (published by Clymer Publications, 222 North Virgil Avenue, Los Angeles, CA90004), and *Chilton's Repair and Tune-Up Guide for Blazer/Jimmy* (published by Chilton Book Company, Radnor, Pennsylvania). In addition, on a normal trip, you should have with you the manufacturer's directory of dealers and distributors. General Motors is well represented in most of Europe, but unfortunately you won't have much luck finding G M dealers in the Sahara. It'll be all up to you!

Maintenance Check Before Departure

Whether your vehicle is new or old, it goes without saying that it should be in perfect condition for the trip you're planning. If the vehicle is new, I recommend that you treat it as if it's a Formula racing car and get a highly skilled expert with all the right facilities to blueprint, balance and port the engine. If you're planning a major expedition like the Sahara one discussed in this article, you'll want to set off in a vehicle that may not be old, but should at least be well tried and tested. If the vehicle is new, and you really are going to tackle the Sahara, try taking it for a hard-driving weekend once you've already done your first 1,000 miles in it. Such a weekend should cover at least 300 miles. The next stage is a week long hard-driving trip. If that works and you're satisfied that you're beginning to sort out the teething troubles of your new prized machine, then you should be ready for the warm-up for the Sahara, a long transcontinental drive. Obviously you won't want to rush into the Sahara with a new vehicle without first carefully testing its strengths and weaknesses, and the progressive expedition testing outlined here seems to me the best way of doing a careful professional job.

1. Change oil and oil filter.
2. Clean air filter and change oil bath.
3. Lube driveshafts, winch, speedometer cable.
4. Lube all locks with graphite; adjust and lube all doors.
5. Clean or replace all gas filters.
6. Inspect undercarriage for fluid leaks, loose bolts on pans and mounts.
7. Rotate all five tyres inspecting for cuts and even wear.
8. Adjust brakes if needed.
9. Check fanbelts and accessory belts.
10. Check adjustment of carburetter.
11. Check sparkplugs. Clean and regap if needed. (Replace as necessary.)
12. Check ignition timing.
13. Check and top up:
    - front and rear differentials
    - swivel-pin housings
    - transmission
    - transfer case
    - overdrive (if applicable)
    - steering box
    - battery
    - brake and clutch fluid
    - cooling system
    - crankcase
14. Check that there are no rattles.
15. Inspect radiator and heater hoses.
16. Check breather vents on both axles.
17. Check all lights.
18. Check wheel balance and front end alignment, and always do so with new tyres and wheels.

Some Useful Tips

To round off, here are some useful ideas whether you're going off tracks for the weekend or planning a major expedition in Africa.

1. If you're trying to attract the attention of rescuers and have no flares, try setting fire to the spare tyre with gasoline. But remember that unless a buddy is expecting you, no one is likely to come out looking for you in the Sahara if you break down in the middle of nowhere. Always make sure, therefore, that there's someone who'll come to look for you if you haven't reported by a certain date.

2. Air-conditioning causes radiator water temperature to rise. One way to create the opposite effect if radiator temperature is getting too high is to turn on the heating—not very pleasant in tropical heat, but it may save your radiator from blowing.

3. Normally run about 2lb more pressure in the front tyres than in the rear. The manufacturer's recommendations should be looked at in terms of the load instructions you should get with your set of tyres. The front of the vehicle should usually be only slightly heavier than the rear when the load is balanced.

4. If you deflate the tyres for off-road driving, remember to reflate them again when you return to the highway. Don't deflate tyres too much off road with a heavy vehicle. For the vehicle described in this article, 20psi is likely to be your very lowest off road tyre pressure.

5. You should know the maximum weight supportable by each wheel at maximum tyre pressure. You should inflate the tyres below the maximum tyre pressure unless GVW is close to tyre maximum. The tyre manufacturer should supply a chart showing optimum pressure for different loads on different terrains (usually limited to on/off road).

6. Never drive a deflated tyre over sharp rocks.

7. To get the correct tyre pressure, measure pressure when the tyres are cold before use.

8. Remember that an imperial gallon is about 25 per cent greater in volume than a US gallon. The weight of one imperial gallon of petrol is around 9lb, and the weight of one imperial gallon of water is around 10lb. These figures are extremely important on a long desert trip when it comes to calculating GVW without the help of a truck weighing scale.

9. If your radiator is gathering a lot of insects, you'll help cooling by cleaning them off from time to time. You'll help keep temperatures down by not mounting spare tyres, jerry cans or other pieces of equipment in front of the radiator grille.

10. Do not mix two different types of engine oil if you can help it.

11. When filling up with engine oil, bring the level up to half way between the high and low marks on the dipstick, and no further.

12. Rotate your five tyres every 4000 miles (5000 for radials).

13. Any expedition into the Sahara will rely heavily on a compass for accurate positioning. Always take compass readings well away from the vehicle to avoid distortion by the vehicle's magnetic field.

14. Regardless of what type of vehicle

insurance, *carnet* indemnity cover you have for your trip (and make sure you do have the right kind for every place), there is always the risk that you'll injure or kill someone. Most people travel on foot in the Third World, are not used to a lot of traffic, and may have more confidence in your ability to avoid them than is justified. I suggest you have public liability insurance valid worldwide for £125,000.

15. Try to get clearly written guarantees with everything you buy for your rig, and return warranty cards to the manufacturer for registration.

A Special Recommendation for Camper Tops for 4WD Vehicles

Four Wheel Campers, Inc., 2890 W. 62nd Ave., Denver, CO 80221.

Four Wheel Campers manufacture an outstanding telescoping camper top for Blazer, Bronco, Scout, Ramcharger and other US off road vehicles. They're designed to fit the vehicle's standard mountings when the vehicle's standard top is removed. Very little increase in vehicle weight (only 230lb on the Blazer) means minimal additional fuel consumption, normally the main drawback with camper tops on 4WD vehicles. And of course with little extra weight, the suspension system isn't put to unnecessary strain. The advantage of the telescoping top is a low centre of gravity while you're trucking along, so you get none of the road-holding problems normally associated with camper tops. But when the top is up, you get over six feet of height inside. The welded aluminium frame gives a unit with superior strength. The upper expandable walls are made of Dacron webb which is not affected by rain or moisture as canvas is, so overall life is increased and maintenance is reduced. These campers have insulated roofs and lower side walls for living comfort. You get a family-sized camper with four-wheel-drive mobility, a walk-through interior to or from the driver's compartment. The unit comes fully equipped with kitchen built-ins and adult sized double beds, one of which sits over the driver's cab when the roof is in the 'up' position. No alterations are required to the original vehicle. Standard equipment includes: 9,000 btu automatic furnace, two burner surface mount stove with safety cover, single stainless steel sink with hand water pump, 20lb LP gas supply, nine gallon water supply, adjustable roof vent, removable centre table, sliding storage doors, screened sliding windows, overhead lighting, shag carpet trim, Herculon upholstery with 4in foam cushions, full privacy curtains, cherry wood interiors with natural wood counter and table top. Optional equipment: porch light, additional storage, gas/electric refrigerator, 12 volt electric water system, carpet or vinyl floor covering, extra interior lights (12 or 110 volt). WEXAS considers these tops (priced around $2,500) as amongst they very best buy in 4WD aftermarket equipment available anywhere. The Four Wheel Camper top is a must for comfortable long distance off-roading, especially if you're planning a long expedition to foreign parts. The Sales Department at Four Wheel Campers will reply promptly to enquiries and send you a free copy of their latest folder. Write or call (303) 426-0131.

Camper Top Equipment for a Trip Across the Sahara

I've assumed that for my trip across th

Sahara, my vehicle is going to be equipped with a camper top from Four Wheel Campers, described in the previous section. (The closest alternative I've seen in the telescoping lightweight range is built by: Fiber Kraft Mfg., 10501 West 48th Avenue, Wheatridge, CO 80033). Some of the equipment I'm going to list could be left behind since it comes close to being in the luxury bracket. However some of the world's toughest explorers (especially mountaineers) learnt years ago that making things too uncomfortable for yourself not only lowers morale but can in the end be downright dangerous. A happy, comfortable and healthy expedition is less likely to make fatal wrong decisions. So here's our list of equipment to go on the camper top or to be fitted inside it:

Battery powered mosquito buzzers (2) fitted to the walls. (From: Good Time Recreational Supply).

Insect screens for all the windows. Make sure that when windows are open (you'll be sleeping with them open to keep cool at night), no insects can get in through or around the insect screens, or around the edges of the doors.

2 burner LP gas stove (butane since propane is not available in North Africa) with thermocouple to cut gas if flame blows out. If possible, get the type of stove designed to double as a heater.

40lb capacity LP gas tank with pressure regulator, gas gauge, safety blow-off valve (to prevent over-filling) and master shut-off valve. The tank should be AGA-approved and carry an ICC, ASME or API tag. Tanks should be painted (preferably with aluminium or white paint) and have a valve guard. Tank must be welded to the base that carries it, but not to the vehicle. The master valve of an LP tank should be closed if the tank is empty to prevent water from entering. (Recommended LP gas gauge from: Marsh Mfg. Co., 7349 Coldwater Canyon Avenue, North Hollywood, CA 91605).

Sink with hand pump, easily filled tank of at least 27 gallons capacity, built-in water purification system (e.g. Everpure CI-RV with spare replacement cartridge filter). Anti-algae tablets may be a useful added precaution to keep water drinkable for days and even weeks without refill or a chance to empty the tank completely and refill. (Recommended hand pump is the one made by Shurflo). (Alternative water purifier is the one made by Ogden, or the PD451 from Viking.)

A reinforced fibreglass water tank is preferable to a steel one. It should be located over or in front of the rear axle. Supply hoses should be of the guaranteed tasteless type. (Thetford Mfg. Co. make fibreglass tanks).

Foamed plastic or fibreglass matting for insulation for floor, walls and roof. Good insulation not only keeps heat in, but keeps cool air in if you are using the air conditioner, thus reducing the frequency or needed output of the air conditioner. (Any cooling should come from the vehicle air conditioner. A separate air conditioner for the camper top would be a ridiculous luxury in such a confined space, and would require its own generator. Evaporation coolers would be useless in the Sahara since they require large quantities of water to work, and only function well where humidity is very high. The heat in the Sahara is extremely dry).

Cab to cowl foldaway shock absorbers

if you feel that the weight on the cab roof from the cabover section of the camper top is going to be excessive when (if) two people are sleeping in the cabover section.

Cab to coach weatherproof boot for waterproofing between cabover section and roof of the driving cab.

Fixed porch light (with yellow bulb to keep insects away). The door itself should be made of aluminium and be sturdy enough to resist a superficial attack by thieves. It should be fitted with a 200 degree viewer. (Available from Western Auto). The outside light is not a decoration. It's there to help you see if anyone is outside when you put it on and look through the viewer. Midnight visitors are not an uncommon event if you unwisely choose to park conspicuously and in close proximity to a town in an Arab country. (It is a good idea never to spend more than one day in the same place, and to try to park towards nightfall.)

Storm windows (double glazing) are a luxury you might like to consider for extra insulation to help reduce the need to use the air conditioning.

All the windows should have a solar shield coating to reduce the amount of light and heat reaching the interior of the vehicle, and to reduce glare and keep the vehicle shady inside. (E.g. Scotchtint Solar Control Film from Pullman Campers, 8211 Phlox Street, Downey, CA. This windowfilm reduces heat/cold loss 75 per cent and glare 82 per cent according to the manufacturers. Also available from Viking). An alternative window glass tint is silver shade screen from Viking. This gives a one-way glass effect (useful as an anti-theft device) and reduces glare. A choice of colours is available (e.g. bronze). I don't know how this screen compares with a solar shield coating for insulation.

Extra storage cupboards in addition to those supplied with the camper top. For a long journey, a lot of food (mostly lightweight dehydrated) will be carried, and should be packed well out of the way until needed.

The roof could be used for a luggage rack which could also carry the two 5ft sand ladders and lightweight everyday equipment packed with tarpaulins on top and underneath. However anything on the outside of the vehicle is an invitation to thieves. Never leave the vehicle unattended in any populated place when the roof rack is loaded. Some drivers prefer to avoid the worry by dispensing with a roof rack altogether. With good packing, it is usually possible to fit most items inside the vehicle, leaving only the following lockable items outside: accessory driving lights, two jerry cans, spare wheel, two sand ladders, the vehicle's four wheels. A roof rack must be large and sturdy, must fit the vehicle and be bolted on if possible. If it is attached to the rain gutters, the clamps should be tested for tightness every day. Ensure that the camper top itself is properly painted or anodized for protection against the weather.

The camper top should have a roof vent big enough so that you can easily climb out of it in an emergency. It should therefore have a simple but very strong interior locking catch.

The vehicle should be equipped with a readily available full first aid supply (see *The Independent Traveller's Guide to Health*) as well as some more sophisticated medical facilities that the crew knows how to use in emergencies. It should also contain vaseline, ear drops, dramamine (for travel sickness), a sting relief spray such as Hobson's (from Don Gleason's Camper Supply).

A wall-fitted magazine rack, well-stocked, is a useful item for a little extra luxury.

Ray-o-Vac 200 wall-mounted light fixture operating on four size D heavy duty batteries. (From: Don Gleason's Camper Supply.)

Camper Top Kitchen Equipment

Selection of plastic bags (for waste disposal, etc.) and wraparound sealers.

GI inflatable canvas water bag (or a quart insulated canteen with shoulder strap available from Indiana Camp Supply—useful as part of a survival kit).

5 gallon collapsible plastic water container with spigot. (From: Good Time Recreational Supply).

Collapsible canvas bucket with watertight top (for wells and use as washing machine—the clothes slop around in it as you drive).

Plastic scourers.

Supply of dishcloths and dish-wiping rags.

Set of 4 plastic containers with lids for food.

One saucepan (medium size) with lid.

One kettle-type saucepan with lid and spout. (From: Don Gleason's Camper Supply).

Tin opener.

Corkscrew.

4 x knives, forks, spoons, plates (all stainless steel).

Coleman folding oven for baking bread. (This oven heats up by sitting on top of a hot plate). (From: Don Gleason's Camper Supply).

Aerosol kitchen fire extinguisher. (From: Edgewood National, Inc.).

Plastic bottles 2 x 1 pint and 1 x 1 quart. (From: Good Time Recreational Supply).

4 x mugs, teaspoons (stainless steel).

2 bottle stoppers, slide-on type.

1 wooden spoon.

1 bread knife.

Supply of tin foil.

Large supply of paper towels.

Plastic kitchen funnel.

Large supply of matches and waterproof matchbox; and/or disposable lighters.

Supply of salt (granular, not salt tablets).

Supply of freeze dried and other food for trip. (From: Indiana Camp Supply).

Large supply of paper plates to save on water, but bury all refuse.

Washing liquid for dishes.

One small frypan. (From: Don Gleason's Camper Supply).

General Trip Equipment to be stored in Camper Top

Short handle axe 20oz. (From: Western Auto).

18in machete. (From: National Surplus Center).

2 razor blades.

Can of spray type petroleum distillate such as WD40.

1 dozen wire clothes hangers.

100ft solid braid nylon ⅜in rope (for wells, clothesline, etc.). (From: Don Gleason's Camper Supply).

Can of spray type insect repellent (e.g. Cutler insect repellent from Don Gleason's Camper Supply).

Washing powder for clothes.

Beach rug with rubberised bottom.

Roll 2in wide plumber's tape.

20ft ½in all-weather tasteless plastic hose for filling water tank.

Tube of Plasticel to caulk leaking windows, or Butyl rubber sealant. (From: Viking).

Rubber cup 'plumber's friend'.

One folding hunting knife. (From: Herter's).

Tie-down nylon strap 10ft x 1in. (From: Western Tire Center).

Teflon pipe sealant tape ½in x 118in. (From: Viking).

Chamois leather.

Sponges.

6 elastic tie-downs, length 40in. (From: Dick Cepek).

Ball of strong twine.

Ray-o-Vac interior fluorescent table lantern with two light elements. (Powered by batteries). A similar lantern is made by Coleman.

4 beach size towels.

Halazone water purification tablets (in case the water purification system breaks down) (or Hobson water purification tablets, available from Don Gleason's Camper Supply).

Maps and a map container file divided into compartments to hold them.

Moneybelt to be worn by crew member(s) when carrying money, passports, etc., to banks, police stations, customs posts.

Soap and soap dish.

Toothpaste.

Pair of kitchen scissors.

Plastic pan and soft bristled hand broom or brush. (From: Viking).

Cleaning liquid for oily hands.

One dozen clothes pegs.

Telescope or binoculars.

Supply of cassettes for the cassette player.

Good quality compass (available from National Surplus Center), set square, protractor, dividers, ruler. (New techniques of desert navigation have developed in recent years and can be extremely useful. A sundial mounted to the vehicle and a sextant properly used could make all the difference between going the right way and getting lost).

Elastic bands.

Sewing kit.

Safety pins.

Waterproof matchbox. (From: National Surplus Center).

Large supply of toilet paper.

Presents for local people in North Africa. (This is an essential part of successfully completing any trip in the Third World. I recommend pens, pencils, exercise books; and a Polaroid camera is a good idea if the local people are not superstitious about having their photo taken).

Bedding.

Waterproof flashlight with red flasher and a set of spare batteries. (Heavy duty D batteries—7 amps).

Dictionaries and guide books.

Vitamins and minerals to supplement the daily diet.

Frostpak icebox from Koolatron Industries, 56 Harvester Avenue, Batavia, NY 14020. These iceboxes have a capacity of 4½ gallons and draw 2 to 4 amps. They plug into the lighter socket or can be wired direct to the battery. Any more elaborate fridge would draw too much power from the batteries to be really acceptable.

12 volt oscillating fan for hot night cooling.

2in x 10yd roll grey racer's ductape. (From: Fredson).

Windup razor. (From: Haverhill's, 585 Washington Street, San Francisco, CA 94111).

There are plenty of other things you'll be taking along, but most of these are personal belongings rather than part of the camper top equipment list. They include: dental floss, waterproof watch, books to read, Kleenex and handkerchiefs, personal clothing —including underwear, socks, trousers, shirts, handkerchiefs (good for many uses other than blowing your nose), tie (for the formal invitation that may crop up; store the tie rolled up in a jar with a lid), dress (for that same occasion), jacket, coats, raincoats, gloves, bathing suit, shoes, sweater, belt, parka with hood—sun tan lotion, a barrier cream for strong sunlight, moisturizing cream, toothbrush, comb (for looking presentable when faced with difficult bureaucrats at border crossings), pocketknife, camera, film, photographic accessories, footwear, sunhat, sunglasses, passport, visas, traveller's cheques, inoculation certificates, car papers, insurance papers, money, traveller's cheque receipts. (Valuables should be kept in the lockable strongbox under the bonnet, which itself can be opened only from the interior of the vehicle), airmail writing paper and envelopes, pens.

Petrol Costs

In countries where fuel is imported, costs of petrol and oil are usually higher than in countries that have their own vehicle industry and are fuel exporters.

Petrol is most expensive in Europe. In Asia, costs are lower, partly because some Asian countries are self-sufficient in petroleum. In the oil-producing countries of North and West Africa, prices are quite low; elsewhere in Africa somewhat higher, about on a par with Asian prices, i.e. between 50 per cent and 80 per cent of European prices. In Latin America, petrol costs roughly half as much as in Europe.

In certain European and North African countries (consulates or tourist offices will advise which), heavy taxes are imposed on fuel in order to restrict domestic consumption. At the same time, the governments wish to encourage tourism. So tourists are offered petrol coupons, to be used instead of or as well as money in paying for petrol.

Diesel fuel prices are generally about ten per cent lower than petrol prices.

Land-Rovers

by Jack Jackson

Despite some weaknesses, Land-Rovers are the most durable and reliable four-wheel-drive small vehicles for expedition and overland use. Their spartan comforts are their main attributes: most of their recent copies are too softly sprung and have too many car-type comforts to be of any real reliability in hard cross-country terrain. Though somewhat underpowered, their low fuel consumption and general availability of spare parts around the world make them a popular choice for expedition work.

No vehicle will remain in mint condition in cross-country use, but in the UK at least, Land-Rovers can currently be resold after a year's hard work for almost the original price or even at a small profit. Having driven some fifty different Land-Rovers over the last ten years in difficult terrain in Africa and Asia, I have discovered in them various weaknesses. A discussion of these may help you in your choice and use of the various models.

Which model? The short wheelbase is usually avoided because of its small load-carrying capacity but remember that in off the road use, particularly on sand dunes, a short wheelbase is a dis tinct advantage. Hard top models are best for protection against thieves. When considering long wheelbase models it is best to avoid the standard six cylinder models, including the one ton and the forward control. All of these cost more to buy, give more than the normal amount of trouble, are harder to find spares for, and recuperate less on resale. The six cylinder engine uses more fuel and more engine oil than the four cylinder engine and the carburetter does not like dust or dirty fuel, often needing to be stripped and cleaned twice a day in very dusty areas. The electrical fuel pump always gives trouble. The forward-control, which is now discontinued except in Spain, turns over very easily and as with the six cylinder, Series II and Series IIA Land-Rovers, rear half shafts break very easily if the driver is at all heavy-footed.

The one-ton and Series III six cylinder models with their stronger Salisbury rear differentials wear the drive plates instead of the half shafts and these are much easier to replace. But you can never find spare parts for the one-ton EMV front differential outside the UK. Both the one-ton and forward-control models go through rear prop shaft universal joints very quickly and their balloon tyres wear out. It is generally agreed that the four cylinder models are under-powered but the increased power of the six cylinder model does not compensate for its disadvantages.

An eight cylinder Range Rover engine was tested in the Sahara in a station wagon body with balloon tyres but to many people's disappointment was not put into production. An eight cylinder engine is now in production for the forces in a 101 inch chassis and with the weight more evenly distributed between front and rear axles than on the normal models. This vehicle is getting some very good reports though at present it is only available with a canvas top. A new civilian Land-Rover is through pre-production testing and in theory is now available.

The current Range Rover is not spacious enough nor has it the load-carrying capacity for true expedition use.

Any four cylinder hard top or station wagon Land-Rover will suit you if you do not intend to go off the road much when heavily loaded. If you buy a new Land-Rover run it in for a few months before setting off on your trip. This allows the wet weather to get at the hundreds of nuts and bolts keeping the body together. If bolts rust in a bit it will save you a lot of time later on. If you take a brand new Land-Rover into a hot climate you will regularly have to spend hours tightening nuts and bolts that have come loose, particularly those around the roof and the windscreen.

Series II and Series III Comparisons. Most regular users of Land-Rovers believe that the later models of the IIA four cylinder petrol Land-Rover were as good as they could get for reliability. Apart from the usual lack of power their only serious weakness was the differential. However the front and rear differentials are interchangeable and the job does not require any special tools. This problem was accentuated in the six cylinder model. Spares were readily available throughout the world and replacement half shafts could easily be carried and changed without special tools. IIA gearbox troubles were minimal so long as the driver learned to double-declutch, because first and second gears were not synchromeshed. The wheel studs on early IIA models had a habit of stripping out complete with the thread, and the exhaust valves tended to burn out on bad petrol, but again repairs for these problems are not difficult to make in the field. If much driving on bad roads is anticipated, a reinforced front axle is necessary on the Series IIA models, but in general this model gives very little trouble.

The rear axle and differential on the Series III model is bigger and stronger, but it is not interchangeable with the front one if things go wrong. These differentials soon develop a lot of slack and require a differential case spreader to get them out, and gauges to reset them. I have never known a half shaft to go on this new differential. Instead the drive plate on the road wheel end goes. This is much simpler to replace as long as all the little metal shavings that spread out into the bearing are washed out.

There are two very important problems with these larger axles. Models from 1971 and early 1972 had a weakness where the axle casing entered the differential casing and on corrugated roads these tended to collapse at around the twenty thousand mile mark, so it is best to avoid buying a model of this age. Also these differentials, unlike those of the IIA model, are not fitted with a recessed drain plug. This is a very simple and not very costly point, but when driving on rough roads the stones thrown up constantly strike the differentials and quickly batter the drain plug to a shape which no spanner or even Stillson wrench will fit. Even worse, the drain plug can become unscrewed and all the oil run out before you become aware of it.

The Series III gearbox is all synchromesh and is developing a poor reputation for reliability. Its problems usually show up within ten thousand miles. It is very common to find after a short mileage that you cannot engage first gear and at twenty thousand miles third gear tends to jump out when vou use it to brake going downhill. On models up to 1976 there is a plastic bush bonded onto the gear change end of the gear change lever where it meshes with the gear selector shaft. Unfortunately this breaks up when used in hot countries, making first or second gear difficult or impossible to engage. When this happens the whole gear lever has to be replaced as the bush is not available as a separate item. For this reason many gearboxes on the Series III have been known to need a new gear change lever every ten thousand miles or even considerably less. Rover have now reverted to the Series IIA type gear lever with a rubber 'O' ring in a groove; this does not give any trouble, so if you have a gearbox with the plastic bush type gear lever it is best to change it to the one with the 'O' ring.

Early Series III engines tended to break timing chains. Series III gearboxes constantly blow the oil seal between the main gearbox and the transfer box and oil is then constantly pumped from the main gearbox to the transfer gearbox. The main gearbox therefore needs constant topping up. The rear output oil seal on the transfer box often fails from the increased pressure and oil level.

The alternator fitted as standard to Series III models is not as reliable as it should be and its fragile diodes cannot be repaired in the field. It is best to get an exchange unit after 40,000 miles or before any big trip and also carry a spare ring and brushes.

The Series III windscreen frame is much weaker than that on the Series II. It usually has to be reinforced at all four corners if much travel on bad roads is envisaged. The new bracket where the windscreen hinge fits to the bulkhead is also very weak and the inner windscreen tie bolt breaks regularly, so some spares should be carried. As the Series III big-end nuts have nylon inserts instead of tab washers you must carry some spares for they cannot be re-used if you take them off. The oil filler cap tends to vibrate off and disappear, so carry a spare. The same cap also fits the radiator. Keep an eye on the door hinge pins which tend to vibrate upwards and out. The Series III also has nylon inserts on the bonnet hinges, which regularly vibrate out and disappear. Either make some aluminium replacements of your own or carry plenty of the nylon ones. The plate holding the horns also regularly breaks off, complete with horn, on the Series III. It is better to remove it completely and fit an air horn instead. Since the catch which holds the bonnet closed will also break, especially if the spare wheel is carried on the bonnet, add some webbing or, better still, spring metal bonnet catches.

The Series III chassis tends to be weaker than that of the IIA. If it will be getting really heavy use it pays to plate it. Plate all the cross-pieces with a T-shaped plate. Also plate the area by the front spring hangers in the vertical plane. I have seen several Series III chassis completely break off on both sides just behind the front spring hangers. Regularly check the four studs on the bottom of the swivel pin on the Series III; as these often come loose on bad *piste* and can fall out, despite the fact that the nuts have tab washers.

Rebuilt ex-military Land-Rovers are quite a good buy and there are still a few Series I Land-Rovers around, but they are very basic and spares for them are easier to find in scrapyards abroad than in the UK.

Tyres. Radial tyres are tougher, run cooler and last longer than crossply tyres. Being more flexible, they also give the springs, the chassis and you an easier ride. They are very expensive outside England so it is best to take spares with you. Their one weakness is their soft flexible side walls which are likely to be cut when driving amongst large sharp stones. When encountering sharp stones it is best to try and drive straight over them so that only the heavy part of the tread comes into contact with the sharp edges.

For many years the Michelin XY radial was the best all-round tyre for Land-Rovers unless you spent a lot of time in mud or really soft sand. These, however, are now being phased out and the XZZ is the best replacement. If you intend a lot of work in soft sand then the Michelin XS is better for flotation; but do not use these on stones as they will cut up very easily. Tyres other than Michelin tend to crack up if exposed constantly to strong sun. Michelins are made of a tougher rubber which gives them longer life and better resistance to the sun. You pay for this with poor traction on wet tarmac. The spare carried on the bonnet is especially susceptible to cracking up in the sun, so if any other make of tyre is carried make a fitted cover to protect it. A spare wheel and tyre carried on the back door will split the door as soon as the going gets rough and also makes the door heavy to use. The alternative place for the spare has to be the roof.

Make sure that you get the correct Michelin inner tubes for Michelin tyres, as they definitely last longer. Always try to get the ones with short valves as long valves can be ripped off if you get a puncture and drive on for a while without noticing. Avon make a similar looking tyre to the Michelin XY called the Avon traction radial. It is much lighter in weight and cannot take the punishment it is sure to get. Though it lasts reasonably well on the ont wheels, if used on the rear the metal wire around the tyre will soon reak up. Unless it is cut by sharp ones a good average life for a Michen XY or XZZ radial tyre is 40,000 miles. Most other tyres will give 20,000 miles or less. For maximum tyre life a loaded Land-Rover on Michelin tyres needs 42-45 psi pressure in the rear and 28-30 psi in the front tyres.

Roof racks. These need to be strong to be of any use. Many on the market are very flimsy and will soon break up on badly corrugated *piste*. Weight for weight, tubular section is always stronger than box section and it should be heavily galvanised.

To extend a roof rack or to put jerry-cans of water or petrol over or even beyond the windscreen is absolute lunacy. The long wheelbase Land-Rover is designed so that most of the weight is carried over the rear wheels. The maximum extra weight allowed for in front is the spare tyre and a winch. It does not take much more than this to break the front springs or even bend the axle. In any case, forward visibility is impossible when going downhill with such an extended roof rack. A full length roof rack can be safely fitted but it must be carefully loaded, remembering that Rover recommend a total roof weight of not more than two hundred pounds and that a good roof rack (full length) weighs almost two hundred pounds itself! Expect damage to the body work and reinforce likely points of stress. A good roof rack design will also have its supports positioned to be in line with the main body supports. It will also have fittings along the back of the vehicle to prevent it from juddering forward on corrugated roads. Without these fittings, holes will be worn in the aluminium roof.

Nylon or terylene rope is best for tying baggage down as hemp rope doesn't last too well in the sun. Hemp is also hard on the hands from all the absorbed dust and grit. Rubber roof rack straps are useful but those sold in

Europe soon crack up on the sun. You can use circular strips from old Michelin inner tubes and add metal hooks to make your own straps: these will stand up to constant sunlight without breaking.

Tall drivers will find more room in a standard hard top Land-Rover than in a station wagon, because the seat back can fall back further on the standard model. On the station wagon you can solve this by moving the top handrail behind the front seats back by 2in, with $\frac{1}{8}$ steel plate from the door pillar to the handrail mounting holes. The seat back can then fall back the maximum amount, held correctly by the stops at its base.

Other extras. Ex-military plastic water cans are the best buy. Both these and metal jerry cans for fuel can be obtained from the government surplus stores. The cans should fit into the holders very tightly or vibration will soon make them full of holes. Jerry cans carried on the roof rack should be protected from vibration by placing a thin sheet of plywood between the sides of the cans and the bars of the roof rack. It is surprising how many people still carry the cans upright on the roof, untie them and lift the heavy weight down to fill the tank. I always put the cans on their backs with the filler spout pointing to the rear of the roof, then fill the cans from the fuel pumps *in situ,* and use a plastic syphon tube to fill the vehicle's tank without moving the petrol cans from the roof. This means that each jerry can carries only 18 litres instead of 20 but the saving in effort makes it worthwhile.

Stone guards for lights are very useful, but you need a design which allows you to clean the mud off the lights without removing them (water hoses do not usually exist in a desert) and such a design is hard to find. Air horns must be fitted in such a location that the horns do not fill with mud, e.g. on the roof or within the body. Horns can be operated by floor-mounted dip-switch. An isolator may be located on the dashboard to prevent accidental operation of the horn.

A heavily loaded vehicle will need one-ton springs fitted onto the rear and diesel springs or stronger fitted to the front. It is perfectly possible to fit one-ton springs to Series IIA and Series III Land-Rovers, though you may come across mechanics who tell you it is impossible. When you do this of course, you must keep a careful eye on the spring hangers, shackle pins and bushes. Generally the bushes go and you will have to replace them most often. Since I began to fit one-ton springs I have found that I rarely have had to replace one. So instead of carrying many complete spare rear springs I carry the main leaves only as spares in case of emergency.

Land-Rover springs are remarkably expensive. Independent spring-makers will make you the same or stronger springs for two-thirds the price.

Sump guards are unlikely to be required because the chassis and the front axle take most of the knocks. The petrol tank at the rear is more likely to be clouted and a rear prop shaft guard is probably more useful but this tends to clog up when you are driving over grass and scrub.

A good powerful spotlight fitted on the roof rack will be invaluable when reversing and will provide enough light for pitching a tent. Normal reversing lights will be of no use.

The steering lock on any Land-Rover used in a dusty area regularly

jams. If you don't remove the steering lock be sure to carry spare keys. If it gets very sticky with dust it is better to leave the key in the lock and never remove it. Do not oil steering locks but use the special dry graphite lock lubricant.

Spares — All vehicles using petrol.

Engine—3 fan belts. 1 complete set of gaskets, 4 oil filters (change every 3,000 miles/5,000 kilometres), 1 complete set of radiator hoses, 2 exhaust valves, 2 valve springs, 1 inlet valve, fine and coarse valve grinding paste and valve grinding tool, 1 valve compressor, 1 fuel pump repair kit (if electric type take complete spare pump), 1 water pump repair kit, 1 carburetter overhaul kit, 2 sets spark plugs, 3 sets contact breaker points (preferably with hard fibre cam follower as plastic types wear fast and close up in the heat), 2 rotor arms, 1 distributor cap, 1 set high tension leads (older wire type), 1 ignition coil, slipring and brushes for alternator, or brushes for dynamo, 2 cans spray type ignition sealer for dusty and wet conditions.

Extras for diesel engines—delete spark plugs, contact breaker points, rotor arms, distributor cap, high tension leads and coil from above and add:—1 set injectors and cleaning kit, 1 set injector pipes, 4 extra fuel filter elements.

Brakes and clutch—2 wheel cylinder kits (1 right and 1 left), 1 flexible brake hose, 1 brake bleeding kit (or fit automatic valves), 1 brake master cylinder seals kit, 1 clutch master cylinder seals kit, 1 clutch slave cylinder seals kit. It is important to keep all these away from heat.

General spares—2 warning triangles (compulsory in most countries), 1 good workshop manual (*not* car handbook), 1 good torch or fluorescent light and lead, 1 complete set of bulbs, 1 extra tyre in addition to the spare (making two spares), 3 extra inner tubes, 1 large tyre repair outfit and tyre pump (engine type Schrader pump is very useful if you have a petrol engine), 1 set tyre levers, 1 good jack and wheel brace, 1 (at least) metal petrol can (e.g. jerry can), 1 grease gun and a tin of multipurpose grease, 1 tow rope (nylon, terylene or wire, not hemp), 1 gallon (4.5 litres) of the correct differential gearbox oil, 1 large fire extinguisher suitable for petrol and electric fires, 1 set electrical jump leads, 10 push fit electrical connectors (of type to suit vehicle), 1 universal joint for propeller shaft, ½ litre can of brake fluid, 1 small oil can and a general oil for door locks, etc., 1 starting handle (if available), 2 sets of complete spare keys kept in different places, 1 small isopon kit for repairing petrol tanks and body holes, 2 tubes general adhesive (i.e. Bostik or Araldite Rapid), 1 tin hand cleaner (neat washing-up liquid will do in emergency), 1 radiator cap, antifreeze if passing through cold areas, windscreen wiper rubbers for use on return journey (store away from heat).

A good tool kit containing—wire brush to clean dirty threads, socket set, torque wrench, ring and open ended spanners, hammer and small cold chisel for large or stubborn nuts, a 12 inch Stillson pipe wrench and a self grip wrench, insulating tape, 10ft of electrical wire, an assortment of nuts and bolts of the same type and thread as already on the vehicle, a set of feeler gauges, a small adjustable wrench, a tube of gasket cement such as red hermetite, large and small standard and Phillips screwdrivers, a pair of pli-

ers, good accurate tyre pressure gauge. *Extra spares for a Land-Rover used off the road*—1 Schrader tyre pump if petrol engine, 8 wheel bearing oil seals, 1 box spanner for wheel bearing lock nuts, 1 rear gearbox oil seal, 1 rear differential oil seal (different for Salisbury and Rover differentials), 1 rear spring main leaf with bushes complete, 1 front spring main leaf with bushes complete, 4 spare spring bushes, 1 set spring shackle plates (4), 1 set spring shackle pins (4), 1 set shock absorber mounting rubbers, 2 door hinge pins, 1 strong hydraulic jack, 4 swivel pin bottom nuts and bolts, 1 screw jack (to use on its side when changing springs and/or bushes), 1 high lift jack in case you are bogged down, 2 prop shaft universal joints with grease nipple, 2 steering ball joints (preferably the old type with grease nipple), 2 rear axle U bolts, 1 front axle U bolt, 2 spare padlocks, 2 bypass hoses, 1 clutch plate, 3 brass nuts for exhaust manifold studs, 16 yards of terylene or nylon rope strong enough to upright an overturned Land-Rover or give a good tow in an awkward place, a hard wood support for jack on soft ground.
Specific to Series IIA Land-Rovers—1 set rear axle half shafts—heavy duty.
Specific to Series III Land-Rovers—1 complete gear change lever (or replace with groove and rubber ring type if you have welded bush type), 4 nylon bonnet hinge inserts (or 2 home-made aluminimum ones), 2 windscreen outer hinge bolts (No. 346984), 2 windscreen inner tie bolts, 2 rear differential drain plugs, 1 set of big-end nuts, 1 rear axle drive plate, 2 radiator caps, 1 timing chain.

Motor Manufacturers' Concessionaires and Agents

You're planning a long journey in your vehicle, which is made by one of the large motor companies—say, the Vehicle Division of General Motors Ltd., or Chrysler-Talbot, or Ford. You know which countries you'll be passing through. You have prepared your vehicle to the best of your ability and have all the appropriate tools and spares. Still, there's the possibility that something will go wrong with your vehicle *en route* that it will be beyond your means to put right. What to do?

Regrettably, the answer is not simple. You can get in touch with your national or local office or plant of the motor manufacturer before you leave, specify the countries you'll be visiting and ask for a list of concessionaires and agents in those countries. We recommend that you do that. But even this is not foolproof. Suppose your vehicle is made by General Motors. Is it an Opel, a Bedford, a Pontiac, an Oldsmobile, a Cadillac or an Australian Holden? If it's US-made, you'll find that General Motors have divided the country up into anything from 18 to 28 different zones and you'll have to locate the office address for that zone. Suppose on the other hand you have a Matra Rancho made by Chrysler France. The Chrysler Corporation is currently in a state of flux. It has plants in thirteen different countries, but in some of these Chrysler now has only a 49 per cent share, with PSA Peugeot-Citroën holding a 51 per cent share. In these cases, plants are likely to give out lists of Peugeot-Citroën concessionaires, which may or may not be able to service Chrysler vehicles. Your Matra Rancho is likely to have Simca parts and so your best bet for servicing is a Simca dealer, not a Chrysler dealer.

Then there is the time factor. In the case of Chrysler, which is rapidly changing, no list even of zone or national headquarters is going to remain valid for long. So a list of concessionaires obtained from one of these headquarters is bound to be even less reliable. Add to this the political and therefore economic upheavals taking place, particularly in many Third World countries, and it will be clear why you can't expect a guarantee of service for your vehicle. And even if you had an up-to-date list at the time of departure, how accurate would it be six months later?

By all means ask around; get lists of dealers and agents along your route. But once you're on the road, work and pray for a smooth run, and if a breakdown does occur, then hope that you'll be lucky enough to find one of those inventive and resourceful mechanics who inhabit the Third World, who don't know what a concessionaire is but do know, as John Steele Gordon puts it in *Overlanding,* how to make 'the radiator hose of a 1953 Chevrolet serve as an exhaust pipe for a 1973 Volkswagen or vice versa'.

Truckin'

by Bez Newton

It may be the fact that I have a Spanish gypsy grandmother that has made truckin' so attractive a life-style to me personally. I like nothing so much as the feeling of independence and self-sufficiency that truckin' brings, free —as long as my old VW is willing—to wander where I will, like a snail with my house on my back. Over the years I have derived great pleasure from converting vans—Kombis, Bedfords, Commers—using interior space as economically as possible, to suit the requirements of planned trips—across Africa, around Australia, through Central America. I carry not only basic motor tools but also my carpentry set so that modifications can be made *en route* as they suggest themselves. I prefer gas lighting to running down the battery! I always carry a spare battery anyway, and have, on occasion, when space allowed, toted along a small (Honda 250) petrol generator. I try to ensure that there is adequate room for spare water and petrol tanks—on the roof? I have known truckers who grew their own vegetables up there!

Locks on engine compartments, spare tyres and tanks stowed outside the vehicle and a 'secret' lockable compartment inside the van for papers and money are 'refinements' available on some vehicles built specially for the serious expedition-goer, but the budget traveller must be prepared to improvise. I have found, in general, it is best to buy in the continent to be trucked (avoiding shipping a vehicle whenever possible) and go prepared to do one's own conversion.

A wide range of good truckin' vehicles (including 4WD models) is available in the United States and all western countries, but the expense of shipment and a maze of customs and import regulations often make buying at home before one sets out a very uneconomical proposition. European and American travellers to Australia and New Zealand, where the cost of vehicles—both new and secondhand—is high, might still like to consider the pros and cons of taking the home product with them overseas. From Europe this is a fairly easy matter and not too expensive. The vessels of the Baltic Shipping Company (European and American agents: CTC Lines) maintain a fairly regular schedule between Bremerhaven/Southampton and Auckland/Sydney and carry up to twenty vehicles (in the hold and, tarpaulined, on deck) with little formality for as little as £290 ($US620).

Formality starts at the other end where vehicles have first to be quarantined and steam-cleaned—might take as much as four days. So don't expect to drive off the docks as soon as your ship arrives in Sydney or Auckland. Depending on the type of visa held by the owner certain other (expensive) regulations will have to be complied with.

To take the Australian example —and other western countries have similar regulations—a resident or temporary resident (and anyone visiting Australia for a fairly long period will most likely have been issued with this latter type of visa) may not import a vehicle he has owned overseas for less than 18 months without paying 33 per cent duty and sales tax (at Australian valuation). No vehicle bearing foreign plates may be driven in Australia for more than one month. Re-registration varies in cost from state to state—in New South Wales, for example, it is $A177 ($US203)—and no lefthand drive vehicle may be registered in Australia except in the ACT and Northern Territory. Because conversion (expensive) will be insisted upon, customs officials are unlikely to admit the vehicle in the first place. Furthermore, all Australian states have very stringent laws on permitted exhaust emission and minimum safety requirements. To avoid facing all these hassles on arrival—and possibly having one's vehicle immobile in a customs pound throughout one's stay—travellers are advised to check very carefully with the consulate or high commission of the country concerned before arranging for shipment.

Shipping costs are linked to volume in cubic metres, not to weight. It is cheapest therefore to strip the vehicle of equipment mounted on the outside, storing it instead inside, though you should remember, on the other hand, that insurance for shipment of the vehicle does not cover contents. Unaccompanied vehicles are prime targets for theft and negligent handling. A vehicle shipped on the vessel on which the owner is travelling is classed as accompanied baggage and attracts a cheaper rate than a vehicle being shipped alone. Additional costs may include various licences and taxes, marine insurance and a shipping agent's fees. An agent handles the paperwork and other aspects of shipping cargo and his efficiency and familiarity with the whole subject are often well worth the fee. Your local automobile club will be able to put you in touch with a shipping agent.

It is perhaps unfortunate that Australia is one of the few countries mentioned in this article where off-the-beaten-track driving is still very much a possibility and where one would like to arm oneself beforehand with a good (four-wheel-drive) vehicle. As in the United States and Europe a wide range of such vehicles is available on arrival, but the following rough guide to prices illustrates how uneconomical—even taking into account the possibility of resale—buying a vehicle Down Under is. The cheapest 4WD vehicles available in Australia—New Zealand prices are higher— are the Suzuki three cylinder (two-stroke) and four cylinder models, the Puch Höfflinger (made by Daimler of Austria)— a two cylinder motor—and the four cylinder Daihatsu, all priced around $A4500 ($US5175), but these babies hardly make good truckin' vehicles. Be prepared instead to shell out ten or fifteen grand for the Toyota six cylinder, International Harvester, Land-or Range Rover, Willis Jeep or Ford (of Australia) F100 pick-up truck, or up to $A15,000 for Australia's own eight cylinder Holden Sunseeker panel van and much more for the Mercedes Unimog.

In the USA the most popular van-type vehicles are the Ford Club Wagon and Chevy van, the Chrysler Dodge Sportsman, the Volkswagen Station Wagon and the Volkswagen 'Thing', made in Mexico and with the ruggedness common to four-wheel-drive vehicles without actually being one.

In Europe, new vehicles are readily available cheaply in their country of manufacture: Volvo and Saab (Sweden), Land-and Range Rovers (Britain) and Mercedes and Volkswagen in West Germany. They can be bought free of local taxes by overseas residents—delivery arranged in some cases before leaving home—and driven around Europe on temporary plates (in Germany, the oval Z-plate; in Britain, Q-plates; in France, the red *targe temporaire)*. But for real economy truckin' a secondhand vehicle can be easily picked up.

The principal entry points to Europe for young American and Australasian visitors are Amsterdam, London and Athens. Amsterdam has become the centre of the converted Kombi trade. Old Dutch and German delivery vans (1600cc VWs) are bought up and comfortably converted for mobile living by Dutch dealers. They—and visitors wishing to sell redundant vans before flying home—hang around the American Express offices on the Damrak in Amsterdam. The money that changes hands is usually American dollars or English pounds, although almost any 'stable' currency is acceptable. A Kombi that will (unless one is very unlucky) serve for a 5,000-or-so mile trip can often be purchased for as little as £500 ($US 1200, $A800). One advantage of driving in Europe is that a vehicle need be registered only in the country of origin, so that, outside Holland, (where registration is cheap anyway—Florins 12 or about five dollars) an unregistered vehicle bearing Dutch plates is perfectly legal, providing of course that it complies with local regulations about the state of the tyres, lights, brakes, etc. While Dutch registration fees are low, insurance premiums are high. The cheapest place for vehicle insurance in Europe is Britain where foreign-registered vehicles can be insured through the A.A.

The London market-places for secondhand Kombis (and other truckin' vans, particularly British-registered Bedfords, Ford Transits and (Commers) are near Australia House and the National Film Theatre. The sale of foreign-registered vehicles in the UK is (strictly) illegal. (In fact, such vehicles can be temporarily imported for one year only.) But both these laws seem to be waived when it is obvious that the sale is being made by a non-resident to a non-resident who is about to leave the country anyway. London prices are inflated in April and May when visitors to Europe arrive and are likely to reach their lowest when they leave in September and October. As homeward flight dates draw near, the summer's truckers will lower their asking price dramatically. It goes without saying that in buying secondhand in this way one should have at least an elementary knowledge of the mechanics of the automobile!

The vicinity of the American Express office in Sintagma Square in Athens is the southern European market-place for second hand Kombis. Greek regulations make the change of ownership more difficult than elsewhere, for the registration number of an automobile temporarily imported into Greece is entered in the owner's passport. He must, of course, leave in the same vehicle.

When any vehicle is sold in Athens the general routine is for the purchaser and vendor to visit the consulate of the country of origin of the vehicle and have the change of ownership officially stamped. They then drive together to the nearest border point (say, Gevgelija in Yugoslavia) where the importation certificate is cancelled in the vendor's passport. The vehicle can then be driven back into Greece, if desired, the new owner's passport being stamped with a new importation certificate.

It is impossible in this short article to give hints on all the vagaries of truckin' in so many different countries. Overall I've discovered that a left hand drive vehicle is more convenient than one with the steering wheel on the right, that there is little or no economic advantage in choosing a diesel motor, and that a reasonably small vehicle is to be preferred. In Portugal, for example, a daily tax is imposed on large vehicles at the point of exit—when Portuguese money may well be short! Some ferry companies cannot or will not take larger vehicles, and some by-ways are likely to be narrow and more easily negotiated in a smaller vehicle. This is why most truckers starting out in Europe (where VW spares are readily available) will choose the Kombi, which, because of its box-like shape and engine at the rear, is small yet reasonably spacious for living. *(Editor's Note:* But see Henry Myhill's opinion *of the Kombi in Van Conversion Land Outfitting.)* Even without the raised roof which one will see on many conversions, household chorse can be done quite comfortably within the van back, which is generally fitted with a Camping Gaz Stove, sink and ample cupboard space. It might be as well to mention that Camping Gaz (which is readily available throughout Europe) can also be used to run a small fridge. The Camping Gaz company make fridges which can be powered from the car's battery when on the road and from the gas cylinder when stationary. They also make a small gas heater which is useful for truckin' in colder regions.

The end of the line for most Europeans truckers is North Africa, but with a few modifications even an old VW Kombi could attempt the drive through Africa. As a result of unstable politics, the Algeria-Mali route across the Western Sahara is now most commonly used. A high proportion of vehicles using this route are 4WD, and between Bidon 1 and Bidon 5, two-wheel-drive vehicles are allowed to proceed only in convoy with 4WD vehicles. Sometimes it is necessary to wait a few days to join such a convoy.

A popular route for truckers bound for Australasia is overland through Asia to Kathmandu or southward through India to Colombo, from where a vehicle can be shipped to Fremantle. An alternative route involves two or three sea crossings: Madras to Penang, Singapore to Darwin or Singapore to Djakarta, Djakarta to Darwin.

Van Conversion and Outfitting

by Henry Myhill

There is much to be said for expeditions converting their own van or vans to their own requirements. So much of the equipment built in by manufacturers is superfluous to expedition needs—wardrobes for unwanted suits and dresses, or large windows which invite curiosity and reduce insulation. Insulation, indeed, is the most important single operation in such a conversion, rendering the van a refuge from extreme heat or cold. Flame-retardant polystyrene roofing tiles are easy to apply. Finishing off with a 'second skin' of ply or hardboard will add to their life and improve appearances. Insulation should extend to the floor, where a sheet of chipboard cut to shape can be neatly covered in vinyl. (Only on an expedition to Spitsbergen might one consider carpeting.)

The absence of side windows not only increases insulation, but makes the van less conspicuous when camping in civilisation, and less of an attraction for thieves. But it must not mean inadequate ventilation. Fresh air can be obtained through the back window opening, by a replaceable opening in the floor (a useful safety measure in any case where gas is carried), and above all by a rooflight.

I suggest the rooflight not only because it is much cheaper than a full elevating roof, but because it preserves insulation and, properly located, it enables the roof to be used for roof racks (to carry expedition stores) and for a roof tent. Two of these at either end of a roof would enable a single 1500cc van to sleep up to eight.

Interior fittings of professionally converted motor caravans can be very complicated. But journeys off the beaten track have a way of spilling out the contents of even the best fastened roof cupboards, swinging open the doors of china cabinets, and dislodging even the most tightly packed volumes from bookcases. There is much to be said for a simple arrangement of sturdy boxes: one perhaps for tinned food, one for cooking utensils, one for mechanical spares, one for expedition photographs and literature, and so on. These can then be arranged to form beds at night, seats and (one on top of another) working surfaces during the day, and can be taken outside in good weather.

The ideal van for conversion measures at least six feet across. Beds can then lie transverse instead of lengthwise. This rules out the popular VW Kombi or Microbus, which has the additional disadvantage of an air-cooled rear engine which takes up too much floor-space, and which performs unevenly at altitude. Spares are available worldwide, though expensive; but the ability of mechanics to improvise grows the further one moves away from the beaten track. The Leyland light vans (current production model is the Sherpa) are not quite wide enough, but have everything else going for them. The Ford Transit and the Bedford pass on every count.

Unfortunately none of these is four-wheel-drive. This feature is really essential only for expeditions which are bound for the Sahara and are literally going to leave the track altogether. There are limitations as to what can be done with the inside of a Land-Rover, even one with a long wheelbase. Here, a roof tent would be particularly valuable. The one (rather expensive) way of motor caravanning in comfort in such a vehicle is with the Suntrekker demountable caravan on the back of a Land-Rover pick-up.

A converted motorised caravan is really the only kind of caravan suitable for overland or off road use. A luxury caravan has been towed from Algiers to the Cape. But as a general rule the trailer caravan, with its deliberately lightweight construction, does not respond well to long rough journeys. Two additional arguments against it are that its wind resistance adds from 20 per cent to 35 per cent to the fuel bill and that ferry fare structures are such that a car plus caravan (or even a car plus the more manageable trailer-tent) can cost more than twice as much as a car alone.

Renting a Car Overseas

by Richard Harrington

Renting a car on arrival in a foreign country can be very easy—or very difficult, depending on the country, the amount of planning and the size of your budget.

Naturally it will cost you more to rent from one of the big international companies. On the other hand, the vehicles are likely to be newer and better serviced if you use Avis, Hertz or Budget, the three most international rental companies. Back up will also be better in the event of breakdown or an accident, and you may be able to drop the vehicle off at a different place from the pick-up point.

Time as well as cost will influence your decision about whether to rent from one of the big companies. Do not expect the service you get from the Big Three to be uniform in all countries, however. You'll find it varies from excellent to abominable, and no one of these firms, on average, is any better or worse than the next one. This is because local offices are in fact franchises owned locally and the degree of control exercised by Avis, Hertz and Budget will tend to be minimal at times.

So if you like to reserve a roof rack with your vehicle when you make your reservation, don't expect that part of the message to have reached the order form at the other end. I generally use Avis and found that only once in eight rentals did the request for a roof rack get through to the country I was visiting. On the other hand, you may be able to get one fixed up if you request it on arrival.

Most people pick up rental cars on arrival at foreign airports, and this is where the big international companies are usually located (with usually a town office as well). However you pay for this convenience in rates that are distinctly higher than you could negotiate with a small local rental company nearby in town.

Not all airport rental locations are controlled by the Big Three. You will frequently find that local rental firms also have an airport desk. In general, however, their services are no better than those of the Big Three, and their costs are generally about the same. There may be distinct disadvantages too. With the big companies, you can pay with vouchers bought before the trip, avoiding the need to carry too much money in traveller's cheques. If the voucher is of the maximum value type (I'm familiar with those of Avis) then you won't have to pay a hefty deposit in case you write the vehicle off. This is true also if you pay by credit card (assuming they will accept your particular brand of card). However, if you insist on paying in cash, be warned that you will probably have to leave behind a large deposit, possibly the equivalent of a couple of weeks' rental.

A word of warning. A lot of airport rental concessionaires practise various types of currency rip-off. And this applies equally to the Big Three. You'll generally find the airport currency exchange desk pretty close to the car rental desks. If the rental firms accept traveller's cheques, it's usually at an extortionate rate of exchange. If however they won't accept traveller's cheques and you're not paying by credit card, you will have to pay in cash. So the chances are that you will have to change your money at the only nearby place, the money changer's next door. Here, rates of exchange are likely to be much higher (you'll pay as much as 10-15 per cent more) than at the same bank in town. You may even get the feeling, occasionally, that some of the extra profit is getting passed back to the car rental concessionaire. You could well be right.

Watch out if you have to pay a deposit in hard currency or traveller's cheques and then get refunded in the local currency—again at a rip-off rate of exchange—when the vehicle is returned intact. You may then be faced with problems turning a heap of local currency back into your own hard currency 30 minutes before take-off. What's more, it may be illegal to export local currency, and you could be faced with no alternative but to buy $500 worth of airport souvenir junk.

So if you do pay a deposit in hard currency, insist that you get your deposit back in the same currency and make sure they put it in writing at the beginning. If you pay your deposit in traveller's cheques, you have to countersign them. Sometimes these are not cashed but are held over until the vehicle is returned. Then your traveller's cheques are returned to you. That's tricky as you will want to use them again, and will have a problem explaining how it is that they are already countersigned.

Try not to return a rental vehicle too close to flight departure time if you're leaving it at an airport. Checking the vehicle back in can take five minutes or two hours—longer if you're in a place where everything, including the airport, closes for lunch. Remember that in some countries, flights may continue to take off during the lunch hour, but the car rental desks may be closed for lunch. Then you realise why they insist on a deposit or some indemnity system. However, if you've paid a deposit that more than covers the rental charge, you could end up severely out of pocket if you simply leave the vehicle and catch your flight in this situation. Forget about writing to them once you reach home—there's not the slightest chance you'll get a reply.

The ABC Airline Guide in the UK and the OAG Guide in the US publish monthly updated (more or less) car rental guides—country by country—for the big rental companies. Anyone can subscribe to these, but they're expensive. Better to look at one at your friendly local travel agency. The car rental companies are usually pretty efficient if you 'phone them before leaving home to make a reservation.

I always used to wonder how Avis in New York could confirm, say, a Toyota model such-and-such in Fiji right there and then on the 'phone. The truth is that they have little more idea than you or I do if that vehicle is available. They assume that there will be some sort of vehicle in that class available, or a more expensive class. The order is telexed by Avis to Fiji and they reserve the Toyota if they can. Otherwise you get another vehicle in the same class, or failing that, a vehicle in a more expensive class at the same rate as the Toyota.

Getting a bigger car can, however, have disadvantages. It may make you look wealthier and therefore make you more theft prone, it will almost certainly guzzle more petrol, and it may be uncomfortably large for the country's narrow roads, especially if you plan to explore tracks and back-country roads.

Don't be fooled by all kinds of low cost advertising come-ons from rental companies. Add on vehicle insurance, third party insurance, mileage charge, gasoline, local taxes (very common) and the ultimate cost per mile can leave a big hole in your pocket.

Unlimited mileage deals may be the best bet if you plan to do a lot of driving. Fortunately most companies charge you the cheapest way at the end of the trip, according to the miles you've done. It may be worth stretching a six day trip into seven days, however, to take advantage of a weekly rental rate. This won't generally save a great deal, however, against seven times the day rate. The saving is likely to be in the region of 15 per cent.

In some very poor Third World countries it may be possible to rent a vehicle with local driver for only a little more than the cost of the vehicle. I rented a car in Bali which cost $16 per day, or $17.50 with driver. Admittedly he wasn't much of an interpreter, but at least he knew the way round all the back roads that never found their way onto any map. So with luck you may get a guide and interpreter thrown in as well. In some countries, local regulations and conditions prevent outsiders from driving rental cars. You have no alternative but to take a 'with driver' deal. However these are likely to be countries where labour costs are cheap, so it isn't like hiring a chauffeur-driven car at home. The cost differential is likely to be minimal.

In countries such as Colombia, Ghana and Guatemala, the roads are badly marked, rutted or muddy and your best bet is to rent a car with driver. A driver comes in useful too in Japan and

Singapore where an international licence and returnable deposit, and a special operator's licence, respectively, are required before you can drive a car yourself. In Japan the road system—no numbers—is especially hard for a foreigner to follow. Panama and Venezuela offer good roads and cheap petrol. Western Malaysia is a driver's dream, but the east is a nightmare, especially during the monsoon (October-January). Don't hire a car indiscriminately, without thinking whether you really need it. In some places, like Hong Kong, traffic congestion is such that you're better off getting about by public transport or on foot.

Four-wheel-drive vehicles may be available for rental in a few countries but this is uncommon. Where they are to be found (e.g. East Africa) the cost is likely to be extremely high, and you will almost certainly have to spend time in town making the rental agreement. The Big Three, at least in most countries, are not interested in off-road drivers. They're not too keen either on your taking their saloon car into the 'boonies', but there's no rule that says you can't, provided you don't get bogged down and leave it there. That could get expensive. If you're paying for the rental by credit card, the contract you sign will normally allow the rental company to charge the entire value of the vehicle to your card should you disappear with it or write it off without covering insurance.

If you plan to rent a car in the Third World, my advice is that you investigate thoroughly beforehand what options are available to you, and rent a vehicle that will be suitable for the terrain you plan to cover.

Check it over carefully when you pick it up. You don't want the gear lever to come off in your hand or the brakes to fail miles from anywhere. And check to make sure the windows shut properly and the doors and trunk lock. If you must leave valuables in the vehicle at any time, a good old-fashioned trunk is the best place to lock them—and much safer than the hatchback vehicles now prevalent everywhere.

There are plenty of pitfalls to renting vehicles, but a little forethought should cut down on the hassles.

Staying Put

Accommodation

Camping
1: Do You Really Need It?

by Anthony Smith

The first real camping I ever did was on a student expedition to Persia. There I learned the principle of inessential necessities. We were travelling by truck and therefore could pile everything on board, every considered need on which to lay our hands. The truck could transport these items, and we only had the problems of sorting through that excess of gear whenever we wished for something. Later the expedition travelled by donkey as this was the most practical method of visiting the outlying districts in our study area. Miraculously the number of necessities diminished as we realised the indisputable truth that donkeys carry less than trucks. Later still the expedition travelled on foot, this being the only practical method of prolonging the journeying after our donkey-drivers had failed to coerce higher rates of pay from very empty student pockets. Amazingly the necessity number diminished yet again as a bunch of humans realised they could carry far less than donkeys and much, much less than trucks. The important lesson learned was that happiness, welfare and ability to work did not lessen one iota as the wherewithal for camping decreased in quantity. In fact there was even argument that the three blessings increased because less time was spent in making and breaking camp.

This lesson had to be learned several times. In a subsequent year I was about to travel from Capetown to England by motorbike. As I wished to sleep out, provide my own meals and experience a road network that was largely corrugated dirt, I found no difficulty in compiling a considerable list of necessities. We must all have made these lists—of corkscrews, tin-openers, self-heating soup—and they are great fun, with a momentum that is hard to resist. 'Why not a spare tin opener?' says someone. 'And more medicine, and another inner tube? And isn't it wise to take more shirts and stave off prickly-heat?' Fortunately the garage that sold me the bike put a stop to such idiot thinking. I had just strapped on a sack containing the real essentials, such as passport, documents, maps, money and address book, when a passing mechanic told me that any more weight would break the bike's back. (It was a modest machine.) Thus it was that I proceeded up the length of Africa without a sleeping bag, tent, groundsheet, spare petrol, oil, tools, food or even water, and never had cause for regret concerning this wealth of lack. Indeed I blessed the freedom it gave. I could arrive anywhere, remove my one essential sack and know that nothing, save for the bike itself, could be stolen. To have possessions is to be in danger of losing them. Better by far to save the robbers their trouble and start with nothing.

A sound tip is to do what the locals do. If they sleep out with nothing more than a blanket it is probable that your frame can do likewise. If they can get by with a handful of dates at sunset it is quite likely that you too can dispense with half a hundredweight of dried egg, cocoa, vitamin tablets, corned beef, chocolate—and self-heating soup. To follow local practice, but to attempt to improve upon it, can be disastrous. Having learned the knack of sleeping diagonally in a Brazilian hammock, so that my body was as horizontal as if it were in bed, I decided one thunderous night to bring modern technology to my aid. I covered myself with one of those metallic 'space' blankets to keep out the inevitable downpour. Unfortunately, during sleep, the crinkling thing slipped round beneath me and I awoke to find my body afloat in water collected by that wretched blanket. Being the first man to drown in a hammock is a poor way of achieving immortality. I looked, of course, at my Indian travelling companion. Instead of fooling about with sub-lethal blankets he had built a fire longitudinally beneath his hammock. Doubtless kippered by the smoke, but certainly dry, he slept the whole night through.

Camping in Antarctica is not a matter of having a good night's sleep as against wakefulness. It is also not a matter of proving, as the old saying goes, that any fool can be uncomfortable in camp. Any fool camping down south is quickly a dead fool. A blizzard can arrive within a very short time, and a traveller who cannot erect his tent in time can die even more rapidly. Essentially the standard two-man Antarctic tent is a four-legged pyramid. These four poles are already within the tent canvas when the tent itself is on the sledge. Pitching it is solely a matter of spreading the legs and jabbing them in the snow. (If there is wind it will even assist you after the first two legs are in place.) Provision boxes on the upwind side help to fix the canvas and there is just room inside for the currently vital boxes (food, radio, stove and light) between the two beds. Unlashing the sledge, erecting the tent and throwing the necessities inside is a matter of a minute or so, if need be.

One trouble with our camping notions is that we are confused by a lingering memory of childhood expeditions. I camp with my children every year, and half the fun is not quite getting it right. As all adventure is said to be bad planning, so is a memorable camping holiday one in which the guys act as trip wires, the air mattresses fart into nothingness and even the tent itself falls victim to the first wind above a breeze. Adults are therefore imbued with an expectation that camping is a slightly comic caper, rich with potential mishap. Those who camp a lot, such as wildlife photographers, have got over this teething stage. They expect camp to be (almost) as smooth and practical and straightforward a business as living in a house. They do their best to make cooking, and sleeping, and washing, and eating no more time-consuming than it is back home. The joy of finding grass in the soup or ants in pants wears off for them about the second day. It is only the tempor-

ary camper, knowing he will be back in a hotel (thank God) within a week, who does not bother to set things right within a camp.

I like the camping set-up to be as modest as possible. I have noticed that others disagree, welcoming every kind of extra. If it will not rain, and is not cold, for example, I see little need for a covering. This may be laziness on my part but it is no Spartan longing to suffer. A night spent beneath the stars that finishes with the first bright shafts of dawn is hardly punishment—but some seem to think it so, removing the natural environment as much as possible.

I remember a valley in the Zagros mountains where I had to stay with some colleagues. I had thought a sleeping bag would be sufficient and placed mine in a dried-up stream, which had piled up sand, for additional comfort. Certain others of the party erected large tents with yet larger fly-sheets (however improbable rain was at that time of year). They also started up a considerable generator which bathed the area in sound and light. As electricity was not a predominant feature of those wild regions a considerable quantity of moths and other insects, idling their time between the Persian Gulf and the Caspian Sea, were astonished at such a quantity of illumination and flew down to investigate. To counter their invasion one camper set fire to several of those coils for repelling insects, and the whole campsite was shrouded in noxious effluent. Over in the dried-up stream I and two fellow spirits were amazed at the camping travesty down the way. We were even more astonished when, after a peaceful night, we awoke to hear complaints that a strong wind had so flapped at the flies that no one inside a tent had achieved a wink of sleep.

The most civilised camping I have ever experienced was in the Himalaya. The season was spring, and tents are then most necessary at the lower altitudes (where it rains a lot) and the higher ones (where it freezes quite considerably). Major refreshment is also necessary because walking in those mountains is exhausting work, being always up, as the local saying puts it, except when it's down. We slept inside sleeping bags on foam rubber within thick tents. We ate hot meals three times a day. We did very well—but then we did not carry a thing. For the six of us there were thirty-six porters at the outset, the number being reduced as we ate into the provisions these men were carrying. I laboured up and down mighty valleys, longing for the next refreshment point and always delighted to see the already-erected tents at each night's stopping place. Personally I was burdened with one camera, the smallest of notebooks and nothing more. The living conditions, as I have said, were excellent, but what would they have become had I been asked to carry all my needs? It is at that point when neither trucks, nor donkeys, nor incredibly hardy mountain men are available, that the camper's true necessities are clarified. For myself, I am happy even to dispense with a toothbrush if I have to carry the thing all day long. Just a blanket will do if that is what the locals use. My body may not be like theirs in the early days; but, given encouragement, it can become half as good as the weeks go by.

2: Yes, You Really Need It

by Jack Jackson

These snippets of advice apply to campers for whom weight of equipment is not of the first importance. Backpackers and campers travelling light should consult the article on *Lightweight Equipment*.

Expeditions planning to have a fixed base camp for some time will find it very useful to have a large mess tent where the party can all congregate for meals and during bad weather.

Inflatable tents are not good on expeditions. Dust, grit, thorns and sunlight destroy the rubber and if blown up in the cool of the day these tents have a habit of bursting when sunlight warms up the air in them too much.

The pegs normally supplied with tents are of little use on the hard sunbaked ground in hot countries, so have some good thick strong pegs made for you from ¼ inch iron or else use six inch nails. As wooden mallets will not drive pegs in, carry a normal claw hammer and you can also use the claw to pull the pegs out again. In loosely compacted snow standard metal pegs do not have much holding power, so it is useful to make some with a larger surface area from one inch angle alloy. Even this does not solve all the problems because any warmth during the day will make the pegs warm up, melt the snow around them and pull out so that the tent falls down. The answer is to use very big pegs or ice axes for the two main guys fore and aft and then for all the other guys dig a hole about nine inches deep, put the peg in horizontally with the guy line around its centre and compress fresh snow down hard with your boots on it to fill the hole.

If you sleep without a tent, in some areas you will need a mosquito net. Several are on the market but they are not usually big enough to tuck in properly, so get the ex-army ones which have the extra advantage of needing only one point of suspension: your camera tripod or ice axe will do for this if you do not have a vehicle or tree nearby.

Since tents take heavy wear, carry some strong thread and a sailmaker's needle for repairs plus some spare ground sheet material and adhesive. Tents which are to be carried by porters, on donkeys or on a vehicle roof rack are best kept in a strong kit bag or they will soon be torn.

When you are staying in the same place for several days and do not want to get up too early remember if you are in a sunny area to try to pitch your tent in some shade (NB the sun rises in the east!) or the heat of the sun will drive you out soon after it rises. If you are sleeping without a tent, flies become a nuisance soon after sunrise, so use a mosquito net if you wish to sleep on.

In cold places, you should not sleep directly on the ground but use some form of insulation. Air beds have disadvantages: inflating them is hard work when you are tired; sharp stones, thorns and sunlight all work against them; and you will certainly spend a lot of time patching them up. If you decide to use one be sure it is rubber, not plastic, and then only pump it up half way. If you pump it any higher you will roll around and probably fall off. On cold nights you will find that your perspiration condenses on the cold rubber and you will wake up in a cold puddle of your own sweat. So put a blanket or woollen sweater between you and the mattress. Camp beds are not very good either. They tend to be narrow, collapse frequently, soon break up and tear holes in your groundsheet. Worst of all on cold nights the air circulates underneath the bed and as your body weight compresses the bedding you will soon be cold unless you are lying on several layers of blankets.

The best mattress for this sort of camping is polyether or rubber foam. Unfortunately the majority of those on the market are cheap and not dense enough. Even the best name brands such as Dunlop are not adequate. However one brand stands out above all the others in density and thickness. This is made by P.T.C. Langdon and called Spatzmola. Spatzmola foams are not easy to find and may have to be ordered. In the UK these can usually be got at Pindisports or ordered from other camping suppliers. Though they are twice the price of the others they are by far the most comfortable and are highly recommended.

All foams tend to tear very easily but if you make a washable cotton cover which fully encloses them they will last for several years. Foams are bulky, so are best wrapped in strong waterproof covers during transport. Personally I use only half a Spatzmola, enough for hips to shoulder, and use a climber's closed cell foam insulating mat called Karrimat, known in the USA as insulite, for my legs and feet. This cuts down on bulk. One advantage of foam mattresses is that the perspiration that collects in them evaporates very quickly when aired so they are easy to keep fresh and dry. Remember to give the foam an airing every second day. Karrimat make mats of any size to order. Karrimat also comes in a 3mm thickness and this is good to put under your groundsheet for protection against sharp stones or on ice because the tent groundsheet can otherwise stick to the ice and you can tear it when trying to get it free.

On a long overland trip, you can combat changing conditions with a combination of two sleeping bags. First get a medium quality, nylon covered down filled sleeping bag and if you are tall make sure that it is long enough for you. This bag will be the one you use most often for medium cold nights. Secondly get a cheap all synthetic bag, ie one filled with artificial fibre. Make sure that it is big enough to go outside the down bag without compressing it when fully lofted up. These cheap, easily washable bags are best for use alone on warmer nights and outside the down bag for very cold nights. If you decide to take only one sleeping bag, choose the best and ensure that it is large enough to allow you to wrap yourself in a blanket *inside* the bag. (Sleeping bags need to loft up, ie swell up, in order to keep the body heat in. Putting blankets around the outside would prevent this and limit the effectiveness of the bag.)

In very cold conditions, the goose down sleeping bags made by Fairy Down of New Zealand are to be recommended. These contain a few feathers to keep the bag lofted up and warm when it is damp. Wool and artificial fibres retain their insulating properties when wet; down does not.

In polar and high mountain areas the golden rule when travelling is never to be parted from your own sleeping bag, in case a blizzard or accident breaks up the party.

The aluminium tables and chairs on the market today are covered with light cotton. This rots quickly in intense sunlight. Look around for nylon or terylene covered furniture or replace the cotton coverings with your own. Full size ammunition boxes are good for protecting the kitchen ware and also make good seats.

When buying cooking utensils, go for the dull grey aluminium ones. Shiny chromium aluminium pans tend to crack up and split with repeated knocks and vibration. Billies, pots and pans, plates, mugs, cutlery, etc., should be firmly packed inside boxes or they will rub against each other and end up as a mass of metal filings. A pressure cooker guarantees sterile food, saves on fuel and can double up as a large billy, so if you have room it is a very good investment.

Whistling kettles are difficult to fill from cans or streams, so get kettles with lids. For melting snow and ice it is best to use billies. Big strong aluminium billies for base camp use are best bought at Army & Navy auctions or surplus stores.

A wide range of non-breakable cups and plates is available, but you will find that soft plastic ones leave a bad after-taste so it is best to pay a little more and get melamine. Stick to large mugs with firm wide bases that will not tip over easily. Insulated mugs soon become smelly and unhygienic because dirt and water get between the two layers and cannot be got out. Many people like metal mugs but if you like your drinks very hot you may find the handle too hot to touch or burn your lips on the metal. Melamine mugs soon get stained with tea and coffee. Melamine cleaners are available but Steradent is a perfectly adequate and much cheaper substitute. Heavyweight stainless steel cutlery is much more durable than aluminium for a long expedition.

Two types of gas are available to campers. Low pressure gas is usually referred to as Calor Gas or butane gas in the UK and by various oil company names worldwide such as Shellgas and Essogas. Though available worldwide they have different fittings on the bottles in different countries and they are not interchangeable. High pressure gas, commonly called Camping Gaz, is available in Europe. Enterprising campsite managers in Turkey and Iran have discovered ways to fill both low and high pressure gas bottles from their own supplies. Stand well clear whilst they do it as the process involves pushing down the ball valve with a nail or small stone and then overfilling from a bottled supply of higher pressure. This can cause flare-up problems when the bottle is used with standard cooking equipment, so if you use this source of supply it is advisable to release some of the pressure by opening up the valve for a couple of minutes well away from any open flame before hooking it up. Refillable Camping Gaz containers with a side overfill valve are obtainable in France to special order.

In general high pressure gas is widely available in Europe but harder if not impossible to find elsewhere. Low pressure gas on the other hand is less easy to find in Europe (except in England) but the only sort you are likely to find in the Middle East. Even if it is called Gaz outside Europe it will still be low pressure. High pressure burners can be used with high pressure gas only, but low pressure burners can be used with either if the correct pressure regulator is fitted. Therefore it is better to have a low pressure cooker and a bottle of each sort of fuel so that you can stock up on whichever is more readily available in the area where you happen to be. The six pound gas cartridge or the ten pound gas cylinder are the best sizes to carry. Gas is without doubt the easiest and cleanest fuel to use for cooking. Where you use a pressure reduction valve on a low pressure appliance, there will always be a rubber tube connection. Make sure that you carry some spares of the correct size rubber tubing.

Lighting stoves is always a problem in cold climates or at altitude. Local matches and Russian matches never work unless you strike three together, so take a supply of Bryant & May matches. The best answer seems to be a butane cigarette lighter kept in your trouser pocket where it keeps warm. Remember to carry plenty of refills.

Space blankets, very much advertised by their manufacturers, are on the evidence not much better than a polythene sheet or bag. Body perspiration tends to condense on the inside of them, making the sleeping bag wet so that the person inside gets cold. In hot or desert areas, however, used in reverse to reflect the sun, they are very good during the heat of the day to keep a tent or vehicle cool. At base camp a plastic sheet or space blanket can be spread over a ring of boulders to make an effective bath.

When buying equipment, be especially wary of any shop that calls itself an expedition supplier but does not stock the better brands of equipment. All the top class equipment suppliers will give trade discounts to genuine expeditions or group buyers such as clubs or educational establishments, and some, such as Pindisports and Field & Trek (Equipment) Ltd., in the UK, have special contract departments for this service.

Youth Hostelling

John Carlton and Diane Johnson

The off-beat traveller will inevitably meet others going to stay at a Youth Hostel and may well want to know what they are all about. The first point is that, although the facilities offered are designed primarily for young people, there is no age limit for staying at Youth Hostels (except in Bavaria, Germany) and they are used by 'the young in heart' of all ages.

The second point is that Youth Hostel facilities are provided by a club run not for profit, but to help young people to travel, to know and love the countryside and appreciate other cultures, thereby promoting international friendship. Each country runs its own hostels independently (usually by committees from its own membership) and the national Youth Hostel Associations of each country are linked through the International Youth Hostel Federation. The Federation (a United Nations sort of organisation) lays down basic standards for its members, but each National Association's interpretation of these does inevitably vary in the light of its own local culture. Theoretically, membership of an Association is necessary for all individual members of hostels but this rule is lax in some countries outside Europe. However membership is very worthwhile for the traveller who thinks he might be using hostels, and those resident in England and Wales should apply to YHA, Trevelyan House, 8 St. Stephen's Hill, St. Albans, Herts. AL1 2DY, Tel. 0727-55215 (or in person to YHA Services Ltd, 14 Southampton Street, London WC2E 7HY). In England and Wales the annual subscription is currently £2.65 for those over 21 or £1.60 for those between 16 and 21. Scotland and Northern Ireland have their own Associations and their headquarters' addresses are 7 Glebe Crescent, Stirling FK8 2JA and 93 Dublin Road, Belfast BT2 7HF, respectively. American residents should apply to American Youth Hostels, Inc., National Campus, Delaplanc, Virginia 22025 and Canadian residents to Canadian Hostelling Association, National Sport and Recreation Centre, 333 River Road, Vanier City, Ottawa KIL 8B9, Ontario. It is possible to obtain YHA membership at an Association office (and sometimes at a hostel) outside one's country of residence, but one then pays a much higher fee.

Once a member, you can stay in any of about 4,500 hostels in 49 countries throughout the world. Basically a Youth Hostel will provide a bed in a dormitory of size varying from 4 to 100 beds. There are toilet and washing facilities and a communal room where members can meet, all at a cost of the local equivalent of from 50p to £2 for the night. In most countries members will find facilities to cook their own food. Cooking utensils and crockery are provided but not always cutlery. Surprisingly in some countries the Youth Hostels offer cheap meals cooked by the warden or staff in charge.

A feature of Youth Hostel life is the sheet sleeping bag—a sheet sewn into a bag with a space for a pillow. Any traveller intending to make much use of Youth Hostels should have one, though there are some hostels at which sheets may or indeed must he hired, so as to protect mattresses. Most hostels provide blankets and consider that these are adequately protected by the traveller's own sheet sleeping bag. In this respect, as in others, Youth Hostel customs vary from country to country.

A full list of the 4,000-odd Youth Hostels of the world (published annually) can be obtained at YHA Services Ltd, 14 Southampton Street, London WC2. Ask for the International Handbook—Volume 1, Europe and the Mediterranean, and Volume 2, The Rest of the World. Each volume costs £1 (£1.15 by post). As well as listing the addresses and facilities of each hostel, the handbook summarises the local regulations for each country, including prices, lower age limits, facilities for families, etc. However all the information given is subject to correction as circumstances change during the year and of course prices will inevitably rise in timc.

We attempt below to make a quick survey of the Youth Hostel facilities of the world.

Europe (including many countries in Eastern Europe but not Russia) is well covered by hostels and the wide variation in their characteristics reflects the local culture of each country. The British Isles hostels are perhaps now unique in expecting a small domestic duty from members before departure, but this does help to emphasise to members that they are part of a self-helping club. This idea may be less apparent in some countries, where the Youth Hostel is often run, with the agreement of the National Youth Hostel Association concerned, by the local municipality as a service, and relations between members and staff are strictly commercial. The club atmosphere is stronger also in France, Holland and Greece. For hostel atmosphere try Cassis situated in an isolated position on the hills overlooking the *calanques* of Marseilles, 20 miles from the city. In West Germany, where the Youth Hostel movement started in 1909, Youth Hostels are very plentiful—mostly large well-appointed buildings, but lacking members' cooking facilities and largely devoted to school parties. Scandinavian hostels are also usually well-appointed, many having family rooms, and there is therefore more emphasis on family hostelling. Spain still has one-sex hostels, but these seem to be rapidly disappearing. Iceland has six simple hostels.

In North Africa there are hostels in Morocco, Tunisia, Egypt and the Sudan. These too reflect the local culture. Try calling at Asni, a hostel in a Moroccan village 40 miles south of Marrakesh on the edge of the High Atlas mountains. Here the warden has three wives and will talk to you with great charm in French.

The Kenyan YHA has nine hostels, two of which are on the coast. One is at Malindi and the other is at Kanami, about 15 miles north of Mombasa, which has an idyllic setting amongst the coconut palms a few yards from a deserted white sandy beach. The new Nairobi hostel is a meeting point for international travellers and at Nanyuki the hostel is close to one of the routes up Mount Kenya. Kitale Hostel near the Ugandan border is part of a farm with accommodation for eight people and the one room serves as a dormitory, dining and common room.

The rest of Africa is devoid of hostels until one reaches the south. Lesotho has one hostel, Mazeru, which is well worth a visit. Local young Basutos use the hostel as a youth centre, so travellers have a chance to meet with them. There is a South African YHA but because of the country's apartheid policy it cannot be a member of the International Federation; this ruling also applies to the Rhodesian Association which has two hostels. The membership cards of white members of the Association in the International Federation will, however, be accepted in these countries. The South African YHA Handbook can be obtained from YHA, PO Box 4402, Capetown. The hostel at Camps Bay is in a beautiful situation looking out across the Atlantic Ocean with Table Mountain almost immediately behind it. Unfortunately Port Elizabeth Hostel, a lovely old Victorian House, is very scruffy by South African standards but in a good position near Kings Beach.

Israel's YHA consists of some thirty hostels, the smallest having 100 beds. All provide meals, many have family rooms, but the members' kitchens are poor. Orphira hostel in Southern Sinai is fairly new, with superb snorkelling and diving close at hand. However it may become part of the Egyptian YHA if the peace negotiations succeed. Although there have been hostels in Lebanon for many years, owing to the civil war there, no recent information is available about them. Syrian hostels are small and reasonably equipped. Many hostellers travelling to or from India meet at the one in Damascus.

Travelling eastwards, one finds three hostels in Iran, one in Kabul, Afghanistan, and a good network in Pakistan. The Pakistani hostels are quite well kept and there are also a number of Government rest houses open to hostellers, as are some schools in certain areas during school holidays. Indian hostels tend to be mainly in schools and colleges and are therefore only open for short periods of the year, although there is a large new permanent hostel in Delhi. Some hostels do not provide any kind of bedding, even mattresses. Sri Lanka has several hostels including one in Kandy and one in Colombo; here, too, Government rest houses and dak bungalows provide alternative accommodation at a reasonable price.

The Philippines, South Korea, Malaysia and Thailand all have some hostels of which the Malaysian ones are particularly well organised. In Thailand, some hostels listed in the International Handbook appear not to exist. The Bangkok hostel, however, certainly does. None of the five Hong Kong hostels is in the city itself.

Japan has the most extensive network of hostels outside Europe, numbering 600. There are two kinds of hostel—western style with the usual bunk beds, and the Japanese style with a mattress rolled out on the floor. Television is a common feature. Several hostels are on the smaller islands of the country such as Awaji, an island in the Inland Sea. Japanese food is served in most hostels—a bowl of rice, probably served with raw egg, fish and seaweed and eaten with chopsticks.

Australia has over a hundred hostels with most of them in New South Wales, Queensland and Western Australia. Distances between them are great. The smaller, more remote hostels do not have a resident warden and the key has to be collected from a neighbour.

New Zealand has hostels throughout the country on both the North and South Islands. They are fairly small and simple with no meals provided but have adequate cooking facilities. Many are in beautiful country, such as the hostel near Mount Cook.

The Canadian and United States Youth Hostel Associations are the only ones now imposing the restrictions on motorists that were at one time applied by all Associations. The North American Associations ask that those using cars should stay at least two nights at a hostel and not use their car in the intervening day. There are not many hostels in North America, considering the size of the continent. There are a few hostels in some of the biggest cities. (In the USA a city hostel will often turn out to be a YMCA offering rooms to YHA members at reduced rates). Most are in isolated areas of scenic interest not always accessible by public transport. There are, however, chains of hostels in New England, Colorado and the Canadian Rockies. A feature of United States hostelling is the 'Home Hostel' where accommodation is offered to members in private houses. The Canadian Association also runs a large number of temporary city hostels in the summer.

In Central and South America, Youth Hostelling has not yet caught on seriously, although there are a few hostels in Mexico, Argentina, Chile and Uruguay.

Although in the poorer countries of the world you can obtain other accommodation as cheaply as in the local Youth Hostel, members have the advantage of being able to look up an address in advance at points all over the world. They can then stay at the local branch of their own 'club' finding (albeit minimal) common standards of accommodation and be sure of meeting and exchanging experiences with fellow travellers.

Finding a Place to Stay

by Richard Harrington

If you can afford to stay in a luxury class hotel, or have the inclination to, then you're probably reading the wrong book. Most people, however, are not in that income/expenses bracket, so I'll confine this article to those with modest or limited budgets.

Finding a place to stay really comes down to three things: cost, comfort and security. The three don't always match in the way you would expect, and higher prices are no guarantee that your belongings won't get stolen, or that you'll pass a comfortable night. If you spend a lot of time on the road, and you're into vehicle or tent camping, a reasonable lodging with water (better still hot water and maybe a shower) every three or four days offers a chance to get clean and 'civilised-looking' again, which is much appreciated and can smooth your path at borders, if timed to coincide with crossings.

Maybe your budget will run to hostels and even cheap hotels every night. There is a lot to be said for maximum comfort, even in the wilds, if you can afford it. But remember this: nothing breeds friendships of the road better than adversity, and the cheaper you travel, the more interesting people you're likely to meet. Money only seems to put up barriers between people. Poverty breaks them down.

Guide books list principal places of accommodation in main towns in Third World countries. Some of the books listed in the Directory section of this handbook will also list (i.e. describe how to find) useful cheap lodgings—at say one to five dollars per night—that don't find their way into the regular guidebooks.

Don't be afraid to check at several places to find the one that suits you best. *Always* ask to see the room, check whether any meals are included in the price, whether there's a government tax to be added to the bill, whether hot water is available, whether taps and shower (if they exist) work, whether the door has a lock that works, whether your prospective room could easily be entered by a window and so on. Also check on check-out time, usually negotiable in small lodgings.

Even in Communist countries, it pays to 'shop around'. In Peking or Hanoi a 'tourist' hotel may charge even for an average room three or four times as much as the little hotel nearby with no tourist facilities. In Russia, one saves a lot by going 'sixth class' for meals and hotel rooms.

Throughout many African and Asian countries cheap board and lodging is provided by 'rest houses'. Built with no windows and cool stone floors, those in Iran, for example, offer welcome relief from the burning sun. Stone benches around their walls are spread with thick Persian carpets, upon which one sits cross-legged, to enjoy delicious local food. Dak bungalows throughout India, erected for travelling officials who used to go from

village to village dispensing justice and administering the law, can be rented now very cheaply for the night on a first-come, first-served basis. Some are quite plain, others are opulent delights of Victorian comfort.

It's hard to avoid using terms like *hotel, hostel, pension,* etc., but the types of lodgings you'll encounter may defy all classification. The smaller and more out of the way the place, the more likely you are to become effectively a private guest in the house of someone trying to earn a little money—or better still, simply kind enough to offer good old-fashioned hospitality for nothing. Don't abuse such kindness, and if your host won't accept money in return, try to lend a hand with the work in the field, or offer something in return. A personal item that means little to you (e.g. an old shirt) could be a treasured luxury to a peasant farmer in Iraq, dressed only in rags.

In general, you'll find that the cheaper the lodgings, the friendlier your host will be, and if you have his/her friendship, this will often (but not always) be an excellent protection against theft. But as the lodgings get more expensive, they are less likely to be owner-occupied. Here you may run into all the problems that are a feature of airports and banks in Third World countries—the problem of the under-educated petty functionary given a little status and maybe a uniform.

If you want to test this approach, try rolling up to a top international hotel in, say, East Africa, on foot, dirty and unshaven, with a rucksack, and see what reaction you get from the door-man and the clerk in the lobby. Ask how much the cheapest room is and you'll almost certainly be told the hotel is full, whether it really is or not. Of course, in supposedly top class hotels like this, reservations may be meaning-less unless you look extremely rich and presentable, or slip a sizeable tip to the clerk at the desk. A reservation sent from abroad and confirmed is likely not to exist in the hotel management's eyes, and heaven forbid you should produce a prepaid hotel voucher from your travel agent at home. Chances are pretty good that you'll have to pay again, even if your reservation did get through.

In some Third World countries, a booking and even a bribe—*baksheesh* in Asia, *mattabiche* in Zaïre—can't get you a room. Anger achieves nothing and is often taken as a sign of weakness. The best thing may be to brazen it out; make yourself at home in the lobby and refuse to budge until someone comes up with a bed for the night.

In a hotel in the under $5 per night bracket, it's likely that room price will be negotiable. Even a little of the local language goes a long way to keeping prices down and making friends, so it might pay to invest in a record/cassette language course before you leave home. Ignore the local language and it will surely cost you dearly.

In more expensive accommodation, standards tend to be more inter-national. Down the price scale, more local standards will prevail. So try to avoid ethnocentricity—seeing things only through the eyes of your culture, and regarding other ways as 'inferior'. Your ways may seem strange to them.

Don't cheat by leaving without pay-ing. A group of young Australians did this after staying for two weeks in a *los-man* (cheap simple accommodation) in Bali. The owner of these simple dwell-ings costing about £1 per night had provided meals as well. Not only did the Australians up and away one night, but some of them subsequently boasted to their friends about how they had ripped off the natives. Bali is still one of the friendliest places on earth, but at Kuta Beach, with its hordes of cheap-living Australians, goodwill is getting in short supply.

Do not assume that every hostel manager, hotel keeper and *losman* owner is out to fleece you. He may be open to bargaining, but that's probably part of his culture. Don't get carried away with your own wonderful negotiating ability and try to screw another 50 cents out of the poor guy. Did you haggle like that when you bought your secondhand Land-Rover to make the trip? Give in occasionally and let him make his profit. He'll still not be as rich as you are.

Travelling in the Third World is all about the people you'll meet. The places you see may be fascinating, but the things you'll remember most about a place are the people you met there —locals in the bar, fellow travellers from the western countries, stall owners in the local market—and the owners and staff of the places you stay in.

The best experiences you're likely to have will be far off the beaten track. Try to get away from cities to the rural areas. In the remotest places, most people will be friendly and honest and you're unlikely to be without somewhere to stay if you're desperate. In a few places, like the valleys of the Peruvian Andes, you'll find the people hostile and xenophobic. These poor Indians have been ripped off by everyone for centuries, so their reaction is understandable. But in general the poorest people will also be the kindest. They may feel honoured to be your host, so repay their hospitality in every thoughtful way you can. And don't be too paranoid about your possessions being stolen. Theft is usually common only near cities and towns.

One of the best nights I ever spent was on the earth floor of a Fijian chief's hut. It would have been bloody uncomfortable but for the soporific effect of the *kava,* a Melanesian alcoholic beverage with a mighty impact. That sort of accommodation experience I wouldn't swap for all the Hilton hotels in the world.

Getting Straight

Documents, learning, preparation

Choosing Maps

Ingrid Cranfield

Note: Addresses of mapsellers/mapmakers mentioned in this article are to be found in the Services Directory section of this book.

Women often claim, and men just as often prove, to be incompetent when it comes to map reading. Yet maps are simple tools and their efficient handling depends only on the understanding of a few basics.

First, there is scale, the measure that relates distances on the map to corresponding distances on the ground. A small scale map gives a broad overview of a sizeable area, while a large scale map shows a limited area in greater detail. One source of confusion about scale is the phrase 'on a large scale', which, in common parlance, means extensive: yet to show a 'large scale' feature on a map may require a *small* scale. One way of remembering the difference between small and large scale maps, therefore, is to note that features on a small scale map *appear* small, those on a large scale map, large. For most practical purposes, maps at a scale of about 1: 1,000,000 or smaller are generally considered small scale; those of, say, between 1: 20,000 and 1: 1,000,000 large scale. Scales larger than 1: 20,000 are used on town plans, maps of individual properties or installations and the like.

Map readers may find it useful to make a mental note of one scale and the measure it represents; all other scales can then be compared with it. Thus a scale of 1:250,000 (more accurately 1: 253,440) means that a quarter of an inch on the map represents a mile on the ground. Metrication in mapping has meant a transition from scales representing round distances in miles (1: 63,360 = 1 inch to 1 mile; 1: 253,440 = ¼ inch to 1 mile) to scales using multiples of ten (1: 50,000, 1: 250,000 and so on). The official body that produces maps of the United Kingdom is the Ordnance Survey, which in the last five years or so has been phasing out its maps based on the mile and replacing them with metric maps. The standard unit for a moderately large scale O.S. map is now no longer 1: 63,360 but 1: 50,000.

The choice of scale in a map naturally depends on the purpose for which the map is intended. A motorist planning a route cross-country will probably find a map at 1: 500,000 quite satisfactory. A rambler, eager to note smallish features in the field, be they *tumuli* or pubs, will be well advised to acquire a map, or, more frequently, several adjoining map sheets, drawn on 1:50,000 or 1: 25,000. Many official mapping authorities base their map series on a national grid, a network of lines which divides the country into small units and represent the edges of individual map sheets. An index for the series (available for consultation at the map retail outlet or in some cases printed on the back of each sheet in the series) shows which map sheets are needed to cover the area of the purchaser's interest.

Another basic distinction to note is that between topographic and thematic maps. Topographic maps (which include most O.S. maps) show the general nature of the country: the lie of the land, the location and extent of built-up areas, some indication of land use and land cover (forests, marshes, farmland), the courses of roads, railways and other lines of communication, the presence of waterways and any other salient features, whether natural (e.g. mountains, sea-cliffs) or man-made (e.g. airports, quarries). A trained geographer, it is said, can study a topographic map of any area and 'read off' from it a large amount of information about the way of life of the area's inhabitants: where and how they live, in what pursuits and how successfully they earn a living, and something of their social, cultural and religious customs.

A thematic map focuses instead on a particular aspect of the country: relief, communications, climate, land use, population distribution, industry, agriculture. Thus a road map gives—or should give—detailed information on the road network and on associated features and amenities (e.g. petrol stations, motorway exit points, mileages between towns); but it may give little or no indication of relief, built-up areas or features of interest to travellers. Similarly, a 'tourist' map will show attractions for the sightseer—castles, museums, lakes, archaeological sites, parks—but will most likely skimp on information on the exact road pattern, sizes of towns and other features of, presumably, peripheral interest to the tourist. Thematic maps can be extremely useful provided they are chosen with care, bearing these limitations in mind.

All maps employ symbols and it is a good policy to familiarize oneself with the symbols used before taking a map into the field. Some mapmakers use representational symbols, i.e. simplified drawings of features; others use abstract or geometric symbols, e.g. triangles of different colours to represent different products at industrial sites. The representation of relief is the subject of much variation: methods include hill-shading, which simulates the appearance of the terrain as it might look from the air, spot heights, and contours (lines joining points which are the same height above sea-level). Recently the Ordnance Survey, having lately altered the symbols used on some of its maps, was compelled to issue a warning bringing to users' attention the vital distinction between the symbol for parish boundaries and the rather similar symbol for tracks (both broken lines): confusion between the two might have had disastrous results.

Apart from scale and content, what criteria should be used in selecting an appropriate map? Legibility is one: a balance should exist between the provision of information, especially of place names, and the prevention of a cluttered look; and a clear type face and size are of course most important. The map's main features should make a strong and unequivocal impact, e.g.

colours should be graded logically to convey the correct impression of a variation in altitude, concentration or any other scale or continuum. Language and place name variants may have a bearing on the map's usefulness in some circumstances. One series of excellent Italian road maps of northern Yugoslavia, for instance, gives Italian instead of Slovene place names, which may cause the user some difficulty in pinpointing locations. Poor physical design in a map—the ease with which it folds, whether or not it is waterproof, the presence or absence of a folder or cover—may cause disproportionate inconvenience in the field.

Official map publishing continues to bear the stamp of colonialism, so that, for example, the best maps of former British colonies and territiries are published by Britain's Directorate of Overseas Surveys, and former French possessions, e.g. Chad, Algeria and many other countries in north and west Africa, are covered by series published by the Institut Géographique National in Paris. Excellent maps of their own countries are produced by many of the old Commonwealth and other rapidly developing industrial nations (e.g. Australia, Canada, Israel); and the USA is in the course of being mapped at a scale of 1: 25,000, a series which numbers many thousands of sheets.

Most governments indeed have their own civilian (often called geological) survey as well as a military agency, both producing maps that are useful for explorers, walkers and overlanders. Two mapping agencies in the USA are worth noting. The US Army Topographic Command produces excellent detailed road maps, including maps of Southeast Asia and the Middle East, and contributes maps of the USA and elsewhere to the 1: 1,000,000 and 1: 250,000 series, which are produced according to internationally agreed standards. The National Ocean Survey publishes Operational Navigation Charts covering much of the world. These are meant for airline pilots but can be useful to route-planners as well; and for some remote—or politically inaccessible—areas they are simply the most detailed maps available in the west.

The publications of commercial map producers however often rival those of the national authority in quality. In Britain there are a number of fine map publishers such as George Philip and Son, Geographia and John Bartholomew. The latter for many years published a series of maps of Britain at a scale of ½ inch to 1 mile, which obviated the need for the Ordnance Survey to do the same. In the USA the National Geographic Society has a prolific cartographic output; NGS maps always carry a profusion of place names. Map publication is not of course the exclusive province of specialists. Road, tourist and other maps are put out for the purposes of promotion or information by, for instance, petrol companies and tourist offices. In western countries, petrol companies' maps are usually given away free and are of a high standard; in the Third World, they are generally less reliable, despite the fact that the user has to pay to get one. Automobile clubs usually produce maps of their own and neighbouring countries, though often these are available to members only. The Australian and the American Automobile Associations produce excellent maps of their areas; in Britain the Automobile Association and the Royal Automobile Club sell maps of all

Ǝurope, Asia and Africa. The road naps of the Argentine Automobile Club are said to be exceptionally good. Motoring organisations in Europe can often supply detailed maps of the Sahara routes.

The Michelin Tyre Company has a eputation for excellent route maps, specially of Africa. One authority on he Sahara notes, however, that even hese maps contain some quite serious rrors. Michelin maps of Mali show nain routes incorrect, tracks comletely misaligned, track classification ut of date and emphasis on the wrong lace names. As a safeguard, travellers hould always carry large scale maps as ell. Michelin are of course not the nly company guilty of perpetrating artographic errors. Mapmakers often ke the easy way out by referring to kisting sources rather than surveying e ground and so mistakes in reading, opying, interpreting, printing and ven in wishful thinking are perpetued. Why, for example, is Nome in laska so called? Because somebody isread 'Name?' scrawled on a map.

For the prospective map purchaser Britain wishing to know which maps e available to cover his field of intert, a trip to one of the public map libries, e.g. at the Royal Geographical ociety or the British Library, is commended. Here he may consult aps, discover where to buy them, ake his own notes or copies from the aps, or arrange for them to be photopied for his private use. (Note: the ws of copyright generally prevent aps from being photocopied for production. Also note: if a photocopy a different size from the original, it is o a different scale.) Maps may be rchased from specialist outlets (e.g., London, Edward Stanford Ltd., 12—14 Long Acre, WC2) or from other stockists (e.g. Harrods).

The traveller proposing to stray from the beaten track may well expect to find 'his' area mapped poorly or not at all, and be surprised, even dismayed, to find that this is not so. The increasing use of aerial and satellite photography in mapping makes it virtually certain that the mapmakers have visited or surveyed the area before him. If however he looks upon maps as the informative, reassuring, perhaps even lifesaving devices they are, he will surely find that the security afforded by possessing a good map more than outweighs the mild disappointment of having been, so to speak, pipped to the post.

Using Maps (and other Navigational Devices)

by John Lindstrom

Many people are afraid to get off the beaten track. They lack the confidence, tools or imagination to go anywhere on the map. But the point is this. *Trails* go to a few selected places. *You* can go anywhere you want to!

Just because it isn't in your guidebook doesn't mean it isn't a fantastic place to visit. In fact, if it isn't in the book, it might be considerably less desecrated than those places that have received the kiss of PR.

The three basic navigational devices for wilderness exploration are map, compass, and altimeter.

The altimeter will accurately tell you how high you are, assuming that barometric pressure has remained constant. For instance, if the weather stays clear and dry throughout your trip, you may assume that your altimeter has stayed accurate, without further adjustment. This is true because altimeters and barometers are constructed on the same principle. Both devices react to the surrounding air pressure. And air pressure varies with elevation (gravity) and with weather. A joke acknowledges the principle. 'Hey. . . . either our campsite has gone up . . . or it's going to rain.'

If, on your trip, a low pressure system (bad weather) moves across the landscape, your altimeter will *not* tell you your true elevation. Instead it will indicate that you are higher than a kite. Under such circumstances, you must reset the altimeter at some checkpoint on the map where the elevation is given, and where you know you are. When bad weather changes to good weather, you must reset also.

Altimeters are perhaps the least important of the three devices mentioned. People who enjoy climbing 2,000 foot rock faces may contest this, since an altimeter may be the only item that tells them just where they are on the face. (This is because a sheer rock face isn't shown with much refinement on a map. Handholds don't show. And an overhang can't be shown. Showing an overhang on a map is like going back in time.)

Two items remain; the old woodman's pair, map and compass. They are probably equally important, although this depends . . .

If you enjoy slogging through swamps or hiking through corn fields, a compass is more useful than a map, or at least as useful, because the terrain lacks visually prominent features. In mountainous terrain, it may well b that a map is more useful than a compass. Here's the reasoning:

If the weather's good, you'll be moving and exploring. You'll be able to se features of the landscape and easily locate them on your map, pin-pointin your whereabouts. A compass won't b so necessary.

If the weather's lousy, misty cloudy, you'll have to navigate b compass . . . but who the hell wants t explore in such weather? Instead yo may be inclined to hole up until the su shines.

A map is a miniature, symbolic pic ture of the land. When you ponder map, you are omnipresent, like Go You are hovering above the Eart looking down at her features . . . show in a special code. How do you make flat sheet represent a steep hillside You invent the 'contour line'. A line an infinite series of points. A contour an 'iso-elevational' line. On a contou map, all points on any given line are th same elevation abve sea level.

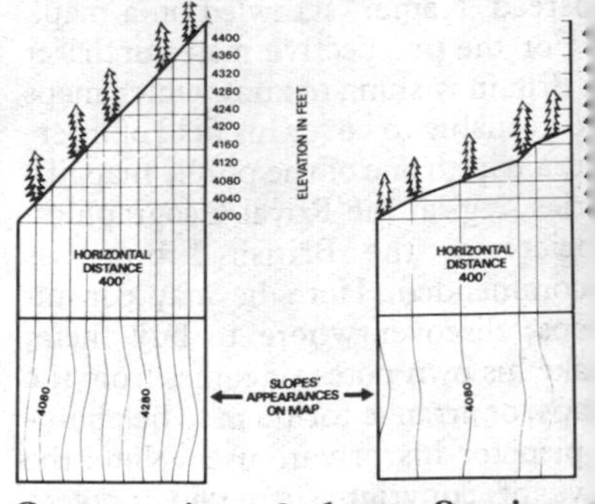

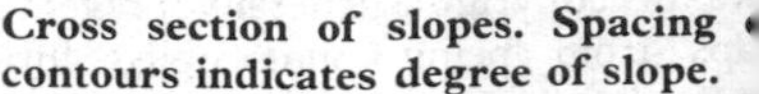
Cross section of slopes. Spacing contours indicates degree of slope.

When a series of contour lines a close together, the terrain is stee When the lines are far apart, the terra is gentle. Every fifth line is heavier a will have the elevation marked. You can count heavy lines and multiply by 200 feet (in this example) and determine elevation to be gained in a given distance to be travelled.

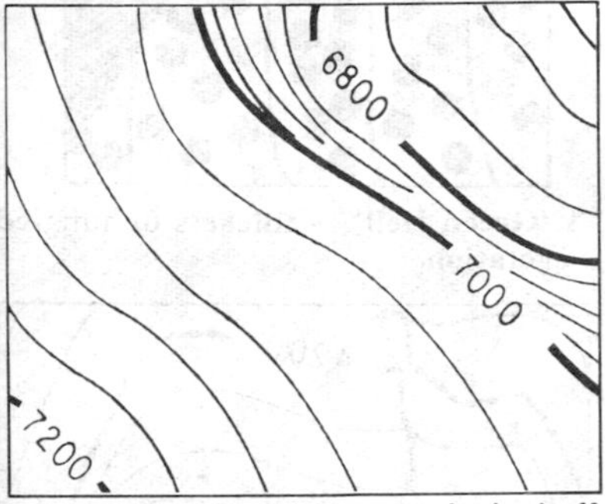

Sheer cliff: contour lines are pinched off or bunched. Map appears 'browner'.

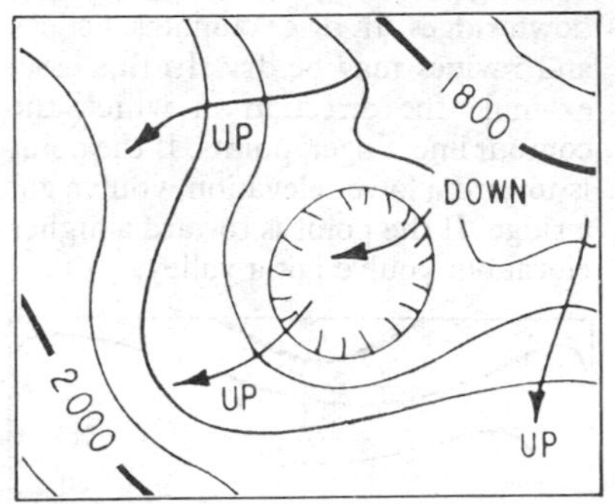

Depression: Hachured line indicates terrain goes lower.

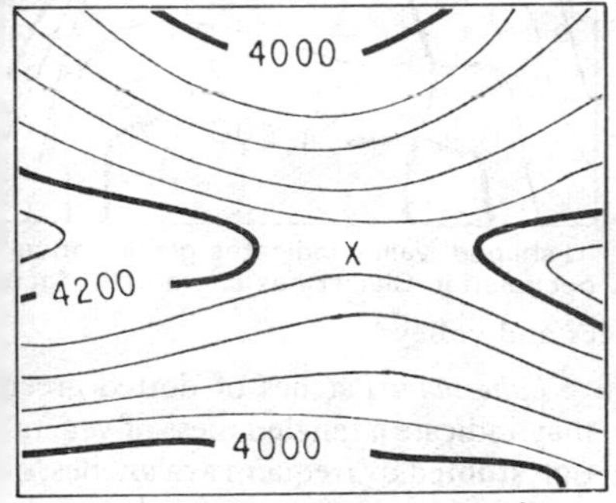

Saddle in ridge: Could be a campsite. 'Fingers' point at each other.

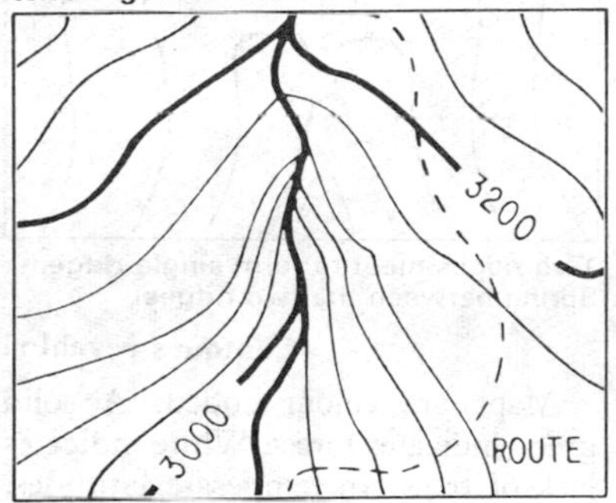

Narrow stream valley probably with waterfall. Not passable. Circumvent.

Examples of contour patterns.

This example has been marked in 40 foot contour intervals, which is typical for mountainous terrain, as depicted on the most detailed maps now being produced by the US Geological Survey. Such maps are drawn to a scale of 1: 24,000 and are called '7½ minute maps'. (60 minutes = one degree of latitude or longitude.) On such a map, approximately 2½in equals one mile of horizontal distance. You are also likely to encounter maps from an earlier era, with a scale of 1: 62,500, called '15 minute maps'. They usually have a contour interval of 80ft. On such maps, one inch equals almost exactly one mile. For flat land, 5, 10 and 20ft intervals are common. In the UK the Ordnance Survey is the body responsible for official mapping of the country. The standard scales on maps for field use are 1: 50,000 and 1: 25,000. On the 1: 50,000 maps, the contour interval is 50m and the heavier contour is marked with a figure representing the elevation to the nearest metre. Contour lines reveal specific things about the landscape as well as the above-mentioned general slope of the terrain.

How do you tell a ridge from a valley, since each produces contour lines in the shape of pointing fingers? A valley will often have a blue line indicating water. Water doesn't run down ridges. In drier climates, valleys and ravines may be dry. In this case, examine the direction in which the contour line 'finger' points. If the point is toward a lower elevation, you've got a ridge. If the point is toward a higher elevation, you've got a valley.

finitely 'polynary'. Actually, there is one other coding.

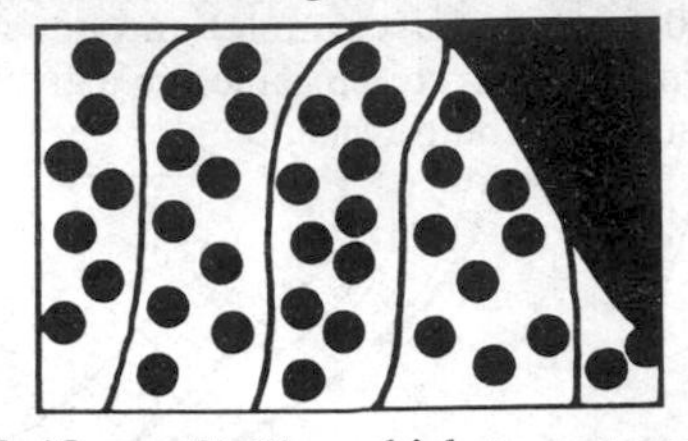

A 'Green Hell' — thickets or tangled vegetation.

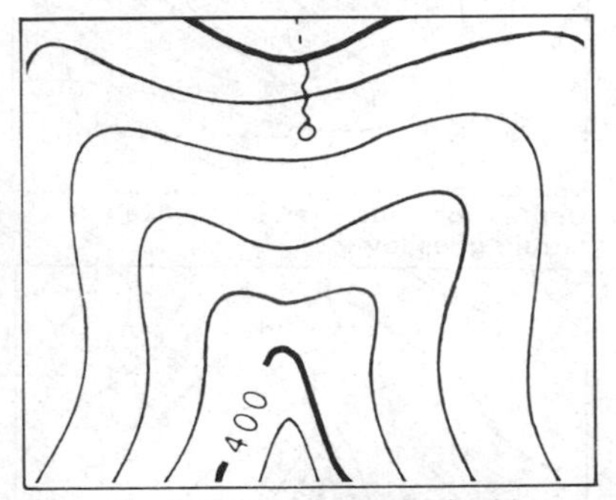

Two ridges meet to form single ridge. Spring between the two ridges.

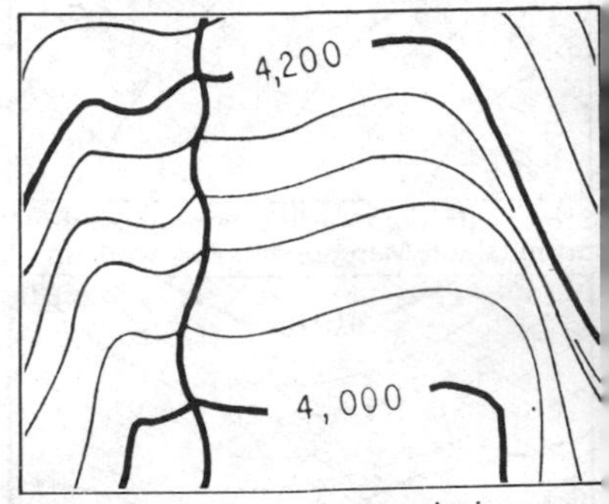

'U-shaped' valley indicates glacier once occupied it. Glacier has carved cliff face

Contours reveal ridges and valleys.

Maps are colour coded. A solid green indicates forest. White indicates lack of trees. In temperate latitudes, white at low elevations indicates fields, pastures, burns and clearcut forests, or deserts. White at higher elevations indicates alpine meadows or barren rock. At high latitudes, such as in Alaska, white likely indicates tundra and meadows even at low elevations. Ice is also white.

You will discover a certain arbitrariness in the colour shading. For instance, the map may show solid green, but you may find that the forest is actually subalpine, and very easy to walk through. Green/No-green is a binary system . . . and the countryside is de-

Take note: Patches of dotted greer may indicate a tangled mess of vegeta- tion stunted by frequent avalanches, o thickets like the famous 'slicks' of the Great Smoky Mountains in the easterr USA. From a distance, they look lik nice meadows. In reality, they are th explorer's nemesis, a 'Green Hell' Avoid them unless you're on a trail o perhaps traversing *downward* throug the mess because you have no othe choice.

Remember: Open meadows ar *white*—lovely for high elevation tra verses. Blue means water. Glaciers ar white patches, with contour lines i blue. Dotted blue lines are intermitten streams.

Once you can interpret the symbolic code of the map, you have about *half* the knowledge necessary to abandon yourself safely to the wilderness. Knowing something about the biological, hydrological, and geological systems of an area is equally necessary.

A few examples:

1. In the northern hemisphere, the sun's azimuth in the sky is toward the south, instead of straight up. A result of this solar system geometry is that steeply north facing slopes spend a lot of time in the shade. If such slopes are growing tall timber, the understorey can be very sparse. You may discover that you can climb up through this dark cathedral with case and peace, digging your boot toes into moist ground; sweating, but not fighting. Choose such a slope over another one that faces south. South means sun, brush and a struggle. By contrast, in dry climates, south facing slopes may be more open because of lack of moisture, while north facing slopes will be tricky thickets.
2. A 45 degree meadow slope may have a few little trails, thoroughfares for marmot and other game, and some well-seated rocks and occasional branches of alpine scrub to hang onto.

A 45 degree granite slab that is wet could be treacherous, especially under full pack.

The map will probably not distinguish between these two features! To cross or not to cross such a slope is a judgement you make from knowledge.

3. A contour map, no matter how refined, is only an approximation of the actual landscape. Standards for 'river' and 'creek' don't exist. Be prepared for the 'hidden cliff discrepancy': in some areas mapped by aerial survey without thorough field-checking, a cliff may exist on the ground but go unnoticed by the mapmakers. This happens particularly in areas of dense forest cover, where the top of the slope is drier, supports shorter trees, the bottom of the slope is very shady and damp and supports larger trees and the cliff itself supports only brushy trees. The apparent land surface is a smooth unbroken slope and will be shown as such on the map. An adventurer has said, after rappeling over an unexpected cliff, 'It's very clear. . . .the hillside didn't read the map.' (Similar to 'That mushroom apparently didn't read the book', uttered by an ill mycophagist.)

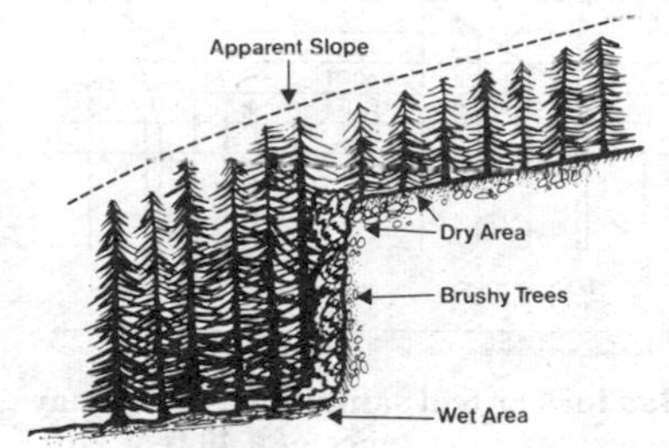

Cross section of slope which has northern exposure.

Remember: the symbol comes *after* the real thing.

4. The 'wandering stream discrepancy': a stream, especially in a glacial region, may wander as well as change volume at different times of day and from year to year. The map simply cannot keep up with these changes in stream patterns.

This article's intention is to praise maps, not to debunk them. However, one last detailed point will be presented here to show that maps are fallible:

An identical slope profile can be

'synched' or 'unsynched' with the contour lines. This is chance, but produces two very different map pictures. The two map pictures do not give the same impression. What we've got is a sampling error. The lefthand version clearly shows the cliff. The righthand version simply implies a slight steepening of the slope. From a statistical standpoint, the odds are good that a 40ft cliff will not be clearly revealed. The two cases here are the extremes. Usually, the map will give an intermediate picture of such a cliff, neither as poorly nor as well shown as the two cases here.

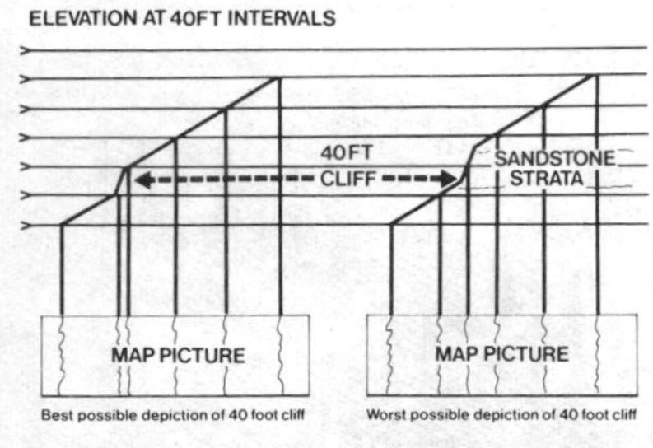

Bad luck or bad sampling? A map may conceal the existence of a 40 foot cliff.

So, a 40ft contour interval map may well not depict a 40ft cliff. This may not seem fair, but it is true. In a region where sedimentary strata abound, frequently not tipped very much, you could come across an unexpected cliff that would require ropes to descend and other additional gear to ascend. In other regions you may encounter a sill of volcanic rock.

This is the challenge! . . . off the trails. Take a map and go. Give your elbows some room. Maybe I'll meet you in the woods. Maybe. . . .

Desert and other Land Navigation

by Gerald Dunn

There can be little doubt that anybody who intends to move extensively 'off track' in desert country for longer than a few days should be trained in astro-navigation, or his party should include one who is.

This is not to say that for a beginner, who proposes a brief sortie or a direct crossing by a marked and charted route, the heavy expenditure and rigorous study which this preparation may demand is necessarily justified. In this case good 'pilotage and dead reckoning' will usually suffice.

Nevertheless, since the most experienced navigator can get into difficulty if his equipment is accidentally damaged, and since the most meticulous planning of the less experienced can be upset by unforeseen misfortune, it is a wise precaution for both of them to cultivate a simple back-up system which can be brought into play in the event of trouble. Obviously, simple methods must not be expected to yield results of great precision—but provided that this limitation is kept in mind it is surprising how accurate these simplified methods can prove to be.

The term 'pilotage' embraces all those arts concerned with the direct observation and interpretation of fixed and visible landmarks. This is by far the most important part of navigation in a desert context and the most important element of it is good log keeping. To illustrate this—desert tracks are sometimes well defined and easy to follow. At other times they may consist of a tangled skein of inter-twining tracks two miles wide which may well conceal an important track junction in their mesh and lead to confusion. Sometimes these junctions can be obliterated altogether. It is therefore a good idea to compose a scenario in advance, noting all junctions and other landmarks which you can expect to encounter with their distances from the starting point of a run, and tick them off the list one by one as they come up. Then in case of doubt you can refer back to the log and the last significant feature seen, and so re-interpret your position.

It is manifestly impossible in a short discourse to cover the whole field of pilotage—so I must advise you to research the matter at your district library. The bibliography of this subject is enormous—any recognised standard work will serve you well. It is all empirical common sense and contains no mysteries.

Dead reckoning, on the contrary, though its principle is extremely simple, often causes alarm in the beginner. The term denotes the art of keeping track of your whereabouts in the absence of any terrestrial or celestial marks or reference points whatsoever. But remember, it is only when you divert from a marked route, or find yourself off tracks in undifferentiated terrain, intentionally or otherwise, that the need for dead reckoning makes itself felt. In the vast majority of desert crossings this need does not arise. It is comforting, however, to possess this art in case it should.

The essential principle of D R amounts to this: that if you are certain of your location at the start of a day's journey—say you are at a township or settlement, or near to a distinctive geographical feature which you can identify on the map—and if you keep account of your movements throughout the day, you can plot these details onto the chart at nightfall and know precisely where you are. The line thus drawn on the chart will be a faithful representation of your track on the ground. This bland assertion, which may satisfy the mariner who has no dunes or ravines to deflect him from his chosen course, presents to the desert-goer a little more difficulty. The scale of his chart is quite small, and his day's route may have twisted this way and that, with perhaps a hundred or more legs, some of them extremely short, forced on him by obstacles in his path of which the mariner knows nothing. Now if the dead reckoner is going to depict all these squiggles in miniature on his chart there is every chance of a great many minute errors of placement, which ride pickaback upon each other and could create a false notion of whereabouts. Moreover, the chart will soon become an untidy scrawl and may cause confusion as a result.

I therefore commend you to experiment with a novel method of D R log keeping which you will not find in the navigation text books. I have borrowed it from analytical geometry. It is known as 'vector analysis'. The accompanying drawing shows how every change of course and the distance travelled in any direction can be recorded in the form of a vector with respect to the starting point of a morning's trek. Back-reading this log we can see that this traveller set out from point X at

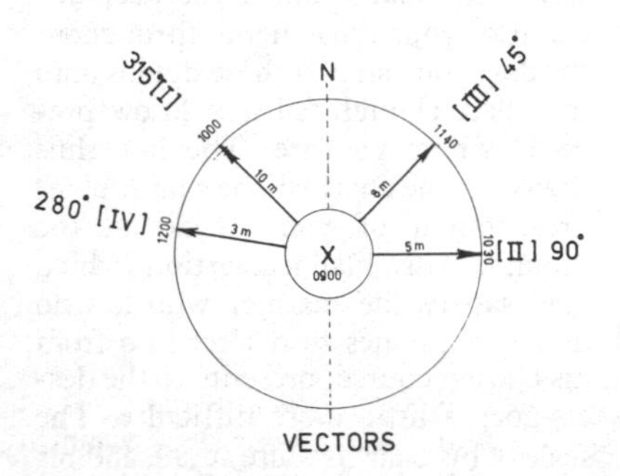

Vector analysis. May be drawn free-hand as journey proceeds — no need for scale or angular accuracy. Note all movements are shown as though from the starting point X.

0900 hrs on heading 315° (true). At 1000 hrs after he had gone ten miles he turned onto Course (II) bearing 90° and held it for five miles. At 1030 hrs he switched to Course (III), bearing 45° for eight miles until at 1140 hrs he turned westward on Course (IV) bearing 280° for a further three miles. Let us suppose he then halted at noon at a point designated Y in order to witness the sun's transit and to write up his log. The traveller now makes a 'resolution' drawing to the largest convenient scale possible in the manner shown—say 1 cm = 1 mile—and all that he transfers to the chart is the equilibrant course line after measuring its length and bearing.

This method of logging is more compact and visual than narrative logging. But of course there are column-ruled log sheets available through the chandlery trade if you prefer them. Don't forget to include in your inventory the necessary paper, pencils (don't trust pens), a straight-edged ruler (preferably scaled to the same ratio as the chart, which you can easily make up for yourself), a simple school protractor, and some kind of drawing surface on which to spread the map sheet—perhaps the fore-hood of your vehicle will serve. Compasses or dividers are useful but not essential.

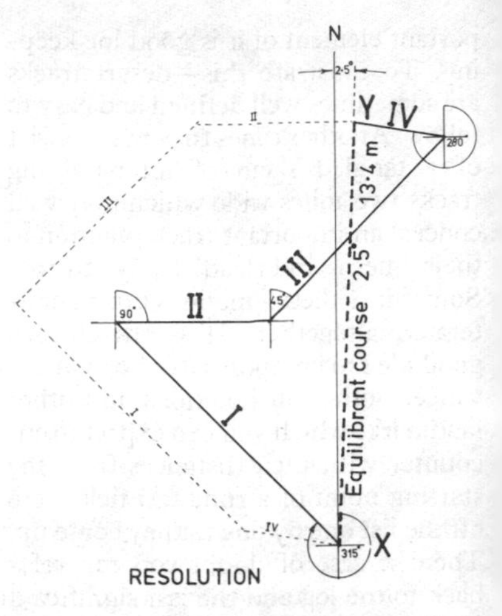

Resolution. Draw to largest convenient scale as accurately as possible. Equilibrant line only is transferred to the chart. Note that vectors may be resolved in any sequence. Light broken line represents same vectors taken in clockwise order. All other permutations arrive at same point Y.

D R is always tedious but it is a satisfactory method provided you do not trust it for longer than two or three days. Its danger lies in the aforementioned accumulation of minor errors as time goes by. But it is very unlikely that three days will elapse without observing a landmark which can be identified on the map—in which case the D R position can be corrected afresh. It is only when D R becomes untrustworthy that a navigator falls back on astro-navigation, using the sun and stars for reference. Good log keeping is the paramount thing. Drivers accustomed to the high standard of route signposting in Europe and North America and encountering desert for the first time could get themselves into serious trouble without close attention to this chore.

Whilst it is very unlikely that careful D R will let you down there always exists the possibility of becoming hopelessly lost. This is where astro-navigation comes into the picture. Since this article is not a full scale treatise on that difficult subject let me ask a question instead. Did you know that any numerate child after half an hour's instruction can easily and quickly ascertain his approximate geographical position anywhere in the world with no more than a good watch, a compass, a suitable pocket sun dial and a few memorised facts about the sun's apparent movements? This will locate him somewhere in a 'circle of uncertainty' whose radius should be as little as ten miles or less. My own average error by this method is just under five miles and is never greater than fifteen miles adrift. An instrument of my design with which this can be done is in the collection of the National Maritime Museum at Greenwich, England, the premier navigational authority of the world. Since the apparatus is small enough to fit in a jacket pocket it occupies no stowage space in a vehicle.

Now I don't suppose this level of resolution would satisfy a professional navigator, but reflect on this thought—that if the circle of uncertainty does not exceed the circle of normal vision in open countryside, and if this method is coupled with extra sharp look-out, ought it not to fulfil nearly all navigational requirements? No mathematical skills, nor almanac, nor books of nautical tables are required, and the operation is extremely rapid as I shall now try to demonstrate.

The abbreviated procedure for finding longitude is as follows:

(i) Read local SUN time.... The sundial does this automatically.

(ii) Note Greenwich SUN time.... Your deck watch is set to this kind of time.

(iii) Find the difference in MINUTES between (i) & (ii).

(iv) Divide this difference by four.

That is all there is to it. The result is your longitude in degrees east or west of Greenwich.

Obviously these instructions need a little explanation. They refer always to SUN TIME—which is seldom the same as MEAN TIME—because whereas Mean Time is constant Sun Time varies with the season. The diagram entitled Equation of Time shows the variation through the year. On any day you can find the difference between the two times and apply the appropriate number of minutes to Greenwich Mean time to convert it into Greenwich SUN time. Your deck watch is then set accordingly to G S T, and from then onward the clock and dial between them present a clear expression of longitude throughout daylight hours. You could memorise the diagram if you wished, or take it with you on your journey, but I wouldn't

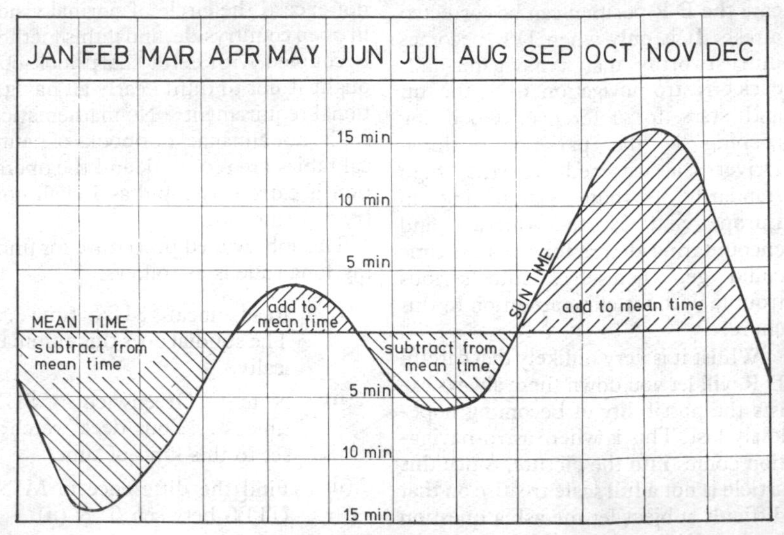

Equation of time. When navigating by sundial set a deck watch to Greenwich SUN Time, instead of GMT. Clock & dial now express your longitude directly.

advise it. The sun's going rate does not change very much in the course of a week, so I should set the deck watch before departure to the sun time appropriate to the third or fourth day of the trek and leave it alone throughout the trip. The watch will be a little over-tuned at the beginning and a little under-tuned at the end. But since you cannot read a sundial of pocket size closer than to the nearest minute there is little to be gained by straining out this small gnat. The inexactitude of the instrument itself will swallow it quite happily.

Another word of advice is never to rely upon a single observation. Take five or more readings and average out. This will miraculously reduce error. Here is a specimen work-out taken from my own records.

21 Aug 1976	Dial Time	GST
	09 32	09 30
	09 34	09 33
	09 49	09 45
	09 57	09 52
	10 08	10 05
	5) 49 00	5) 48 45
	09 48	09 45

The difference is 3 minutes, which when it is divided by four declares my longitude to be 0° 45′ distant from the Greenwich Line. In fact I was ° 52′ distant from the Greenwich Line. So my error was 7′ of arc (which at my latitude is equivalent to (7 x cos 51° 45′) = 4.34 nautical miles. In fact the 45′ line, on which the instrument informed me I was standing, passes through Osea Island in the Blackwater Estuary and I can clearly see

Osea Island from where I am now sitting.

Before dropping the question of longitude I should mention that when you are close to the Greenwich meridian there can be some doubt whether your longitude is east or west. A significant band of the Sahara Desert falls within the area of this difficulty. Mariners settle this doubt by a mnemonic doggerel—

Greenwich time best—Longitude west

Greenwich time least—Longitude east

In the specimen work-out Greenwich Time was earlier (or less) than Dial Time, so the longitude is east.

An even easier way to estimate longitude, though not a very accurate one, is to note the GST at the moment when the sun passes through your meridian. It is then noon at your site. Compare this with GST and the difference is the longitude. You can also adapt this trick to hold a meridian course from day to day. Ascertain the time at which the sun *should* transit your chosen meridian and then take steps to make sure that it does so. If the sun arrives on your meridian too early you are too far to the east (by the number of minutes divided by four). And likewise if the sun arrives late you are too far west and should bias your course to correct it by tomorrow noon.

Having established your longitude you draw a line on the map to correspond. This only tells you that you are somewhere on that line. For the time being you must assume your D R latitude is correct and cross your meridian accordingly to provide a 'fix'. Your position is at the junction of the calculated lines of longitude and latitude, or close thereby.

Not every sundial is capable of finding latitude. My own dials are generally provided with such means. But since they vary so widely in form and geometry I cannot do more than describe the latitude drill movements with one type only. First the instrument is set to the appropriate date. This takes care of the sun's declination for that day. The instrument is then offered up to the sun at the moment of noon (when the sun bears true south or north of my position) in such attitude that when a sighting arm is raised until a shaft of light through an aperture runs true along a datum line, this operates a pointer which, in turn, indicates the latitude of that place on a scale. No figuring of any kind is involved. It is quite automatic.

The basic idea of latitude finding is that the height of the noon sun in degrees above your rational horizon is the same as your angular distance from the nearer pole of the earth. From this (your co-latitude, or polar distance) it is easy to deduce your distance from the equator—which is, of course, your latitude. Unfortunately this basic fact is true on only two occasions in the year—namely at the equinoxes. During the remainder of the year the sun drifts north and south of its equinoctial path as far as the limits of the tropic zone, as indicated by the drawing. This drift is named the declination of the sun. The whole art of latitude finding by solar observation consists in discounting the declination—to restore the sun, so to speak, to its proper position on the equinoctial path. The general proposition to be remembered is therefore—

Noon height declination = Co-latitude

Declination tables for every minute

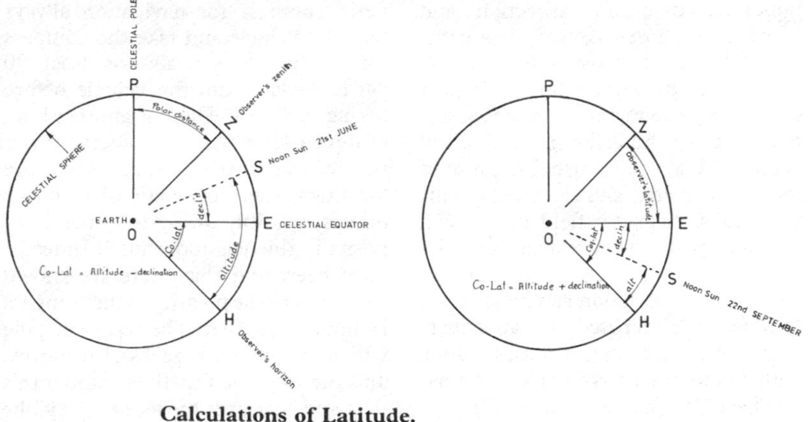

Calculations of Latitude.

$\angle POE = \angle ZOH = 90°$

$\therefore POZ = EOH =$ Co-latitude or Polar distance

$ZOE = 90°$ - Co-lat = Latitude

of the year are included in all almanacs. My suggestion is that in preference to carting a heavy book across the desert you would do better to extract from your district library only the figures which will apply to the period of your journey. Indeed if your journey will occupy ten days or less, recognising that the declination will not change very much during half that time, it may be sufficient to calculate the declination applicable to the fifth day and commit this single figure to memory. If this correction is applied to all noon heights during the trip it should serve well enough. Alternatively you could ask the mathematics or geography teacher at your local school to work out for you the correction you must apply to the noon height during the period of your journey. The physical science department of the local museum will probably be happy to do this for you free of charge.

The marine sextant is used at sea for taking the height of the sun, but it is not so well suited to use on land because it requires a clear distant horizon line, which you will never have in a desert. An artificial horizon can be carried but it is a nuisance and rather difficult to use. A surveyor's theodolite is best but it is a clumsy and expensive shipmate. There are many other curious ways of ascertaining the height of the sun from which I give only one of the more bizarre methods that are open to the forlorn traveller who lacks any better means. If all else fails you can try a bit of jungle trigonometry but it is easier to employ simple graphics. By this I mean placing any regular object, such as a match box, on a level surface and measuring its height and the length of its shadow. Draw these to the largest scale possible and close the triangle with a hypotenuse line. Measure the enclosed angle with a protractor—this angle is the height of the sun.

A very important part of navigation is good compass work. By this I broadly mean the ability (by any means or none) to determine direction and maintain a chosen course. The common magnetic compass is the maid-of-all-work in this regard. But as I expect everybody knows this instrument behaves very erratically in mechanical vehicles. The vehicle itself, because of its ferrous metals and electrical circuitry, has a magnetic field of its own which disturbs the terrestrial flux in its vicinity and so deflects the compass needle. This deviation can be reduced (but never eliminated) by an expert compass adjuster, but it must be done with all intended cargo in place. Disasters have been caused by a metallic tool or other utensil being carelessly placed too near the compass bowl.

Only recently I made a fool of myself. Noticing an index error of five degrees in my hand bearing compass, I puzzled for many hours to find the cause. A few days earlier whilst stripping out my sail boat for the winter I had slipped a small emergency compass into my inside jacket pocket and forgotten to remove it. The two compasses were acting upon each other. Even steel rimmed spectacles can deviate an optical compass held up to the eye.

The moral is that if you use a magnetic compass for navigation always halt the vehicle and take the compass for a nice long walk—at least 30 yards—away from the vehicle before taking bearings. There is another kind of deviation which can sometimes put the evil eye on the compass. There are localities where magnetic disturbance occurs, possibly owing to mineral deposits in the neighbourhood. Indeed I have been told that there are known cases where the polarity of the compass is almost reversed. The magnetic pole which the compass seeks also moves, and the isogonic flux lines wander this way and that over the face of the globe so that variation differs quite sharply from place to place—often over quite short distances.

In spite of its crimes the compass is still a very useful aid but it does need understanding. This is why many people prefer to take advantage of the plentiful sunlight that desert climates usually afford at most seasons of the year and use a sun compass instead. This is not a bad idea because all sun compasses embody a sundial of one kind or another which can be used for position finding as well as for direction. There are many kinds of sun compass. I have designed and built

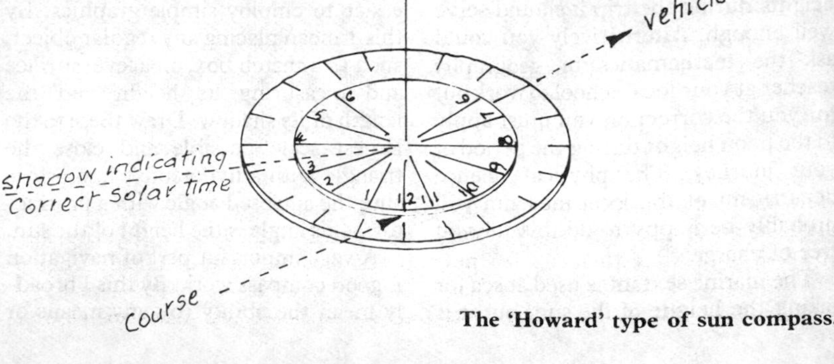

The 'Howard' type of sun compass.

three different types for my own use. Their supreme advantage is that all bearings are TRUE without any correction for variation or deviation. The handicap is that they will not function at night or when the sky is overcast. One of the simplest forms of sun compass is the 'Howard' type, illustrated. A circular dial is graduated anti-clockwise in degree intervals. The zero point is the 'heading' along which direction the journey is required to proceed. The heading line is set along the lubber line of the vehicle. Another blank circular disc is laid upon the first, and the operator graduates this for himself in hour divisions somewhat in the manner of a sundial (except that the intervals are calculated by azimuth instead of by hour-angles). In the centre is a vertical pin whose shadow, falling across the hour disc, will indicate solar time. The noon mark of the hour disc is set to correspond with the required bearing on the lower disc. Steering is now a matter of causing the clock face to register correct local solar time. This raises two questions.

The first is how to graduate the hours. The information can be got from the almanac, remembering that the direction of shadow fall varies greatly with place and season of year. To get over this difficulty I have developed a graphical hour-azimuth reckoner which predicts the true bearing of the sun visually at any time or location on the earth, and this device is clapped to the back of the main disc for reference when required. Thus an operator can prepare an hour disc for himself in a few minutes to suit his own geographical location at any season of year. You will appreciate that as a journey proceeds and your longitude and latitude change, it may be necessary to make up fresh hour discs from time to time, say when your location changes by one degree in any direction—roughly 60 - 70 miles.

The second question is how to keep track of solar time so that the dial can be compelled to show this. The easiest solution is this. At the start of a day's journey take the Howard out of range of the magnetic field of your vehicle and orient it with a hand held magnetic compass. It will now indicate the Local Sun time. Set an independent deck watch to this time and it will continue to register solar time throughout the day. Now all you have to do is to steer so that the two clocks agree. Since a sundial can only show correct sun time when it is properly oriented it must follow that if a dial is known to be showing correct time it must, of necessity, be oriented. This being so the vehicle must be headed on the course indicated by the noon mark.

This may sound very complicated, but like so many other things it is simple in practice. In any case it is always an amusing stratagem to sneak up on a troublesome problem from the rear and kick it in the backside.

However, in order to eliminate such difficulties I have designed the universal sun compass illustrated. Its sundry members can be tilted to any required angle with respect to each other so that steering a course becomes a matter of simply holding a shadow on a datum line. If the vehicle yaws the shadow will fall off the line and must be steered back into place in the same way as a compass needle is kept on the mark.

But pause to consider another thought. If the skies are open is it really necessary to use a mechanical compass of any kind at all? The firmament itself, especially by night, provides a

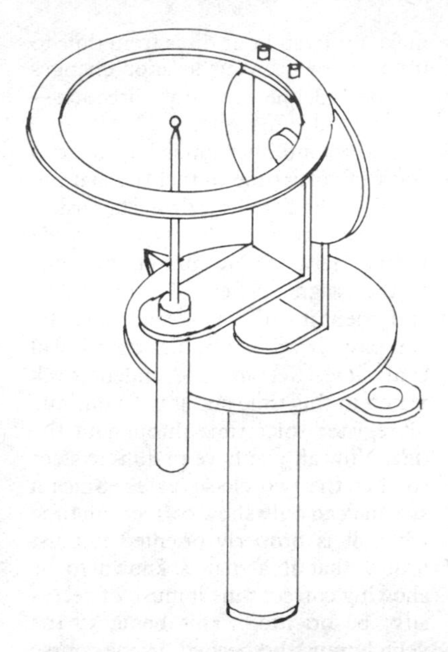

Universal sun compass. One of the author's creations with longitude, latitude, and local time finding capability. May be operated manually or vehicle mounted.

splendid compass ready made. Experienced mariners spend very little time gazing at the compass. A helmsman will sometimes put the vessel on course and then 'hang a star in the rigging', as they say; or by day he may fix a shadow to the deck and then steer to keep it in place for a quarter of an hour or so. During this time the sun or star will move a little. So the helmsman resets the course and repeats the process. Others steer by the wind, or by the pattern of the waves, or even by the wake sometimes.

Some of these tricks can be adapted to desert travel. It is well known, though it is something of a fallacy, that if you direct the hour hand of a clock towards the sun the meridian will run half way between the said hour hand and the noon mark on the clock face. Actually this is true only when you are fairly close to the meridian line of the time zone to which civil time is linked in your district, and when the sun's diurnal error is nil. But it is worth knowing. A derivative of this idea is to visualise yourself as being at the centre of a huge clock face, uptilted towards the equinoctial. Knowing, as you should, that the sun traverses 15 degrees each hour along this track, and provided you know the south point (or north), it is fairly easy to estimate time by eye, say within ten minutes of correct. You can then mentally estimate the required angle between your chosen heading and your shadow at any time of the day and so maintain quite a decent direction irrespective of the sun's movement. There is some evidence that homing and migratory birds may sometimes guide their flight by this method.

It is also worth a pause to ponder the relative necessity for accuracy in course keeping. Many people imagine that if they stray a little off course from a given heading they will run into trouble. The real truth is that a helmsman who can hold a vessel on course within five degrees either side is rated good at his job. Perhaps you had not realised that if you have a fault of one degree in your compass you will have to go 57 miles before you are more than one mile off course. Elementary trigonometry shows this to be so, because the cotangent of one degree of arc is 57. In fact, if it were possible to go bee-line through the heart of the Sahara Desert, from the Mediterranean coast to Accra, on the South Atlantic coast, with an index error of one degree in your compass, you would miss that city by only 30 miles. I am not advocating sloppy compass work. I intend only to suggest that slavish bond-service to precision, for its own sake, can sometimes be an extravagance and lead to needless terror.

I have inspected many desert charts in an attempt to estimate the average scatter of geographical features likely to be encountered in the world's deserts. Experts may dispute my conclusion that charted features are generally not much more than thirty miles apart. Certainly it is true that, in parts of Cyrenaica for example, there is absolutely nothing to be seen for hundreds of miles. But I think we may exclude such awesome regions from this discussion because they are not likely to attract small expeditions—if only because they lack nearly all military, commercial, or scientific interest—and they are no place for the tourist either. If my assessment is accepted (made in the knowledge that general hypotheses are seldom of much value) we are left with a desert model over which physical features are spread at mean intervals of thirty miles. Now suppose you wish to travel from one such landmark to the next, but there is an index error of five degrees in your compass. You wonder to what kind of danger this will expose you? Apply the cotangent equation. Your cross-track error after thirty miles will not be more than two and a half miles. You can hardly avoid finding your target, even with so serious a defect in your instrument.

Potential navigation difficulties can often be foreseen and circumvented in the route planning stage. By way of example, it would obviously be frivolous for a beginner (as also for a master navigator) to choose a route such that his survival is going to depend on finding an unmarked isolated water well in the midst of a featureless wilderness. Not even a professional could be certain of achieving this near-impossible feat. It could be worth a thousand mile detour to avoid so dangerous an undertaking. If you doubt this remember that in earlier days travellers from the Eastern States found it better to go to California by way of Cape Horn than to face the hazards of the direct route—and the Horn passage in those days was no buggy ride.

A further useful rule to observe in finding the way through wild country is to set a clear objective for each stage. As you begin to approach a destination—an inconspicuous oasis perhaps—make up your mind well in advance as to the time at which you expect it to appear on your horizon. If it does not appear on time, halt at once and think out your reckoning afresh. If reconsideration confirms your previous estimate the probability is that your objective is quite close at hand, and may be concealed in a fold of dead ground, or that it should come into view after a few miles more. Conduct a careful square search of the surrounding countryside making full use of any high ground in the vicinity—even standing on the cab top of the vehicle extends your rational horizon marvellously. Aim to expand your circle of visibility to correspond with your circle of uncertainty aforementioned. Don't omit to look backwards as well as forwards. You may have just bypassed the objective. But above all, do not carry on in the vague hope that it will turn up. This may cause you a lot of bother harking back to it if you have inadvertently overshot.

With regard to this problem of overshoot, the Eskimo people evolved an interesting technique for reducing this risk. Sir Francis Chichester modified it on his record-breaking first flight over the Tasman Sea, and bomber pilots applied it to a radio flight guidance system during the last war. It embodies the principle of 'intentional error'. To use it you do not shape a bee-line course towards your objective, because if you bypass or overshoot there is no way of knowing if it lies on your left or right hand side. But if you deliberately aim off one way or the other this problem is halved. An extension of this idea is to 'run down the latitude' of the destination point but keeping appreciably to the north or south of it. Or if you are clever at longitude you may try to put yourself onto the meridian of the point, but well off to the east or west thereof. Then in either event you will have some notion of the hand on which your destination is situated. If you are lucky enough to find (indeed it is worth contriving that it should be so) that your objective is conveniently on, or near, a natural fault, or a dry water-course, you should aim to pick up this feature at some distance from the target. Then when you come upon it the said feature will lead you home. The amount of aim off should in all cases exceed the radius of your uncertainty circle.

It is not often that bad navigation *per se* leads to catastrophe in the desert. Cool thinking overcomes most obstacles. It is more frequently defective planning, or foolhardiness, but most often panic which causes trouble. The fear of being lost is a universal human preoccupation from the cradle to the tomb. Like other primordial sentiments it can best be disarmed by intelligent self discipline.

If interested readers have difficulty obtaining pocket (or mounted) sundials, sun compasses, star dials and so forth—which are rather scarce—and would like to have such an instrument designed and built I can accept a few commissions from time to time. I have been creating these things for many years and have not so far made two alike. Alternatively I may be able to provide working drawings for people to make their own, or have them fabricated by a local craftsmen. Some of these instruments are pretty things to possess in themselves—even if only as display ornaments, conversation kindlers, trophies, heirlooms, or commemorative pieces.

Expedition Insurance

by Philip Harrison

In 1976, a small expedition went out t Ecuador to climb Mt. Sangay, an ac tive volcano. The expedition consiste of six members, and they were all o the slopes of the volcano when it er upted, killing two members of the tear and seriously injuring two others. Th full details of the expedition and th rescue of its members by the Ecuado rean Air Force can be found in Richar Snailham's book *Sangay Survive* (Hutchinson, 1978); but what is not re vealed is the insurance battle wit which I was myself involved. It turne out that each expedition member ha made his own insurance arrang ments—or lack of them in one ca —and the baggage and transpo cover to and from Ecuador was i adequate. Even the commenceme date of the cover was wrong.

I want to show by this example how important it is for every expedition leader to make sure that his or her expedition is fully protected in every possible way before they start, because once the accident happens it is too late! Insurance cover is one of those intangible things that is always the first to be discarded when financial problems arise. Instead it should be given high priority, considering that goods are expensive to replace these days and life is irreplaceable! There are three main areas of insurance cover that all expedition leaders should consider and I will deal with them in order of priority.

First there is Life Cover. This is a difficult one to quantify but here are some suggestions. If the expedition is purely scientific, the destination is low ground in Western Europe, and travel is by scheduled services, then the figure can be reasonably modest, and it may be that the expedition members' own policies can be extended to cover the period of the expedition. If, on the other hand, the expedition is adventurous, and is visiting more remote parts, for example the Himalaya, then the risk is correspondingly greater and the figure for life cover must take account of this. Overall, the leader should consider the personal situation of the expedition members. For example a married man or woman with children should have very high cover, whereas an unmarried expedition member can probably manage with less cover, as there is no one immediately dependent on him.

Second comes Personal Accident Cover. This is very important and the one most likely to be called on, but again the figure to be settled upon depends on the type of expedition and where it is going to operate. If the objects are mainly scientific and the place is near civilization and doctors, then a modest cover should be sufficient. If the expedition is going into more dangerous territory, where medical help will be difficult to find, then the cover must be proportionately higher and a figure of £5-6000 ought to be considered. This may seem high, but one must not forget that an injured man or woman in a remote place may have to be airlifted out and this can prove very expensive. I once heard of a university expedition in Norway, from which a man, injured on a glacier, had to be lifted out by helicopter. A large bill for hiring the helicopter was sent to the expedition leader who had to meet it out of his own pocket, as the expedition's insurance cover was inadequate to cover such an eventuality. The cost of recovering an injured expedition member can be as high as medical expenses, but the latter can be reduced if the expedition has a doctor among its members. Some everyday health insurance policies will cover the insured whether he is at home or away. If the members' existing policies are valid for all geographical areas, it may only be necessary to extend them slightly.

Third comes Baggage and Transport Cover. Most expeditions travel by scheduled airlines if possible and standard baggage cover is available for this part of the journey. If the expedition is taking some very expensive equipment then these pieces should be itemised and covered separately. It is important to read the small print of any cover offered and make sure exactly what is included in the total figure. Most baggage cover does not include things like wireless sets, scientific equipment, special cameras, etc., or baggage stolen from a vehicle where there are no signs of forcible entry. Some domestic policies have clauses that protect property away from home. At the same time, make sure that the cover extends from the assembly point in the country of departure to the base camp, and for the return journey home. If you lose an expensive wireless set after arriving in a country but before reaching the expedition's starting point, you will be disappointed if you put in a claim and find that the cover extends to the airport only of that country.

For details of insurance required by motorists, see *Documentation for the International Motorist*. But note that most countries require a traveller to take out third party insurance as he enters the country. Third party or liability insurance covers a motorist for judgements found against him. Third party insurance is recommended for Mexico and Central and South America and can be bought, for Mexico, from a Mexican insurance company at border towns or in advance through the automobile club, or, for Central America, in border towns in Guatemala. Although it is not required in South America, a driver involved in an accident may find himself in gaol unless he has it. In Asia, only Turkey requires this insurance. In Africa, the countries of the Mediterranean, the east and the south require it, the others do not. The best places to buy insurance for Africa are London, Capetown and Nairobi. Third party insurance is required in Australia and New Zealand and can be bought on arrival. If a vehicle is to be shipped, marine insurance will be needed to protect it from loss or damage while it is in the custody of the shipping company.

All the above may sound expensive, and an example of the costs of a big expedition can best be summarised by the Scientific Exploration Society's latest venture, Operation Drake. This is a round-the-world expedition involving a large number of people, including scientists, and visiting a number of countries, some remote. The figures are as follows: Temporary Life Cover £10,000 for death due to natural causes. Personal Accident Cover £10,000 for accidental death or loss of one or more limbs or eyes or permanent total disablement. Medical Expenses £5000 or Repatriation Costs up to £5000. Baggage Cover £500 per capita but with a limit of £100 on any one article. Valuable equipment is itemised and covered separately. The cost per person per annum is £210, but is proportionately less if a member of the expedition is involved for a limited time only; for example £65 covers a member for three months.

I realise that this is an extreme example and that special rates could be negotiated because of the size of the expedition. So how does the small expedition go about arranging cover for its members? If a school or university expedition is visiting Iceland or Scandinavia, for example, both popular for this kind of expedition, then it is worth asking the travel agent in the first instance, as the agent may be able to make special arrangements. For instance, Oxford University has its own travel club which is able to arrange cover at very reasonable rates. Otherwise, the travel agent should be asked to recommend a broker who might be able to help, or else approach somebody such as (in the UK) the Young Explorers' Trust. For the more specialised expedition, particularly on that is going to a remote area, it is best to approach a broker who specialises in High Risk Cover. In most (British) cases this can be negotiated only through Lloyds of London. A broker is a 'must' here, as only he will know which syndicate at Lloyds to approach, but this will prove to be the cheapest type of cover in the long run. He may be able to find the firm that will offer comprehensive insurance for a whole expedition. Also Lloyds can arrange cover all over the world, which most companies are normally unable to do. There is usually a geographical limit for which UK brokers can issue insurance. In the USA brokers are normally tied to just one insurance company.

If a claim on the policy or policies does arise, it is important that the insurer be informed immediately, by cable if possible, and full details should be sent as soon as possible. Some sort of claim form will have to be filled in, and the sooner this is done, the sooner payment will be made.

If an expedition has a large amount of baggage and stores to be transported, it will probably prove cheaper to send it by sea, the more so if it is in a container. This is of course feasible only if a container port is within reach of the expedition site. Insurance cover for this type of transport is well established, and will probably prove cheaper than insuring individual items. Remember that the container is insured from its place of departure to the port in the destination country only, so that extra insurance cover will be needed at either end, in particular to the expedition's base camp. One point to note here is that all container companies have fixed dates for sealing containers and for departure, so it is important for the expedition to work out its timetable very accurately, to ensure that the stores will have arrived in the destination country by the time they are needed.

For the duration of a long expedition or trip, you may be able to suspend your normal insurance cover, so as to achieve some saving on premiums.

One final point. Most expedition budgets seem to include a figure for insurance cover, but I am sure that this is a 'guesstimate' in many cases. Before completing the expedition's budget, find out what the insurance premium is likely to be, so that an accurate figure can be included, and the expedition members' contribution adjusted accordingly to cover this. A lot of expense and worry in the future can be prevented if this item of insurance cover is properly dealt with at the beginning.

Money Problems

by Richard Harrington

Most travellers will carry their money in the form of traveller's cheques. Companies issuing traveller's cheques usually advertise the ease with which lost cheques can be replaced. Often it's a lot easier in advertisements than in reality, especially if you're travelling well away from big cities. Just changing the cheques in the first place can be difficult. And you can be charged anywhere from the correct rate of exchange to as much as 20 per cent over the correct rate if you're not careful. Big well-known banks are no guarantee against rip-offs, but usually offer better rates than small street-corner *bureaux de change*.

Here are a few tips. Use only strong currency cheques. While you can get cheques in a strong currency like Japanese yen, Swiss francs or Deutschmarks will suit most travellers best, or else dollars or pounds sterling unless these are going through one of their periodic crises. Get a mixture of big and small denominations. While you don't want to be carrying too much cash around at once, you don't want to have to change cheques every day of the week if it's going to take two hours in a bureaucracy-minded bank each time. So make it a compromise between a lot and a little when you change money.

In choosing traveller's cheques, opt for a well-known company whose cheques are known to most banks. The obvious choices are American Express, Thomas Cook, First National City Bank, Bank of America. On the other hand, the big companies tend to charge 1 per cent of face value, ostensibly for insurance. Smaller banks usually have to waive the 1 per cent charge in order to persuade people to use their cheques. That's cheaper, but you could get caught, as I did, with Bank of New South Wales travellers cheques in Hawaii. They were in US dollars but the shop where I tried to cash them refused to accept them, claiming they had never heard of New South Wales, Australia.

Deak Perera, Inc., of 630 Fifth Avenue, New York (and at the airport), issue travellers cheques, but they also provide bags of small coins of different currencies at cost plus a premium. These coin packs are designed to help the busy traveller who needs money for 'phone calls and taxis on arrival somewhere and doesn't have time to change money at the airport, where the local bank may be too busy, or closed, or changing money at 20 per cent over the going rate in town.

To the really off-beat traveller, cheques may not be of much use. It goes without saying that only banks and medium-large hotels and shops will be willing to accept them. So be prepared to take more currency with you if you move away from towns in Third World countries. And if you lose your travellers cheques, remember that even American Express cheques aren't usually replaced within 24 hours, even if they do have an agent close to hand. The 24 hours is usually *working hours* for a start (excluding weekends) and is the time it could take from moment of verification at point of issue of the cheques back in your own country, not from the time you notify the local agent. So in, say, the average West African city, allow up to a week for replacement of American Express cheques, and be prepared for hassles. In Europe, allow 2 working days. With all other brands of cheque, double these figures and add some.

In fact, on the beaten path, you may even prefer to pay by charge or credit card. Among charge cards, only American Express and Diners Club are universally known. Of credit cards, Visa is by far the most international. English cards like Access and American cards like Mastercharge are spreading in overseas acceptance, but of all these cards American Express is still the most universally accepted. But that's in major cities and airports, and only to buy goods and services. Elsewhere in the Third World, forget it. It won't help buy manioc in a market in Benin.

There may be ways of reducing the

amount of money you take abroad. Hotel vouchers may be useful but you could have problems getting them honoured. Some countries, e.g. Morocco, issue petrol vouchers at special discount rates to tourists, but the problems associated with getting these vouchers may be such that you will prefer to pay the going rate at the pumps.

Four years ago, Ghana decided to encourage more spending by tourists by introducing the *cedi* voucher system. You have to spend a minimum of so much a day in hard currency (unless you are exempted as a student, etc.) and can only enter the country with *cedi* vouchers for a substantial minimum amount bought through a Ghanaian Embassy or High Commission before leaving home. Some Third World countries use this tactic and variations on it to keep out poor tourists or to encourage high levels of hard currency income, or both. The Himalayan Kingdom of Bhutan is a typical case. Here travellers can enter only as part of a pre-arranged group that has paid a minimum of £50 per day for ground arrangements. Not for the impecunious! Many countries insist that you show an exit ticket when you arrive, and sufficient funds to maintain a high level of living (spending?) while there.

But beware of showing too much money. The police/immigration could pass the word on, especially if you've given an address on your entry form, and you could become a primary target for theft if the money was mainly in cash. On the other hand, if you don't declare all the funds you have on arrival, and if more is discovered by a search when you leave, the difference, or all the money could be taken (stolen) by the police, ostensibly as a fine. This isn't likely, but could occur in some South and Central American countries. However the genuinely wealthy are usually immune as corrupt police fear the influence of the rich (they've learnt that money buys jets and political influence) and prefer to pick on travellers who look young and poor, and probably are. So that's a very good reason for smartening up at borders to look wealthier than you really are.

One good idea for your travels is to leave with say £25 in £1 notes in your money belt. (If you wear one, don't ever delve into it publicly. Instead of a money belt, you could use an old pocket on a tape; as fastener use Velcro, which opens with a noise and will alert you to any attempts at thieving. Put it around your neck by day, inside your trousers, preferably between your legs, at night.) These notes make useful sweeteners in West Africa and South and Central America when it comes to solving bureaucratic problems. But don't let the official see that you have more than you're offering him. He may want it all, or arrest you for attempted bribery.

It won't help much in most local markets to offer payment in foreign currency, but in larger shops and hotels you may solve the dilemma of whether to run the risk of changing money on the black market for a better rate of exchange, by simply paying cash in hard currency at a negotiated rate equivalent to a black market rate. That way you don't break the law (the recipient might if he stuffs the cash under his bed instead of banking it) and still get the benefit of a black market rate of exchange, one which could give you as much as three to four times the official rate, in some countries. The only problem with this approach is that it can involve large quantities of hard currency for extended periods and that adds up to a security risk.

On the other hand changing money on the black market can be risky. I was once haggling on a street in Ceylon (now Sri Lanka) for a good rate on two Australian pounds. A van pulled up and four armed men jumped out and dragged into the van the unfortunate money changer, who had been offering me three times the official exchange rate. Black market dealings can land the traveller as well as the money changer in jail. And the money changer himself may be out to rip you off. Be suspicious of any move a street money changer makes.

An old con trick runs like this. You agree with the money changer to a rate of exchange which seems to you to be highly acceptable. You count the bundle of notes he's giving you. He takes back the bundle for a second. An accomplice then diverts your attention by moving in and trying to sell you something, and meanwhile the money changer switches the bundle of notes to a bundle of worthless paper wrapped in one real note. The deal is completed and by the time you discover the trick, money changer and accomplice have disappeared into the crowd. Your greed has been rewarded. Console yourself with the thought that the thief will never be as materially rich as you are.

In some places, you may get by without money—maybe bartering shirts and odd western artefacts (knives, spoons, diving masks, etc.) for local goods, food or lodgings. This tends to happen nearer the end of a trip, and may be a good cheap way of obtaining local souvenirs before flying home.

A wise traveller will check on a country's latest money regulations before he goes there. And he'll keep all exchange receipts for transactions in that country. In fact, he may be breaking the law locally if he doesn't. It will be up to you to judge whether you should risk sidestepping the restrictions the country may have set up to maximise its hard currency income at your expense. If you plan a trip to say Kenya and you plan to take about £400 in cash for an independent 10-day trip, then I would do the following. If renting a vehicle, use a prepaid Avis voucher so that that cost is taken care of before you leave home. Carry £30 in £1 notes for tips/bribes, £100 in cash for purchases when you should gain a few per cent (but not a lot in Kenya) on the official exchange rate, especially if buying from Asian traders, and £270 in your chosen brand of traveller's cheques (2 x £50, 6 x £20 and 10 x £5) for hotel and some meal payments.

When it comes to changing money in banks, you may notice that usually you get a worse rate of exchange (generally between 1 per cent and 3 per cent worse) for cash than for travellers cheques. I have never figured out the reason for this. Maybe cash is such a security risk that the cost of insuring it has to be offset by a poorer rate of exchange. Anyway it's another point in favour of using travellers cheques.

Some countries limit the amount of money you can take out. Very few limit the amount you can bring in. If you are taking a local currency out, make sure it's legal to do so. And if it's not a 'hard' currency don't expect to be able to exchange it once you get home. Some 'semi-hard' currencies may be exchangeable at home but at extremely poor rates of exchange, well below the rate you might have got on leaving the country the money came from, assuming you could exchange back unused currency there. Coins are rarely changeable at home, and then at terrible rates.

Whatever your level of travel and wherever you go, a little foreknowledge and a lot of precaution are good investments.

UK Passports and Medical Requirements for Travellers

A UK passport is valid for 5 or 10 years and is obtainable through any Employment Exchange or from regional passport offices:

Passport Office Addresses

The Passport Office
Clive House
70-78 Petty France
London SW1H 9HD

Branch Passport Office
5th Floor
India Buildings
Water Street
Liverpool L2 0QZ

Branch Passport Office
Olympia House
Dock Street
Newport NPT 1XA

Branch Passport Office
55 Westfield Road
Peterborough PE3 6TG

Branch Passport Office
1st Floor
Empire House
131 West Nile Street
Glasgow G1 2RY

Foreign and Commonwealth Office
Passport Agency
1st Floor
Marlborough House
30 Victoria Street
Belfast BT1 3LY
(Personal applications only).

British subjects must apply for a passport on a special form obtainable from travel agents, passport offices or any main post office. The application must be countersigned by a bank manager, solicitor, barrister, doctor, clergyman or someone of equal standing who knows the applicant personally. The application should be sent to the passport office for the applicant's area of residence. Two full-face photographs must accompany the application. The fee for a passport is £11/£22. Three weeks should be allowed for the application to be processed.

For travel within Western Europe (excluding the German Democratic Republic and East Berlin), you can travel on a British Visitor's Passport. This costs half as much as a full passport, but is valid for 12 months only. British Visitor's Passports and application forms for them are available from any main post office from Monday to Friday. The British Visitor's Passport is not available from passport offices, and is only available to UK citizens for holiday purposes of up to 3 months. It is not renewable.

If you lose your passport, tell the local police first. Then contact the nearest British embassy or consulate, by telephone or telegram. The telegraphic address of any British Embassy is PRODROME, of a High Commission UK REP and of a British Consulate BRITAIN, followed by the name of the town.

Normally the Consul will issue a new passport, valid for up to 12 months, as soon as possible after checking with the issuing office in the UK. For this reason, it is highly advisable to keep a separate note of your passport number and its date and place of issue. The new passport is valid in all the countries in which the original was valid, and costs the same. On expiry, it can be extended to a full 10-year passport when you return to the UK, at no extra charge, if your original passport has not been found.

In case of emergency or great urgency, the Consul can issue instead a Single Journey Emergency Passport, which enables the holder to return to the UK but will be confiscated on his return. Or the Consul may issue a passport that is valid for, say, a month—-without checking with the issuing officer—and this permits the holder to return to the UK via several other countries.

Medical Requirements for UK Travellers

The Department of Health and Social Security, Alexander Fleming House, Elephant and Castle, London SE1, publishes a leaflet *Notice to Travellers* which contains up-to-date inoculation requirements for travellers. This may also be obtained by telephoning the Department on (01) 407 5522, ext. 6711. The main offices from which the leaflet may be obtained in Wales, Northern Ireland and Scotland are, respectively: Welsh Office, 17th Floor, Pearl Assurance House, Greyfriars Road, Cardiff CF1 3RT; DHSS, Dundonald House, Upper Newtownards Road, Belfast BT4 3SF; Scottish Home and Health Department, St. Andrew's House, Edinburgh EH1 3DE. The Health Control Unit, Terminal 3 Arrivals, Heathrow Airport, Hounslow, Middlesex TW6 1NB (Tel (01) 759 7208) gives similar information.

Sources of advice on prevention of tropical diseases are the London School of Hygiene and Tropical Medicine, Keppel Street, Gower Street, London WC1E 7HT (Tel. (01) 636 8636), and the Liverpool School of Tropical Medicine, Pembroke Place, Liverpool L3 5QA (Tel. (051) 709 7611).

US Passports and Medical Requirements for Travellers

Application for a new passport must be made in person to:-

1. a postal employee designated by the postmaster at selected post offices;
2. a passport agent at one of the Passport Agencies, located in Boston, Chicago, Honolulu, Los Angeles, Miami, New Orleans, New York, Philadelphia, San Francisco, Seattle, and Washington, DC (see below for addresses);
3. a clerk of any federal court;
4. a clerk of any state court of record;
5. a judge or clerk of selected probate courts.

The applicant must present:-

1. proof of US citizenship, i.e.
   (i) a certified copy of his birth certificate under the seal of the official registrar; or

(ii) a naturalisation certificate; or

(iii) a consular report of birth or certification of birth; or

(iv) his previous passport;

2. two recent identical photographs that are good likenesses (Note 1);

3. identification, e.g. a valid driving licence with signature and containing a photograph or physical description.

The application:

1. is usually processed in ten days or less;
2. costs $10.00 + $3.00 'execution fee'.

The passport:

1. is valid for five years from the date of issue;
2. can be valid for more than one person (Note 2);
3. comes in two sizes, the standard 20 page size and the 48 page size (Note 3);
4. can be supplied with detachable pages (Note 4).

Application for a passport renewal by a passport holder can be made, in person, at any of the places listed above, or in certain circumstances by mail, together with a completed form DSP-82 'Application for Passport by Mail', (available from tourist agencies and the places listed above), the previous passport, two new photographs and $10.00. (Note 5).

Note 1: Photographs need not be taken professionally, as long as they are clear and show the applicant full-face and with no hat against a plain white or light background. Photographs should measure 2in x 2in. Since photographs are often required for visa applications and other formalities, it is advisable to have quite a number printed. Colour photographs are permitted on US passports but since other countries often insist on black and white photographs for visas, etc., it is best to have these. Composite photos of the bearer and his family are no longer allowed.

Note 2: A wife can be included on her husband's passport, but not a husband on a wife's, and unmarried children under the age of 12 included on the passport of a parent, brother or sister. However, no-one so included on another's passport can travel without the bearer. On the other hand, the fewer the number of passports in a group of travellers, the faster are formalities at borders sometimes completed. Applicants should weigh these factors against each other.

Note 3: Travellers intending to visit Third World and other countries, where visas are required and copious entry and exit stamps entered into passports, should ask specifically for the 48-page passport.

Note 4: An accordion sheet of extra pages can be also issued, to provide additional space in a valid passport, by any consulate or passport office abroad. Detachable pages can be used for visas of countries which are at odds with others on the traveller's route (e.g. for an Israeli visa, when the holder proposes to visit Arab countries). The pages can be removed after use.

Note 5: When a passport expires, so too do all the visas and permits contained therein. These must be reapplied for after obtaining the new passport. The same is truc if a passport is lost and replaced.

For convenience: print the holder's name and the passport number on a label and stick this to the front cover of the passport to ensure instant identification and save time at borders.

For safety: keep the passport with you at all times, if possible in a leather or cloth pouch strung around your neck. Apart from customs officials, police, passport agents, consular officials, and, in many European countries, hotel clerks and train conductors, very few people are authorised to handle or inspect passports. Do not give your passport to any unauthorised person. It may be a good idea to take a xerox copy of your passport with you abroad, especially if you intend to visit Eastern Europe, where an official may casually borrow your passport and disappear while he stamps your visa in it.

Visas and Tourist Cards: A visa is an official permit to enter a country, granted by the government of that country. Visas are usually stamped in the passport and are valid for a particular purpose and stated time. Visas can be obtained from the country's consulates either in the USA or abroad. Ordinarily, they should be obtained in advance; certain countries even require that the applicant obtain his visa from the consulate nearest to his place of residence. Some countries require a tourist card instead of a visa; these are available from consulates in the USA, from some travel agencies and airlines, or at the border.

Visa and tourist card requirements change frequently according to the vagaries of political circumstances. To check the current situation, consult form M-264 'Visa Requirements of Foreign Governments', available at any Passport Agency.

For the address of the nearest consulate or consular agent, consult the *Congressional Directory*, available in most libraries.

US Passport Agencies

Boston	Room E 123, John F. Kennedy Building, Government Center, Boston, MA 02203. Tel (617) 223-3831
Chicago	Room 331, Federal Office Building, 230 South Dearborn Street, Chicago, ILL 60604. Tel. (312) 353-7155
Honolulu	Federal Building, 335 Merchant Street, Honolulu, Hawaii 96813. Tel. (808) 546-2130
Los Angeles	Hawthorne Federal Building, Room 2W16, 1500 Aviation Boulevard, Lawndale, Los Angeles, CA 90261. Tel. (213) 536-6503
Miami	Room 804, Federal Office Building, 51 Southwest First Avenue, Miami, FL 33130. Tel. (305) 350-4681
New Orleans	Room 400, International Trade Mart, 2 Canal Street, New Orleans, Louisiana 70130. Tel. (504) 589-6161
New York	Room 270, Rockefeller Center, 630 Fifth Avenue, New York, NY 10020. Tel. (212) 541-7710
Philadelphia	Room 4426, Federal Building, 600 Arch Street, Philadelphia, PA 19106. Tel. (215) 597-7480
San Francisco	Room 1405, Federal Building, 450 Golden Gate Avenue, San Francisco, CA 94102. Tel. (415) 556-2630
Seattle	Room 906, Federal Building, 915 Second Avenue, Seattle, WA 98174. Tel. (206) 442-7945
Washington	Passport Office, 1425 K Street, NW, Washington, DC 20524. Tel. (202) 783-8170

Medical Requirements for US Travellers

The US Public Health Service (USPHS) is the main source of information for the traveller on medical requirements. There are two centres:

The USPHS, National Communicable Disease Center, Atlanta, GA 30333, deals with health requirements

AMERICA FOR PEANUTS
USA CAMPING TOURS
From New York to Los Angeles, from Alaska to the Rio Grande —3, 4, 6 and 9 week Adventure Overland Holidays coast to coast across America for the 18-38 age group. Departing from New York, Los Angeles or Anchorage, you'll visit famous cities, beautiful National Parks and see the REAL AMERICA — friendly people; mountains; cowboy rodeos; deserts; with the flexibility and informality of a small group. Travel is in sturdy Dodge minibuses; camping is at American style sites, most with full facilities. 3 weeks approx £190. 6 weeks approx £350 9 weeks approx £485, connecting flights e.g. London New York London, from £135 via Laker Airways.
Free colour brochure from:
TREK AMERICA LTD
62 Kenway Road, (Dept. WH), London, SW5. 01-373 5083
B A T A

and animal and plant quarantine regulations for the US and other countries; and publishes a small booklet entitled *Health Information for International Travel,* which is available on request.

The USPHS, 330 Independence Avenue, SW, Washington, DC 20201, provides information on vaccinations and other immunisations required for visitors to foreign countries and in some cases will also administer the necessary shots.

Documentation for the International Motorist

In all cases you will need the following documents:

Personal
Passport
Driver's licence

Vehicle
Current registration document and plates

and, depending on your country of departure and your destination(s), you may additionally need:

- Birth certificate
- Extra passport photographs
- International Certificate of Vaccination and Personal Health History Card = 'yellow card' (1)
- International Driver's Licence (2)
- Translation of Driver's Licence (3)
- Permits to travel in restricted areas (4)
- International Student Identity Card or International Scholar Identity Card (5)
- Automobile Club membership card (6)
- Pink slip or notarised letter from title holder giving permission for you to take the vehicle out of the country
- *Carnet de Passages en Douane* (7)
- Nationality plaque
- International Automobile Certificate (8)
- Petrol coupons (9)

(1) Available from your local health authority, together with advice on vaccinations recommended.

(2) Many countries now no longer require an international licence. The national or state licence is sufficient. US drivers travelling abroad may choose to take out (i) an Interamerican Driver's Licence, issued by the American Automobile Association or other recognised automobile clubs, for use in Central or South America. The address of the A.A.A. is

American Automobile Association
World Wide Travel,
1712 G Street N.W.,
Washington, D.C.20006,
USA;

(ii) an International Driver's Licence, issued by the AAA for use in the rest of the world. Neither licence is valid within the USA itself. UK drivers can apply to the Automobile Association for an International Driver's Licence.

(3) Required by some countries.

(4) e.g. the Sahara region of Algeria, the Suez Canal zone of Eygpt.

(5) Both issued to full-time students and valid for one calendar year. The

Student Card is for college students, the Scholar card for high-school students. The card, created by the International Student Travel Conference (ISTC), an association of student travel bureaux in some 40 countries, entitles the holder to discounts at concerts, theatres, museums, historic sites, student hostels and restaurants, tours, trains, buses, ships and Student Air Travel Association (SATA) flights, ski-lifts (Switzerland) and clothes (Eire).

A faked card often secures the same benefits and can be easily made with a college crest, a photograph of the holder, some suitable information, a signature and a laminated finish.

(6) Nearly all countries have an automobile club. Most are members of the Alliance Internationale de Tourisme (AIT) or the Fédération Internationale de l'Automobile (FIA) and honour the membership privileges of other member clubs. A club card entitles the holder to discounts on maps, accommodation, shipping of vehicles, insurance, repairs and other services.

(7) Similar to a passport for the vehicle. Permits temporary importation of the vehicle into the country visited without the need to post a bond or pay duty. It is a guarantee by the issuing club that it will pay any duties applicable if the owner violates another country's customs regulations, e.g. by selling the vehicle illegally. Often a *carnet* is not required by the countries visited: yet to be caught out when one *is* required usually means leaving a large monetary deposit or being turned back. Most travellers therefore do have to have them for an expedition of any length.

To obtain a *carnet*, US motorists should fill out an application for International Customs Documents for Automobile and submit the form to the AAA or to their local auto club together with an 'irrevocable letter of credit' (or, better still, a 'time certificate of deposit', which permits their money to earn interest while they are out of the country) in the Association's name. In Canada the letter of credit can be up to 375 per cent of the value of the vehicle and in the USA up to a maximum of $3,000. You need to have on deposit at the issuing bank funds sufficient to cover the amount of the letter of credit or time certificate of deposit. These cannot be withdrawn until the letter of credit is surrendered by the automobile club. The *carnet* is returned to the issuing body on expiry.

In the UK *carnets* are issued by the Automobile Association or the Royal Automobile Club for about £20 and a Bank Guarantee is drawn up. Motorists may take out double indemnity against having to repay an amount to AA or the RAC. This costs around £175 on a vehicle valued at £2,500.

Under certain circumstances, the Automobilclub von Deutschland, D6 Frankfurt, Niederrad Lyonerstr.16, Fed. Rep. of Germany, will issue a *carnet* without a letter of credit. As no funds on deposit are required, a *carnet* obtained here can mean substantial savings for a party setting out from West Germany.

Correct use of the *carnet en route:* Each page of the *carnet* is divided into three sections. On entry into a country the top third is surrendered to the customs official and the lower two thirds stamped. On departure, the middle section is surrendered and the bottom third (the stub or *souche*) stamped again. The holder should ensure that this is done properly, otherwise he may have trouble en route or when he comes to return the *carnet* to the issuing body. If you can, tell the customs official, on entering a country, by which exit you intend to leave, as he may have to forward the records to that border post.

(8) A translation of the vehicle's registration into the languages of the Central and South American countries (Spanish, French, Portuguese and English). It is issued to any traveller driving his or her own vehicle in Central or South America, upon application to the A.A.A. World Wide Travel office, or to the A.A.A. Mexican Border offices in Laredo, El Paso and Brownsville, Texas, or Nogales, Arizona. The cost is $US 2.00.

(9) These are issued to visiting motorists in countries where there are restrictions on the residents' use of petrol. Motoring organisations can advise which countries issue petrol coupons.

Laissez-Passer, Visas, Permits, Restricted Areas

by Jack Jackson

Ten years ago travel in the Third World for a westerner was relatively easy with few restrictions, whereas travel in Europe was full of customs and police checks. Nowadays the position is reversed. Europe is easy but in most Third World countries the paperwork and permissions required go on growing every year.

Allow plenty of time for obtaining visas; usually in London you need three to five days per visa if applying in person, and up to two weeks if applying by post; also send your Vaccination Certificate with your application. Some countries will also ask to see your airline tickets or a receipt for them.

If you are travelling on, or overland, many countries will not issue a visa unless you already have a visa for the next country *en route,* so get all your visas in the reverse of the order in which you will use them.

Visas usually state specifically the Port of Entry into the country so overland travellers should state clearly their intended Port of Entry and stick to that route.

Visas obtained in your country of residence will be full Tourist Visas (different types of visas are issued to people on business or working in a country). If you travel overland, most countries have a small consulate in the nearest town across the border where you may obtain a visa, e.g. Meshed or Peshawar for Afghanistan; Amman for Saudi Arabia, but these consulates usually issue only a Transit Visa which quotes a fixed route, is valid for one week only, and requires that you exit at a different border point. A Transit Visa can sometimes be extended to a full Tourist Visa in the capital city, but not always.

Fanatical Muslim countries are hard on women travellers. Libya will not give visas to unmarried women under 33 years of age or to men under 26 years of age. Saudi Arabia will not give visas to unmarried women at all and even married women find it almost impossible to get one. For Transit Visa use only, girls travelling with a man can 'obtain' Muslim marriage licences in Amman (Jordan).

If your passport gets filled up and you still have a valid visa in it, a new passport can be tied to the old one with a seal, thus retaining the use of that visa. British Nationals who renew their passport abroad should ensure that the new passport has written into it that 'the owner has the right of abode in the United Kingdom', otherwise they can be refused entry on their return.

At borders or airports, you may be asked for ambiguous taxes, some legal, some not so legal, but the man behind the desk is all-powerful so you do not have any choice.

Many countries where monetary systems are unstable and which therefore have flourishing black markets now produce a currency form on which you declare all your monies, jewellery, cameras, tape recorders, etc., and this is checked again when you leave against bank receipts for any money you have changed. Algeria, Afghanistan, India, Nepal and Sudan are examples of this. Obviously all traveller's cheques must be changed in a bank anyway, but if you neglect to declare all your cash and change some of it on the black market you must understand the risk. Most countries are not too thorough in their search when you leave, but Algeria is one that is. A word of warning on currency forms: with groups, border officials naturally try to cut down massive form filling by filling in one form for the group leader. This can make life very difficult later if you want to change money at a bank but do not have your own individual form and the group leader is not around; so try to get individual forms.

Some countries e.g. Bhutan, Ghana, Zaïre, have followed the Eastern European system of making you change a certain amount of money each day. Other countries purposely delay you whilst waiting for permits, e.g. Nepal now keeps you waiting in Kathmandu for your Trekking Permit to make you spend more money there. As most trekkers are limited for time, a straightforward Tourist Tax would be more acceptable.

Many countries insist that you register with the police within 24 hours of arrival and often a fee is charged for this. Usually if you are staying at a hotel the registration is done for you by the hotel and the costs are included in your room charges, but if you are in a very small hotel, camping or staying with friends you will either have to do it yourself or pay someone to do it for you. As this often entails fighting through a queue of several hundred local people at the Immigration Office—with the chance that you have picked the wrong queue anyway—baksheesh to a hotel employee to do it for you is a good investment. Most of these countries require that you register with the police in each town at which you stop, and in some cases, e.g. South Sudan, you have to report to the police in every town or village through which you pass. In smaller places the registration process is usually much easier.

Many countries require that you get permission from the central government to travel outside of the major cities. This is so in the Yemen Arab Republic, the Sudan and Nepal. Usually for this permission you go to the Ministry of the Interior but if a Tourist Office exists it is wise to go there first. Any expedition or trekking party will have to do this anyway; only Nepal has a separate office for Trekking Permits.

Most countries have restricted or forbidden areas somewhere. To visit Sikkim, Nagaland or Bhutan you must apply for a permit to the central government in Delhi and for Nuristan or the Wakhan Corridor you must apply in Kabul. In some other restricted areas permission remains with the local officials, e.g. Kabul or Herat for the Afghanistan central or northern routes; Tamanrasset or Djanet for the Algerian Sahara, Agadez for the Niger Sahara; the District Commissioner in Chitral for Kafiristan.

Much of Africa and Asia have large areas of desert or semi-desert. Restrictions on travel in these areas are formulated by the governments for travellers' safety and take account of such obvious things as ensuring that travellers have good strong vehicles carrying plenty of water and fuel and are spending the nights in safe places. Unfortunately officials in these out-of-the-way places tend to be the bad boys of their profession. Forced to live in inhospitable places they are usually very bored and often turn to drink and drugs. Hence when a party of westerners suddenly turns up they see this as a chance to show their power, get their own back for the old colonial injustices, hold the travellers up for a day or more, charge them baksheesh, turn on a tape recorder and insist on a dance with each of the girls and suggest that they go to bed with them, and if there is a hotel locally, hold them overnight so as to exact a percentage from the hotel keeper. Unfortunately, your permit from the central government means nothing here. These people are a law unto themselves.

The police in Djanet (Algerian Sahara) really have it tied up: you cannot get fuel to leave without their permission and to get that you have to spend a lot of money with the local tourist organisation and hotel as well as fork out baksheesh to the police themselves. Recently in Southern Sudan the police at Kadugli insisted on being paid 50 Sudanese Pounds (£80) by each person in the party before allowing them to go into an area for which the central government in Khartoum had already given them permission. Further south at Tonga a drunken policeman tried to arrest a member of the party because he did not like his face. Fortunately the District Commissioner was in residence and, being sober, overruled the policeman.

Local officials also have a habit of taking from you your central government permit and then 'losing' it. This makes life difficult both there and then and also with local officials later on in other areas. It is therefore best to carry ten or so photostats of the original government permission (phototstat machines are always available in capital cities) and *never* hand over the original. Let officials see the original but always give them a photostat instead.

If you are travelling as a group, all officials will want a group list from you, so carry a dozen or more copies of a group list made up of names, passport numbers, nationalities, dates of issue of passport, dates of expiry of passport, dates of issue and numbers of visas, and occupations.

Some countries, e.g. Sudan, Mali, Cameroon, require that you get a photography permit. These are usually available in the capital only, so overland travellers will have problems until they can get to the capital and obtain one. As with currency declarations, officials obviously like to save work by giving one permit for a group, but it is best to get one per person. I have known several instances where big-headed students have made citizen's arrests of people taking photographs—who then had to spend a couple of hours at the police station waiting for their leader with the photo permit to be located.

Most countries will extend a full Tourist Visa two or three times for a fee but only when the present visa has almost expired and normally only in the capital city. Once you have had a Tourist Visa extended, you will normally also need an exit visa before you can leave. This takes two to three days to get, so apply early. Some countries require an exit visa anyway, e.g. Yemen Arab Republic. Do not overstay a Tourist Visa without renewing it: this can involve a heavy fine, an appearance in court (Iran) or in times of political unrest, a spell in prison (Afghanistan).

Where enmity exists between countries, e.g. between South Africa and Tanzania, or between Israel and the Arab countries (now with the exception of Egypt), passport-holders having visas or entry stamps from one country will virtually always be refused entry into the others. It is possible to get around this problem in advance by applying for two passports, one to enter each of two mutually hostile countries, or for loose-leaf pages in a passport which, once they bear the offending visa, can be removed prior to entering the next country.

Do as much as you can before you leave home in the way of obtaining visas and permits, but be prepared, throughout the Third World, for delays, harrassment, palms held out and a large dose of the unexpected.

Learning a Language for your Trip

Adapted from an article in *Holiday Which?*, and reproduced by kind courtesy of the Consumers' Association.

Can you tell a tarte from a Torte, or even a tortilla? A casa from a Kasse or even a cassata? Or is your contact with the locals confined to schoolboy French, gesticulating hands or English pronounced slowly and clearly at the top of your voice?

Adult education classes in holiday languages are booming, and in every newspaper or bookshop you can see claims for 'almost-instant-packaged-language-learning' using the 'latest-wonder-method'. So how do you start?

We've looked at the main ways members used to learn a language—local education authority (LEA) classes and teach-yourself methods—and at the main advantages and disadvantages of each.

The main ways of learning are:

Group classes: Lots available, run by LEAs, private language schools or cultural institutes—such as Institut Francais. They can be anything from one evening a week to intensive courses all day for several weeks. Ask at your local library, the cultural section of an Embassy, or look up 'Schools-Language' in the Yellow Pages. Cost: from about £4 for a term's LEA class, about £14 for a term at a private school, to over

£1,000 for an intensive course of several weeks.

Private tuition—either by private language schools (see above), or by finding a tutor locally through such sources as the Institute of Linguists, 24 Highbury Grove, London N5 2EA, regional branches of teachers' associations such as the Modern Language Association (head office at same address as Institute of Linguists). Some LEAs also have private classes. Cost: from £4 to £8 per hour.

Correspondence courses—offered by some colleges, listed by the Council for the Accreditation of Correspondence Colleges, 27 Marylebone Road, London NW1 5JS. Some also have Summer Schools and language laboratories. Cost: from about £17 to £31.

Teach-yourself—using books, or a combination of books, cassette tapes, records, radio and television; you might be able to borrow these from libraries. Cost from nothing if you borrow a book, to over £250 for a full programme of cassettes and learning books.

Language laboratory courses—offered by LEAs, particularly in larger polytechnics or technical colleges, and by private language schools (see under group classes for how to find out about these). These may be flexible 'use-the-lab-when-you-want' schemes, or fixed classes, or as a supplement to other, mainly group, courses. They vary from simple tape-recorders with headphones to computer-controlled systems, with individual booths connected to a master console. In some laboratories students can work at their own pace, recording and then listening to their own voices. In others, pace is controlled by the teacher, so a student can't always record his answers and then play them back. In all language laboratory work, a lot depends on how much supervision the teacher is able to give, and how good the course material is. Repeated drills can quickly become boring.

Residential short courses—details in a calendar published twice-yearly by the National Institute of Adult Education, De Montfort House, De Montfort Street, Leicester (50p inc postage). Cost: from about £9 to £22 for 3 days, £94 for 15 days.

You can get information on these ways of learning and details of how to find out about them in *Information Guide 8, Part-time and Intensive Language Study: Notes on facilities available*, 10p, from the Centre for Information on Language Teaching and Research (CILT), 20 Carlton House Terrace, London SW1 5AP, Tel. (01) 839 2626. Their library has directories and lists of courses. They publish bibliographies of course materials, and will give advice on the type of course to suit you, and where you can get it. Write to them, giving them as many details of yourself and your needs as you can.

Every method has advantages and disadvantages. Which you choose may depend on things like how much time you have, whether you can get to a language centre, whether you can do day or evening classes, what LEA facilities are like, and what you are prepared to spend.

Many people seem to find a combination of several ways of learning a language best and most enjoyable—such as group classes which use audio-visual aids and/or a language laboratory, group classes plus a BBC TV or radio course, or a teach-yourself system with individual tuition or perhaps a short residential summer course. However, it sometimes takes quite a bit of extra effort.

Even within one method of learning, systems of teaching may vary. For example, many courses use 'direct method' teaching—which means that English is banned, and grammar is not taught formally. You are immersed in the new language and learn to repeat and imitate. While this is a popular method for learning holiday words and phrases, it may have drawbacks in that you can only learn a language superficially without grammar—although it might hold your interest better in the early stages.

There are many local education authority courses, with differing levels, aims and approaches, such as 'conversation', 'refresher', 'intensive', 'audio-visual', 'examination' and even ones called 'French-for-the-family', 'Rush-Hour class' or 'Short course for the holidays'. There are day and evening classes, usually for one to two hours, once or twice a week, running from 10 to 32 weeks with breaks during school holidays. You're more likely to find the exact class to suit your needs, particularly past the beginners stage, if you live in or near a large town. In some areas, you have to join an exam course, such as one for GCE 'O' level. If the language you want to learn isn't offered, try to make up a group yourself. LEAs will usually try to find a teacher if there are about 15 people wishing to study a subject.

A large proportion of LEA courses use audio-visual aids such as TV, videotape, radio, slides and cassettes or records. Some classes use only the foreign language, others a lot of English. Some are conversational, others concentrate on grammar. Some students find their classes excellent for learning vocabulary, others are disappointed. Quite a few classes use BBC material.

The level of ability in classes varies a lot, particularly in those other than for beginners, and also in less populated districts, or for less popular languages. Too wide a range of ability can make for uneven or slow progress, especially when pace is determined by the slowest members of the class.

The standard of teaching also varies, as does the amount of individual help given. In a two-hour session there is not much time for individual help. LEA enthusiasm for using native teachers, often proudly claimed, is not always matched by their ability to select a competent one. 'We were taught by a French woman whose English was difficult to understand' and 'Teachers were Spaniards rather than teachers' are typical complaints.

The fall-out rate in LEA courses is high, a fact that some students find to their advantage, since teaching becomes more personal and individual. However, classes can be closed if they fall below a certain size (usually eight or nine—somctimes as many as 14).

Cost: Courses range from £4 to £21, and vary according to length of course, number of hours per week, and area. At larger centres such as Polytechnics and Colleges of Art they usually cost more, from £14 to £21—they may include language laboratory facilities and NUS membership. OAPs, under 18s and those doing more than one evening class pay less.

Overall students enjoy their course very much. Indeed classes become social occasions. How well you do depends on how much work you put in.

There is a difficulty in achieving any real competence in the language in only one lesson a week, and it is highly advisable to supplement this method by as many short intensive courses as possible. Two or three days' concentrated practice can bring quite surprising progress.

No one Teach-Yourself method seems to stand out as better than others for all things. The main problems are:

i) The self discipline needed. Quite a few people who have used the Linguaphone system, for example, have four 1 it very easy to discipline themselves to do the work. You need to set aside regular study periods and resist the temptation to skip over boring bits of grammar or those aspects which give you trouble.

ii) The lack of opportunity to have pronunciation corrected. There is no opportunity to ask questions and you are apt to learn wrong pronunciation and have no one to correct you. More simply there is no one with whom to converse. With the Linguaphone Audioactive system you can record cassettes and send them in to have them corrected.

iii) Mistakes may go unnoticed, and so be repeated. Some of the Linguaphone systems include a correspondence service for marking work.

The main advantages are:

i) You can work at your own pace, and after the course is finished you can always start again.

ii) You can listen to records and cassettes whilst doing other things.

Recommended Books

Living French, by T. W. Knight, £1.25 (tape also available £7.86 inc. p&p from Tutor-tape Co. Ltd., 2 Replingham Road, London SW18; *Beginners' Italian,* by Ottario Negro, £1.25 (tape £6.78 from above): and the *Teach Yourself* series in several languages, from 60p to £1.25. All these are published by Hodder and Stoughton.

Users have also recommended Hugo's Simplified System series such as *French in Three Months* and *Italian in Three Months,* 95p, and *German Made Simple,* by Eugene Jackson and Adolph Geiger, £1.50, published by W. H. Allen. Some have used *Phrase Books,* 75p, published by Collins.

Teach-Yourself systems

Linguaphone have four different systems from Sonodisc (with cassettes, in French, German, and Spanish, £18) to systems at £63.50 (records or cassettes in many languages), £78.50 (records or cassettes, in French, German, Spanish and Italian) and an Audioactive course at £87.50 (cassettes, in French, German, and Spanish) and £350 including portable language lab. (cassettes in French, German, Spanish, Italian, Russian and Arabic). They all include an optional correspondence course: the Audioactive course includes a pronunciation tutorial service. They also do a Travel Cassette, in French, German, Spanish and Italian, which includes captioned picture cards covering emergency holiday situations; £3.95.

The Stillitron cassette system, in French, German, Spanish, Italian and Arabic, includes a 'visual feedback' device which corrects multiple-choice questions, and a multiple-choice exam service. The system, £250, can be used at various Stillitron centres in the UK. You can buy a portable language lab., £145.

World of Learning's (359 Upper

Richmond Rd. West, London, SW14 8QN. Tel. (01) 878 4314). PILL cassette system in French, German, Spanish, Italian and Afrikaans costs £45.50, Russian £49.50. A cassette recorder costs £36.56, a portable language lab £140.40. They also do a French Conversation Course at £39.50.

Reader's Digest do a record system, *At Home with French*, £17.95.

BBC material

Television or radio courses usually start at the beginning of October. You can find out about these from: the September Radio Times supplement *Look, Listen, Learn;* or from your local library; or from Educational Broadcasting Information, Adult Education (30/BC), BBC, Broadcasting House, London W1A 1AA, Tel. (01) 580 4468. There are combined radio and television courses and short 'crash courses'. You can get past BBC programmes in the form of cassettes and records, as well as the accompanying textbooks, from BBC Publications, 35 Marylebone High Street, London W1M 4AA (also by mail order) or from your local bookseller. Costs range from about 50p to £10. For example, the 25 lesson beginners' German course *Kontakte* would cost £3.75 for three books plus £4.20 for three records or £7.44 for three cassettes, including VAT. A 30 lesson Italian course *Amici, buona sera!* costs £1.75 for a book and £3.13 for two records. Not all courses have cassettes.

How long does it take? Experts reckon that an adult can expect to learn 1,000 to 3,000 words during the first year. One student who went on holiday to Leningrad after two terms of weekly group classes (plus up to four hours homework a week) said 'I obviously didn't know very much, but was able to read names of shops, goods and so on, exchange common courtesies, ask the way and do shopping; it added an extra dimension to my holiday. It was great to be able to read the Cyrillic script. It meant one could confidently go around Leningrad on one's own; one didn't have to stay with the group for fear of getting lost.'

How long *you* will need before you could cope as well depends on many things: the skill of your teacher, your motivation, how much time you have, and your ability to learn languages, or knowledge of other languages.

A few points which might help you:

1) Have a goal at the end of the course, such as a planned trip to the country where the language is spoken.

2) Group classes can be a stimulus, particularly if you are encouraged to chat to your fellow pupils in the language.

3) Lively teaching makes a lot of difference.

4) Audio-visual aids help, particularly when combined with other methods.

5) You need to study regularly, (CILT recommends you do homework for double the time you spend in classes each week,) and to be very enthusiastic and self-disciplined.

In the end it all depends on your commitment. Nothing compensates for laziness and nothing replaces steady daily practice.

Getting it Together

Fit and ready to travel

The Independent Traveller's Guide to Health

Drs. Peter Steele and John Frankland

The unprepared and the unwary are those for whom expeditions abroad can end in misery and expense. You are off on the trip of a lifetime. If you are a wise traveller, you prepare documents, check equipment and carry spares in case of emergency. It is logical to take as much care of your health for if this lets you down you may be throwing away a great experience and risking huge cost. At least one party member and preferably all would profit from a formal course in first aid as organised by the St. John's Ambulance or Red Cross. Much of this can be forgotten but the basic principles are essential as is the ability to improvise, since accidents never happen in close proximity to your first aid supplies.

This article aims to guide you when visiting out-of-the-way parts of the world where certain health hazards exist. Whatever your reasons for travelling and wherever you plan to go, the medical problems you meet vary little; but disregard them at your peril.

The art of travelling is learned by building on the experience of previous journeys, but the objective always seems unattainable when poring over a map laid out on a comfortable sitting-room carpet. We have never lost the feeling of awe that goes with undertaking a new journey—and we hope you won't either.

Preparations

Beside your travel documents, do not forget: Form E.III or your medical insurance policy (British Travellers) (see later). If you are taking any medicines or drugs regularly take an adequate supply with the dosage and pharmacological name marked on the bottle since proprietary names vary in different countries. If you have any current or past significant illness get medical advice on how sensible an undertaking your trip represents. Advice on how any relapses or complications might be handled and whether any particular drugs should be available for these should also be sought.

Personal medical information which can be imprinted on a bracelet or medallion; blood group, allergy, diabetic or steroid treatment dosage. This is safer than a card carried in the pocket and may be life-saving in an emergency. A doctor's letter setting out any special medical problems (with translations into appropriate languages).

Spare spectacles and your lens prescription.

Tooth fillings tend to loosen in cold so a dental check-up may save agony later. If you suspect piles, seek an examination. Feet should be in good shape as much may be expected of them.

If you are not in the peak of condition build up your fitness with regular graded activity over some months before departure and seek medical advice if this brings on any health problems.

Insurance

Falling ill abroad can be very expensive. The European Economic Community (Belgium, Denmark, France, West Germany, Ireland, Italy, Luxembourg and the Netherlands) have reciprocal arrangements with the British National Health Service. The detailed workings of the scheme are laid out in Form SA.28, issued by the Department of Health and Social Security. This explains how the certificate of entitlement to medical treatment (Form E.III) may be obtained. It is not available to self-employed or unemployed persons. (Students beware!)

Elsewhere the cost of consultation, medicines, treatment and hospital care must be paid for by the patient. As this could be financially crippling, full health insurance is a wise precaution. A typical 'package deal' insurance includes baggage, personal accident and medical expenses up to £1,000. Routine dental treatment is not usually included in the policy. Insurance covers the cost of medical treatment abroad or of flying the sick person home. This may be safest and cheapest in the long run. Don't forget to insure every member of the party.

If you incur medical expenses, present your policy to the doctor and ask him to send the bill direct to your insurance company. Many doctors will demand cash and the level of their fees may alarm you. Keep a reserve of traveller's cheques for this purpose, insist on a receipt and the insurance company will reimburse you on return. Do not expect the medical standards of your home country in your wanderings. Some practitioners include expensive drugs routinely for the simplest of conditions and multiple vitamin therapy, intravenous injections and the inevitable suppositories may be given unnecessarily to run up a bigger bill. Be prepared to barter diplomatically about this, to offer those drugs you are carrying for treatment if appropriate and even to shop around for medical advice.

If you are going to take part in 'high-risk' sports you will need a more specific insurance cover. Consult a company which specialises in this field and will give you an individual quote. If you have difficulty in obtaining expedition insurance, a quote from a Lloyd's broker may be the answer.

Immunization

Immunization can protect you from certain infectious diseases that are common in countries abroad but rare at home. Your local District Community Physician's Department (UK) or Public Health Service Office (USA) will advise you on the inoculations necessary for a particular country you may wish to visit and how to obtain them, either at a clinic or through your family doctor. Do not leave it to the last moment as a full course can take up to three months. Carry with you an International Certificate of Vaccination.

Immunization is not obligatory in Europe or North America, but it is wise to be protected as follows:

a) Typhoid, paratyphoid A & B and tetanus (TABT). Two injections are given one month apart. An unpleasant reaction, with a sore arm and headache, is not uncommon and you must avoid alcohol for 24 hours. After a wound from a dirty object or an animal bite, you should obtain a booster dose of tetanus toxoid. Full protection against tetanus is essential to anyone roughing it in the tropics where this nasty illness is more prevalent.

b) Cholera, yellow fever and possibly smallpox. The world is now smallpox free and the only need for a certificate is to satisfy frontier bureaucrats in developing countries. At the time of writing, the WHO is expected within the next twelve months formally to declare that smallpox has been eradicated and then perhaps even this requirement will then cease. Meanwhile travel agents have the best up-to-date information on countries that still demand production of a certificate. Cholera certificates are still definitely needed to cross many frontiers—two injections a minimum of ten days apart. This injection can be combined with TAB (TABChol). The certificate is valid for six months only, so on a long journey have this protection just before departure. Yellow fever immunization is needed for Central and South America and Central Africa and for travellers in other countries who have journeyed through or come from these areas. In the UK this can be given only in designated centres (address from your local Community Physician); all other immunizations can be done by your General Practitioner. A single injection is necessary. There are no reactions and the certificate is valid for a period of ten years.

If smallpox vaccination is also needed get the yellow fever injection first and the smallpox can be given a week later—if you do it the other way round there has to be a three week interval between the two.

c) Most will have had childhood immunization against diphtheria, polio & tuberculosis (BCG). However these last two can be real hazards in developing countries, especially off the tourist track, so check your immunization status and consider a poliomyelitis booster dose.

Three months should be allowed for a full course of immunization, but in emergency, a 'crash course' of smallpox, TABT, yellow fever (and cholera) can be given in 15 days.

d) Check whether your route includes a malarial area, using an up-to-date reference. This problem has reappeared in some areas previously cleared. The embassies of some countries may deny they harbour malaria when this is not now the case. In the UK anti-malarials can be bought without prescription from a chemist. Paludrin taken one tablet daily from the day before exposure and continued for 28 days after leaving the malaria zone is effective and probably easier to remember than Daraprin taken one tablet weekly over the same period. Fatal malaria can be contracted from just one bite of the Anopheles mosquito so the above precautions should be observed even in the case merely of a brief aeroplane stop in an affected area. The 28 day post exposure treatment is most important as the disease can otherwise develop weeks after exposure.

Above 7,000 feet one is most unlikely to contract malaria. Many overseas communities have their own legends concerning the disease which are often fanciful, eg that a local herb or a quantity of alcohol is the best preventative. Listen politely but ignore all of these and obsessively take your own medication.

Infectious hepatitis (jaundice) is a real hazard to travellers through areas with absent or primitive sewage systems. Perhaps more return with this ailment than all the exotic diseases. An injection of 750 mgm of human gamma globulin gives reasonably effective passive immunity for six months against the commonest strain and is highly recommended. In the UK this has been available free on the NHS to all at-risk travellers since 1976.

Where vaccines are contraindicated for a particular patient, this should be noted on his International Certificate and he would be wise to secure written permission to enter a country without the relevant vaccination, from its consulate.

Advice

Now you are ready for the journey: rucksack packed, pockets crammed with documents and wallet bulging with travellers' cheques. Have a happy, healthy holiday. The following advice aims to help you avoid illnesses commonly met abroad, most of which can be treated by yourself in the first instance. The commonest medication you will need will be 'stoppers and starters', antacids for upset stomachs and simple dressings for minor trauma. If however, the condition rapidly worsens or does not improve within 24-48 hours, you should consult a doctor or other medical practitioner. In some areas, the most highly trained person around will be the local pharmacist who will both dispense medication and perform medical treatments. Your consulate or embassy and the nearest office of the Peace Corps usually have a doctor attached to them or near at hand and will be able to refer you to him. In the remotest of places sponsored mission hospitals may offer an excellent and devoted service, probably with English speaking staff, and may be worth seeking out. When your trip is over, if you do not wish to ship home drugs and dressings, they will be more than glad to accept them.

Several different drugs can be used to treat any one illness. Those recommended here are included in the suggested medical kit at the end of the article. Approved names of drugs are generally used; proprietary preparations are set in italics. Unless otherwise stated, a medicine should be taken four times a day. Children's doses are usually half the adult dose.

Traveller's Diarrhoea

Gippy Tummy, Delhi Belly, Kathmandu Quickstep—traveller's diarrhoea has as many names as patent remedies. It strikes most travellers at some stage in their journey, making more trouble than all the other illnesses mentioned here put together. The causes are usually untraceable but may include gluttony, change in climate and an upset in bacteria that are normal and necessary in the bowel. Infection with disease-causing organisms carried in water and food is less common.

Much of the pleasure of travelling abroad comes from eating local food and drinking wine; it is hardly worth going all the way for beer, fish and chips. Be moderate to prevent the tummy upset that will spoil your trip, and even make you into a useless member of the party.

Water warrants the utmost care; carelessness may be very costly. In some clean hotels and restaurants, the water is safe to drink. Stream and river water is likely to be polluted unless it comes directly from a hillside spring, and glacial mud or mica in alpine rivers is especially irritant to the gut. If in doubt, sterilize all drinking water.

Boiling briskly for a few seconds kills most organisms (including amoeba cysts and infectious hepatitis virus). So drink tea or coffee.

Water-purifying tablets. Chemical treatment is less effective than boiling, takes longer and leaves a taste of chlorine, but in hot climates where high fluid intakes are mandatory, using such tablets (Puritabs UK, Halazone Tabs USA) may be the only practical way of treating water whilst on the move. Take an adequate supply.

Filters get rid of the murky colour of suspended organic matter but purify water less surely than boiling.

Food and Drinks: see *Food and Water: Advice for the Long-Distance Traveller*.

Hygiene: Lavatories abroad are often dirty. You may have to squat and keep your balance by holding on to the walls. Wash your hands carefully with soap as soon as possible afterwards. Take your own toilet paper as newsprint is rough and fragile. At campsites, dig a latrine hole well away from the tents and your water supply.

Brush your teeth in clean water only—chewing gum is a useful temporary cleaner.

NB Cholera: 1973 saw a worldwide epidemic, notably in parts of Italy. The cholera organisms come only from the human intestine and are spread by faecally contaminated water, not by direct contact or inhalation. Raw shellfish collect the bugs so are particularly dangerous. A sudden onset of profuse watery diarrhoea in an epidemic area calls for immediate attention. Unfortunately the immunizations against this disease are not too effective.

Treatment: The illness usually clears up on its own in two to three days. You may also vomit and, because a lot of body water is lost, you may feel groggy. Go to bed and drink unlimited fluids (at least a pint an hour). Avoid eating —except dried toast and peeled grated apple gone brown (pectin). Under expedition circumstances, sufferers may have to continue to work and may choose to continue to eat, so that anti-diarrhoeal tablets will be needed.

Lomotil (four tabs initially and then two tabs four times daily as required) or Codeine Phosphate Tablets 30gm—two tabs four times daily as required—are probably as effective and compact a medication as any.

Antibiotics, though fashionable, should not be used blindly since they kill normal bacteria, which are protective, as well as poison-producing ones.

Dysentery

If diarrhoea does not stop within 24 hours on this treatment, or if blood appears in the stools, consult a doctor since you may be suffering from dysentery. If you cannot find help, the best drug to start with is Cotrimoxazole *(Septrin, Bactrim)*. Bacillary dysentery starts suddenly with acute diarrhoea, fever and malaise. Amoebic dysentery causes slimy mucus and blood and warrants laboratory investigation and treatment. Its severity builds up slowly over several days.

Worms

Worms are common in tropical countries. They cause an itchy bottom and can often be seen in the stools. Take one Peperazine *(Pripsen)* sachet.

Indigestion

Any antacid will ease gut rot, indigestion and perhaps a hangover. Tablets are more portable than mixtures and just as effective. Magnesium trisilicate acts as a laxative and as some diarrhoea often coexists (aluminium hydroxide *(Aludrox)* tablets may be more suitable. Everyone will need them on most journeys even if only to cure indiscreet eating and drinking. *Maxalon* (Metaclopramide) tablets four times daily will help suppress vomiting.

Constipation

Drink plenty and eat fruit. If this fails, take two laxative tablets (*Senokot*).

NB Beware the person who feels sick, has no appetite, a dirty, coated tongue and pain in the belly. If the abdomen is tender, particularly in the right lower quarter, suspect appendicitis and visit a doctor. If no doctor is available rest the patient, give fluids only with antibiotics and pain killers. Evacuation must then be considered.

Waterworks

Urinary infection is more common in women, and begins as frequent passing of urine with burning pain. Drink a pint of water hourly with a tablespoon of bicarbonate of soda and take an antibiotic if it does not improve in a day.

Women travellers have the extra burden of coping with menstrual problems *en route*. If on the pill it is totally safe to start the new pack immediately after finishing the current pack without the usual six or seven day interval. This will avoid menstrual loss or certainly minimise it, saving quite a nuisance. If however a woman on the pill in a mixed party gets diarrhoea for more than one day the pill may not be absorbed and she should assume that she is not protected for that cycle. An unwanted pregnancy in a remote spot will cause much distress!

To complicate the issue the menstrual cycle of both non-takers and takers of the pill may become irregular or stop temporarily whilst travelling and adventuring. This is harmless and needs no medical intervention. Women on the pill are more at risk from thrombosis and thrush.

General Infections

Antibiotics must not be eaten indiscriminately, but if you develop an infection with a high fever and rapid pulse when you are away in the wilds on your own, blind therapy with a broad spectrum bug-killing drug may be justified. Cotrimoxazole *(Septrin, Bactrim)* or Amoxycillin *(Amoxyl)* should be taken for a full five-day course. In malarial areas treatment with Chloroquin Phosphate Tablets 250gm—four immediately, two in six hours, then one twice daily for two more days—must also be considered, even if regular prophylaxis has been taken, and this three-day treatment should abort most attacks. If it does not, then seek medical advice. If there is fever, as well as the above treatment, supportive measures such as rest, shade, frequent sponging with cool water, two Aspirin tablets every four to six hours and an adequate intake of fluids are essential.

Local Infections

Eyes: If the eyes are pink and feel gritty, wear dark glasses and put in chloromycetin ointment. A few drops of Amethocaine will anaesthetise the cornea so you can dig out a foreign body. Homatropine dilates the pupil and relieves spasm but will temporarily blur the vision.

Ears: Keep dry with a light plug of cotton wool but don't poke matches in. If there is discharge and pain, take an antibiotic.

Sinusitis gives a headache (felt worse on stooping), 'toothache' in the upper jaw, and often a thick snotty discharge from the nose. Inhale steam with Tinct. Benz. or sniff a tea brew with a towel over your head to help drainage. Decongestant drops may clear the nose if it is mildly bunged up, but true sinusitis needs an antibiotic.

Throat: Cold dry air irritates the throat and makes it sore. Gargle with a couple of Aspirins or table salt dissolved in warm water; or suck antiseptic lozenges.

Teeth: When brushing teeth is difficult, chew gum. If a filling comes out a plug of cotton wool soaked in oil of cloves eases the pain; gutta percha, softened in boiling water, is easily plastered into the hole as a temporary filling. Hot salt mouthwashes encourage pus to discharge from a dental abscess but an antibiotic will be needed.

Feet take a hammering so boots must fit and be comfortable. Climbing boots are rarely necessary on the approach march to a mountain; gym shoes are useful. At the first sign of rubbing, put on a plaster.

Blisters: Burst with a sterile blade or needle (boiled for three minutes or held in a flame until red hot). Remove dead skin, spray with Tinct. Benz. Cover the raw area with zinc oxide plaster and leave in place for several days to allow new skin to form.

Athlete's Foot can become very florid in the tropics so treat this problem before departure. The newer antifungal creams, eg *Canestin,* are very effective and supersede antifungal dusting powders but do not eliminate the need for sensible foot hygiene. In very moist conditions, eg in the rain forest, on cave explorations or in small boats, macerated feet can become a real and incapacitating problem. A silicone based barrier cream in adequate supply is essential under these conditions, eg *Conotrane* ointment.

Skin sepsis: In muddy or wet conditions most travellers will get some skin sepsis on small wounds. Without sensible hygiene these can be disabling especially in jungle conditions. Cuts and grazes should be washed thoroughly with soap and water or an antiseptic solution; 5 per cent Mercurochrome Aq. dabbed on cuts and grazes is an excellent antiseptic as the skin remains dry (creams will leave it greasy and attract dirt). Other suitable antiseptics are potassium permanganate and gentian violet. Large abrasions should be covered with a vaseline gauze, eg Jelonet or Sofratulle, then dry gauze, and kept covered until a dry scab forms, after which they can be treated daily with Mercurochrome solution and left exposed. Anchor dressings are useful for awkward places, eg fingers, heels. If a cut is clean and gaping, bring the edges together with *Steristrips* in place of stitches.

Sun

The sun can be a stealthy enemy. Sunlight reflects strongly off snow and light-coloured rocks; its rays penetrate hazy cloud and are more powerful the higher you climb. Until you have a good tan, protect yourself with clothing and a hat. An ultraviolet barrier cream *(Uvistat)* screens the skin but with excessive sun it merely acts as fat in the frying process. Rationing sunlight is cheaper and more effective.

If you are planning to travel in hot weather, train for it by exercising in the heat beforehand and/or spending a few sessions in a sauna bath. This way your body will learn to perspire at lower temperatures and the network of capillaries in the skin will increase so that more blood can travel to the skin. Enzymes in the body will also change, allowing you to make more physical effort while producing less heat. On the trail, stop frequently to rest, and drink and eat before you need to so as to replace all the salts necessary to prevent cramps and weakness. Salt tablets, part of the White Raj in a pith helmet image, are needed less than most imagine. In the tropics most people will produce almost salt free perspiration after acclimatisation, especially if conditioning is gone through, and generous salt supplements on food will keep a satisfactory salt level in the blood. In the first week of exertion in the heat it may be reasonable to offer them (eg *Slow Na* tablets one three times daily) but after this it is generally not necessary. However some will feel better if they take them and it is perhaps unfair to deny them this probable placebo response.

Sunburn: Calamine soothes shrimp-pink prickly hot skin; if you turn bright lobster you are severely burnt and should obtain a steroid cream.

Heat exhaustion and heatstroke ('sunstroke'): If you develop a high temperature and feel ill after being in strong sun, cool yourself with cold water sponging or ice packs, take ample fluid, drink slowly, and take Aspirin to lower your temperature and relieve headache. This, together with salt and rest, is the treatment for heat exhaustion, which is a fairly common condition that can occur in or out of the sun, eg after heavy work in shaded but hot and humid conditions. Heat exhaustion can be due to a) simple faints precipitated by heat, b) water loss, c) salt deprivation or d) psychological factors.

Rarely a more serious condition occurs, mainly in elderly or ailing people. On a humid day, an overheated body may attempt to cool by massive sweating, with little effect, for it is the evaporation of sweat that cools, not the sweating alone. Excessive water loss will eventually cause the body's heat regulating mechanism to break down and inhibit any further sweating. The patient's temperature may rise to 105°F or more. Collapse from heatstroke warrants urgent medical help, as there is danger of damage to internal organs and the brain and a 25-30 per cent death rate. Meanwhile keep the patient cool, by immersion in cold water if possible or in a well ventilated place. Try to reduce his temperature and keep it from rising again above 101°F.

Snowblindness is caused by an ultraviolet burn of the cornea, resulting in intense pain and swelling of the eyes. It can be prevented by wearing dark glasses or goggles; horizontal slits cut in a piece of cardboard will do in emergency. Amethocaine drops will ease the pain enough to reach camp. Then put Homatropine drops and chloromycetin ointment in the eyes and wear dark glasses or cover with eye-pads and a bandage if the pain is severe.

Travel Sickness

If you suspect travel sickness could upset you take antihistamine tablets starting one hour before the journey. All of these can make some people drowsy, so if you need to be alert, eg to drive, find one on which you are safe. *Piriton* (Chlorpheniramine) 4gm three times daily is usually as good as any.

Nature's Annoyances

From flies and mosquitoes, bees, wasps, ants and hornets; from fleas, ice and bedbugs; from sea urchins and ellyfish; and from a host of other creepy-crawlies we pray deliverance. Repellent sprays and creams (usually based on Dimethylphthalate) last only a few hours but are essential for those prone to a severe reaction to insect bites, who will generally already be aware of this. All will be bitten, causing fierce reactions and distress in a few. Remove any stings. Calamine cream or otion or *Anthisan* (Mepyramine Malate) cream will help. If distress remains, antihistamine tablets, eg *Piriton* gm three times daily, with Aspirin as painkilling adjunct, should be used. Good hygiene is necessary to prevent arge reactions to multiple bites from becoming affected. If this happens, rest, antihistamine tablets, antibiotics and clean dressings will usually effect a cure.

Anthropods: Lice, fleas and bedbugs are kept at bay by ICI Louse & Insect Powder.

Mosquitoes are usually only a bother at lower altitudes. A net makes sleeping more comfortable but does not guarantee protection from malaria.

Snakes: Clean the area of the bite (not by urinating on it). Try to identify the snake: if possible, kill it and take it with you to the hospital for identification. In the USA, where rattlesnake and cottonmouth snakes prevail, sucking the wound is favoured in a victim who is well covered in fat. To prepare for this, sterilise a knife in a flame, make a cut into each fang mark about ½cm long and ½cm deep. Suck the wound, spitting out the venom, for about 15 minutes. If more than half an hour has passed since the victim was bitten, do not cut or suck the wound, as this may do more harm than good. In South America, Africa and Asia, sucking is in any case useless, since cobra venom is the main hazard and this is not easily absorbed by suction.

Keep the victim quiet, do not move the affected part, eg in the case of a bite on the foot, do not allow him to walk even one step. In any case, the victim should be carried to hospital, if possible, instead of walking. Meanwhile apply a cloth or elastic bandage between the wound and the heart to slow circulation of venom; loosen bandage for a minute or two in every fifteen. Give no stimulants, eg alcohol, as this dilates the blood vessels and accelerates circulation of the poison, but give a sedative.

Local hospitals probably carry a serum against the bites of common local snakes. Polyvalent antivenin is available but it is expensive, difficult to get hold of and itself very hazardous because of the risk of inducing shock. If given it should be injected, but not more than three hours after the bite and *never* unless the features of poisoning develop—despite assurances from some quarters that its use is mandatory. Watch the patient for signs of allergic shock: shivering, rapid heartbeat, low pulse. If ice is available, wrap it in a cloth and pack it around the affected part. In case of infection, give antibiotics. Consider also giving pain-killing medication and antihistamine. If the patient survives the first twenty-four hours after being bitten, he will probably recover, though some deaths do occur after this interval. It is worth mentioning, however, that only 30 per cent of victims of the most venomous snakes die.

Dogs: Rabies exists in most countries with a few fortunate exceptions such as Britain and Australia. To help limit the spread of this awful disease, abide strictly by the anti-rabies regulations and never smuggle animals home from abroad. The new human Diploid cell 'Merriaux' anti-rabies vaccine has completely superseded the old painful multi injection Duck Embryo vaccine but will probably not be available in primitive countries and may not be available in the USA either. Anyone handling bats, small mammals, etc., should have this before departure (cost in the UK c.£20). The British should cease to be dog lovers abroad and should leave well alone any animal displaying abnormal behaviour—especially exceptional tameness. A bite from a dog or any other mammal always warrants a doctor's advice on the prevalence of rabies in the district and the advisability of vaccination for a victim not already so protected. At the very least, an anti-tetanus booster is recommended. If at all possible, capture the animal that has attacked a victim so that it can be tested for rabies. Wash the wound as soon as possible with soap and water and follow this with three per cent solution of hydrogen peroxide. In the absence of water or hydrogen peroxide, wash with any sterile liquid—beer, cold tea or coffee or any carbonated drink will do.

Scorpions: Only a few species have a severely poisonous sting. As with snakes, prevention is better than cure. Carry a stout stick to test the nature of anything you can't identify. Wear thick boots and watch where you put your feet. Before donning clothes and boots in the morning in scorpion territory, ie dry country, shake them out. If bitten, the treatment is rest, analgesics, antihistamines and probably a course of antibiotics. Tarantula-type spider bites come in this category. Whip scorpions are harmless.

Leeches are most troublesome during and shortly after the monsoon in the tropics. You do not feel them bite and may only notice a bootful of blood at the end of the day. Open sandals let you see them early and insect repellent discourages them—a lighted cigarette or salt makes the leech drop off.

Wasp stings—vinegar. *Bee stings*—antihistamine ointment.

Bilharzia is widespread in parts of Africa so avoid swimming in slow-flowing rivers and lakes where the flukes breed.

Poisoning: Try to make the person sick by sticking fingers down his throat. Under ideal conditions, the treatment of choice in children is syrup (not fluid extract) of ipecacuanha when there is a risk of toxicity, provided that treatment is given under medical supervision within four hours of ingestion and that the poison is not corrosive, a petroleum distillate, or an antiemetic. For adults, support of vital functions should be the primary concern of those administering first aid, followed by stomach washout. Though giving a salt solution is no longer the preferred treatment, it will be the best available under most expedition or trip conditions.

Injury

Nature is a wonderful healer if given adequate encouragement.

Superficial wounds: see *Skin sepsis* above.

Deep wounds: Firm pressure on a wound dressing will stop most bleeding. If blood seeps through, put more dressings on top, secured with absorbent crepe bandage, and keep up the pressure. Elevate the part if possible.

On trips to remote spots at least one member of the party should *learn to put in simple sutures*. This is not difficult—a friendly doctor or casualty sister can teach the essentials in ten minutes. People have practised on a piece of dog meat and on several expeditions this has been put to good use. Pulling the wound edges together is all that is necessary; a neat cosmetic result is usually not important.

Burns: Superficial burns are simply skin wounds. Leave open to the air to form a dry crust under which healing goes on. If this is not possible cover with *Melolin* dressings. Burn creams offer no magic. Deep burns must be kept scrupulously clean and treated urgently by a doctor. Give drinks freely to replace lost fluid.

Sprains: A sprained ankle ligament —usually on the outside of the joint —is a common and likely injury. With broad Elastoplast strapping a 'stirrup strapping' can often allow the sufferer

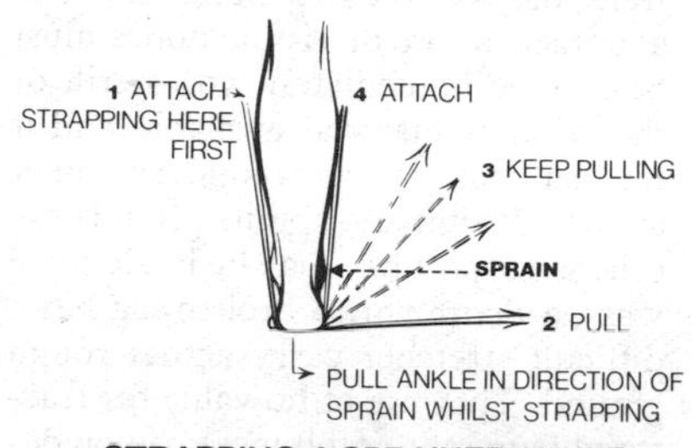

STRAPPING A SPRAINED ANKLE

to continue walking (see diagram). Put two or three long lengths from mid calf on the non-injured side under the foot and, with tension applied and the ankle twisted towards the injured side, attach along the calf on the injured side. Follow this with circular strapping from toes to mid calf overlapping by half on each turn. The first aid treatment of sprains and bruises is immobilisation (I), cold, eg cold compresses (C), and elevation (E) (remember ICE).

If painful movement and swelling persist, suspect a fracture.

Fractures: Immobilise the part by splinting to a rigid structure; the arm can be strapped to the chest, both legs can be tied together. Temporary splints can be made from a rolled newspaper, an ice-axe or a branch. Pain may be agonising and is due to movement of broken bone ends on each other, and needs full doses of strong pain-killers.

The aim of splinting fractures is to reduce pain and bleeding at the fracture site and thereby reduce shock. Comfort is the best criterion of efficiency of a splint but remember that to immobilize a fracture when the victim is being carried, splints may need to be tighter than seems necessary for comfort when he is at rest, particularly for transport over rough ground. Wounds at a fracture site or visible bones must be covered immediately with sterile or the cleanest material available and if this happens start antibiotic treatment at once. Pneumatic splints provide excellent support but may be inadequate when a victim with a broken leg has a difficult stretcher carry across rough ground. They are of no value for fractured femurs (thigh bones). If you decide to take them get the Athletic Long Leg splint which fits over a climbing boot where the Standard Long Leg splint does not.

Pain: Pain killers fall into three strengths for different grades of pain:

Mild: Aspirin (lowers the temperature but can irritate the stomach). Dose: Up to 4 x 300mgm tablets initially, then repeat two tablets four hourly as necessary. Paracetamol is a useful alternative. Dose: Up to 4 x 500mgm tablets, then repeat two tablets four hourly as necessary.

Moderate: Pentazocine *Fortral* is probably the best for parties without a doctor who do not wish to impinge on scheduled drugs regulations. Dose: Up to 4 x 25gm tablets or 2 x 50gm capsules four hourly as necessary.

Strong: Pethidine, Morphine—available on special prescription only and there is a risk of trouble with the law if you are caught with these drugs in your possession. Further, they are potentially hazardous (eg they may depress breathing) and *should be used only by those with previous special instruction in their use.*

Unconsciousness

The causes range from drowning to head injury, diabetes to epilepsy. Untrained laymen should merely attempt to place the victim in the coma position—lying on his side with the head lower than the chest to allow secretions, blood or vomit to drain away from the lungs. Hold the chin forward to prevent the tongue falling back and obstructing the airway. Don't try any fancy manoeuvres unless you are practised, as you may do more harm than good.

All unconscious patients from any cause, particularly after trauma, should be placed in the coma position until they recover. This takes priority over any other first aid manoeuvre.

Fainting: Lay the unconscious person down and raise the legs to return extra blood to the brain.

Exposure/Exhaustion Syndrome

Hypothermia occurs when the temperature of the central core of the body falls below about 35°C owing to the combined effect of wind, wet and cold. Exhaustion and low morale worsen it. If someone behaves in an uncharacteristic manner—apathetic, stumbling, swearing, uncontrolled shivering—be on your guard. He may suddenly collapse and die.

First priorities are to stop and shel-

ter the victim in a tent, lean-to or polythene bag, and to re-warm him by skin-to-skin contact, by dressing him in dry clothing and by putting him in a sleeping bag, in close contact with someone else if possible. Then give him hot drinks, but no alcohol. If his condition does not improve you may have to call help and evacuate him by stretcher.

Those travelling in areas where exposure is likely should read up the features and treatment of this very real hazard (eg James Ogilvie: 'Exhaustion, Exposure' in *Climber and Rambler,* Sept./Oct. 1977).

High Altitude Ills

Up to 12,000ft you have little to fear—no more than on an ordinary mountain walking holiday. If you are not shaping up too well, reconsider the wisdom of climbing higher for you are entering the realm of the high, thin, cold, dry air. Slow ascent is the secret of easy acclimatization to altitude. Breathing and heartbeat speed up; a thumping headache and nausea make you feel miserable. At night, sleep is elusive (*Mogadon* one tab.) You may notice a peculiar irregularity in the pattern of breathing (Cheyne Stokes respiration) when, for a short period, breathing appears to have stopped and then gradually increases in stepwise fashion until it eventually falls off again. The normal output of urine may be diminished and very dilute.

The unpleasant symptoms of acclimatization usually pass off in a few days, but they may develop into Acute Mountain Sickness. This rarely starts below 15,000ft so is unlikely in the Alps, but may occur in Africa, the Andes or the Himalaya.

If you begin to feel more ill than you would expect for your own degree of fitness and acclimatization, go down quickly and stay down rather than battle on for glory—and end up under a pile of stones on the glacier. Acute Mountain Sickness can quickly develop into High Altitude Pulmonary (lung) Oedema, or Cerebral (brain) Oedema, (known in the USA as HAPE—high altitude pulmonary edema, and HACE—high altitude cerebral edema). This is swelling due to abnormal retention of water. Women are more susceptible in the days before their periods. This is a potentially lethal disease, the cause of which is not understood, but it can affect all ages, the fit and unfit, those who have risen quickly and those who have not.

If someone suddenly feels, and looks, puffy in the face, goes blue round the lips, has bubbly breathing and even pink frothy sputum, evacuate him urgently to a lower altitude. Oxygen (if available) and a diuretic drug Frusemide *(Lasix)* may help to clear water from the lungs, but they are no substitute for rapid descent which has a miraculous effect. Those who have suffered once are likely to do so again, and should therefore beware.

Thrombosis: Persistent deep calf tenderness and slight fever and pain—more than muscular ache—may indicate a vein thrombosis. Women on the pill are especially at risk. You should rest, preferably with the legs bandaged and elevated, and start an antibiotic. This is a serious illness, so descend and seek medical advice.

Piles, which commonly trouble people at high altitude, are probably due to raising the pressure inside the abdomen by overbreathing while carrying heavy loads. A haemorrhoidal suppository *(Anusol)* gives temporary relief.

Dry cough is eased by inhaling steam. Codeine Phosphate 15mg dampens it. In a violent bout of coughing, you can fracture a rib. The agony may make you think you have had a heart attack but the chances of this are slim.

Frostbite should not occur if you are clothed properly and take commonsense precautions. If you get very cold, rewarm the part quickly against warm flesh (someone else's if possible). Do *not* rub it or you will damage the skin and cause further wounding which nay become infected. Drugs which di-ate the blood vessels (Vasodilators) nave no specific action against frostbite lthough they make you feel a warm low inside. This can be very dangerous as you are losing heat from the rest of your body and you may be tipped nto exposure.

If a foot is frozen, it is better to walk n it back to a low camp where you can apidly rewarm it in water 42-44°C. hereafter the victim must be carried.

Dehydration

The sedentary dweller from a moder-te climate may well find that tropical emperatures plus the need for a high vork rate will cause weakness and sub-ptimal performance due to dehy-ration, despite an increased fluid in-ake. In deserts, in small boats and also t high altitude dehydration can be a eal risk.

Owing to immobility from any cause nd also illness, particularly if fever or iarrhoea are present, the fluid intake nay fall to a level where dehydration an develop. In a temperate climate round 1500ml (2.6 pints) of fluids aily are adequate but working hard in the tropics may cause this volume of perspiration in just 1½ hours.

Dehydration is best expressed as a per cent loss in body weight, 1-5 per cent causing thirst and vague discomfort, 6-10 per cent causing headache and inability to walk and 10-20 per cent delirium leading to coma and death. Drinking sea water or urine in survival situations causes only a more rapid deterioration.

To plan expedition fluid requirements, assume that an average unacclimatized man working out of doors in extreme hot/wet or hot/dry conditions will drink 7-9 litres (12-16 pints) of fluids per day. 'Voluntary dehydration', symptomless initially, is common if drinking fluids are not within easy reach and palatable.

In temperate climates the average diet contains an excess of salt which is excreted in sweat and urine. Over two to three days in the tropics adaptation reduces the amount of salt in sweat and urine to negligible levels. During the first two weeks dehydration may be accompanied by salt depletion so that supplements are of value, but generally the treatment is simply rest and an increased fluid intake until the urine volume is adequate (around one litre or 1½ pints per day) and visibly normal or pale in colour.

In early days of heat exposure a definite self discipline in achieving a sufficiently high fluid intake is necessary. Those treating ill patients must watch and encourage this aspect of their treatment. The most obvious features of marked dehydration are sunken eyes and a looseness of the pinched skin. If these cannot be corrected by oral fluids a serious situation is developing and medical aid should be sought as intravenous fluids are likely to be needed.

Immersion in Water

Prolonged immersion in all but tropical waters carries a life threatening hazard of hypothermia which is probably a bigger risk than drowning. The amount of subcutaneous fat will affect survival time considerably but a naked man of average build will be helpless from hypothermia after 20-30 minutes in water at 5°C (41°F) and 1½-2 hours in water at 15°C (59°F).

If thick clothing is worn these intervals will be increased to 40-60 minutes at 5°C (41°F) and 4-5 hours at 15°C (59°F). Thus if a ship is to be abandoned or a small boat is threatened, warm clothing should be donned with a waterproof outer suit if one is available. Cold can cause dilation of blood vessels in the hands and feet and thus increase heat loss here so that mitts and footwear are also desirable as is protection for the head and neck.

Some flotation aid such as a lifejacket, wreckage, an upturned bucket or even air trapped in a waterproof coat should be sought. When in the water float quietly instead of swimming. With the stress of cold water combined with a threatening situation swimming is a normal reaction but, because of its stirring effect on the surrounding water and despite the heat it generates, swimming will merely accelerate loss of body heat. Swim only if no flotation aid is available, if threatened by a sinking ship or if rescue by others is not possible and land is within reach. Whilst waiting for rescue float quietly as all exercise will accelerate cooling.

Suggested Medical Kit

For British travellers our normally generous National Health Service does not supply medication for trips abroad and all drugs needed for this purpose have to be purchased—usually at a considerably greater expense than anticipated. Friendly doctors, usually General Practitioners, may, if feeling generous, defray this expense by donating drug 'samples'. Travellers should otherwise ask their doctor to sign the entire list as a private prescription. American travellers should also obtain drugs on prescription; in some cases it is illegal to buy or possess them without prescription. Take with you the drug prescriptions supplied with the drugs or photocopies of their descriptions (with indication, recommended dosage, etc.). In the USA all drugs are listed in *The Physician's Desk Reference*, published annually. American travellers within easy reach of Mexico may consider buying drugs there: many more drugs are available without prescription than in the USA and they are considerably cheaper.

Leader's Kit (for 4-6 persons for 2 months)

Basic

Dressings	1 Elastoplast dressing slip 36″
	50 Assorted size Elastoplast dressings
	1 Zinc Oxide strapping plaster 1″ x 5 yards
	2 Elastoplast strapping 3″ x 5 yards
	1 pack Steristrip ¼″ x 4″ – to hold small wounds together
	1 Bandage Crepe 3″
	1 Bandage Cotton 2″
	1 Bandage Triangular
	12 Gauze squares plain
	6 Melolin dressings 4″ x 4″ — place next to dry wounds
	4 Jelonet or Sofratulle

	Dressings 4" x 4" — place next to moist wounds
	1 small cotton wool pack (compressed)
	1 wound dressing No. 15
	1 Netelast Dressing (Head size) — to retain dressings
Cleansing	1 bar soap
	50 ml Dettol or TCP solution — to wash wounds
Instruments	1 pr. each Scissors blunt/sharp
	1 Forceps — blunt end
	1 Scalpel blade (sterile)
	4 Safety pins
	2 Thermometers (1 'subnormal' if low temperature likely)
	Disposable syringes and needles if injectable drugs included
	Paper and ballpoint pen to note drugs given
	2 x 3.0 chromic catgut sutures

Medicines

Pain Killers (mild) see text	100 tabs. Soluble Aspirin 300 mgs.
	50 tabs. Paracetamol 500 mgs.
(moderate) see text	20 Caps. *Fortral* 50 mgs. (Pentazocine)
(strong)	10 tabs. Pethidine 50 mgs.
	or 4 Inj. Pethidine 100 mgs—See warning in text.
Antibiotic	40 Cotrimoxazole (*Septrin, Bactrim*)-2 tablets night and morning for five days
	2 ampoules Triplopen
	2 ampoules sterile water—
	1 injection i.m. Triplopen and sterile water to patient with an infection who can't swallow.
Antihistamine	40 tabs. Promethazine (*Phenergan*) 25 mgs. — 1 at night
	50 tabs. Chlorpheniramine (*Piriton*) 4 mgs. — 1 3xdaily as needed
Sleeping	20 tabs. Nitrazepam (*Mogadon*) 5 mgs. — 1 or 2 at night
Sedation	20 tabs. Diazepam (*Valium*) 2 mgs. — 1 3xdaily as needed
Diarrhoea	½lb Kaolin powder — 2 teaspoonsful in water as necessary
	40 tabs. Codeine phosphate 30 mgs. — 2 4xdaily as needed
	100 tabs. *Lomotil* (Diphenoxylate) — 4 immediately then 2 4xdaily as needed.
Constipation	10 tabs. *Senokot* — 1-4 at night as needed
Indigestion	30 tabs. *Aludrox* or Magnesium Trisilicate — 1 or 2 anytime
	10 tabs. *Maxalon* — 1 up to four times daily as necessary for vomiting
Salt	20 tabs. *Slow Na.* (Slow Sodium) 1 3x daily for 2 days
Anti-worm	4 sachets *Pripsen* (Piperazine) — 1 sachet, repeat in 10 days
Eyes and Ears	5 ml *Neo-cortef* eye/ear Drops — apply four times daily as necessary
	2 mimims amethocaine — to anaesthetise the eyes
	2 mimims homatropine — to rest an inflamed eye
Teeth	Oil of cloves — apply to aching tooth
	1 gutta percha temporary filling
Skin: Sun	2 tubes *Uvistat* 1 *Lipsyl* 1 tube Calamine cream apply when necessary
Insects	3 tubes *Flypel* or equivalent—apply four above items
	3 tubes *Anthisan* cream — apply 3xdaily
Fungal infections	1 tube *Canestin* ointment — apply 3xdaily
Barrier cream	1 60g *Conotrane* ointment — apply when necessary
Bruises and sprains	1 x 14g *Lasonil* ointment — apply 3xdaily
Antiseptic	1 10ml 5% Mercurochrome Aq. — apply daily

Infestations	1 ICI Louse and Insect powder — dust with this daily as necessary
Haemorrhoids	10 *Anusol* suppositories — use twice daily
Diuretic (high altitude only)	20 tabs. *Lasix* (Frusemide) 40mgs. — 2 tabs 4 hourly
Anti-Malarials	
Preventative	tabs. *Paludrine* (Proguanil) — 1 daily
or	tabs. *Daraprim* (Pyrimethamine) — 1 weekly from entering malarial area until 28 days after leaving the area.
Treatment	40 tabs. Chloraquine phosphate — 4 immediately, 2 in six hours then one twice daily for 2 more days
Directory	IAMAT or INTERMEDIC or other directory of English-speaking doctors.
Individual Kit	
Dressings	1 Plaster strip 12" x 2½"
	1 Zinc oxide 1" x 3m
	4 *Melolin* 2" x 2"
	1 Crepe bandage 3"
	1 Razor blade
	Paper and ballpoint pen
	10 *Fortral*
	20 Phenergan (25 mg)
Diarrhoea	60 *Lomotil*
Sun	1 tube *Uvistat*
	1 *Lipsyl*
Throat/Skin/Eyes/Ears	12 Aspirin
	1 tube (4 gm) *Neo-cortef* eye/ear drops
	100 Water purifying tablets

If away from base for long, appropriate quantities of antibiotics, Chloraquine phosphate, and *Paludrine* or *Daraprim*.

Personal drugs sufficient for the whole trip or a list of sources from which they can be obtained in the countries to be visited.

Store your medical kit in a waterproof and dustproof container that is lockable and kept locked. Transfer liquids and tablets to plastic bottles and attach labels with Sellotape as stuck-on labels can come off in wet heat. Certain items may be stored in plastic pouches with zips. If storing the kit in a vehicle, keep it away from the floor, which is likely to be too hot, and keep it easily accessible at all times. The doctor or medical officer of the party should be familiar with the contents of the kit, knowing instantly the exact location and application of each item. The kit therefore needs to be kept permanently in impeccable order. Where possible one person should issue supplies, as open house will encourage rooting about, the opening of new supplies when part used ones are available and general chaos. The same person should oversee stocks of drugs and dressings in case more need to be obtained.

Coming Home

On your return from a long trip, have a check-up — urgently, if you still have symptoms of illnesses suffered during your absence. Report to your doctor where you have been and any relevant medical incidents. Blood, urine or stool samples may need to be tested by a specialist in tropical medicine or other appropriate field.

Drugs referred to — Alternative names used in other countries

The person in charge of the medical kit should understand the appropriate useage and dosage of all the drugs carried. It also helps if brief notes on this are on the labels of all drug containers.

Pharmaceutical Name	*Commonest UK Brand Name*	*Commonest Brand Names in other countries*
Aluminium Hydroxide	Aludrox	Maalox (USA) Actal (Various countries)
Chloraquine Phosphate	Avlaclor	Aralen Hydrochloride (USA)
Chlorpheniramine	Piriton	Allertab, Chlormine, Histadur, Telmin (USA) Chlortrone (Canada)
Chlorimazole 1%	Canestin Cream	Lomotrin (USA)
Codeine Phosphate	Codeine Phosphate	Paveral (Canada)
Cotrimoxazole & Sulpha	Septrin or Bactrim	Septra (USA)
Diazepam	Valium	E-Pam, Serenach, Vival (Canada)
Dimethicone & Hydrargaphen	Conotrane Oint.	Versotrane (Canada)
Diphenoxylate & Atropine	Lomotil	Diarsed (France)
Frusemide	Lasix	Lasilix (France)
Heparinoid & Hyaluronidase	Lasonil oint.	Hyazine (USA) Wydase (Canada)
Mepyramine Maleate	Anthisan Cream	Antical Cream (USA) Statomin Cream (USA)
Mercurochrome Aq 5%	Mercurochrome Aq 5%	Mercurescein (Canada)
Metoclopramide	Maxolon	Maxeran (Canada) Reglan (USA)
Mexenone	Uvistat Oint.	Uvicone (Australia)
Neomycin & Hydrocortisone	Neo-Cortef Eye/Ear Drops	Neobiotic (USA) Herisan Antibiotic (Canada)
Nitrazepam	Mogadon	Remnos (Various countries)
Paracetamol	Paracetamol	Capitol, Dolanex, Febrigesic, Nebs (USA) Tempra (Canada)
Penicillins (Mixture of 3)	Triplopen Inj.	Benapen (S. Africa)
Pentazocine	Fortral	Talwin (Canada & USA) Sosenyl (Various countries)
Pethidine	Pethidine	Demeral (USA & Canada) Physadon (Canada)
Piperazine	Pripsen	Antepar (USA) Piperzinal (Canada)
Prognamil	Paludrine	Paludrinal (Canada)
Promethazine	Phenergan	Promethopar (USA) Zipan (USA) Histantil (Canada)
Pyrimethamine	Daraprim	Fansidar, Maloprim (Various countries)
Slow release sodium chloride	Slow Na	Neutrasil (Canada) Trisomin (USA)
Standardized Senna Extract	Senokot	Colonorm (Germany)
Vaseline Gauze	Jelonet Gauze or Sofratulle Gauze	Petroleum Gauze (USA)

Seasonal Travel

by Richard Harrington

An inexperienced traveller may not think too much about seasons of the year before he takes off. He knows it's always hot in Indonesia and cold in the Arctic. The seasoned traveller, on the other hand, plans his trip very carefully around certain times of the year.

Airlines, hotels and tour operators have off-peak seasons for very good reasons, and adjust their prices downwards in these periods to attract people who, if they could afford it, would prefer to travel in high season.

Let's look at the factors that determine the seasons for outside visitors. An obvious one is weather: rainfall, humidity, temperature. Climates that are no trial to local peoples may have devastating effects on those ill-adapted and arriving from more temperate regions.

For reasons still little understood, certain tropical regions of the globe are subject to seasonal monsoon rains, cyclones, hurricanes and tornadoes. For most people, these are non-travel seasons. On the other hand, travel deals may be so attractive in these periods that you may decide to make the trip because you know you could never afford it at any other time. Or you may be a surfer and choose to travel in the stormiest seasons of the year, knowing that these are usually the times for the biggest and best waves.

But weather isn't the only factor. Big game and birdlife may be more spectacular in certain seasons. Endemic diseases may be more easily caught at certain times of the year. In Arctic Canada the air is clear and crisp and cold in winter. In summer it warms up, but you may get eaten alive by mosquitoes.

The going may be physically impossible, or almost so, in certain months. Few have dared to move on the Arctic Ice Cap during the continuous night of freezing winter. Yachtsmen crossing the Atlantic from west to east avoid a winter crossing on the northern route, and those sailing on the Pacific circuit from North America to Hawaii, Tahiti and New Zealand try to make the trip from Papeete to New Zealand before the summer cyclone season begins. In the jungles of the African West Coast and of 'Africa's armpit'—Cameroon, the Congo Republic, Zaïre, etc—the going is very rough and extremely unpleasant during the rainy season, from May to August. The best time to start a Trans-Africa crossing from London would be September/October when the height of the Sahara summer and the rainy season farther south have both passed. Autumn and, better still, spring are the best times for a Sahara crossing. No European traveller would willingly expose himself to the Sahara in midsummer.

Man too can create seasons which affect the traveller. In Muslim countries, the festival of Ramadan lasts several weeks and can inconvenience visitors. If you plan to visit an Arab country in the August/September period, bear in mind that local people will do without food from sunup to sundown. Various services will be disrupted or unavailable. Meals will be hard to obtain outside tourist areas. And you may get woken before dawn as whole families noisily eat a hearty meal before the sun puts an end to the revelry. A meal shared with an Arab family at sunset during Ramadan is a treat to be remembered. In the Hadj season, when Muslims from West Africa to Indonesia flock to Mecca on pilgrimage, air services along the necessary routes are totally disrupted for regular travellers.

In the West Indies, Guyana and Brazil, Carnival is a time of the year when normally poor people are given the chance to forget their worries and feel rich. Cities like Port of Spain in Trinidad and Rio de Janeiro in Brazil have become magnets for tourists, but are to be avoided if you have business to do there and are not interested in drinking, singing and dancing in the streets.

Major festivals in Europe and elsewhere always attract culture-seekers and other visitors. Those who are not interested in such occasions but don't plan well enough to avoid them can end up finding no place to stay, getting 'bumped' off overbooked aircraft and paying over the odds for everything because all the prices in town have been doubled for a week. So for certain countries, especially in Central and South America, a look at the festivals calendar should be part of your planning.

Jet lag combined with 'climate shock' can really hit you hard going from a temperate country in winter direct to the tropics. If you can afford it, and are going from London to Fiji in January, you would be advised to make a stop and get a tan halfway, say in Hawaii, before reaching the all-out heat of the tropical South Pacific in mid-summer.

A little known fact is that since hot air is thinner than cold and hot air rises, the air at altitude will be even thinner when hot than when cold. While the heat itself, in the high altitude cities like Mexico City, La Paz, Addis Ababa or Nairobi, is unlikely to be overwhelming, the rarefied air may leave you exhausted for several days if you don't take it easy when you get there. If you're susceptible to altitude (some people are far more susceptible than others, and fitness is no indicator), remember that winter is probably the best time to go. The heat is also likely to be less overwhelming then.

Before you go anywhere, look carefully at the temperature, humidity, altitude and rainfall you might encounter. See *Climate and its Relevance to Travel.* Read a *geography* book about the place you'll visit. You may learn things the tourist brochures and propaganda guidebooks won't tell you for fear of putting you off.

Checking the seasons will affect your choice of clothes for the trip, the amount of money you take, maybe even the choice of film for your camera —slow speed Kodachrome for bright summer pictures, high speed Ektachrome for not-so-bright winter conditions. It can also affect your shopping. You won't find any trouble getting summer swimwear and suntan lotion in London in January, but try buying swimwear in mid-winter in some other parts of England. Not too easy, which is why the airport shops do a roaring trade. In the end, much of it comes down to commonsense—and your budget.

Air Travel and Your Biological Clock

by Richard Harrington

Even the healthiest of us know what a tiring experience it can be to fly across time zones in a modern jet aircraft. A little knowledge about how our bodies function around the clock will make it easier to understand what exactly happens to our metabolism when we subject ourselves to the severe stress of long-haul flying, a process we may not be able to avoid, but one in which we may be able to reduce the negative effects with a few careful precautions.

Various functions in our bodies each follow their own changing rhythms around the clock, all of them ultimately synchronised with each other over a period of 24 hours. Our heart rate and breathing rate change on a circadian (literally 'around the day') basis, reaching their slowest point during sleep, when our body temperature also drops. At the same time, our ECG and EEG patterns show marked differences from those of our waking state. Our bodily functions respond differently in different cycles—the day/night cycle, the wake/sleep cycle and the female menstrual cycle to name only the most obvious. These cycles affect our appetite, our thirst, our intellectual and our emotional performance.

Before November 18, 1883, there were no different time zones dividing up the world into 24 sections between the two poles, and at that time no one travelled fast enough to notice the effect on the human body of moving latitudinally around the globe. It is doubtful, in fact, whether it is possible to travel fast enough on the ground for changes in our biological subsystems to be measurable, though doubtless slight changes do occur when for example someone crosses the US non-stop by train.

The reaction of one individual's body to a long journey by jet will vary considerably from another's, and the state of the individual's health and his age have a lot to do with his ability to withstand the stress of long distance flying across time zones. Most adults will still be able to notice changes in their own rhythms, especially the waking/sleeping cycle, three or four days after a jet trip that has crossed at least five time zones non-stop. Physiological tests can usually measure a degree of discoordination in certain cycles a week after such a flight, as the body still struggles to bring all its rhythms back into complete harmony in a pattern suited to the new time zone.

Research has shown that to fly east to west (with the sun) is more stressful than to fly west to east. The ability of individuals to withstand flying stress varies enormously, and age and health are not the only factors involved in this variation. Some individuals suffer very easily from jet lag, while others experience relatively little fatigue after the same long journey. Experience and frequent travel don't inure the traveller to jet lag, so that a frequent air traveller will actually find his ability to withstand travel fatigue diminishing over the years as he gets older and his health declines.

Jet lag will manifest itself in different ways on arrival. Efficiency will be inhibited, reaction time and decision time will slow down, urinary output will be affected, appetite and sleep patterns will be altered. Jet lag cannot be cured by popping a few pills, but one thing you can do to help yourself is to leave on your trip fully rested, to try not to get up earlier on the day of departure than you would normally, and to rest for 24 hours (even if you can't sleep) on arrival before exerting yourself. Remember that if you fly up or down within the same time zone, your biological clock will not be affected. It is crossing into another time zone, and readjusting your bodily time cycles within the new time zone, that is so exhausting. A point worth remembering is that it takes less time for your body to readjust on returning 'home', than it does on the outward trip.

Obviously the more time zones you cross, the more your circadian rhythms will be affected. So let's take a look at the variable factors that can affect the extent to which you are affected: (i) time of arrival and departure; (ii) length of flight; (iii) duration of flight; (iv) length of stopovers, if any; (v) direction of flight; (vi) stress of a rough/crowded flight; (vii) comfort of flight (1st class or economy class, width of seats, chance to stretch out on adjoining seats); (viii) your age and physical condition; (ix) climatic conditions on arrival and ease of acclimatisation.

Flying is physically a lot more stressful than a lot of people realise. And there's more to the problem than time zones. Modern jet aircraft are artificially pressurised at an altitude pressure of around 10,000ft. That means that when you're flying at an altitude of say 40,000ft in a Boeing 747, the cabin pressure inside is what it would be if you were outside at a height of 10,000ft above sea level. That means if you normally lived at an altitude of 10,000ft and took off from your local airport at an altitude of 10,000ft, you'd feel very much at home in the air. However most people live a lot closer to sea level, and to be rocketed almost instantly to a height of 10,000ft (so far as their body is concerned) takes a considerable amount of adjustment. Fortunately the human body is a remarkably adaptable organism, and for most individuals, this experience is stressful but not fatal.

You may ask why cabin pressure is equivalent to 10,000ft and not to normal sea level pressure. The answer is that, when flying at altitude, a modern jet with a sea level cabin pressure would have to have extremely strong (and therefore heavy) outside walls to prevent the difference between inside and outside pressure from causing the aircraft walls to rupture in mid flight. At present, no economically viable lightweight material is available that is strong enough to do the job. Another problem is that if there were a rupture at say 45,000ft in an aircraft with an interior pressure equal to that at sea level, there would be no chance for the oxygen masks to drop in the huge sucking process that would occur as the air inside emptied through the hole in the aircraft. A 10,000ft equivalent pressure at least gives passengers and oxygen masks a chance if this occurs.

Inside the cabin, humidifiers and fragrance disguise all the odours of large numbers of people in a confined space. On a long flight you're breathing polluted air. You've been shot up at high speed from sea level to the top of a

10,000ft mountain, and at the end, you're shot down again equally abruptly. What can you do to help your body survive such an onslaught? First you can loosen your clothing. The body swells in the thinner air of the cabin, so take off your shoes, undo your belt, tilt your seat right back, put a couple of pillows in the lumbar region at the small of your back and one behind your neck, and whether you're trying to sleep or simply rest, cover your eyes with a pair of air travel blinkers (ask the stewardess for a pair if you haven't brought any with you). Temperatures rise and fall notoriously inside an aircraft, so have a blanket ready over your knees in case you nod off and wake up later to find that you're freezing. Japan Airlines and Singapore International Airlines have recently introduced *couchettes* on certain routes. These are a great asset, but they're expensive—usually the first class fare plus a supplement. When I look at all that space wasted over passengers' heads in a Boeing 747, and all those half empty hand baggage lockers, I often wonder why the aircraft manufacturers don't forget about the propriety of cabin interiors and arrange things so that comfortable hammocks can be slung over our heads for those who want to sleep—or better still, small *couchettes* in tiers like those found in modern nuclear submarines. Personally I'd prefer such comfort, whatever it might do to the tidiness of the cabin interior. After all, air travel is now just another form of mass transportation, and is anyone really concerned with appearances on a long haul flight any more?

It's tempting on a long flight to feel you're not getting your money's worth if you don't eat and drink everything going. Stop and resist the temptation, even if you're in first class and all that food and drink seems to be what most of the extra cost is all about. Most people find it best to eat lightly before leaving home and little or nothing at all during the flight. Foods that are too rich or spicy, and foods that you're unaccustomed to, will do little to make you feel good in flight. Neither will alcohol. Some people claim that they travel better if they drink fizzy drinks in flight. That may be true for certain individuals. I prefer to take along a bottle or two of pure fruit juice and drink that during the flight in small doses.

Walk up and down as frequently as possible during a flight to keep your circulation in shape, and don't resist the urge to go to the toilet (avoid the queues by going before meals). The time will pass more quickly, and you'll feel better for it if you get well into an unput-downable novel before leaving home and try to finish it during the flight. This trick always works better than half-heartedly flicking through an in-flight magazine.

You may try to find out how full a plane is before you book, or choose to fly in low season to increase your chances of getting empty seats to stretch out on for a good sleep. If you've got a choice of seats on a stretched DC8, for example, remember that there's more leg room by the emergency exit over the wings. On the other hand, stewardesses tend to gather at the tail of the plane on most airlines, so they try not to give seats away there unless asked. That means you may have more chance of ending up with empty seats next to you if you go for the back two rows (also statistically the safest place in a crash).

You might travel with your own pillow, which will be a useful supplement to the postage-sized pillows supplied by most airlines.

Finally, if you're going to sleep in flight, put a 'DO NOT DISTURB' notice by your seat and pass the chance of another free drink or face towel every time your friendly neighbourhood stewardess comes round. You probably won't arrive at the other end raring to go, but if you've planned it wisely to arrive just before nightfall, and if you take a brisk walk before going to bed, you just might get lucky and go straight to sleep without waking up on home time two hours later.

Food and Water: Advice for The Long Distance Traveller

by Ingrid Cranfield

The following notes are for travellers who intend to visit countries that are off the beaten track, and who will therefore need to draw on resources to be found *en route*. Wilderness backpackers, who can buy food in advance and carry with them much of what they need for the duration of their trip, should turn to Mary Schantz's article *Food and Cooking on the Move*.

Here are some pointers for buying food in the countries of the world where western standards of hygiene and provision of commodities do not apply.

Canned, powdered and dried foods are usually safe to eat, though, because they are usually imported, they are more expensive than at home, and especially so outside major cities.

Staples such as flour, grains and cooking oils are nearly always safe.

Fresh foods: Meat, poultry, fish and shellfish should look and smell fresh and be thoroughly cooked, though not overcooked, as soon as possible after purchase. They should be eaten while still hot, or kept continuously refrigerated after preparation. Clean fish yourself if possible. Eggs are safe enough if reasonably fresh. Milk may harbour disease-producing organisms. The 'pasteurised' label in underdeveloped countries should not be depended upon. For safety, if not ideal taste, boil the milk before drinking. (Canned or powdered milk may generally be used without boiling for drinking or in cooking). Butter and margarine are safe unless obviously rancid. Margarine's keeping qualities are better than those of butter. Cheese, especially hard and semi-hard varieties and aged cheeses with a rind, are normally quite safe; soft cheese is not so reliable.

Vegetables for cooking are safe if boiled for a short time. See that on vegetables for peeling and fruits with a peel, the peel is intact. Wash them thoroughly and peel them yourself if you plan to eat them raw.

Moist or creamy pastries should not be eaten unless they have been continuously refrigerated. Dry baked goods, such as bread and cakes, are usually safe even without refrigeration.

Always look for food that is as fresh as possible. If you can watch livestock being killed and cooked or any other food being prepared before you eat it, so much the better. Don't be deceived by plush surroundings and glib assurances. Often the large restaurant with its questionable standards of hygiene and practice of cooking food ahead of time is a less safe bet than the wayside vendor from whom you can take food cooked on an open fire, without giving flies or another person the chance to recontaminate it. Before preparing bought food, always wash your hands in water that has been chlorinated or otherwise purified, e.g. with iodine. In restaurants, well-cooked meat, fish, poultry, vegetables and eggs, cheese, unpeeled raw fruit, coffee, tea and bread are nearly always safe; so are fruit drinks if the fruit is pressed in front of you. Cold foods, especially those that can't be peeled, such as salads, are a risk, as are any foods that have been left standing for some time after cooking, when they will be a breeding ground for microbes. Ice cream is especially to be avoided in underdeveloped countries.

Water: Assume all water is unsafe for drinking unless you are perfectly sure it is otherwise. You have no natural immunity to local diseases carried in the tap water: dysentery, typhoid and cholera among them. Hotels and restaurants and even foreign airlines almost all use ordinary local water which could be contaminated, however much the owners and staff tell you otherwise. If it is hot enough to make you withdraw your hand, it is probably safe.

Water from streams, lakes and other untreated sources should be assumed unfit to drink except in high mountain areas where there is no habitation upstream. In desert and semi-arid areas, you may commonly find sulphurous streams which have water that is unpleasant to taste but not harmful, and rarely find arsenic springs, which are very much the reverse! Sea water is undrinkable unless distilled, but perfectly all right for cooking.

You can purify water for drinking by boiling it vigorously for at least ten minutes or by using water purification tablets or solutions. Pharmacists can usually supply tablets that purify with chlorine, such as Halazone, or those that release iodine, such as Potable-Aqua. Tincture of iodine (2 per cent concentration) is equally effective, though more troublesome to use. One drop added to a glass of water, the mixture swilled around to wet all surfaces that you may touch with your lips, and left to stand for twenty minutes will render clear water safe for drinking. (Strain first and double the dosage for cloudy water.) For very long term use, avoid iodine, which can cause a temporary, if harmless, enlargement of the thyroid gland. Another method is to add two drops per quart of ordinary bleach (5.25 per cent solution of hypochlorite) to clear water (twice as much for cloudy water), agitate and leave it to stand for thirty minutes.

Water purifying devices usually act only as filters, which improve the water quality but don't thoroughly sterilise it. Alcohol does not sterilise water for drinking, though undiluted alcoholic beverages such as wine and beer are normally safe.

Water in the form of carbonated bottled drinks is usually safe, if bottled, diluted or transferred to other containers in large reliable plants. Open them yourself or have them opened in front of you. Even this isn't a foolproof guarantee, because unscrupulous vendors have been known to water such drinks as Coca-Cola and sell them in resealed bottles. If this story makes you nervous . . . don't even buy carbonated drinks in a Third World country. Uncarbonated drinks can be dangerous, as the sugar content permits rapid multiplication of bacteria. Don't buy drinks from a jug.

Ice is no safer than the water from which it is made. Do not boil mineralised water as the taste is unpleasant and the process may only concentrate dangerous minerals. Once purified, water remains safe indefinitely as long as there is no chance for recontamination from the outside.

For brushing your teeth, use for preference water that has been made safe or some carbonated drink. For washing and bathing, tap water is usually sufficiently safe. Swimmers should consider the possibility that water in rivers, lakes, ponds and rice paddies, drainage ditches and irrigation canals and even oceans near the mouths of rivers may be inhabited by a freshwater snail that is infested by a parasitic flat worm. Bathing in infested water (or falling into infested water and failing to towel yourself dry quickly) may cause these parasites to enter the bloodstream, where they can do untold permanent damage to internal organs. The disease is called schistosomiasis, bilharzia or snail fever and the remedy is both unpleasant, involving toxic and potent medicines, and uncertain.

To check whether the parasite is present, consult the country's public health agency, *not* its tourist offices.

Hookworms, leeches and other snail-borne parasites picked up from water or moist ground may cause a variety of other ills from the mild to the rather serious. Leeches can be removed by rubbing them with salt or by touching them with a lighted cigarette or match. Don't pull them off, as this will only increase the bleeding.

Last word: Take reasonable care, but don't get hysterical about precautions with food and water. Expecting to stay healthy is half the battle; and your vaccinations will make strong allies.

Food on the Move: Planning, Packing and Preparation

by Mary Schantz

'Good food even makes up for rain and hard beds. Good fellowship is at its best around good meals.' Whoever so aptly said this really said a mouthful. Wholesome, mouth-watering food, and plenty of it, can make even the worst of camping conditions more bearable. But no matter how fantastic the scenery, soft the beds or fine the weather, the enjoyment of any outing wanes quickly when the food is lousy. The problem is: How do you insure you'll have good food on your trip off the beaten track?

There is no one 'right' menu for a camping trip. We all have slightly different tastes in food. Besides, there is an almost endless number of menu possibilities. The choice is yours. So, what should you pack along? Here are a few points you'll want to consider when choosing the right foods for you: *weight, bulk, cost per pound:* Obviously, water-weighted, tin-plated canned goods are out. So are most perishables. Especially if you're going to be lugging your pantry on your back, you'll want only lightweight, long-lasting, compact foods. Some of the lightest, of course, are the freeze-drieds. You can buy complete freeze-dried meals—just add boiling water and wait five minutes. Most are that easy to prepare. They have their drawbacks, however. First, they're mighty expensive. Second, even if you like these pre-packaged offerings, and many people don't, you can get tired of them very quickly.

I've found a much more exciting and economical alternative is to buy dry or dehydrated foods at the supermarket and combine them to create your own imaginative dinners. Dried beans, cereals, potato flakes, Bisquick (pre-leavened biscuit mix), meat bars, crackers, dry soup mixes, cocoa, pudding, gingerbread and instant cheesecake mixes are just a few of the possibilities. But don't forget to pack a few spices to make your creations possible—chilli, curry, garlic and onion powder, Italian seasoning, cinnamon, salt, pepper and whatever else you fancy.

Nutritional value: Because hiking, biking, canoeing and all-around camping are pretty strenuous activities, you'll need to fuel yourself with plenty of high-energy foods (around 3,000 calories/day in summer and more in winter).

Fats contain more than twice as much energy per pound as proteins and carbohydrates do. So include plenty of hard cheese and margarine (both of which will last for weeks if kept out of the sun), and foods like nuts and peanut butter that stick to your ribs. Powdered milk and eggs are great lightweight sources of complete protein that can be added to almost anything you make and will help balance any meal. Dried fruits, pasta, rice and flour are good high-carbohydrate fuel foods, as are the quick-energy fruit drinks and sugar snacks. If you're worried about vitamins and minerals, pack vitamin pills. And drink high-in-C fruit drinks.

Quantity: Most people tend to work up a big appetite outdoors. About 2 to 2½pounds of food per person per day is average. How much of which foods will make up that weight is up to you. You can guess pretty accurately about how much macaroni or cheese or how many pudding mixes you're likely to need.

Unless you're a fantastic angler, don't count on fish to supplement your diet. And unless you're well versed in an area's flora, don't count on eating nature's salads. Except for the common berries or easily-identified greens, why risk stomach aches, indigestion or the effects of more serious poisoning?

Palatability: Last but not least, what do you like? If you don't care for instant butterscotch pudding or freeze-dried stew, don't take it along. You'll probably like it even less after two days on the trail. And if you've never tried something before, don't take the chance. Do your experimenting beforehand. Don't shock your digestive system with a lot of strange or different new foods. Stick as closely as possible to what you're used to in order to avoid stomach upsets and indigestion. And make sure you pack a wide enough variety of foods so that you won't be subjected to five oatmeal breakfasts in a row or be locked into an inflexible menu plan.

Packaging Your Food: After purchasing your food, the next step is to repackage it. Except for freeze-dried meals or other specially sealed foods, it's a good idea to store supplies and spices in small plastic freezer bags. Just pour in your pancake mix, salt or gingerbread mix, drop an identifying label in if you want to take all the guesswork (and

fun) out of it, and tie a loose knot. Double-bag when in doubt.

Taking plastic into the wilderness may offend one's inner sensibilities, but believe me, it works well. On a month-long expedition recently into rainy South eastern Alaska, I learned just how handy these lightweight, flexible, recyclable, *moistureproof* bags really are.

Preparing Great Meals: Although cooking over an open fire is great fun, many areas don't allow and can't support campfires. So don't head off without a stove. When choosing the stove that's right for your needs, keep in mind that size, weight and *reliability* become more of a concern the further off the beaten track you go.

Note: Butane or propane gas, with which many of the easy-to-use cartridge stoves are fuelled, 'freezes up' in below-freezing temperatures. This makes these stoves impractical for year-round use. White gas (refined petrol), kerosene (paraffin) and alcohol-fuelled stoves, though they require pumping, priming, etc., are more reliable. Moreover in Asia you will not find fuel available for butane or propane stoves; there, paraffin is the only fuel you can be sure of finding. (See also *Lightweight Equipment.*)

Aside from a stove, you'll also need plenty of matches, a collapsible water jug, water purification tablets (just in case), a nesting set of aluminium pots, a non-stick frying pan with cover, a few knives, pot grips or holders, a scrubbing pad, and a heavy bag in which to store your soot-bottomed pans. You'll also need individual eating utensils: spoon, cup and bowl will do.

Now you're ready to whip up some memorable camping meals. On a typical day you might serve up the following fare:

Breakfast

Orange juice reconstituted from powder
Coffee, cocoa or tea, with honey and milk
Freshly baked cinnamon coffee cake:
1½ cups Bisquick
½ cup flour
½ cup brown sugar
¼ cup milk powder
1 tablespoon melted margarine
2 tbspn. cinnamon

Mix ingredients together with just enough water to make a stiff dough. Shape into a large patty, place in greased frying pan, and sprinkle with lots of raisins, nuts, cinnamon and sugar.

To bake—place frying pan on top of lukewarm coals or over low flame on stove. Then shovel hot coals or build a small twig fire on top of the frying pan cover to create an even-baking oven. For best results, use a relatively flat lid (one that you don't mind getting sooty) with upturned edges. Or put small stones around the edge of the cover to help hold your coals or fire in place. Bake slowly till coffee cake is done—about 15 minutes. Better than a reflector oven—you can bake breads and cakes this way.

Lunch

Crackers and cheese (or peanut butter or jam)
Hot vegetable-beef soup
Cheesecake—simple to make:

Just melt margarine in frying pan, mix in crumbs and press onto bottom of pan. Then mix reconstituted powdered milk with the cheesecake mix and pour into crust. Let set. Make this *before* you begin lunch.

Between Meals

Plenty of fruit drink to avoid dehydration
Snack:
Mix together chocolate drops, peanuts, sunflower seeds and raisins, or any other fruit-nut-confectionery combination you like. Excellent for quick energy.

Dinner

Onion soup appetiser
Spaghetti and meat bars:
1½ cups cheddar cheese, sliced thinly for easy melting
½ cup tomato base powder, and enough water to make a rich tomato paste
½ cup raisins
1 or 2 meat bars, or a handful of bacon bits, oregano, garlic and onion powder, salt and pepper to taste.

Melt cheese into the tomato sauce, toss in raisins and crumbled meat bar or bacon bits. Season, simmer and serve over anything: spaghetti, noodles, baked fish and rice.

Dessert

Stream-cooler lime jelly with rehydrated fruit, or lemon pudding with coconut topping

Here are two bread recipes. One uses yeast and the other does not.

Basic Yeast Bread

1¾ cups water
1 packet active dried yeast
4 cups flour
2 teaspoons salt
2-3 tablespoons sugar

After dissolving yeast in lukewarm (not hot) water, add salt and sugar and set aside for 10 minutes. Then add 2 cups flour and beat hard for several minutes. Add the rest of the flour and mix well. Knead bread on a flat rock or in a frying pan for 5-8 minutes, flouring your hands and the kneading surface as often as necessary to avoid sticking. Then shape dough into a loaf, butter the loaf top, and place in a well greased pan. Cover the pan and set it in a warm place (about 85°F—if cold outside, place pan near the fire or over a pot of boiling water). Let rise to double its size (about an hour). Then bake the bread by placing the pan over hot coals (or low flame) and shovelling more hot coals (or building a small fire) on the lid. Bake for 30-45 minutes. Bread is done when it turns a golden brown and sounds hollow when the bottom of the pan is thumped.

Easy Whole Wheat Bread

2 cups whole wheat flour
1 cup biscuit mix
1 tablespoon baking powder
1 teaspoon salt
3 tablespoons brown sugar

Mix all ingredients together, shape into a loaf, and place in greased frying pan. Bake till bread has finished rising and turns a dark golden brown. Bake either by 'frying' (holding pan over low flame and flipping it over to cook top just as you would a giant pancake), or by 'baking' (setting pan on low heat and covering pan lid with hot coals). The bread rises further when baked than fried.

Some things about camp cooking I've learned the hard way:

cook on low heat to avoid scorching

taste before salting—bouillon cubes and powdered bases, often added to camp casseroles, are very salty. Don't overdo with more salt!

add rice, pastas. etc, to boiling water—not to cold—to avoid sticky or slimy texture.

add freeze-dried or dehydrated foods (onion flakes, freeze-dried corn, etc.) early on in your recipes to allow time for rehydration.

add powdered milk and eggs, cheese and thickeners (flour, wheat germ, etc.) to recipes last when heating.

when melting snow for water, don't let the bottom of the pan go dry or it will scorch. Keep packing the snow down to the bottom.

add extra water at high altitudes when boiling. Water evaporates more rapidly as the boiling point drops.

Cleaning Up

Soap residue can make you sick. Most seasoned campers, after one experience with 'soap sickness of the stomach,' recommend using only a scouring pad and water when it comes time for the clean up. Boiling water can be used to sterilise and is good for removing grease residue. Cold water is best for removing the remains of your glued-on pasta or cheese dinners. Soak and then scrub.

Use those recyclable plastic bags to store leftovers and to carry out any litter. Leave your wilderness kitchen clean—and ready for your next feat of mealtime magic!

Appendix

Here's a sample list of lightweight, inexpensive, versatile, easily-packable, mostly non-perishable foods to feed four or five for two weeks off-the-beaten-track:

Food, in pounds:

4	macaroni
3	noodles
2	rice
2	spaghetti
1	pearl barley
1	navy beans
1	split peas
½	wheat germ
5	wheat cereal
5	oatmeal
8	Bisquick (biscuit mix)
5	whole grain flour
2	cornmeal
2	potato flakes
1	green beans (freeze-dried)
1	onion flakes
½	corn (f-d.)
½	peas (f-d.)
½	carrots (f-d.)
5	raisins
2	prunes
1	apricots (dried)
1	apples (dried)
1	mixed fruit (dried)
2	tomato base powder
½	beef base powder
4	soup mixes
1	bacon bits (vegetable protein)
5	meat bars
5	peanut butter
5	peanuts
1	sunflower seeds
2	walnuts
½	coconut
½	popcorn
10	margarine
5	powdered milk
1	powdered eggs
10	hard cheddar cheese
5	cheese or plain crackers
8	brown sugar
5	cocoa
5	honey
½	tea
1	instant coffee

5	fruit drink crystals
1	jelly mix (5 boxes)
1	pudding mix (5 boxes)
1½	gingerbread mix (2 boxes)
1	instant cheesecake mix (2 boxes)
2	chocolate slabs or drops
	Spices, in ounces:
16	salt
2	pepper
1	oregano
1	onion powder
1	garlic powder
½	curry powder
½	chilli powder
½	dry mustard
¼	basil
4	cinnamon
¼	ginger
¼	nutmeg
4	baking powder
2	baking soda
2	vanilla
¼	yeast (5 packets)

Note: This list is just a suggestion. If you choose to use it, be sure to tailor it to your tastes. In other words, if you want powdered scrambled eggs and bacon bits for breakfast every morning, you're going to need more of both. And less oatmeal!

Climate and its Relevance to Travel

by Gilbert Schwartz

Prospective travellers may carefully prepare for their trip by consulting guide books, by choosing the most desirable accommodation, by designing an appropriate itinerary, and by thorough preparation in general. But the traveller frequently fails to investigate the most important ingredient affecting the failure or success of the trip . . . the weather.

Well, maybe there isn't much you can do to guarantee good weather but you can do some things that will minimise disappointment.

Be sure to do some homework. Look into reference books on the subject. One source of this type of information is my own book, *The Climate Advisor: A Complete Reference Guide to Climate and Weather in the United States, Canada, Mexico, and The Caribbean* (Climate Guide Publications, P O Box 323, Sta. C, Flushing, N Y 11367). The book is a non-technical reference based on objective and scientific weather information. Use it to help select the most favourable times and travel locations. Remember, when interpreting climate information, some statistics are necessary but they could sometimes be misleading. Look for comparisons. Especially compare the prospective location with an area at home or an area with which you are very familiar. For example San Francisco, California, has a temperature range for July from a maximum average of 64°F (18°C) to a minimum of 53°F (12°C) with no precipitation. This becomes more meaningful when it is compared with New York City which has a range of 85°F (29°C) to 68°F (20°C). New York has 3.7in of precipitation and, on average, 11 days during the month have rain of one-tenth inch or more.

So, in spite of the fact that California has a reputation for being warm and sunny, if you're planning a trip to San Francisco in the summer, don't forget to bring a sweater! The average temperatures are cold and the winds are a brisk 11mph, windier even than Chicago, the 'windy city'.

Up-to-date weather conditions and forecasts may be obtained from various sources. A current weather map, which is based on information furnished by government as well as private weather services, is the main source for getting a general picture of weather patterns over a large area. These weather maps show conditions around the country at ground level. Elements which are of particular interest to travellers and may be shown on the map include temperature, pressure change, wind speed and direction, cloud type, current weather, as well as precipitation.

Of course, the weather information and projected forecasts must be interpreted. You may do well to alter your itinerary and stay clear of areas that project undesirable or threatening weather conditions. Especially keep alert for severe weather conditions such as storms, heavy rains, etc. For example, you should remember that when travelling in mountainous regions, flash floods can strike with little or no warning. Distant rain may be channelled into gullies and ravines, turning a quiet streamside campsite into a rampaging torrent in minutes.

Incidentally, there is excellent literature available through the US Government Printing Office prepared by the National Weather Service that could prove to be valuable. For example, information on lightning safety, flash flood safety, hurricane and tornado safety, as well as publications containing summaries and other data pertaining to weather and climate are available. Write to Superintendent of Documents, US Government Printing Office, Washington, DC 20402, for a list of publications.

After you have had an opportunity to review reference materials on climate and sources for weather forecasts you may do well to become acquainted with the meaning of some basic weather elements and learn how they may affect your travel preparations.

Perhaps the most important weather element is temperature. Temperature is important because it is a good indicator of body comfort. The ideal air temperature is about 80°F (27°C).

Temperatures are generally colder to the north and at higher elevations. On the average, temperatures decrease by 3.5°F (1.7°C) for every 1000 foot increase in elevation up to 30,000 feet.

Wind, which is air in motion, is another important weather element. Winds are caused by pressure gradients, the difference in pressure between two locations. Air moves from an area of high pressure toward an area of low pressure. The greater the pressure gradient, the faster the wind. Sea breezes form when cool high-pressure air flows from the water onshore to the low pressure area created by the warm air over the land. On a clear hot summer day, the sea breeze will begin at mid-morning and can blow as far inland as 10 miles with wind speeds of 10 to 15mph. In the evening the process is reversed. An offshore land breeze blows at a more gentle speed, usually about half the speed of the daytime onshore wind.

A somewhat similar situation occurs in the mountains and valleys. During the daytime, the valley floor and sides and the air above them warm up considerably. This air is less dense than the colder air higher up so it rises along the slopes creating a 'valley wind'. In the summer, the southern slopes receive more sun and heat up more, which results in valley winds that are stronger than their north slope cousins. At night, the process reverses. Downslope 'mountain winds' result from the cold air above the mountain tops that drain down into the valley.

Winds are also affected by such factors as synoptic (large area) pressure differences and by day-night effects. The sun produces maximum wind speeds while at night winds near ground are usually weak or absent. Wind speed is also influenced by how rough the ground is. Over smooth water surfaces, the wind speed increases very rapidly with increasing altitude and reaches a peak speed at a height of about 600 feet. Over rough terrain the wind speed increases more gradually with increasing altitude and does not reach its peak speed until about 1500 feet.

As we well know, wind, temperature and humidity have a bearing on our comfort. To indicate how combinations of these elements affect the weather we experience, two indexes should be understood: wind/chill factor and temperature/humidity comfort index (THI).

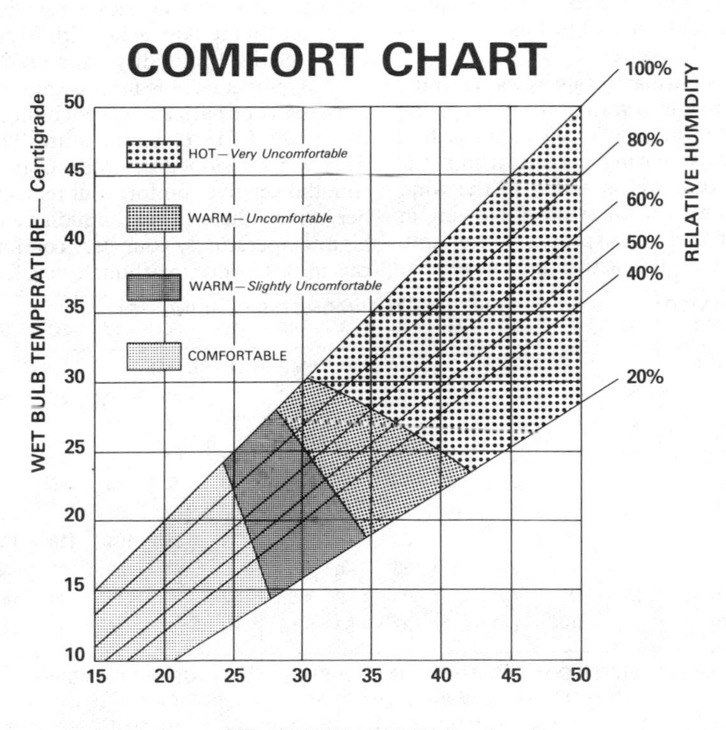

AIR TEMPERATURE—Centigrade

To use the comfort chart, start by locating the vertical line that corresponds to the air temperature. Next, find the sloping line that corresponds to the relative humidity. The intersection of these two lines will put you into one of four comfort categories: (1) hot and very uncomfortable, (2) warm and uncomfortable, (3) warm and slightly uncomfortable, and (4) comfortable.

The wind/chill factor is the cooling effect on the body of any combination of wind and temperature. It accounts for the rate at which our exposed skin loses heat under differing wind-temperature conditions. In a wind of 20mph, 25°F (-4°C) will feel like -3°F (-19°C). This effect is called 'wind chill', the measure of cold one feels regardless of the temperature. Chill increases as temperatures drop and winds get stronger, up to about 45mph, beyond which there is little increase. Thus at 10°F (-12°C), increasing the wind from 0 to 5mph reduces temperatures by four degrees, but a change in wind speed from 40 to 45mph reduces it only one degree.

The wind may not always be naturally caused. For example, someone skiing into the wind may receive quite a chill. If one is moving into the wind, the speed of travel is added to the wind speed; thus if the wind is blowing at 10mph and one's speed is 15mph into the wind, the actual air movement against the body is 25mph. At 15°F (-9°C) this air speed gives a wind chill equivalent to -22°F. This is easily cold enough for exposed parts of the body to sustain frostbite.

A combination of warm temperatures and humidity have a significant bearing on our comfort. Particularly at warmer temperatures, the higher the relative humidity, the less comfortable we are. This is a result of the corresponding decrease in the rate at which moisture can be evaportated from the skin's surface. Since the cooling of the air next to the skin by the evaporation of perspiration is what causes a cooling sensation on the skin, a day with 70 per cent relative humidity and 80°F (27°C) temperature is far less comfortable than one with 25 per cent humidity and 90°F (43°C) temperature. The THI was developed in order to measure this relative comfort. But remember, where there is low humidity and high temperatures, your very comfort can mislead you, for though you feel

ESTIMATED WIND SPEED IN MPH	ACTUAL THERMOMETER READING (°F.)											
	50	40	30	20	10	0	-10	-20	-30	-40	-50	-60
	EQUIVALENT TEMPERATURE (°F)											
calm	50	40	30	20	10	0	-10	-20	-30	-40	-50	-60
5	48	37	27	16	6	-5	-15	-26	-36	-47	-57	-68
10	40	28	16	4	-9	-21	-33	-46	-58	-70	-83	-95
15	36	22	9	-5	-18	-36	-45	-58	-72	-85	-99	-112
20	32	18	4	-10	-25	-39	-53	-67	-82	-96	-110	-124
25	30	16	0	-15	-29	-44	-59	-74	-88	-104	-118	-133
30	28	13	-2	-18	-33	-48	-63	-79	-94	-109	-125	-140
35	27	11	-4	-20	-35	-49	-67	-82	-98	-113	-129	-145
40	26	10	-6	-21	-37	-53	-69	-85	-100	-116	-132	-148
(wind speeds greater than 40 mph have little additional effect)	LITTLE DANGER (for properly clothed persons)				INCREASING DANGER			GREAT DANGER				
					Danger from freezing of exposed flesh.							

WIND CHILL CHART

safe you may be at risk of burning.

Wind chill factor and THI are best understood by examining an appropriate Wind Chill Chart and Comfort Chart. These too may be obtained by writing to the US Government Printing Office.

Lacking the sophisticated instruments and sources for weather data you may still be able to project your own forecasts. Become familiar with basic weather elements such as pressure signs, clouds, wind changes, etc. Learn how these indicators change before the weather changes.

A layman should beware of the climate statistics he sees in many tourist brochures. The climate will almost always be more severe than is evident from the quoted rainfall, temperature and sunshine figures. All-important humidity figures are usually not given (Bali might be empty of tourists half the year if they were), and temperature figures may be averages over day and night, and well below (or above) actual normal maximum (or minimum) temperatures. Or they may represent averages recorded at 0600 or 1800 hours, because these figures will look most attractive to visitors.

Something else you will not find easily is water temperature. Winter sun holidays are now extremely popular. A lot of people do not realise, however, that although daytime temperatures may be in the 70sF, (low 20sC) water temperatures may only be in the 40s and 50s (about 5-15°C) and swimming without a wetsuit impossible.

The sea takes longer to warm up than the land each summer. Conversely it takes longer to cool down in the autumn. Reckon on a lag between sea and land temperatures of about one and a half months. In Tunisia the sea is a lot cooler in March than in October. On the other hand, by March, the air and land temperatures are already rising with the beginning of summer. They will reach their highest point in June/July, but the sea will take until August/September to be fully warmed up.

In winter, comfortably warm water is almost a certainty in the tropics; more doubtful in the subtropics, for which you should find and study year-round water temperatures. You may just decide to go in summer instead, even though it will probably cost more. In short, warm air and warm water don't always go together.

Familiarity with climate information, whether you rely on primary or secondary sources, will go a long way towards permitting you to get the most out of your next trip.

Lightweight Equipment

by P. F. Williams

The main items of lightweight equipment required for an off the beaten track venture comprise a tent, sleeping bag, stove, cooking and eating utensils, plus a rucksack or pack frame and sack. Added to your load will be food, spare clothing and personal gear. Whether you are a hill trekker, hitch hiker or just planning a backpacking holiday by road/air, then on foot, you will have one thing in common—at some time or most of the time you will have to provide the motive power for whatever equipment is to be carried. You therefore need to consider very carefully what gear is to be taken, how to carry it and the best methods of packing to employ. Take note that if your journey is planned across the hills or over uneven difficult terrain then the weight of your load should not exceed one third of your body weight; for young people 30lb is the maximum weight advised. These limitations are necessary as the energy required to cover such ground can equal that used in heavy manual work so you cannot afford to carry non essential gear.

The object of this short chapter is briefly to explain what type of lightweight equipment is needed in order to rough it in comfort.

Lightweight Tents. A well designed lightweight tent should normally have sufficient room for two people to sleep in and, if weather conditions are bad, have adequate space to cook in. Recent years have seen considerable variation in lightweight tent design but many are still based on the traditional ridge tent or single pole pyramid tent. The design of tents influences their performance in bad weather, the ease with which they can be erected and the internal space available for the occupant(s). The one pole pyramid type can be pitched single handed quickly and easily. It is normally lighter than other designs and its front guy(s) take up less ground than many other models that require both front and rear guys. Its shape renders it stable in wind and, being higher, it can be more comfortable to live in during inclement weather. Ridge tents are normally cheaper as they are easier to make. The ridge with no walls weighs and costs less than a walled ridge tent. The steep pitch of the roof sheds the rain well and it will withstand the wind better. However, a walled ridge is more spacious at floor level and gives more headroom. The use of external poles will ensure that maximum space possible is available within a tent. A ridge tent approximately 6ft 6in by 5ft wide and 4ft high will give ample space for two people.

Nylon fabric is commonly used in the construction of the lightest tents and there are many two man designs that only weigh about 4lb. Nylon will not absorb water like cotton fabric so, after a heavy shower, most of the moisture can be shaken off before you pack the tent and move on. However, nylon does not possess the same breathing qualities as cotton and unless adequate ventilation is built into the design of the tent, condensation will form on the walls of the inner tent. If silicone coated walls are used in the inner tent this will allow a transfer of vapour laden air and condensation is not such a problem. Nylon may be affected by the ultra violet rays of the sun which can cause a breaking down of the fabric structure but many modern designs are made of high tenacity ripstop fire resistant nylon fabric which has inbuilt ultra violet inhibitors to prevent the tear strength of the material deteriorating while the tent is in use and exposed to sunlight. Cotton fabric tents are still very popular though heavier than similar designs made in nylon. The average weight of the two man type is about 7lb. Cotton loom fabric of about 3oz per square yard is a well proven material. Cotton possesses the property of allowing air to filter through the fine weave so that little condensation will form in the tent yet the fabric is fully waterproof. However, cotton fibres will absorb water when exposed to very wet conditions so if the tent has to be carried when wet you will find that the weight will have increased considerably.

When selecting a lightweight tent such factors as expected weather conditions, altitude, temperature and any requirement for fine weight limitation should be taken into account. Here are a few points to note:-

a. select a design made by a reputable manufacturer and carefully read the maker's specification for the tent including the work it is designed for e.g. base camp use, mountaineering, a low level bivouac etc.

b. if you intend to operate in wet conditions note that some modern tents are designed so that the flysheet is erected first. The inner tent can then be erected under and protected by the fly sheet. When you strike camp the procedure is reversed and the inner tent packed first.

c. consider the weight of the tent in relation to the all-up weight of the rest of your gear. It may be feasible to split the weight of the tent with a companion, one person carrying the main tent, the other the flysheet and poles.

d. a down to earth flysheet will in bad weather give maximum protection to the inner tent and make the tent warmer in cold conditions. It will also help to keep it cool when the weather is very hot.

e. a built-in ground sheet that extends about 3in up the side, front and back walls will eliminate draughts that would otherwise blow in under the walls. It will give insulation from damp ground and if the area does get very wet the groundsheet will keep the inside of the tent dry.

f. note that some mountain tents are designed for dry conditions and are not suited for prolonged use in very wet weather.

g. fine netting should cover all vents to keep out mosquitoes and gnats.

The Backpacker 2 and the Marriott Kamplite are examples of British two man nylon designs that weigh only 4lb and 4lb 15oz respectively. The American Wenzel Flamezel Mountain tent is a polyester/cotton drill two man design with an all-up weight of less than 7lb. The Skyliner and Shenandoah Pack tents are nylon two-man ridge designs that weigh 5lb. Wenzel also retail the pyramid shaped nylon Tall Trail Tent that gives a 7ft x 7ft floor area, stands 7ft 4in tall and weighs 5lb. Another well known American camping firm, Coleman, sell their two-man Mountain Tent made of 7oz dri-tex. This is a heavier design and weighs 15lb. Your tent may be your home for days, perhaps even weeks. It is therefore important that the design, weight, size and degree of protection it offers suit your purpose and give the comfort you require.

Sleeping Bags. Modern sleeping bags contain such natural fillings as down and feather or man made fillings such as plastic foam, Terylene and Dacron. Down is the undercoating of water fowl and nothing that man can manufacture has the qualities of loft, softness, and lasting resilience possessed by this natural substance. A down bag may be very light yet will keep you extremely warm. Among the synthetic fillings Dupont's new Dacron Hollofil

808 (hollow filament fibrefill) provides about 17 per cent more loft and insulating value than other solid polyester fillings of the same weight. Dacron Hollofil 808 and Hollofil II are used in the American Wenzel and Coleman sleeping bags, some of which are designed for use in cold temperatures and at high altitudes. Wenzel also retail cold weather bags containing the excellent Fortrel Polar Guard insulation. These types of bags weigh between 5 and 7lb, are washable, dry quickly and are much cheaper than bags containing natural fillings.

The degree of insulation afforded is dependent not only on the type and amount of filling in a bag but also its arrangement within the sleeping bag walls. The filling of a sleeping bag has to be contained so that it does not slip or move within the walls and produce cold areas. This is done by dividing the walls into squares, sections or tubes. In some bags the outer wall is drawn by the quilting stitches to the inner wall. The region of the lines of stitches form cold spots because of the lack of thickness and insulation. This type of quilting is unsuitable for use in colder conditions unless you are prepared to wear additional gear when you go to bed such as woollen tights or a wool shirt and socks. However, where two layers of simple quilting are used to make a double wall effect the insulation is excellent. Boxed or walled sleeping bags are normally more expensive than the single walled quilted bag but a much better purchase as the walls are kept completely separate from one another. The filling is contained in areas or boxes by fabric compartment walls stitched between the outer and inner covers. These boxes allow for the full expansion or 'loft' of the filling when it is warmed by the body and therefore give a continuous uniform layer of filling to insulate the sleeper.

A sleeping bag fitted with a draw cord is preferable for a bag will afford maximum warmth only if it can seal in the heat generated by the body. When the cord is pulled, your body from the shoulders down should be as snug as the quality of the bag can make you. Expensive sub zero bags are often mummy shaped with a built in hood to cover the neck and head. Such designs have a draw string that will close the hood to leave only a small space for breathing.

Here are a few points to note when selecting a sleeping bag:-

a. select a design made by a reputable manufacturer and carefully read the maker's specification for the bag and the conditions in which the bag is suitable for use, e.g. light summer use, cold weather, arctic conditions.

b. a down filled bag is best; the weight of the filling will indicate the thermal properties of a bag. 3lb of goose or duck down will make a very snug bag suitable for use in very cold weather. 2lb of similar filling makes a good warm general purpose bag. 1lb of down will provide the warmth for use in summer conditions only.

c. make sure that the length of the bag is adequate. Adult sizes normally range from about 78in to 90in.

d. shapes of bags can vary; some are rectangular, others tapered or mummy shaped. Mummy shaped bags are normally more

suited for low temperatures and tapered designs used for moderate conditions. Rectangular bags are normally best suited for low level summer projects.

e. a waterproof nylon carrying or stuff bag is necessary in order to keep the bag dry and clean.

f. if you do not intend to use a tent but sleep rough, purchase a sleeping bag cover made with a strong waterproof base and polyurethane proofed nylon top that will give extra protection from cold and damp.

g. a wool or brushed cotton inner sheet will give extra warmth and help to keep the bag clean, for otherwise body moisture can be absorbed by the filling.

h. if you are travelling in a vehicle and expect a lot of cold weather, take a foam rubber mattress; if travelling light use an air mattress.

Remember, you will spend six to eight hours daily in your sleeping bag, in bad weather perhaps much longer. The comfort it gives you is therefore very important so buy the best that you can afford.

Stoves. There are various types of lightweight stoves on the market including those fuelled by solid fuel, methylated spirit, paraffin, petrol or gas. However, several factors need to be considered before purchasing a stove and these include price, weight, size and availability and cost of fuel. Information on the fuel capacity, boiling time and the one charge burning time will also give an indication of a stove's overall efficiency.

The most popular solid fuel stove on the British market is the design that, when folded, resembles a small flat light steel box which measures a little larger than a box of cards and encloses the solid fuel blocks. When erected the stove comprises a pot stand with supporting legs and a burner plate on which the fuel blocks are placed. The fuel blocks will light even when damp and partially used blocks can be put out and retained for use later. Heat control is effected by adding or reducing the amount of fuel on the burner plate. The best cooking results are obtained if lightweight aluminium cooking utensils are used as the stove cannot generate pressure like a gas or petrol stove. The solid fuel stove has no working parts that can malfunction. The Hexamine type is a well proven British Service issue that is also available on the open market. Another similar make is the Esbit stove made in West Germany. The solid fuel Tommy stove is another variation of this form of stove in which the fuel is ignited within a round metal stand. The Meta boiling set is a light aluminium set that weighs 3¼oz and includes a saucepan of just over ½pt capacity and a cooking stand. It is specially suited for heating small quantities of liquid.

The Methylated fuel picnic stove is the cheapest and one of the simplest types of lightweight stove available. The Express-type weighs 8oz and measures 4½in in diameter by 3in high. The stove has no working parts but comprises a hollow steel cup with a perforated rim. This stands inside the combined pot stand and wind shield. Meths is poured into the cup to a set level indicated and then a match applied to the centre of the cup ensures a flickering flame that soon settles down to form a very effective gas-like ring of fire emitted from the perforated rim. The disadvantage with the methylated fuel picnic stove is that it is not possible to control the heat output and it uses a fair amount of fuel. A 2oz charge of fuel will burn for 29 mins in calm conditions at 72°F., long enough to heat several dishes; a ½ pint of water will boil in 5mins. The Trangia Storm Cooker 25 is a more advanced design of meths stove and comprises an aluminium cook set with two large frying pans, frypan and pot lifter. The stove will boil a litre of water in eight minutes. The Storm Cooker measures 11 x 23cm and weighs 2½lb.

The Primus type of paraffin pressure stove is a Swedish invention well known throughout the world. It is cheap to run and will give a long burning time. The stove has a high heat output but it is also possible to control the pressure from very low to quite intense heat. The paraffin pressure stove works on the principle of vaporising paraffin by compressing air. The burner of the stove is first heated by a priming of meths or solid fuel set on fire in the cup specially designed for this purpose and located midway up the stem of the stove. A few pumps are then given to the main reservoir and a match applied to the burner enables the paraffin vapour to catch. The stove will then start to roar so that the spreader ring glows red hot. The Primus 96L (½pt), Hipolito (1pt) and Blaxburner (2pt) are popular paraffin pressure stoves of this type. I have used a ½ pint Primus for many years and found it a very reliable stove that is not affected by dampness, rain nor intense cold. However, you cannot afford to neglect a pressure stove. The jet requires to be cleaned from time to time and after much use the pressure washer may need to be renewed. The Primus 96L model is ideal for lightweight projects. It has a ½ pint fuel reservoir and weighs 28oz empty. It will boil one pint of water in just over five minutes. One charge of fuel, which weighs 7oz, will last for approximately 140 minutes. The Optimus 111 paraffin stove is a more recent design housed in a folding steel box. It has a capacity of one pint of fuel and will burn for about two hours on one filling. The stove weighs 4½lb.

There are various excellent petrol fuelled pressure stoves on the British market such as the one pint capacity Optimus 111B which weighs just over 4lb or the smaller one third capacity Optimus 8R which only weighs 1½lb. It will burn for about 45 minutes on a single filling of fuel that weighs 3oz. The well known American camping firm Coleman retail a number of excellent lightweight petrol stoves. Their compact Model 502-700 has a fuel capacity of one pint, weighs 3lb and has an approximate burning time of three and a quarter hours. Another type the Model 576 Peak 1 stove has a fuel capacity of 283 grams, weighs 2lb and has an approximate burning time when set to simmer of three hours 30 minutes. The stove will boil one litre of water in less than three minutes. Coleman produce their own fuel for gasoline camping appliances that is sold widely in the States, even in one store towns. It burns clean and prevents clogging or gumming up and has a rust and corrosion inhibitor. Remember that petrol gives off toxic fumes when burnt so care needs to be taken if the stove is used inside a tent. It is also extremely important that the manufacturer's instructions are read carefully so that you know exactly how to use and service your petrol stove. Also take note that stoves sold in Great Britain are normally intended for use with unleaded petrol.

Generally speaking lightweight gas stoves are not expensive to buy but cost more to run than liquid fuelled paraffin or petrol stoves. The advantage of gas is that it is clean, requires no priming and the pressure can be finely controlled. There are numerous designs of lightweight gas stoves on the market made by such firms as Camping Gaz, Veritas, Tilley and Falk. In the USA Wenzel retail a single propane burner, their Camp Stove 29262. Coleman sell their Model 5418A700 Compact propane stove that has a disposable cartridge, an operating time of two hours 45 minutes and a weight of 5lb. The Bluet burner is another model that is popular in the States. It uses the C200 disposable butane cartridge that will burn for four hours. The stove weighs only 16oz. Generally speaking a gas cartridge with a capacity of 6-8oz of gas will give over two hours of burning time. However, unlike in other types of lightweight stove, as the fuel commences to run out the pressure will drop considerably for quite some time and delay cooking. Take note that when butane gas is used at low temperatures difficulty can be experienced in lighting the stove and the stove may then burn with little pressure. Propane gas however will burn well in cold conditions.

Utensils. At the absolute minimum two vessels are needed for cooking on an expedition, a frying pan and a pot. There are numerous good lightweight canteens on the market that comprise both cooking and eating utensils. When purchasing such equipment make sure that the items are also compact and the utensils will fit neatly inside one another, for a rucksack has little spare room for bulky gear. A two man canteen need not add up to more than 1½lb in weight. Cooking pots should be seamless to avoid possible leaks and a wide base makes for quicker cooking and less fuel used. When considering lightweight frying pans check for stability when on the stove, particularly those types which have long folding handles. If the weight of the handle outweighs that of the light pan and its contents, it will be unstable and could be dangerous unless the handle is continuously held throughout the whole cooking operation. Utensils that incorporate the paint-pot type of handle are not designed for the handle to stay upright away from the heat. You will therefore need a grip handle that weighs only 1½oz with which it is possible to handle these and any other hot utensils with safety. The Hiker (one-man), Gillwell (two-man) and Wanderer (two-man) are well known British lightweight canteens that weigh in the region of 1lb yet comprise a variety of utensils for cooking and eating. On the American market Flaghouse Inc. of New York sell numerous designs of cook sets including United States Army style mess kits made from tough gauge aluminium.

Load Carrying.

There are three types of load carrying equipment to consider—the traditional rucksack, the frame rucksack and the pack frame and sack.

The Rucksack. The rucksack is basically a container bag that has no supporting frame and lies directly on the back of the walker supported by two shoulder straps. Rucksacks are often made of 15oz proofed close weave cotton canvas or 7 to 10oz proofed nylon. They weigh about 2 to 4lb yet can have a capacity of about forty litres or even more. Some modern designs are contoured to shape the back. Large external pockets are useful, particularly if lined with additional proofing such as oil cloth, for the stove and liquid fuel can then be carried separately with no possibility of contamination of food or personal clothes in the main bag. A pull-out plastic sleeve at the top of the sack fitted with a draw cord makes for security and the extra protection of contents in bad weather. A large lid with long straps will ensure that the tent, which is a comparatively heavy item, can be carried in a waterproof cover on top of the sack, held securely by thc tightened lid straps.

The rucksack needs to be packed carefully to ensure that the weight is distributed evenly and that no sharp items can dig into your back. It has the disadvantage that it will cause the back to sweat as there is no circulation of air between the canvas and the body. However, it is normally lighter and cheaper than the frame rucksack or pack frame and sack. When empty it can be used as a mattress to lie on at night and give insulation from dampness and cold striking up through the ground. The Fellman 40 (40 litres capacity, 28oz) and the Yukon pack (13½oz) are two large nylon rucksacks retailed by Blacks of Greenock. Camp Trails of the USA retail the Combo Pack (22¼oz, capacity 3070cu in), made of water repellant, mildew resistant cotton duck. Flaghouse Inc of New York sell Yucca and Duluth packs and Wenzel their Explorer day bag—water repellant cotton—and the Skedaddle pack of Urethane coated nylon.

Frame Rucksacks. Modern frame rucksacks comprise a steel or aluminium frame onto which the sack is built. The frame rucksack distributes the load well and allows adequate ventilation of the back which is not possible with frameless models. The Norwegian Meis design is popular both in the USA and in Europe and has a capacious canvas sack built onto a light aluminium frame that incorporates a waist support. Before purchasing a frame rucksack check that the frame fits the back well and if a waist support is incorporated, as with the Meis, it fits around and does not dig into your back. The frame rucksack should be comfortable to wear, be carried high on the back and have the capacity to take all the gear you need. The Cyclops range of rucksacks by Berghaus of Newcastle-upon-Tyne, England, includes a variety of models that have an internal alloy frame. They are contoured in three dimensions to follow the shape of the back and spread the load so that the weight is positioned high and comfortably on the back. Some of their expedition models have a capacity of over 70 litres.

Pack Frames. The pack frame is capable of supporting large or heavy loads. Most frames weigh only about 1½-2lb and are made of strong alloy tubing and are either welded together or linked by plastic joints. Strengthened V bars are often fitted to resist distortion from heavy loads. However Colemans of Whichita, Kansas, have introduced a new concept in frame construction and produced a moulded frame made from high impact material called Ram-Flex. The material is durable at all temperatures and gives controlled flexibility to move with your body. It has numerous moulded slots that allow considerable adjustment to the supporting straps. The slots can also be used to tie on extra gear. Custom designed sacks are normally attached to frames by pin mountings or by straps but it is possible to carry any reasonable load from the large duffle-type sailor sack to a self designed canvas sack, provided it is strapped securely to the frame and is not unwieldy in shape. It is important that the frame fits properly and some frames are adjustable. The frame should contour the body. The distance between the top harness attachment point and the bottom of the cross band on the frame should equal the distance between the highest prominent neck-bone and the hip-bone. When the loaded frame is carried it will then lie with the cross band resting on the hips. A padded waist belt will help to take the weight off the shoulder and give extra support. Camp Trails of Phoenix, Arizona, produce a wide variety of combination pack frames and sacks. Coleman of Wichita also make a variety of packs to fit their Ram-Flex frames. In Britain Blacks of Greenock produce their Packpacker 2. Tote-em frames are another well known make available on the British market.

Packing the Load. When you walk the body is balanced so that the line of gravity runs down the spine through the legs to the ground. Any weight carried on the back also has its line of gravity that runs parallel to that of the walker. The closer together these two lines can be kept the easier it is to carry a load and walk upright. It is therefore important that the centre of gravity of your load should be kept both high and as far forward as possible. This can be achieved by dividing the pack into three vertical zones. Into zone A, which is the area close to your back, place objects that have the greatest density with the heavier items highest. The tent should be packed at the top of the sack, either held in place by the lid and tightened lid straps of the rucksack or strapped to the poles of the pack frame. In the middle zone B pack items of medium density and in Zone C, furthest away from your centre of gravity, pack the lightest items. If high density items should be packed in Zone C the excess weight will tend to pull you backwards and to compensate you are forced to lean forward from the waist.

In addition to the general principles of zoning your gear there are other factors that need to be considered. Emergency items, such as the *cagoule* and overtrousers that will be needed in the event of sudden bad weather, and emergency rations, should be packed near the top of the sack. If you use a rucksack avoid sharp objects against the back. Last wanted items such as night gear or reserve food should go into the bottom half of the sack. Where the arrangement of items in the sack is not so critical, such as when short distances are to be covered or when hitch-hiking is to be the method of transport, then you may prefer to divide the sack into a night side and a day side with the sleeping bag, sleeping gear, toilet kit, torch, towel neatly packed on one side and day items such as food, stove and reserve clothing on the other side. The general principle of heavier items at the top of the sack will still apply. It is good practice to keep spare clothing and the sleeping bag in light polythene bags for driving rain can penetrate the smallest flaw in canvas. Of course if the sleeping bag is carried externally it should be rolled neatly, placed in a waterproof container and tied below the sack to the pack frame. Above all keep the load as compact as possible so that if you initially travel by train or coach your movements are not hindered by a load that is too wide or has protruding items tied externally that can snag on doors or handles. Such a load will present the same difficulty in the field when moving through brushwood or along overgrown paths.

Once you have satisfied yourself that you have achieved the best way of storing your gear in the sack take note of the actual position of items and always pack them in the same order. In this way your equipment is easy to find—even in the dark.

Equipment Check List

The following list of equipment and personal gear is given as a guide for a walking holiday in foothills during summer.

Tent
Sleeping bag
Hip length air bed or foam mat
Rucksack or pack frame and sack
Stove
Fuel
Cooking utensils
Eating utensils
Knife fork and spoon
Tin opener
Backpacker grill
(to save fuel)
Trowel
Toilet kit
Toilet paper
Small towel
Water purifying tablets
Brillo pads (for cleaning pans)
String (can be used as clothes line)
Shoe cleaning kit
Map
Compass
Watch
Whistle
Torch
First aid kit
Pocket knife
Waterproof matches
Water bottle
Sunglasses
Suncream
Pen or pencil
Candles (to save batteries)
Emergency clothing (e.g. *cagoule* and over-trousers.
Emergency rations (glucose tablets, chocolate, dried fruit, Kendal mint cake etc)
Spare clothing, underclothing and socks
Sneakers
Camera

A Survival Kit for The Motorist

by Richard Harrington

This is a list of items that will help the stranded motorist to survive when he is far from habitation, particularly in the desert or near the sea. It includes some ideas based on the assumption that the traveller has told someone of his intended route and estimated time of arrival at his destination, and that rescue is therefore a possibility.

space blanket (weight should only be a few ounces, and the NASA-developed synthetic material that space blankets are made of means that they make good lightweight sleeping bags if you have to walk out.)

small metal mirror to catch the sun's

rays and help you be spotted from a distance or in the air.

red flare beacon which burns on the ground and can be spotted from the air.

one dozen aerial flares.

orange smoke bomb (visible from further away than a flare).

ACR-4F Rescue Lite (a waterproof flashing strobe light).

two chemical seawater desalting kits (seven pints of fresh water produced by each.)

packet of Tibblin Wet Energy, a substance which is supposed to help reduce your body's fluid requirements by slowing dehydration. I don't know how effective this is. If it is meant to reduce your thirst, or inhibit perspiration (the body's built-in cooling system), then I suspect it could do more harm than good. Check it out before taking it with you.

solid broken-in walking shoes. If you are going to walk out, you'll get a lot further with a strong pair of walking shoes that your feet are already used to, along with a couple of pairs of *soft thick woollen socks* to be worn on top of each other, *however hot it is.* But there will be few occasions where walking out will be advisable in the desert. Even in the middle of winter, you're unlikely to survive more than a few miles because of the inevitability of dehydration and the fact that you can only carry so much water with you.

aerosol type siren to attract attention. The sound should carry a couple of miles in still conditions, in case you are out of other signalling devices and are trying to attract the attention of someone who can't see you although you can see him.

lightweight first aid kit in case you start walking out, and a *vehicle first aid kit,* which should be a lot more comprehensive. One item in it ought to be sterile vaseline gauze to treat burns from the engine or the exhaust. You should also have some Phisohex Sudsing Antibacterial Skin Cleanser in the vehicle kit.

The risk of snake bite is often exaggerated, but you may want to take a *snake bite kit.*

If you're stuck with a broken-down vehicle trying to conserve valuable body fluid in parching heat, don't stay inside your vehicle, where it will be hottest of all. But outside in the shade of the vehicle (and remember at noon there won't be any shade outside) it's still pretty hot. The best thing you can do in that situation is spread out your *folding camp bed* that you've brought along for such an emergency and lie on it without moving. If you had a canvas awning fitted along the edge of the roof of the vehicle before departure, lie underneath that. Make sure that the camp bed is at least two feet off the ground. The reason is that the temperature is at least 10 degrees Fahrenheit less two feet above the ground than on it. And with the awning, you'll be in the shade at mid-day. You'll still be sweltering, but it may give you an extra day or two—maybe the difference between life and death from dehydration. You can probably go for a week or longer without food in the desert, but if the temperature in the shade is over 100°F, you're unlikely to survive more than two days without water, and a lot less if you move around or set off on foot. (See *Survival—Deserts.*)

The solar still may be the single best thing that ever happened to you if you break down and run out of water in the desert. A single still may not produce enough water for even one person for long, far less two or three. Dig a hole three feet deep and three feet square. Put a *plastic bucket* in the bottom and cover the whole thing with a *6ft* x *6ft piece of strong black plastic sheeting.* (You should bring along *two pieces* in case one tears). Weight the sheet with rocks around the edges to hold it in place and place some stones in the centre, causing the middle of the sheet to sink over the bucket without actually touching it. During the night, precipitation forming on the interior of the sheeting runs down to the bottom of the cone over the bucket and drips off into the bucket where it may be gathered early next morning. A solar still built at the bottom of a slope, especially if there is any plant life growing there, is likely to produce more moisture, but don't expect gallons, even at the best of times.

It goes without saying that in the desert you should minimise physical effort in order to retain valuable body fluid. That is why getting stuck in the sand is something you should avoid at all costs. It may even pay to stay stuck until evening. A solar still, if you are going to make one, should also be left to late in the day. Digging a hole in the middle of the Sahara in the middle of the day is a sure way to be dehydrated, even in winter. (Only a fool would try to cross the Sahara in summer.)

If your trip were to take you along a shoreline far from any habitation, one piece of equipment you'd need to get food is a *fishing line* with *sinker and hooks,* or better still a *surf casting rod* with *line and lures. Diving mask, flippers* and a *snorkel, speargun* and *diving knife* might also come in useful in a survival situation, as well as for recreational use.

Finally, if you are going to make it out on foot rather than stay with your vehicle (it might be your only chance and maybe the nearest village is only ten miles away), then make sure you have a *small lightweight backpack* to carry the gear you set out with. This should hold your *½ gallon thermal water container with strap.* And don't forget your compass and map.

A Survival Kit for The Lone Traveller

by Gene Kirkley

Survival—the art of staying alive just one minute longer—until someone finds you and brings you out of your emergency situation, or you can accomplish the same thing on your own.

A list of equipment that would be ideal for a given situation might be totally inadequate for another time or place but, at the same time, there are a few basic items that can form the nucleus of any so-called 'survival kit'. The enterprising traveller who plans to venture off the beaten track should remember that even the best kit will do him no good unless he has it with him when the emergency occurs!

So, in the beginning, this would indicate that any item to be included must be one that can be easily carried along with every other item in the kit. A bag or other container that might be used must lend itself to being readily available. This could be perhaps a *small leather or canvas pouch worn on the belt;* another fine option is a *pint-sized flat can* with a taped-on lid that fits in a coat pocket or just inside your pack. I'll have more to say about that same can later.

One ingredient that does not go into the kit but is considered by many as the most essential is mental attitude. Your acceptance of the challenge to 'stay alive' and your use of whatever knowledge you may have acquired either through experience or study will likely be the determining factors in your survival. It is only human nature to be afraid, but the ability to control fear can mean the difference between a state of disastrous panic and an orderly progression that leads to survival and rescue. Yes, your brain is your most important piece of equipment!

Now for the items that CAN be gathered and carried in some kind of kit. Let's assume that you have worn sensible clothing determined by where and when you ventured off the beaten track. Then body shelter becomes a matter of protection from either the sun, rain or cold temperatures. *A space emergency or rescue blanket* weighing only two ounces that packs in the space of a pack of cigarettes could be the lifesaver in either case. *Plastic garbage bags and leaf bags,* too, can serve as shade, protection from rain, or shelter from cold and wind.

One of the most important skills that the adventurer can learn is how to build a fire under adverse conditions. Not only is fire invaluable as a source of heat, but the soothing effect can help to restore confidence to one who may be near panic. A friend of mine tells of an experience where he was lost in the desert with the thermometer above the 100 degree mark. The first thing that he did was to gather brush and build a fire. He managed to remain calm enough to figure a way out of his predicament. Even if I have to leave out something else, I'll have at least two or three ways for starting a fire in my kit. The best perhaps is a *waterproof* (make sure!) *case* filled with *waterproof matches*. The big wooden kind! My own case that I'm now using has a piece of rough emery cloth glued *inside* the cap to ensure a dry place to strike the match! Instead of the 'flint and steel' that was once the standby for the pioneer and mountain man, I prefer what is called a *metal match.* It's only 2½in long, and it will produce a shower of sparks when scraped with any sharp object such as your knife blade.

Fine steel wool is the best thing I know of for catching those sparks and the beginning of a blaze; the 'flint and steel' folks used a bit of charred cloth. I rather doubt that either would be available to a person in a survival situation, so I like to provide another means for starting that fire. *Fire starter tablets* as well as *fire starter ribbon* that looks much like a tube of toothpaste are available at camping supply stores. Both are good when the tinder is wet, but the best thing I've found for starting a fire under wet conditions is *carbide!* Yes, the same thing once used to light the miner's headlamp. It is readily available in small cans of pea-size pellets. My personal kit contains a 35mm film can filled with them. I start my fire by putting five or six of these little pellets in a small hole under my tinder. By adding a small amount of water, a gas is generated that will readily ignite from just a spark from the metal match. Care should be taken to keep the carbide from getting wet until it is needed, and then be very careful until you understand the power of the

generated gas. But it will guarantee you a fire when you need it!

While we're on the subject of fire, let's add a three inch stub of *candle*. That, too, can be of great help in starting your fire as well as providing some much needed light.

Now that you have the means for building a fire, let's add a *knife* to the kit. Opinions differ widely as to what makes the best knife for survival situations, but I lean toward the sheath knife (not the folding kind) with a four or five inch blade and a sheath that folds down over the handle as added protection against loss. Several brands will fit this description; my own happens to be a Buck. I don't like the folding kind because on more than one occasion, I have been so cold that I could not have opened one. An injury might make it difficult to open the blade, too.

A bit of advice about the knife, and this applies to each item in your survival kit. Don't ever sacrifice quality for economy! Get the best you can afford, and then take care of it. A dull knife can be very frustrating as well as more dangerous than a sharp one. I should mention that likely I'll have a *folding pocket knife* along since I'm almost never without one wherever I may be.

If your travels will take you into the back country, by all means include a *compass* and, if possible, a *map* of the area. I want a good quality compass in a protective case; I have one that has lasted through some pretty rough travel for over thirty years. Two makers of quality ones are Silva and Michaels. Don't get caught without one!

Back to that third item of priority, water. Quite often, unless you happen to be in desert country, there will be some source of water, but likely the quality of it will be questionable. Since it might not be possible to boil the water to insure its safety, I carry a small container of *water purification tablets*. Many of us have used Halazone tablets over the years, but I'm now using a new one called Potable Aqua. To me, the water tastes better. Learn to use one of those plastic bags mentioned earlier as a solar still if no other source of water is available. Moisture you've got to have to keep your body functioning! That pint can as a container for your kit can double as a water container for boiling if necessary or for use in heating food and drink.

Ask a dozen people what they consider to be the best *first aid kit*, and likely, you'll get a dozen different answers. The very minimum that I could accept would contain aspirin, bandaids, antiseptic soap, and ointment. If you need any special medication, be sure to include a few days' supply—just in case. I've added a couple of Darvon capsules obtained from my doctor for use as a pain killer in case of an accident. Hopefully they'll never be needed.

Many, many words have been written on the subject of 'living off the land'; the idea of procuring life-sustaining food from wherever you happen to be in the wilds. This is fine for someone 'living' in the wilds when he is able to supplement the foods from nature. What I'm trying to point out is this: it is very likely that you will use up more energy in obtaining the food than it will provide in return. Sure, if small animals can be snared or trapped or if fish can be caught or if berries can be picked—the list could go on. You must decide whether to go looking for food, conserve your energy or direct your efforts towards finding your own way out. Some *high energy sources* that should be in your kit include sugar cubes, bar chocolate, bouillon cubes, tea bags or packets of instant coffee. Small pieces of hard candy can also provide quick energy; they can give your spirits quite a lift, too!

In spite of what I just said about trying to snare game or catch fish, I do include the means to do both in my kit. A bit of *fishing line* and a few small *hooks* along with a few feet of 18 gauge *wire* take up very little space in your kit, and the added weight is negligible. Even though the caloric content of the natural food obtained may be small, here again your morale gets a big boost from eating a fish you've caught!

Just about every place that I've been in the boondocks, insects have been somewhat of a problem unless it was very cold. You may not encounter any 'deadly' species, but just about any of them can make your life a lot more miserable if you have no protection. Two brands of *repellent* that are available in small plastic bottles are Cutter's and Deep Woods Off. Both are excellent and deserve a place in any survival kit.

Since man does not see too well in the dark, and because we have been used to artificial light all of our lives, a small compact *flashlight* that operates on two AA size *batteries* can be worth its weight in gold. I've used one of these for as much as a week at a time for the necessary lighting in camp on one set of batteries. A little light in the dark is also a wonderful thing for your morale!

Any person in a survival situation needs to be on the alert for rescuers; a method of signalling or attracting their attention is very necessary. One of the simplest ways is with a *whistle*. The lost person may have shouted himself hoarse already. My whistle is of hard plastic; shy away from metal. Very cold metal will stick to your lips and be very painful. A small *mirror* makes an excellent signalling device, especially if it is the kind with a sighting hole in it. *Signal flares* that make either loud noises, bright lights or coloured smoke are nice to have if space and weight permit. And don't forget that green leaves or branches on a fire give off smoke that can be seen for some distance. Make some advance preparations for signalling so as to be ready when you think help is within sight or sound.

A length of *nylon cord* or even an extra pair of *boot laces* could be very useful to you in any emergency situation. Look around you; chances are that you'll find worthy substitutes for the things to which you have become accustomed. Keeping your mind occupied is a great help in avoiding deadly panic.

And that brings me back to the most important thing you will have with you—your brain. Use it! Think positively! Keep telling yourself 'I'm gonna make it!' and very likely you will! Just use what you have to stay alive one minute longer!

Taking it Seriously

Expeditions and other travel with a purpose

Expedition Planning

by Graham Cownie

In our contemporary urbanised existence, the basics of living are taken for granted and survival has become an almost abstract notion only to be read about in books. Now more than ever the thrill of planning and carrying out one's own expedition is attracting more and more independent minded people; giving them the opportunity, perhaps for the first time, of coping with the fundamentals of survival.

This chapter has been designed as a basic introduction to the art of planning and executing such a venture. Obviously, one cannot hope to cover all the problem areas in one short article; instead the following ten sections provide a brief outline of the main points to be considered. The first seven sections deal with the planning stages, the last three concern themselves more with the actual running of the expedition. The real secret is, of course, to know exactly what to expect, and to be aware of every potential pitfall. The importance of reading plenty of background material cannot, therefore, be overemphasised.

Expedition Aims. Establishing the exact aims of the expedition is the most vital step in the planning phase. The nebulous idea that provides the initial impetus for an expedition must be transformed into a more solid and tangible objective. The aims of any expedition may be assigned to one of four categories: (1) *Fun.* A light hearted trip where the sole intention is enjoyment. (2) *Proving or Achieving Something.* For example, the British Trans-Americas Expedition opening up the Darien Gap, or Bonington's successful attempt on Everest's most difficult face. (3) *Discovery.* Few expeditions now make this their main aim since even the most remote areas have been mapped with the aid of satellite photography. However, the more remote parts of the world, such as West Irian in New Guinea, still offer a chance to do some useful survey work. (4) *Scientific.* Expeditions whose priority is scientific experimentation, with all other aspects of the trip being of secondary importance.

Exact classification of the aims of your expedition may at first appear to be rather trivial. It is however vitally important to those who hope to obtain financial support or free equipment. No company will even consider sponsoring an expedition if it has not first received a detailed well organised outline of the expedition's aims and plans. Additionally, without a clear idea of what you intend to do, the chances are that you will eventually embark on a trip with little or none of the right equipment and no specialised expedition members. If you don't know what you are going to do, you won't know how you are going to do it. Bearing these points in mind, decide on your exact aims as soon as possible; don't try too much too soon. In other words don't plan to row the Atlantic until you have at least crossed the Channel!

Team Selection. As often as not other people will have to be found if the planned trip is ever to get off the ground. This is perhaps one of the most crucial stages of planning a venture abroad. The importance of having the right people to go with cannot be overemphasised. Many expeditions have failed owing to wild attempts to coerce dissimilar people into a viable team. Make sure that those you approach are well known to you personally or to another of your close friends who can vouch for their compatibility with the rest of the group. Do not, in desperation, try to persuade friends to come on an expedition, if you have not previously had the opportunity to test their characters. Even the most light-hearted of weekends in North Wales can illustrate the other side of a person's nature. Especially if the weather is bad! Habits that are simply niggling on a small trip become the centre of bitter rows on larger and more serious expeditions. In short, make absolutely sure that the friends you choose will still be friends when the chips are down.

The number of people needed naturally depends upon what you intend to do. There are those who have travelled huge distances over rough terrain, solo, but this form of expedition should be left only to those with considerable experience and who are tough both in body and mind. The psychological pressures on an individual in a strange and invariably hostile environment are enormous and only the most resilient of personalities can hope to cope adequately with the situation. An odd number of people usually means using equipment below optimum performance; eg, three tents for five people is less economic than the same number of tents for six people, and three in a two-man tent is usually very cramped. Expeditions with an odd number of people tend to be more liable to fall subject to frequent bickering; this is not so obvious with a large group but at the other extreme a three-man trip is less likely to prove enjoyable. There is no hard and fast rule which dictates the best number for a particular expedition and the decision is largely a matter of common sense.

If, after a period of search, you are still unable to find enough suitable companions to accompany you on your expedition, you will inevitably have to go about the precarious task of finding them by some other means. This may be done in two ways: (1) by joining exploration and travel clubs and building up new friendships, which is by far the more reliable method though slower than (2) advertising for team members. If you must use the latter method remember to judge applicants on three main points: compatibility, past experience, and qualifications, the most important factor being the applicant's ability to integrate with the rest of the group. Team members should be fit, enthusiastic, calm under stress, loyal to the leader and to the aim of the expedition, and have a good sense of humour. Although past experience of the expedition environment is a useful 'plus' this cannot be preferred to solid qualifications. If you don't have a doctor on the team, for example, then you would be well advised to look for one. The Young Explorers' Trust is an organisation geared to providing medical candidates and is especially useful in this connection.

The leader himself should be self-critical but at all times confident and decisive. He should lead by example and regard the success of the expedition and the well-being of the party as his responsibilities.

Route Planning. Whether the aim of your expedition revolves around reaching a certain destination or whether the idea is simply to prove a route, one of the most critical factors determining the ultimate success or failure of the trip is the route planning itself. Time spent on route planning is never wasted. It is essential to appreciate that the more you know about the route you will be following, the less you will be taken by surprise when things don't go according to plan. Although the art of such planning could fill a whole book there are certain considerations to be taken into account above all else. (1) Climate: in almost every part of the world there is a definite season during which expeditions are mounted; it is usually most unwise to ignore it. Monsoons, winter, or rainy seasons can stop an expedition in its tracks and in certain cases out-of-season expeditions may even be refused permission to make the attempt. For example, the Sahara in midsummer is often 'closed' to allcomers. (2) Logistics: carefully examine the availability of essential stores. Very few major expeditions can hope to be totally self-sufficient so make sure that your vehicles have the range to get from one petrol supply to another. Check the availability of food en route, note the positions of the nearest hospitals, make sure that the water sources are still there at the time of year that you intend to visit. (3) Terrain: take account of the geography of the places you pass through. Don't plan a route regardless of the contours: if you are on foot it may prove exhausting and if you're in a vehicle it may prove impossible. If you intend to traverse a land mass by means of its rivers then choose the direction that gives you the most downstream travel. If you are on foot choose the direction where the prevailing winds travel in the same direction as you do. Check that those depressions you intend to cross aren't filled with water at that time of year. In other words, acquire an intimate knowledge of the terrain *before* you cross it. (4) Politics: take account of political and military restrictions. Many countries pose enormous problems of this nature (especially in Africa). Try to route through those countries most politically aligned with your own; make sure that the borders are open and in sensitive border areas don't wander off the standard routes; you may be shot at or stray into minefields (a big problem in the Western Sahara). See next section also. (5) Transport: make sure that the route is feasible. Assume nothing. Just because the road on the map is intersected by a river don't assume that it will necessarily carry a bridge or that a ferry service will be available. Find out for sure. Again, read as much as you can about the countries you intend to visit; only that way can you be aware of problems before they arise.

Political. Since political problems perhaps represent the greatest pitfall for the modern day explorer, here are some important 'do's and don't's'. First make sure that *before* you go you have all the necessary documents relating to your vehicle, expedition permission, visas, health, restricted drugs (in the first aid kit), equipment, and insurance. Political clearance particularly for the Third World countries should be applied for 12 to 18 months in advance of departure. It is most unwise to ignore formalities and documents deemed only to be 'advisable' as opposed to mandatory. The simple rule is that if there is a form to be filled in—fill it in. In many Third World countries bureaucracy has reached absurd proportions and so the more forms you produce at the border, the fewer you are likely to have forgotten. Plenty of neatly filled forms usually impress customs officials and immigration officers no end. Even in Europe there have been instances where a Student Union card along with a Student Rail card have served in lieu of a passport! Check up on clothing and hair restrictions, and any other Western habits that may get you into trouble. Bribery is also a big problem in underdeveloped countries; no advice can be offered on this subject save the warning not to be indignant when asked for a bribe. It's simply a matter of common sense—and don't do anything likely to irritate an official. Don't argue or fight with the locals, even if you are in the right. Discretion is always the better part of valour. Don't leave your belongings unattended, for if you do they *will* disappear; likewise don't assume that because you are sleeping next to your equipment this is an adequate defence—lock everything up before you bed down for the night. Don't take photographs of anything in the least bit military; you may be arrested for spying as a result. Finally before you go, check that your passport will be valid for the countries you intend to visit, eg, an entry visa for Israel stamped in your passport will preclude you from obtaining entry into Arab states. Rhodesian visas incur similar problems with certain African countries. Never undertake a major expedition without adequate insurance and always make a note of the address and telephone number of the British Government Representative in the countries you intend to visit. If you do this through the Foreign Office it will also give you the opportunity to check up on areas of terrorist activity likely to be encountered—information which the tourist offices and embassies are usually loath to give.

Finance. The financing of your expedition will undoubtedly prove to be one of the biggest headaches of the planning stage. There are two ways to raise the necessary capital: (a) work hard and save! or (b) try to find sponsors. If it is at all possible try to underwrite the costs out of your own pocket. The immediate appeal of instant money from companies and trusts must be balanced against the knowledge that this system incurs obligations and liabilities, ie the need to write reports, take photographs, etc. In addition to this there is the added strain of *having* to complete the expedition when during its execution you would rather limit or modify your aims and objectives. The uncertainty of getting enough sponsorship before your expedition is due to leave, the added initial costs of having to get a brochure printed, the endless letter writing, and most of all the fact that you may have to compromise the quality of the gear you take are all 'cons' that need careful consideration before you opt for outside financial help. Before you even begin to consider this question you first have to calculate a detailed costing covering all the financial aspects of the trip. When doing this there are a number of points to bear in mind. (1) Always pay attention to the smallest detail and cost out each and every item of equipment and spares that you will be taking. Never simply allow a bulk sum to cover, say, kitchen sundries or gaz cooker spares, since their actual cost will invariably exceed the amount you have allocated. Likewise with vehicle spares, buy specific parts before you leave so that you know how much you have spent on repairs even before you go. (2) Don't forget the costs you are likely to incur whilst running the trip e.g. customs charges, added fuel costs due to currency fluctuations, tyre repair services, minimum cash requirements for entering the country, and a small allowance against on-the-spot fines for traffic offences. (3) Remember also the cost of closing down the expedition, especially the cost of having the expedition report printed. No matter how much you budget for your trip, it will always end up costing that bit more—so add on ten per-cent as a safety margin. Provided you have paid sufficiently close attention to detail the final figure should be a fairly accurate indication of the required capital outlay.

Equipment. The rules regarding the selection of equipment are simple and uncompromising: get the best. Quality doesn't cost—it pays. It is always the second rate equipment that can be guaranteed to go wrong just when you need it most. Having said this it is worth noting that the most expensive equipment will not always be the best. A £150 sleeping bag may well represent the finest in down quality but it is not what you want in Borneo. A much cheaper Dacron bag would in this case be infinitely preferable because of its superior warmth-retaining characteristics when wet. Herein lies the point, that you must conduct sufficient background research to enable you to establish the exact requirements for each item of equipment. When you know what your sleeping bag will have to contend with, then is the time to go out and choose one. Buy your equipment as soon as possible. It seems that expedition gear is totally unaffected by politicians and continues to support a price inflation rate way above the official figures! Having most of your gear accumulated at an early stage also gives you the time to examine it in detail; you will then have time to make modifications and strengthen weak points if need be. Each member of the party should supply the leader with a list of his own special equipment requirements well ahead of the departure date but be himself responsible for obtaining it.

Training. As in any activity that requires a high degree of proficiency, training plays an important part in the preparation for your trip. Requirements may broadly be divided into three main areas: physical fitness, technical know-how, and procedure. Regardless of the type of expedition you undertake you will *have* to be fit; jogging, squash, swimming and hill walking all serve to improve your stamina. Whether you are driving a vehicle or climbing a mountain plenty of stamina is of the utmost importance. Technical know-how is also vital. Of course the presumption here is that any major expedition will be based around members already proficient at whatever aspect of adventure they intend to pursue but on a major expedition there are bound to be new items of equipment that will need practice and understanding if they are to be used to full effect. Make sure you get that practice. An emergency is not the time to find out how to give an intravenous injection into an unconscious casualty! Procedures are also important to revise, especially those concerning emergencies. Revise your crevasse rescue methods, revise your deep river crossing techniques, learn how to make a Thomas Stretcher, practise your capsize drill—no matter what type of expedition you plan to do there are always relevant emergency drills to be learned off by heart. All these procedures should, of course, be familiar to you and on a major expedition in a remote area it is all the more important to have them taped.

Health is the most important of the three principles governing the day to day running of the expedition. Ill health is usually caused by one or more of five factors. These are given below, along with the basic precautions needed to ensure that you remain healthy throughout your trip.

Wounds—small cuts and abrasions produced by brushing against thorns, falling on stones, etc. Not serious in the UK but must be treated rigorously in damp tropical climates.

Ingestion—eating and drinking unclean food and water. See *Food and Water: Advice for the Long Distance Traveller.*

Contact—small organisms working their way in through the epidermis after physical contact. Two main offenders, bilharzia and hookworm. Bilharzia in infested water. Hookworm enters through soles of the feet after walking barefoot. An even greater danger is from malaria-carrying mosquitoes. These are most in evidence at dusk and it is vital to have your insect netting set up well before sunset.

Climate—illness such as sunstroke, heat exhaustion, snow-blindness, hypothermia, pulmonary and cerebral oedema. All caused by too rapid an attempt to adapt to prevailing conditions or insufficient protection from the elements. Avoid sunstroke and heat exhaustion by avoiding prolonged exposure to sun and heavy physical work. Avoid snowblindness by *always* wearing goggles whilst on snow in bright conditions. Avoid the two oedemas by not going too high too quickly. Note, the only real cure for this potentially lethal affliction is to retreat to a lower altitude immediately.

Animal Attack—not a very likely cause of injury. See *The Independent Traveller's Guide to Health.*

In addition to the precautions needed against the above five hazard areas, ensure that before you go you have had an adequate course of vaccinations. Be especially careful to make sure that these are well documented on the appropriate forms. In areas where malaria is prevalent it would be most unwise not to take the recommended course of anti-malarial drugs.

Survival. Food, clothing and shelter

are the three prerequisites of maintaining a reasonable level of existence on an expedition. Their order of importance varies, of course, with the climate; for example, in polar regions shelter is of the first importance whereas in the Sahara, food and water are the most necessary items. Despite such minor variations in priority, all three remain essential factors of survival in a hostile environment.

A tremendous amount has been written on the subject of food and water, with all sorts of weird and wonderful advice relating to calorie balanced diets and glucose intake, etc. Most of this may safely be ignored. The essential quality your food must possess is palatability; if you enjoy what you are eating and you have enough of it that is all that matters. If weight is a big problem, don't take tinned food. Dehydrated meals can be quite pleasant if cooked well. 'Vesta Meals' are probably best in this respect. Some brands, supposedly designed for expedition use, are quite revolting. Chocolate, peanuts, raisins, and glucose tablets are all pleasant between-meal 'munchies' and are especially useful on mountains. Since bread will probably be hard to come by take plenty of biscuits. Don't dismiss the food planning as a minor detail. With a little forethought a good selection can make all the difference to team morale. Packets of 'Rise and Shine' orange or grapefruit juice are a change from water and help to hide the taste of chlorine produced by water purification tablets. Hot drinks such as tea or coffee are essential in cold climates so don't forget to take vacuum flasks.

In tropical regions clothing is not of great significance. Its main requirements are that it should be durable. Whether one wears shorts or long trousers is a matter of personal preference: shorts are certainly cooler but long trousers afford more protection against thorns and insects. Don't make the mistake of taking no warm clothing—the nights can be very cool. In areas with a polar-type climate clothing may represent the difference between death and survival. Good quality protection is vital. This must fulfil three requirements: it should be windproof, waterproof, and warm. Waterproofing inevitably involves condensation (although the new 'Goretex' material claims to eliminate this problem) so you might as well get something substantial. Very light nylon waterproofs soon start to leak and are far from adequate. Warmth can best be obtained by wearing fibrepile undersuits with down anoraks and breeches on top. Your head is responsible for a high proportion of heat loss from the body, so invest in thick woollen balaclavas. Dachstein make the best mitts for cold weather. These coupled with a pair of leather palmed nylon overmitts are ideal protection for hands. No matter where you intend to go, a good pair of boots is essential These should have vibram soles and one-piece uppers with the tongue sewn up to the ankle. Spend plenty of time getting a pair that fit you properly; badly fitting boots make walking a painful business. As with other equipment, quality is the key to a good buy.

Little need be said on the subject of shelter: you will inevitably be sleeping under canvas and a first class tent is a must. See *Lightweight Equipment*. For general use, the Vango Force Ten range is unbeatable in terms of quality, efficiency and durability. For more specialised mountain work, the MacInnes Box Tent or the Black's Tunnel Tent can be recommended.

Mobility. Your vehicle is probably your most important piece of equipment. Any major overland expedition worth its salt will not be sticking to tarmac roads, consequently a suitable well-designed vehicle is vitally important. This should be a four-wheel-drive type with a long wheelbase, a permanent 'hard-top' roof, and good ground clearance. The suspension should be stiff, and it is a good idea to have a 'chimney-pot' exhaust system to reduce the risk of the engine stalling when wading shallow rivers. Other useful modifications include an additional fuel filter between the tank and pump, a capstan winch on the front, a sump guard, and an oil cooler. Several companies specialise in preparing vehicles for long overland trips and will make modifications to order. On the whole it is not a good idea to make use of them; doing the alterations yourself will give you practice at working on the vehicle *and* a clearer understanding of how it works. If you *have* to make use of these companies (because you lack the ability to carry out the changes yourself) you should seriously consider recruiting a competent mechanic onto the team.

Although petrol engines are more expensive to run than diesels, they are quieter, more easily understood, and spares are more readily available. Don't use soft-top vehicles—they may be more pleasant in hot weather but their canvas coverings are usually inadequate in torrential rain and are easily ripped. Hard-tops avoid these problems, allow you to use a roof-rack, and afford much better protection against thieving if you ever have to leave the vehicle unguarded. By far the best vehicles to use are long wheelbase Land-Rovers, Range Rovers or Jeeps. Toyota and Daihatsu make adequate alternatives. A good set of spares and tools is, of course, essential. Although a lot may be gleaned from reading books about vehicle maintenance, there is no substitute for a competent mechanic.

Envoi. Before you even attempt to plan an expedition it is as well to get one point perfectly clear. You should *never* approach any venture with the idea of 'conquering' anything. You *cannot* conquer the desert, the jungle or the mountains; the natives you meet are not simply savages who may constitute a threat to your safety. Remote and wild places are to be understood and respected. Do not blindly conduct your expedition to some military schedule; instead stop and remind yourself of the tiny part you play in the grand order. You should not be conducting a battle against the elements. Wild places can teach you so much about yourself if you only stop and think. If you try to become a part of the places you visit then not only will you stand more chance of success but you will also learn to savour and appreciate the wonders of the wild places that so many others clumsily destroy with their unthinking arrogance.

Selecting Travelling Companions

by Dr. John Payne

Over the past 20 years a virtual revolution has been taking place: travelling has become easier and remote places suddenly more accessible. For many young people an adventurous undertaking is often a protest against the rigidity, conformity and lack of opportunity which characterises so much of urban life today. The combination of opportunity and outcry has led, not only to an increase in full-blooded expeditions, but also to an increase in adventure travel: far from wanting travelling to be made easier, many people are trying to find ways of making it more difficult.

Travel of an unusual kind will almost certainly involve a change of life style. Some particular knowledge or skill is often needed—even if it is as mundane as being able to cook, or repair a Land-Rover—and so too is the ability to endure setbacks and disappointments. It is, ironically, those very reversals which, recalled later in comfort, make the whole venture seem worthwhile.

When you are choosing your travelling companions, you will need to be sure of the basic facts: where you are going, what you will do there, and how you propose to get there.

You will need also to be aware that any worthwhile expedition or trip has a dual function built into it: a 'stated objective', which is some exploration of the physical world, and the 'hidden objective' of each individual participant, which will consist of some inner exploration—a chance to learn more about himself, acquire or practise a new skill, exercise his scientific curiosity, learn self-reliance as part of a new group, stretch himself physically. Each participant has a private notion of what he will gain from the trip and it is often true that the more he has sacrificed to go, the more he will hope to gain. This applies to the seasoned explorer as much as to the trainee or student.

On an expedition with a large scientific content, the organiser should always remember that observation and feeling also have an important part to play. The feelers and viewers may not get on easily with the dedicated measurers, but both types have a contribution to make. On an expedition that involves training of young or unskilled people, the leader must ensure that the training is done well and that each part of the expedition is seen to be essential and not merely contrived for the sake of putting obstacles in the way of the less accomplished participants—however much they will later on recall with satisfaction the moments of combined effort and achievement.

An interview is an important part of selecting team members. If you have known the applicants before, you will know if you can get on with them. If you have never met them before, you will have to ask them (and yourself) questions about them. Do they have sufficient experience in the field in which you are interested? Can they climb, canoe, drive, navigate or whatever? Have they camped/mountaineered before? If so, where and in what circumstances? Do they want to go to the same place as you? Is their citizenship likely to be a problem? Are their attitudes to the main object of the expedition, to chores, money, accommodation, very different from your own? What can you flush out about any fads or crankish notions they may have? Do they seem the sort of people who will be dismayed by physical discomfort, or depressed when things go wrong—in short, how adaptable are they? And very importantly, do you think they will get on with the other people who are thinking of going on this trip?

It is a good idea if all the people who are going on the trip meet one another beforehand, so that they can assess each other and try to decide if they will get on well together. If this is to be a democratically organised trip, it is probably best if the majority in a larger group have the right to veto someone who they think will not fit in. In a smaller group of, say, fewer than eight people, all established trip members could have the right to veto any subsequent applicant. This may cause some hurt feelings at the time but, properly handled, it will be less disastrous than having a personality clash when you are too far from home to do anything about it.

Some expedition organisers run training or selection weekends or trips. These are by definition an artificial test and the evidence suggests that such sessions may be useful for testing compatibility or for giving an already selected team the chance to get to know each other, but of little use in providing mock situations which in their nature and effects will bear any relation to real situations that later arise in the field.

People desire to travel for an enormous variety of reasons, but each venture usually falls within an identifiable category, depending on the interests of those who originally conceived the project. These expeditions may be purely adventurous—mountaineering, caving, canoeing, camping—or purely cultural, but whatever the aim, it is better if everyone is in agreement about the project and how it should be carried out. An expedition should be run by common consent, not by dictatorship. It is important to have a mix of personalities—extrovert and introvert, talkative and silent, enthusiastic and energetic, as this will add to the enjoyment of the venture. It is an advantage if everyone is involved in the planning at some stage, as they will then have a share of the responsibility, whatever happens. Mixed groups are theoretically a source of conflict and discontent, but in practice these usually work out well.

It is interesting to note that while many aspects of expedition organisation and equipment and the physical sciences generally have advanced enormously over the past twenty-five years, personnel selection has not made similar strides. Selection of individuals still requires individual judgement and has not yet been successfully replaced by any form of psychological testing or electronic selection. When the New Zealanders were selecting men to spend one year in the Antarctic at Scott Base in the 1950s and early 1960s they found that selection by a panel of experienced men who had extensive practical and scientific experience in Antarctica was the most effective procedure. The panel selected men who gave the impression of being quiet, intelligent, alert, good-humoured, interested, hardworking, experienced in life and moderate in their habits and who did not like women too much! There was a general agreement that a leavening of older men between 30 and 40 was valuable for the stability of the group. Personality testing revealed the majority of these men to be introvert and self-sufficient.

Young people will tend to travel for an adventure, to organise their own lives in their own way for a period and do things that excite them; older people may wish to shake their lives out of their accustomed rut by living spontaneously and differently, and satisfying some long-harboured ambition. In all circumstances they must choose travelling companions who will fit their mood and their endeavour, and so combine with them to make the journey satisfying and enjoyable. If they work together to overcome the difficulties and frustrations, as well as sharing the enjoyment and the excitement, the enterprise may enable them to live more fully in the more mundane life that awaits them on their return.

Survival—Jungles

by Robin Hanbury-Tenison

The key to survival in the tropics is comfort. If your boots fit, your clothes don't itch, your wounds don't fester, you have enough to eat and the comforting presence of a local who is at home in the environment, then you are not likely to go far wrong.

Of course jungle warfare is something else. The British, Americans and, for all I know, several other armies have produced detailed manuals on how to survive under the most arduous conditions imaginable and with the minimum of resources. But most of us are extremely unlikely ever to find ourselves in such a situation. Even if you are unlucky enough to be caught up in a guerilla war or survive an air crash in the jungle, I believe that the following advice will be as useful as trying to remember sophisticated techniques which probably require equipment you do not have to hand anyway.

A positive will to survive is essential. The knowledge that others have travelled long distances and lived for days and even months without help or special knowledge gives confidence, while a calm appraisal of the circumstances can make them seem far less intimidating. The jungle need not be an uncomfortable place, although unfamiliarity may make it seem so. Morale is as important as ever and comfort, both physical and mental, is a vital ingredient. To start with, it is usually warm, but when you are wet, especially at night, you can become very cold very quickly. It is therefore important to be prepared and always try to keep a sleeping bag and a change of clothes dry. Excellent lightweight and strong plastic bags are now available in which these should always be packed with the top folded over and tied. These can then be placed inside your rucksack or bag so that if dropped in a river or soaked by a sudden tropical downpour—and the effect is much the same—they at least will be dry. I usually have three such bags, one with dry clothes, one with camera equipment, notebooks etc., and one with food. Wet clothes should be worn. This is unpleasant for the first ten minutes in the morning but they will be soaking wet with sweat and drips soon in any case and wearing them means you need carry only one change for the evening and sleeping in. It is well worth taking the time to wash these, or even just rinse them out whenever you are in sunshine by a river so that you can dry them on hot rocks in half an hour or so. They can also be hung over the fire at night which makes them pleasanter to put on in the morning but also tends to make them stink of wood smoke.

Always wear loose clothes in the tropics. They may not be very becoming but constant wetting and drying will tend to shrink them and rubbing makes itches and scratches much worse. Cotton is excellent but should be of good quality so that the clothes do not rot and tear too easily.

For footwear baseball boots or plimsolls are usually adequate but for long distances good leather boots will protect your feet much better from bruising and blisters. In leech country a shapeless cotton stocking worn between sock and shoe and tied with a drawstring below the knee and outside long trousers gives virtually complete protection. As far as I know no one manufactures these yet, so they have to be made up specially but are well worth it.

Hygiene is important in the tropics. Small cuts can turn nasty very quickly and sometimes will not heal for a long time. The best protection is to make the effort to wash all over at least once a day if possible, at the same time looking out for any sore places, cleaning and treating them at once. On the other hand where food and drink are concerned it is not usually practical or polite to attempt to maintain perfectionist standards. Almost no traveller in the tropics can avoid receiving hospitality and few would wish to do so. It is often best therefore to accept that a mild stomach upset is likely—and to be prepared (see *The Independent Traveller's Guide to Health*).

In real life and death conditions there are only two essentials for survival, a knife or machete and a compass, provided you are not injured, when the best thing to do is crawl, if possible, to water and wait for help. Other important items I would put in order of priority as follows: a map; a waterproof cover, cape or large bag; means of making fire, lifeboat matches or a lighter with spare flints, gas or petrol; a billy can; tea or coffee, sugar and dried milk. There are few tropical terrains which cannot be crossed with these, given time and determination.

Man can survive a long time without food, so try to keep this simple, basic and light. Water is less of a problem in the jungle, except in limestone mountains, but a metal water container should be carried and filled whenever possible. Rivers, streams and even puddles are unlikely to be dangerously contaminated, while rattans and lianas often contain water as do some other plants whose leaves may form catchments, such as the pitcher plants. It is easy to drink from these, though best to filter the liquid through cloth and avoid the 'gunge' at the bottom.

Hunting and trapping are unlikely to be worth the effort to the inexperienced, although it is surprising how much can be found in streams and caught with the hands. Prawns, turtles, frogs and even fish can be captured with patience and almost all are edible—and even tasty, if you're hungry enough. Fruits, even if ripe and being eaten by other animals, are less safe, while some edible-looking plants and fungi can be very poisonous and should be avoided. Don't try for the honey of wild bees unless you know what you are doing, as stings can be dangerous and those of hornets even fatal.

As regards shelter there is a clear distinction between South America and the rest of the tropical world. In the South American interior almost everyone uses a hammock. Excellent waterproof hammocks are supplied to the Brazilian and US Armies and may be obtainable commercially. Otherwise a waterproof sheet may be stretched across a line tied between the same two trees from which the hammock is slung. Elsewhere, however, hammocks are rarely used and till tend to be a nuisance under normal conditions. Lightweight canvas stretchers through which poles may be inserted before being tied apart on a raised platform make excellent beds and once again a waterproof sheet provides shelter. Plenty of nylon cord is always useful.

The jungle can be a frightening place at first. Loud noises, quantities of unfamiliar creepy-crawlies, flying biting things and the sometimes oppressive heat can all conspire to get you down. But it can also be a very pleasant place if you decide to like it rather than fight it—and it is very seldom dangerous. Snake-bite, for example, is extremely rare. During the 15 months of the Royal Geographical Society Mulu Expedition in Borneo no one was bitten, although we saw and avoided or caught and photographed many and even ate some! Most things, such as thorns, ants and sandflies, are more irritating than painful and taking care to treat rather than scratch will usually prevent trouble.

Above all the jungle is a fascinating place—the richest environment on earth. The best help for morale is to be interested in what is going on around you and the best guide is usually a local resident who is as at home there as most of us are in cities. Fortunately in most parts of the world where jungles survive there are still such people. By accepting their advice, recognising their expertise and asking them to travel with you, you may help to reinforce their self-respect in the face of often overwhelming forces which try to make them adopt a so-called 'modern' way of life. At the same time you will appreciate the jungle far more yourself—and have a far better chance of surviving in it.

Safety and Survival at Sea

by Maurice and Maralyn Bailey

Survival is as old as the world, but, because of the manner of our modern way of living, where we specialise and employ specialists for the necessities of life, relatively few people have to live an utterly basic existence with all the shrouds of 'progressive' materialism and comforts stripped away. Intellectually naked and with only a vestige of civilised living left, survivors discover, or more correctly rediscover, other codes of behaviour to comply with a primitive life style reminiscent of remote millennia. Armed with our basic animal instincts we are all capable of enduring, to a greater or lesser degree, conditions of acute hardships and sailors have long taken more than a passing interest in the consequences of shipwreck.

Our requirements for safety and survival at sea can be summarised under four headings: The Boat; Safety; Damage Control; and Survival.

The Boat. For us safety at sea begins with the boat. A soundly constructed watertight hull with no vulnerable stress points is the first priority. Then must come the security of the mast and the rigging, of the rudder and its fastenings. After this meticulous attention to maintenance will form the basis of safe cruising. If the mast fittings and rigging are oversize they will add appreciably to the crew's peace of mind. The best available ground tackle is a cheap insurance and long distance voyagers will be well advised to carry more than two anchors with ample cable of above average weight. Since the sails will, for the long distance sailor, be the main source of power a second suit of working sails is well worth considering especially when one contemplates the rapid deterioration of Dacron in ultraviolet light. Even on a sail boat everything mechanical and electrical should be thoroughly overhauled annually or before the start of a long passage and ample spare parts and tools should be carried.

Safety. Almost everything in this world that comes under the heading of adventure has its moments of danger and all seafarers must realise this. They must become self-reliant without dependence upon any outside assistance when things start going wrong.

It is impossible to compile a list of safety equipment in any order of priority since it is all, to our minds, necessary to a boat contemplating a long voyage or, in truth, any cruise. There will be many things which have apparently been overlooked and it could be argued, for example, that adequate up-to-date charts are items of safety equipment. We would suggest that another, though less evident, aid to safety would be a self-steering device, electronic or wind activated, which relieves the crew of the exhausting business of steering and allows them to give their full attention to lookout tasks. No excuse can be found for economy with pyrotechnics and every boat should be liberally endowed with up-to-date flares. It is absolutely essential to have rocket-type flares on board and if they go off with a bang, so much the better. White hand flares should be kept handy at all times to warn approaching ships of your presence. The argument for fitting an octahedral radar reflector permanently to the mast in such a way as to present many surfaces to the projected beam is very sound.

Some considerable thought must be given to the various safety harness systems available to ensure the harness selected will be comfortable, easy to get into, quick to adjust and, just as important, easy to get out of in an emergency. Every crew member should have a harness and no one should leave the security of the cabin or cockpit in heavy weather or at night without being hooked on. Whether the crew are equipped with buoyancy aids or lifejackets is a matter of individual preference. We prefer the lifejacket because, in its deflated state, it encumbers the wearer to a minimum.

Lifebuoys should be equipped with automatically activated strobe lights and dan buoys. Lifebuoys come in two versions, horseshoe and circular, of which we prefer the latter. Ample fire extinguishers and a gas detection device must be added to the list.

The important contribution the health of the crew makes towards safety cannot be overlooked. Nothing can upset the efficiency or morale of a crew more than to be constantly beset with worries about the health of one of their number. We cannot warn too strongly of the consequences of the inability of the crew to live amicably together within the restricting confines of a small boat. A trained and rested crew, physically fit and mentally compatible, and following a fair and acceptable routine, should be able to cope with any emergency.

Damage Control. The one really frightening aspect of damage to a small boat at sea is that a fairly small, and potentially repairable, leak in the hull may rapidly exceed the capacity of the pumps and sink the boat. When a leak is discovered, through collision or stress of weather, shock may easily give way to panic and the crew's uncoordinated actions may impede any concerted effort to implement correct damage control.

Any discussion of damage control must include the advantages and disadvantages of installing some emergency buoyancy into the boat itself. Watertight bulkheads with firm sealing doors are an obvious way to contain damage and maintain buoyancy but it is very difficult to achieve this in small boats without encroaching on the living space. The installation of permanent built-in buoyancy by using materials such as expanded polystyrene foam in the bow and stern sections is another possibility. Another, but expensive, method is to install a series of buoyancy bags stowed deflated over 'wasted' space and connected to a remote CO_2 gas cylinder.

With a ruptured hull the one real hope of saving the boat is the use of a quick and effective collision mat or plug to cover the hole. Any damage control procedure must be set in motion promptly, on the understanding, however, that once one is committed to one course of action it will be probably too late to attempt an alternative if the first fails. A sail or canopy lowered over the bows and dragged aft to cover the hole and secured firmly on either side of the boat can be effective. There are available patented collision mats including pneumatic cushion affairs and an umbrella-type device. We carry two of the latter types in addition to a collision mat of our own design. This consists of two pieces of heavy canvas cut as equilateral triangles with 3ft sides sewn together with a piece of carpet sandwiched in between. The edges are roped and heavy thimbles in the corners have ropes spliced on. The ropes are long enough to wrap around the hull.

After a ruptured hull the biggest worry facing the ocean voyager in a sail boat has always been the possibility of being dismasted. From the start it is important to have a clear practical plan for rigging a jury mast and sail. The ship's jury equipment should include tools necessary to clear away damage, prepared patching and shoring implements and copious spares for everything. An emergency steering system must also be worked out before setting sail.

Survival. The best time to learn about survival techniques is before the event and must include a study of past cases and of alternative equipment specifications. Literally survival starts as soon as a boat fails to function in the way it was meant to: in other words, when a sail-boat is dismasted, or a power-boat develops engine trouble, or the hull has sustained some damage which impedes its performance.

After the decision to abandon ship the castaways must be clear in their minds what priority is to be given to items to be salvaged from the wreck. The quicker everyone is off the boat the better but the essential items for survival must be ready to hand and not encumbered by an accumulation of non-essential articles.

There has been a lot of intelligent argument about the merits of using a dinghy instead of a liferaft. The dinghy's main advantage lies in its capacity for propulsion by oar or by sail which would enable survivors to reach safety, especially in coastal waters. However, when compared to a liferaft, it will provide less room for each man, is less stable and it lacks, to start with anyway, the essential canopy to protect the survivors from the sun and from heat loss. Liferafts are a static means of survival and do not lend themselves well to manoeuvrability which is, of course, all to the good if a search for the survivors has been mounted. For ourselves, the liferaft would again be the main instrument of survival but we would make every effort to take the dinghy as well.

Talk of safety and survival must always include the subject of radios. While a good receiving apparatus is a necessary piece of the yacht's navigational equipment, a transmitter is often an expensive luxury for a small boat voyager. A fixed transmitter cannot be taken into the liferaft and must be operated *in situ* and its use is dependent upon the time and the personnel available to operate it in the emergency. The cost of fitting an automatic distress device is justified but even this will take a little time to activate. Small compact radio telephones suitable for the limited space available on board yachts, especially line-of-sight VHF, have too short a range which makes their benefit doubtful in mid-ocean. In addition, transmitters are susceptible to poor earthing. However, every yacht would be well advised to carry a small portable transmitter for broadcasting on the emergency frequencies despite its high cost.

Every yacht should be equipped with a survival pack or 'panic bag' and we think it should contain supplies to sustain a crew for a minimum of four days. The following list would be the *minimum* for four people:

> 16oz cans of water (10); 18oz packs of compressed glucose and vitaminised survival biscuits (6:14,000 cals); vitamin tablets (100); simple first-aid kit; lanoline-based cream for sunburn; anti-sea-sickness tablets (20); electric torch; heliograph; rocket flares, red (4); hand flares, red(4); smoke flares (4); fluorescine dye (2); fishing kit; solar still; knives (2); pocket compass; sponges (2).

Ideally the items should be triple wrapped in polythene and the pack filled out with plastic foam pieces to enable it to float. The pack should be stowed on deck in a waterproof valise as close to the liferaft as possible. Since water is the most essential element of the body's metabolism under survival conditions it will be useful to have several 2-gallon containers filled and handy for quick release on board.

If all else fails and you find yourself adrift on the open sea, water, shelter and food—in that order—will be your main ingredients for survival. Food and water must be rationed and do not eat or drink anything in the first 24 hours. To start with, allow half a pint of water per person per day unless you can be sure of catching rainwater. Any exertion, especially in the tropics, causes a loss of water through perspiration. Everything must be done to reduce this loss by resting in the shade and draping the body in clothes soaked in sea water. Avoid 'immersion foot' by removing footwear, drying the feet and loosely wrapping them in cloth and exercising the feet and toes.

It is important to remember that under survival conditions the intake of food, especially protein, must be modified in direct proportion to the availability of water. For example, one volume of protein food requires two equivalent volumes of water for digestion and elimination of waste. A normal healthy active person requires some 3,000 calories per day but a non-active person in survival conditions could reduce his requirements to about 600 calories per day. If castaways are well nourished initially there will be little cause for alarm if the calorie intake is low to start with. A healthy person is capable of living three or four weeks without food. Thirst can be reduced by sucking a button.

Obtaining sustenance from the sea could be a problem. Large parts of the oceans are biological deserts and the prospect of catching fish and other animals away from the convergence zones and continental shelves will be marginal. Fish always follow dinghies. Brightly coloured fish, fish that 'puff up', fish with teeth like those of humans or with parrot-like mouths and fish covered with spikes or bristles should not be eaten, as they are likely to be poisonous. If there are sharks or swordfish about, throw wastes overboard at night.

A high standard of seamanship with an ability to improvise will compensate for many deficiencies, but when it comes to a serious emergency there can be no excuse for not having the boat and its crew fully prepared.

Survival in the Cold

by Sir Crispin Agnew of Lochnaw, Bt.

Some of the most wonderful areas in the world are cold environments, but living in them poses a constant challenge. Survival becomes a continual battle against exposure or, as it is sometimes called, hypothermia. Exposure occurs when the body loses its heat faster than its mechanism can replace the heat loss. Man needs a constant body temperature of about 36.9°C. If this falls below a certain point, death will occur. In outline, at 33.9°C the muscles cease to work and the victim becomes immobile, at 32.8°C he becomes confused, at 31°C (a drop of only six degrees from normal body temperature) he becomes unconscious and at 28°C he will die.

People who survive longest in the wilds are those who never get into difficulties. Prevention is better than cure, so prepare well. Study the environment and carry appropriate and adequate shelter and clothing, with sufficient food for the whole trip including a survival reserve. An emergency food pack should contain simple sugars which are easy to digest and provide immediate heat generation, but if it is possible to carry a cooker and provide cooked food so much the better. The

route must be within your capabilities and you should note possible shelter and escape routes. We will now consider what must be done if despite all the preparation you are caught out in a survival situation. Three things cause exposure and are therefore the greatest danger to survival. The colder it is the greater the danger, but linked to temperature are the two other factors —wind and wet. Wind carries away body heat by convection and this then has to be replaced by burning more body energy. Scientists have shown a direct correlation between wind and temperature which is called the Wind Chill Factor. The temperature and wind together provide an apparent temperature. One mile per hour of wind is equivalent to between about -18°C (higher temperatures, slower winds) and -19°C (lower temperatures, faster winds.)

The third factor in the equation is the wet. Water is a good conductor and it also destroys the insulation value of clothing, for when it evaporates it extracts heat from the surrounding area and thus lowers body temperature. Physiologists have defined the insulation factor of clothing in 'Clo's'. For normal winter trekking you wear about '2 Clo's' of insulation but if the clothing becomes wet then the 'Clo' factor falls from '2 Clo's' to '.75 Clo's'. Wet clothing increases the speed at which you lose body heat and this is greatly increased if it is windy. Stay dry at all costs and avoid the often fatal downward cold-wet spiral.

You may think that being physically fit will increase your chances of survival if hypothermia sets in. True, it will help a little, but the amount of fat you are carrying is far more important. Body fat reduces heat loss and provides fuel to keep blood temperature raised. Women have a layer of subcutaneous fat and will often survive longer than men as a result.

Good clothing is vital. Even if you have tents and other camping equipment, if your clothing is not good and fitted to the environment then you will be unable to move. Woollens or nylon pile are much to be preferred to cottons or straight nylons because wools and pile retain their warmth much better, particularly when wet. Several layers of clothing are better than one thick layer because they trap the warm air and also give great flexibility in changing temperatures. A suitable combination of clothing for a cold temperature is: (for the top half) a vest, a woollen shirt, a lightweight woollen pullover, a pile jacket such as is made by Javelin, with a windproof anorak and a waterproof nylon *cagoule* for the outer covering; (for the lower half) good boots and gaiters, woollen socks, long woollen underwear, woollen breeches or jeans and waterproof trousers on top. The body loses a lot of heat through the head which should be covered by a woollen hat. Ensure that there is no gap at the stomach where the body's temperature is generated and maintained. Down clothing, sleeping bags and other luxuries should be considered, but their weight must be weighed against their probable use.

In a survival situation you must maintain the body's core temperature as near normal as possible. When the air temperature drops, the body shuts off blood from the extremities (such as the fingers and toes) in order to shorten the circuit and maintain this core body temperature, which is essentially in the stomach area. There is also shivering, which burns up sugars in the muscles and generates heat. You must try to prevent frostbite by keeping the extremities warm—in your armpits or your crotch—but it is better to lose fingers than die. It is essential to seek shelter from the wind. At its most basic this might be the lee of a rock or slope but this should be improved upon wherever possible. Above the snowline it may be possible to dig a snow hole or build an igloo but if these are not available tree or rock shelters can be built. A very simple shelter is provided by a 7ft polythene bag, which keeps out the wind and rain.

You should have fed well before setting off in the morning and continued to eat small snacks at regular intervals during the day to maintain blood sugar levels. If you have done this you will meet any crisis well nourished. A regular intake of food during the survival period refuels the body and helps it to generate heat. Liquid intake is also important because without fluid the body finds it difficult to digest food. Dehydration is one of those factors which lower your body's resistance to the elements. Great care must be taken to keep survival packs as light as possible. Nothing will exhaust a party more than carrying heavy packs, for they may well then be forced to bivouac before reaching their destination.

In the cold, wind and wet you must anticipate and learn to recognise the symptoms of exposure and once they appear take immediate action to prevent the situation from getting worse. The symptoms of exposure can be summed up as 'acting drunk'. A person suffering from exposure may begin to stagger, appear tired or listless, display unreasonable behaviour or have sudden uncontrollable shivering fits. You will notice that he begins to slow down or stumble and he may complain of disturbance or failure of his vision.

If your party is getting exhausted and liable to exposure, stop early, because it is easier to take the necessary action when you still have spare energy. Seek shelter from the elements. Once in your shelter, put on all your spare clothing, have something to eat and make every effort to maintain your core temperature. Huddle together for extra warmth and keep your hands and feet warm by placing them in each other's armpits. There is a great temptation when feeling cold to try and generate heat through violent exercise; resist it because you will merely disperse heat by convection and send warm blood to the cold extremities, which then returns to the core at a lower temperature. Vital reserves of energy will then be used to re-generate the heat. Likewise, do not take alcohol as it creates a false illusion of warmth, sending blood to the cold extremities and overall lowering the body temperature.

Understanding the problem and taking steps to solve it are half the battle for survival, but however good your equipment, clothing and shelter you will not survive unless your mental attitude is right. The will to live is vital. We have many examples in the annals of exploration: Shackleton's party surviving for many months on Elephant Island in the Antarctic, Walter Bonatti surviving on Mont Blanc for over five days in a raging blizzard while some members of his party died on the first day in the same conditions. If you do not have the will to live then you will not begin to take the most elementary necessary precautions. Cultivate determination and it will enhance your survival chances.

Nobody can guarantee you comfort in a really cold environment, but with proper practical and mental preparation, you will probably never be engaged in a real life-and-death struggle.

Bibliography

The Polar World, by P.D. Baird. Longmans 1964.
Safety on Mountains, CCPR Publications.
Mountain Leadership, by Eric Langmuir *Survival*, by George J. Ritchie, *Scottish Mountaineering Club Journal* Vol. XXVII No. 151, May 1960.

Survival—Deserts

by Tom Sheppard

Before examining desert survival, consideration must be given to the kind of circumstances in which a survival situation might arise. The breakdown of a single vehicle on a route carrying relatively infrequent traffic; the demise of one of a group of vehicles necessitating another vehicle's going for help; injury or other problems in a party doing field work some distance from a base camp, again possibly involving some of the party's going for help. Though an alarmist approach is not recommended, a common error in the chronicles of what develop into survival incidents is not thinking ahead or taking the situation seriously enough at an early stage.

Even skeletal preplanning for emergency contingencies can be invaluable in aiding survival and copies of the intended route, rescue plan* or procedures to be adopted should be left with appropriate agencies before departure. Each vehicle or group in a party should be equipped with a heliograph, a small pack of rescue flares and a whistle. Signalling procedures should be worked out—for example, in a convoy each vehicle is responsible for the vehicle behind and a flashing heliograph from a following vehicle could be used to convey 'I have a problem. I have stopped.' A standard procedure for flares is worth working out—for example discharging them only on the hour or quarter hours and preferably only after dark means that all eyes in a rescue party can be trained on the horizon at those times. Many desert routes carry much traffic at night and a flare-equipped lookout can pick his moment to attract their attention from a distance. A fire of old cartons or boxes could also be prepared to attract attention. The universal warning to other traffic on the road of a disabled car ahead is branches or a line of stones laid across the road.

If the party have decided to take an emergency radio beacon (such as SARBE), designed basically for marine use, they should consult the manufacturer on its optimum placing—height above ground, proximity to vehicle sheet metal, etc.—if it is to be used on land. The decision to take such a beacon presupposes that its limited range on land and battery duration still put the expedition within the likely reception radius of regular, known air traffic and that appropriate distress frequencies (UHF or VHF) are also catered for.

*See report on The Joint Services West East Sahara Expedition, Annex Q; Royal Geographical Society Library. Excerpts from expedition Rescue Plan.

A basic appreciation of the physics of heat transfer and the phenomenon of cooling by evaporation is a great help in appreciating nearly all aspects of desert survival*.

Steaming soup in a large soup-plate will cool more quickly than in a jug because there is a larger surface area from which evaporative cooling can take place. Water (steam) is lost in the evaporation process. The human body uses a porous skin and evaporative cooling (sweating) to keep cool (or to maintain a constant 36.9°C body temperature in the face of what would otherwise be a heat build-up due to energy expenditure); in doing so, water is used. The greater the amount of heat to be shifted (or the more the energy expenditure during the heat of the day), the greater the loss of water. If the ambient temperature is above body temperature then to prevent body temperature from rising, water is lost through evaporative cooling even when standing still. In a survival situation there will almost invariably be a water shortage so its conservation is vital. Extracting a severely bogged vehicle will cost less water if the work is done pre-dawn (the sand is firmer then anyway); any walking for help should be done at night, resting in the shade during the day; loose, long sleeved shirts and trousers will allow enough evaporative cooling but not lead to the wasteful loss that shirtless bodies in low desert humidities incur. Remember the slow steady work rate of the desert dwellers and the slightly more humid and therefore evaporation-limiting 'micro-climate' with which their loose robes surround them.

*See 'Exploration Medicine' by Edholm and Bacharach, published by John Wright and Sons, Bristol, 1965.

In temperatures over about 42°C, high winds, sand storms or riding shirtless in a fast moving vehicle will promote heat gain from the air by the human body through conduction (contact with a hotter substance, like putting a kettle on an electric hot-plate); considerable and excessive water loss will result from combating this thermal onslaught and clothing must be donned to limit the heat gain. This is the opposite of the wind-chill factor so well known in winter survival but is avoided by the same means—insulation. The difficulty in determining excessive sweating is that in the exceptionally low humidities prevailing in true desert, sweat evaporates instantly and the skin appears to remain dry all the time—unlike the sweat-soaked shirt and streaming face that manifests itself in hot, wet jungle situations.

Remaining in the shade and limiting heat-load by use of light-coloured clothing, hat and possibly an aluminised 'space blanket' shelter will also reduce the amount of heat transfer the body has to cope with by evaporative cooling. A 'space blanket' will be found a useful aid in general camping as well.

Basically the options in most situations are either to remain with the vehicle awaiting rescue, while making yourself obvious to rescuers, and reduce water consumption, or—and only after very careful thought—walk for help carrying as much water as possible. (A sand-ladder with additional 4in x 48in alloy sheet strips pop-riveted on the longitudinals can be turned into a sledge capable of carrying one or two jerry cans in smooth conditions.)

As a very general guide, observations on a recent Sahara expedition with a daily max/min of 40°/23°C indicated 6.4 litres per day consumption for a 6ft, 73kg man carrying out driving, camp and mechanic duties—no heavy digging. This has been seen to vary from 4.5 litres to 10.5 litres for the same duties when max temperatures are 32°C and 46°C respectively—note the greatly increased rate of consumption as temperature rises beyond 42°C.

Note that in a low energy output, 'sit-and-survive' situation, food is relatively unimportant. A man can last a considerable time without any and indeed dehydration suppresses hunger. Water, however, is a different matter.

The following desert water table shows the number of days of expected survival under two conditions:

Note well the use of the term 'survival'—the person's condition towards the end of the periods indicated would be extremely serious.

The table below is of use when planning the logistics of a situation where maintenance of water balance—and thus full mental and physical efficiency—is required:

Daily Water Requirements to Maintain Water Balance—Rest in Shade at all times

Mean Temp* °C	Litres per 24 hours
20 and below	1
25	1.2
30	2.4
35	5.3

*Mean temperature can be taken as 8°C below the daily maximum in desert conditions.

DESERT WATER TABLE — DAYS OF EXPECTED SURVIVAL

CONDITION	MAX DAILY SHADE TEMP °C	TOTAL AVAILABLE WATER PER MAN—LITRES 0	1	2	4	10	20
1.	50	2½	2½	2½	3	3½	4½
Resting in	45	3	3	3½	4	4½	7
shade at	40	4½	5	5½	6½	8½	12
all times	35	7	8	8½	10	14	20
	30	9½	10½	11½	13	19	30
	25	11	12	13½	15½	22½	34½
	20	12	13	14	16½	23½	36½
2.		TOTAL AVAILABLE WATER PER MAN—LITRES					
Walking		0	1	2	4	10	
only at night	50	1 (40)	2 (40)	2 (48)	2½ (56)	3½ (64)	
and resting	45	2 (40)	2½ (40)	2½ (48)	3 (56)	3½ (64)	
in the shade	40	3 (40)	3½ (40)	3½ (48)	4 (56)	5 (72)	
by day.	35	4½ (48)	5 (48)	5 (56)	6½ (72)	7½ (88)	
(Approx distance	30	7 (65)	7½ (70)	8 (78)	9 (100)	11 (112)	
walked shown	25	8½ (83)	9 (91)	10 (105)	11½ (125)	14 (160)	
in brackets—Kilometres)	20	9 (90)	9½ (98)	10½ (130)	12½ (175)	15½ (225)	

Both tables based on information in UK Ministry of Defence (RAF) pamphlet PAM (Air 225) — 'Desert Survival' 1975.

Keeping out of Trouble

Stumbling-blocks

Mishaps, Major and Minor: What to do

Loss or theft

	To avoid:	*Notify:*	*Provide:*	*Expect:*
Passport	Keep a note of the passport number, separate from the passport itself.	Police and nearest consulate or embassy.	Passport number, details of your travel plans and dates, photos of yourself.	To be issued with an exit visa, an emergency travel document or emergency passport (for which there is usually a small charge).
Travellers' cheques	Ask for printed advice from issuing bank or authority; follow advice and keep it with you. Keep a note of the serial numbers of the cheques, separate from the cheques themselves.	Police and issuing bank or local agent thereof; *or* to Thomas Cook or Wagon Lits office if issued by Cooks; *or* to nearest branch of American Express if issued by Amex.	Details of issuing authority and if possible serial numbers of cheques.	From American Express, during working hours, an rapid refund: *or* from Avis, at weekends and in an emergency, a loan of up to $US100 for American Express cheques *or,* with some delay, replacement cheques from any other issuing body.
Credit cards	Keep a note of card numbers and of the issuing company's address, separate from the cards themselves.	Police and bank or issuing company (telephone, telex or cable from post office, large hotel, etc.)		Old card to be cancelled to avoid fraudulent use; eventual replacement issued.
Money		Police and nearest embassy or consulate; and		Loan from consulate or embassy in an

	To avoid:	*Notify:*	*Provide:*	*Expect:*
Money		telex bank asking them to authorise a local bank to issue money to you.		emergency (a small charge is made for this service, payable later) and your passport to be surrendered as security. Money cabled out to a Third World country takes at least two weeks, and persistent enquiry at the receiving bank.
Vehicle	Don't leave vehicle unlocked!	Police and insurance company.		To be issued with a note of confirmation by police.
Luggage and Valuables	Take with you a copy of your insurance certificate.	Police and manager of hotel, campsite, etc.	Copy of your insurance certificate.	
	Air Travel: Take out extra insurance, available from travel agents, as carrier's liability is limited. Watch to ensure that luggage goes on conveyor belt. Keep receipts.	Airline personnel within four hours of arrival. Ensure that they fill in a 'loss or damage' form.		Carrier to find luggage within three days; or to pay compensation, which, on international flights, is based on weight, not value.
Vehicle documents	Keep a full set of photocopies of driving licence, insurance document (green card or equivalent), registration book, bail bond and other documents, separately from originals. One option: get an International Driving Permit (from motoring organisations) and take this with you *instead* of your driving licence.	Police	Details of driving licence; photocopies of relevant documents.	To be issued with a note of confirmation by police; in some countries a temporary replacement of your driving licence to be issued by the national automobile club on provision of details of your licence.

IF:
you have a road accident—notify police and fill in details on form supplied with your insurance card or, if you don't have one, take down names and addresses of people involved and of witnesses; a photographic record of the incident will also come in useful.
you are stopped for committing a traffic offence—do not remonstrate, pay any fine demanded but insist on being issued with an official receipt. (This helps to ensure that you are being charged at the correct rate for the offence.)
you witness a crime or accident—you are required in some countries to stay on the scene and render assistance. This does not apply to a civil disturbance or commotion, when it is best to keep clear in order to avoid a false arrest or charge.
you get into trouble with the police for a suspected crime—insist on speaking to your consulate or embassy, if the police fail to inform them as a matter of course.
you are the victim of a crime—inform the police and your consulate or embassy.
you need a lawyer—consult your consulate or embassy, who will advise on legal aid and procedures and be able to refer you to English-speaking lawyers and interpreters.
you need to make a telephone call home from a Third World country—the most effective, if expensive, way may be to forget about using a post office or the services of a cheap hotel and actually to book into a Hilton or the nearest equivalent so as to use its telephone system. Even this takes time—one traveller reports a 15-hour wait for a call from the Hilton in Rabat, Morocco.
you fall ill—you will usually be required to pay for some or all of the treatment, even if you can subsequently reclaim expenses. This is possible for British travellers in some countries, provided that they keep receipts for medical attention received and medicament supplied and send these to the Department of Health and Social Security, Overseas Group, Newcastle-upon-Tyne, England NE98 1YX.
a member of your party dies—consider having the body cremated locally, which may be less distressing and is usually cheaper than having the body transported back home. The ashes can be flown home for interment. Inform the police and nearest consulate or embassy, who will provide valuable advice and support.

Editor's Note:
This article draws on material in a feature entitled 'What should you do if . . .', published in *The Sunday Times.*

The Trouble with Travel

by Christopher Portway

It was one of those cleverly-organised pickpocketing operations wherein one villain occupies your attention and, from the cover of a jostling crowd, the finger-man—or is it 'dip'?—lifts your wallet. The dipping in this instance took place on a crowded bus. I should have known better, of course, for I've been the mug in similar goings-on in Kabul and the middle of the Baluchistan Desert. My passport and the remainder of my travel cheques reposed sweatily against my tummy, but my air ticket, credit card and all my local currency were in my buttoned jacket pocket for I was in the ritzy part of town making the initial arrangements for my flight home.

I knew my wallet had flown very few seconds after it had been lifted and made myself very unpopular indeed by yelling to the driver to stop, denying exit to all those wishing to alight and dragging aboard a reluctant policeman on point duty who felt impelled to line everyone up against the bus sides to be searched with their hands over their heads. Meanwhile the two villains, who had escaped before the hullabaloo, simply melted into the horrifying traffic pandemonium resulting from the now uncontrolled intersection.

It was all a wasted effort of course. And it happened in Quito, the alleged law-abiding capital of Ecuador. But I knew my drill for such an eventuality and at least the thieves were considerate in that they had chosen a sizeable city for their nefarious deed. In quick succession I went to the police who provided me with a copy of my statement, to the local office of my credit card company to report the card's loss and I would have gone to the office of my airline to trigger the procedure for obtaining a replacement flight ticket had I been flying home from Quito. But I was departing from Bogotà in neighbouring Colombia, a world away when one is virtually penniless. How I got there without the help of either my embassy or the credit card people is a story outside the confines of this article but needless to say that while in Colombia—the most robbery-prone country in South America, and, probably, the world—I made damn sure nobody lifted my *remaining* belongings. This I did by walking the treacherous streets of its towns with a naked machete in my hand and, back in Britain, my insurance company paid for my losses without a murmur.

Simple theft is the main scourge of the traveller in foreign parts though it can occur the moment he or she has left home. I know an editor who lost his passport and money to a thief in the departure lounge of Gatwick Airport. The funny thing was that he was flying to Bogotà with its terrifying reputation for violent mugging and open snatch and grab! Theft can take many forms and be directed against not only one's person but one's vehicle and/or its contents, and from campsite or hotel.

The open type of vehicle like a jeep invariably constitutes a problem of security. Rubber tie-downs are much easier to use than ropes because they go on and come off so easily. But the light-fingered fraternity will find them equally convenient so such a loaded vehicle should *never* be left unguarded.

It is the heavily populated areas, of course, that pose the greatest threat and on no account should *any* vehicle be left alone and unlocked. It is normally the driver's responsibility to see that the doors and windows are firmly locked and, if the roof rack is loaded but cannot be locked, his absence should be, at most, fleeting. On such occasions it is better to park in a busy street than a quiet one thus giving the passing thief less opportunity to lift anything in peace. When parking for the night try to find a hotel park, garage or guarded parking lot; somewhere where there are other cars which, from a thief's point of view, could be under surveillance. Never leave obviously valuable equipment in plain view within a vehicle. It is an open invitation for theft. At least cover the items if you can't take them with you. Even bolted-down vehicle equipment like wing mirrors can disappear in the night and in those parts of the world where a curious crowd collects to see the strange creatures and their vehicle from outer space watch everything with an eagle eye.

The country is less troubled by theft but never relax your guard. Equipment removed from the vehicle should be kept close to where you camp and valuable items locked in the vehicle—unless it is to be occupied overnight.

Camera equipment is always fair game to the robber. Where snatch and grab thieves are rampant, it is a good idea to hold onto the body of the camera and case itself, not just the strap, when walking in the streets. Remember the more daring razor-blade gentry find it only too easy to snip through a camera strap. The same applies to rucksacks though the only recommendation here is for your companion to walk behind you keeping an eye on *your* sack. But how the rear man guards *his* sack is an unsolved problem unless you are all walking in a circle!

Passports excepted, personal papers and documents are unlikely to have any intrinsic value to a thief and he will usually discard them. But to their owner their value is enormous. If you are travelling alone or in a small group you will probably find it more convenient to keep them on your person, tucked in your money belt, with your other small valuables, next to your skin. A larger expeditionary group, however, might choose to keep such items centrally, in a small safe or locked tin hidden in the vehicle.

But wherever you go outside the country never be without travel insurance which also covers your health and possible curtailment of the proposed tour. Carry out the company's instructions following the successful attentions of a thief and at least you need not be the poorer financially upon your return. Not that obtaining proof of robbery is always easy. When my wallet 'flew' in Kabul I spent a whole day with the Afghan police showing them where and how it was done but they steadfastly refused to sign my prepared statement. Eventually I was driven into going to the British Embassy and paying to have my own signature witnessed. It was worth it, for my insurance company still made good the loss.

Protection by more direct methods such as the carrying of a firearm is not recommended. Even in those countries that do permit it the necessary papers are difficult to come by. But that's not the point. The idea that a pistol under the seat or one's belt is protection is usually nonsense. When walking the Andes of northern Peru a year or two ago I debated upon the idea of arming myself because I had been told that cattle rustlers who kill without mercy inhabited the area. I'm sure they would have got me with or without a gun had I met any. And in many countries a gun is a prize in itself to a violent thief who will make every effort to procure one. In countries with strict gun laws, being found with an unauthorised weapon can bring severe penalties.

So, from guns to 'mutter'. Usually there is no trouble about the temporary importation of such electronic equipment as a tape recorder or radio. In 'awkward' countries make sure the importation is recorded so that the apparatus can be 'exported' when you want to leave. Elaborate equipment should be cleared with a consulate well before your trip commences.

Money transfer is a hassle at the best of times and whenever possible take all the cash you need—in the form of travellers' cheques—for the period you will be away. Many countries make it illegal for a person to receive funds from abroad in anything but local currency. This means that any cash sent will be in 'funny money' and probably of little worth outside that country. But if you do have to have funds sent, there is a procedure. Someone at home deposits the money in a bank which cables an order to its correspondent bank nearest to where it is to pay the money out to you. The person depositing the money at home should ascertain that the country to which he is transferring the cash allows transactions in pounds or dollars and that he pays all the costs (considerable) at his end. A copy of the receipt and order to the recipient should be forwarded so that, if any problem arises, the money can be traced at either end. Where no pound or dollar transactions are permitted and you are absolutely broke your consulate can sometimes help by arranging for funds to be transferred through other channels. But you won't be popular.

The best way of carrying money is—physically—in a money belt and —directly—by travellers' cheque as I have already intimated. But it is a good thing to hold a small reserve of a stable currency for use should you arrive somewhere where there is no establishment willing or able to change cheques. Another tip is to obtain and carry small amounts of money of the country or countries to be visited so that you are not a pauper the moment you arrive. So many times I have arrived in a Moslem country on a Friday to find all the banks and exchange offices firmly closed and, once in Iraq, not even an open bank would accept my Barclays travellers' cheques because they were supposed to be tainted with Israeli business dealings. Thus, up to a point, your choice of bank or agency from where your cheques originate should be governed by the politics and religion of the countries you are visiting.

A word of caution here. Some countries have exceedingly strict currency controls, the Soviet Union in particular as I have found to my cost. You can be severely dealt with for taking even small amounts of her roubles into or out of the country. They, like some other nations, give you a currency declaration form on which it is necessary to declare what monies you have brought into the country, everything you change while in it and what you have left to change back when you leave it. The same principle applies in many of the Communist and Third World countries to goods you bring in. Be sure your currency declaration and exchange records add up or you will be in real hot water.

Dealing in the black market is not always a criminal offence. A number of countries flaunt both an official bank rate of exchange and a parallel or 'free' market rate and currencies can be bought and sold openly to everybody's satisfaction. Don't try it in Russia or the East European people's republics however unless you are quite sure of the person with whom you are dealing. Secret police agents sometimes make tempting offers in an effort to catch out a Westerner.

Bribery is another activity that has no legal defined status. In the less developed countries, especially the Latin American states, bribery is a way of life for the police, the customs authorities, in fact anyone who has you over a barrel in the field of onward travel. But bribery is a subtle art and there can be no laid-down rules. A botched attempt to solve the problem with a handout in countries that are as incorruptible as any human-run authority can be will only land you in more trouble. One has to play things by ear with the help of knowledge or experience gained of the corruptibility of the country you are negotiating.

The last few points I have raised can only too easily land you in the 'stir'. So let us end at the sharp end with the factors surrounding arrest and detention. It has happened to me on a number of occasions. My wife is Czech but before our lives were joined she was simply my girl friend. Simply? Not at all. She lived east of the Iron Curtain; I lived west of it. And in those days it was a case of 'the east is east and the west is west and never the twain shall meet'. Come and get me, she wrote. So I did, making my initial run-in on foot. I got through all the 'nasties' on the firmly-closed frontier but, some ten miles inside the Czechoslovak People's Republic I relaxed for a moment and got nabbed by a border patrol. My crimes under Communist law totted up to a jail sentence of something of the order of 114 years. I actually served about three months, for I became an 'international incident' by virtue of the fact that the British Embassy in Prague had not been *informed* of my arrest. Nobody in the British government minded much that I *was* in jail but they objected strongly to not being *told.* It was a breach of international etiquette and that's what mattered.

The incident serves to emphasise the point that, following arrest, you have every right to be put in touch with the nearest British diplomatic representative. In most countries this is automatic; the consul or whatever will visit you in the lock-up and you can start arguing from there. I've also been apprehended in the Soviet Union, East Germany, France (where somebody 'recognised' me as a terrorist), Yugoslavia and Idi Amin's Uganda but was released before actual imprisonment began. In all cases of arrest I 'went quietly' and I recommend this to my readers. Resistance will, in all likelihood, result in a bashed skull or worse.

Consulates are limited to what they can do for their arrested nationals. They can help in finding a local lawyer and can request but not command fair treatment. Being arrested in a foreign land is a traumatic experience, especially if you don't speak the language. Jails in all countries are frightening places, often lacking sanitary facilities, beds and edible food. Worse is the prospect of never coming out.

Me, I'm a glutton for punishment, mostly on account of my journalistic curiosity. I would never advise my fellow travellers who have the gumption to visit the lesser-visited territories of the world to be otherwise. But wherever you go and whatever may happen, heed the hallmark of a good scout and *be prepared.*

Bureaucratic Hassles Overseas

by Richard Harrington

Whatever the brochures may say, travelling outside the so-called 'developed' world isn't a bed of roses all the way. Often the reverse, in fact. A book could be written about the hassles that the single or group traveller may have to face in Central or South America, Africa, Asia and the Middle East. Here I'll limit myself to a few of the more common problems.

Your troubles may start before you leave home. Just getting a visa from the embassy may be an ulcer-provoking experience. Bureaucracy seems to grow in inverse proportion to the rank of the functionary one is dealing with. One embassy required three photos, three forms and three signatures from my 5-week old son before they'd issue a visa. I won't go into details, but the long lunch hours and early closing at the embassy didn't help much. Their country was much the same way, right down to the Voluntary Tourist Development Fund donation at the airport on leaving—you donated voluntarily or you didn't leave at all.

But let me tell you of the joys of a certain African airport at 3 am. Of being hauled before the Airport Chief of Police and his armed henchmen, dark glasses and all, of being interrogated for half an hour while my passport was scrutinized, only to be told that they had decided to make me 'leader' of our group of transit passengers. Our passports taken, we were dumped on the rubber tyres in the back of a van and herded off across the city (1½ hours): I was to be responsible for ensuring that everyone paid for the compulsory sightseeing trip which our group had joined at our hotel in the belief that the ancient taxis had come to take them back to the airport for the onward flight. At least we finally made it back to the airport in the end. Those of us smart enough to spread a few dollars around accidentally got their passports back just in time to make it to the check-in desk, where a few further dollars secured for me the only seat left on the plane—leaving behind eight bewildered Brits wishing they'd never left home, clutching their confirmed seat tickets, protesting to a now slightly richer check-in clerk, complete with the moustache that is *de rigueur* among all Third World airport bureaucrats, as he gazed through them as inscrutable as the sphynx. Too bad the next flight wasn't for a week. . . .

By now, you'll have gathered that my lingering memories of hassles abroad have a lot to do with airports. That I cannot deny. However at least I have not fared as badly as the passengers transiting through Baghdad who were put up in the local prison for the night. Apparently the overnight hotel in Moscow wasn't much better five years ago. Maybe now the situation has improved.

One extra nuisance for the air traveller is the airport tax, often levied from the passenger about to step aboard the aircraft. Some countrie's exact payment on arrival, others (e.g. New Zealand) on departure, still others, including the USA, at both times. Canada and Australia levy no such taxes.

Make sure your papers are in order before you go and that you can meet all the health and currency requirements. Take along with you receipts or other dated proof of purchase for items of equipment such as cameras, so as to avoid having to pay duty on them. Some countries now demand a minimum expenditure per person per day. In case you think you can cheat, be warned that for the growing number of countries adopting this type of policy, you have to buy currency vouchers before you leave. If you plead convincingly at the Embassy/High Commission that you're going on a serious research expedition and can provide letters from potentates in the country to prove it, you may become an exception. But don't count on it. The guys who wrote the rules for a lot of Third World countries didn't think too hard about exceptions to anything,and you may quickly come to realise that whatever the bureaucratic jungle at home, at least our own system is flexible.

Customs is really a matter of commonsense, as is immigration. Immigration officers are usually superior in rank to customs officers, and therefore entitled to be additionally arrogant. Which they almost invariably are. What about the smiles and the welcomes and the flower *leis?* Forget it. Did you really believe that brochure rubbish anyway? But maybe I should be honest and admit that things aren't always all that bad. My wife and I, having been given the dirt treatment by the customs men on duty in Tangiers, stopped at a building a hundred yards further on to ask the way. It turned out to be the Customs HQ. Since it was the Muslim feast of Ramadan and it was after sundown, the off-duty officers were busily eating their first meal in over 14 hours. To the delight of my wife and myself, they not only showed us the way, but insisted that we stay for half an hour and share their meal. Friendlier customs men I have ever met anywhere, goes to show that it is dangerous to generalise.

My attitude to hassles improved after I spent a year studying social anthropology—the best cure for ethnocentricity and national prejudice. I began to understand why Singapore Airport provided barbers to cut the hair of those whose locks exceeded a statutory length indicated on posters hung around the building; and why miniskirts have been banned in certain African countries; and why so-called hippies have finally been kicked out of Kathmandu and dope has been made illegal there.

I also understand why having a small blond-haired boy of under two years of age was a great ice-melter in most countries—an object of intense curiosity amongst blacks, and of great affection amongst the Latins of the Catholic countries, who have almost a reverence for small children. Only amongst the Arabs is a child of little interest. So if you can manage it, take one along on your travels. If he's under two, his airfare is only 10 per cent anyway.

Ideally you should always wear glasses (not dark ones, which are the prerogative of the police and the refuge of terrorists), a dark suit, a white shirt, a dark tie and carry an umbrella. In practice, this is often not much fun when the temperature is 115°F in the shade, the humidity is 100 per cent, and your luggage weighs 35 kilos. Nevertheless, try to keep your clothes clean, use a suitcase rather than a rucksack if you're not backpacking/hitching, shave, and get your hair as close to a crewcut as is possible without looking like an astronaut. A moustache is better than a beard, but avoid both if possible. Don't try to smuggle anything through customs, especially drugs. Hash and grass may be common in the countries that you visit, but be careful if you must buy any. Remember that a local dealer may be a police informer. Prosecutions are becoming more common and penalties increasingly severe—from ten years' hard labour to death for trafficking in 'hard' drugs and sentences hardly more lenient, in some countries, for mere possession.

In dealing with airport officials, and with police and other government officers (and airline clerks for that matter), adopt an attitude that is neither servile nor arrogant. A friendly smile, even in the face of rudeness, will go a long way, and you should make every effort to be cooperative. As often as not, you'll have to turn the other cheek, but tell yourself that foreigners are like your own countrymen—they're all individuals and you can't judge the behaviour of them all by that of a few. There's no doubt that the white man's colonial legacy has had a lot to do with the hostility that you may encounter. Put it down to the arrogant selfish attitudes of some of your compatriots who have been there before you. It's not for nothing that the most difficult people you'll encounter in Third World countries are the ones who have the most contact with foreign visitors. It's true that every country has an abundance of friendly people willing to welcome you into their homes, and that generally speaking, the further you get off the beaten tourist track, the more hospitable the people get. That in itself is a good reason for heading away from tourist trails. In a country where the immigration officer tosses your passport back at you, and where city youths shout abuse as you drive past in your hired car, you may still find people who will make you wonder how your own countrymen could be so cold. Adopt an understanding attitude and you'll have some tremendous experiences where you really feel you're beginning to connect. Learn the language fluently and it's hard to go wrong.

A few tips to help you on your way: if you must spread around a little 'dash' (West African) to oil the palms that facilitate your progress, do so carefully after checking how to do it properly from a friend who knows the ropes. You may have to do it, say to avoid a few days in a Mexican jail for a mythical driving offence. On the other hand, you could end up in a jail somewhere for trying to bribe an officer of the law—then you might have to bribe a lot more to get out rather than rot for a few months waiting for a trial. The $1 or $5 bill tucked in the passport is the safest approach if you do decide on bribery, as you can always claim you keep your money there for safety. But it may only be an invitation to the officials to search you more thoroughly; and since all officials ask for identity papers, you may go through quite a lot of dollars in this way.

Avoid countries where there is a violent revolution or other civil disturbance taking place. Your own embassy or consulate will almost invariably advise against visiting areas of dubious safety, as it will be their responsibility to help extricate you if you get into difficulties there. On the other hand, the embassy of the country concerned will usually play down the degree of disturbance. The most reliable reports on a situation are to be had from fellow travellers and local newspapers. If a disturbance develops while you are in a foreign country, report to your embassy for advice and to register your whereabouts, in case you should be advised to leave.

Before entering a country, it is as well to check with its consulate across the border what the current situation is with regard to visa requirements, closed border posts and the location of border checks, since formalities are sometimes to be completed at a town some distance from the border post itself. At a border you are likely to have to face immigration (i.e. passport) control, health control, a currency check, a police check and customs. In some countries spot checks are made once past the border and in others, especially communist countries, internal checkpoints exist. You will get an idea of what these procedures may involve in each region from the articles in the *Troublespots and Bureaucratic Hassles* section. Generally the same rules apply to leaving the country as to entering it, except that there are certain items that may not be exported without a licence or, in some cases, at all. The main ones are money, antiques and other valuable works of art, and protected wildlife — animals or parts thereof, e.g. horns, hides, etc.

Don't carry firearms for protection. You'll probably get caught with them on arrival (assuming they haven't been picked out by airport security X-ray checks on departure) and in most countries, importing firearms without a licence is illegal. It's very unlikely you'll ever be in a position where you'll need them. You may get robbed but if you give what you have quickly, you're unlikely to get hurt. Shoot someone and I don't fancy your chances of survival. On the other hand, learn karate and use it if you have to. It almost certainly won't provoke the use of guns.

Always carry your passport with you for identification purposes. If you're driving a vehicle, keep yourself clean-shaven and tidy as you'll by stopped frequently by the police in many countries—if only as a routine check. Be courteous and have all your car papers and your driving licence in order. Be careful to camp at night where you're not likely to get picked up by the police. It's a good idea to have a local friend travelling with you. He'll not only speak the language, but will probably be able to sort out any problems with officials.

When asking directions, make your question as clear, simple and specific as possible. Answers vary from the sullen and monosyllabic to the voluble and inventive. If in doubt about the accuracy of the replies ask two or three different people—and keep on asking patiently until some of the answers tally.

There's no excuse for failing to research the countries you intend to visit. Talk to people who have lived in or visited them and find out what problems you may have to encounter. If you go prepared and adopt a sympathetic, understanding frame of mind, you should be able to manage without trouble.

Communicating: Breaking the Barriers

by Jon Gardey

The barriers to communication off the beaten track exist just because of who you are: a visitor from another civilisation. It is necessary to show the local people that underneath the surface impression of strange clothes and foreign manners is another human being, like them.

The first step is to approach local inhabitants as if you were their guest. You are. It is their country, their village, their hut, their lifestyle. You are a welcome, or perhaps unwelcome, intruder into their familiar daily routine. Be always aware that they may see very few faces other than those of their family or the other families in the village. Their initial impression of you is likely to be one of unease and wariness. Be reassuring. Move slowly. If possible, learn a few words of local greeting and repeat them to everyone you meet in the village. It is very important to keep smiling, carry an open and friendly face, even if you feel exactly the opposite. Hold out your hand, open, in a gesture that says to them that you want to be with them, a gesture that includes them in your experience. Hold your body in a relaxed, non-aggressive manner. Try to take out of the first encounter anything that might anger them, or turn to shyness their initial approaches to you. If they offer a hand, take it firmly, even if it is encrusted with what you consider filth. Be as close as possible, don't hold back or be distant, either in attitude or voice. Coming on strong in an effort to get something from a local person will only build unnecessary barriers to communication.

Begin with words. If you are asking for directions repeat the name of the place several times, but *do not point* in the direction you think it is, or suggest possible directions by voice. Usually the local person, in an effort to please his visitor, will nod helpfully in the direction you are pointing, or agree with you that, yes, Namdrung *is* that way, if you say so. It may be in the opposite direction. Merely say Namdrung and throw up your hands in a gesture that indicates total lack of knowledge. Most local people are delighted to help someone genuinely in need, and, probably after a conference with their friends, will come up with a solution to your problem. When *they* point, repeat the name of the place several times more, varying the pronunciation, to check if it is the same place you want to go to. It is also a good idea to repeat this whole procedure with someone else in another part of the village, and frequently along the way to check for consistency.

In most areas it is highly likely that none of the local people will speak any language you are familiar with. Communicating with them then becomes a problem in demonstration: you must *show* them what you want, or perform your message.

If you are asking for information more difficult than directions, use your hands to build a *picture* of what you need in the air. Pictures, in the air, on the sand, on a piece of paper, are the only ways to communicate sometimes, and frequently they are the clearest. Use these symbols when you receive blank stares in answer to your questions. Use sound: noises with your voice, or objects that you have in your possession that are similar, or of which you would like more.

Not all of your contact with local people will be to get something from them. Don't forget that you have a unique opportunity to bring them something from your own culture. Try to make it something that will enrich theirs; try to give them an *experience* of your culture. Again, show them what it looks like: postcards, magazines. Let them experience its tools. If you have a camera let the local people, especially the children, look through the viewfinder. Put on a telephoto so they can get a new look at their own countryside. Take along an instant print camera, photograph them, and give them the print. Most important: become involved. Carry aspirins to cure headaches, real and imagined, that you find out about. If someone in the village seems to need help, say in lifting a log, offer a hand. Contribute yourself as an expression of your culture.

Setting out to take pictures, I usually try to make myself acceptable to the local people as a person first, a photographer second. I do this in some situations by leaving my camera in its bag, and getting down on the ground with them, drinking tea, asking questions. When a rapport has been established (and, incidentally, I have had an experience of them without a camera between us) I bring out the camera and the taking of pictures is usually accepted. A useful tool is a device that photographs at a right angle to the direction the camera is pointing. Be invisible. Try not to intrude into people's lives any more than necessary for the result. And sometimes it is necessary to stop photographing when there is a protest. If arms are raised and there are angry looks, stop. And smile.

A bribe for a photograph or payment for information is justified only if the situation is unusual. A simple request for directions is no reason for a gift. If the local people do something out of the ordinary for you, reward them as you would a friend at home. The best gift you can give them is your friendship and openness. They are not performers doing an act, but ordinary people living out their lives in circumstances that seem strange to us. I have found myself using gifts as a means of *avoiding* contact with remote people, especially children, as a way of pacifying them. I think it is better to enter and leave their lives with as much warmth as I can give, and I now leave the sweets at home. If you are camped near a village, invite some of the local people over to *share* your food, and try to have them sit among your party.

On some of the more travelled routes, such as Morocco, or the main trekking trails in Nepal, the local children, used to being given sweets by passing trekkers, will swarm around for more. I suggest that you smile (always) and refuse them. Show them pictures, your favourite juggling act. Then give them something creative, such as pencils.

If a local event is in progress stand back, try to get into shadow and watch from a distance. You will be seen and noticed, no matter what you do, but it helps to minimise your presence. If you want to get closer, edge forward slowly, observing the participants, especially the older people, for signs that you are not wanted. If they frown, retire. Respect their attempts to keep their culture and its customs as free as possible from outside influence.

The people in the remote places are still in an age before machines, live their lives close to the earth in comfortable routine. Where you and I come from is sophisticated, hard and alien to them. We must come into their lives as gently as possible, and, when we go, leave no marks.

Editor's Note: In less remote areas where the local people have had more experience of travellers, you must still observe the rules of patience, open-mindedness and respect for the life-style of others. But you will encounter people with more preconceived notions about foreigners—and most of those notions will be unfavourable. Here are a few pointers for cases like these:

In some Third World countries, arrogance and inflexibility can prove your undoing. Officials must be distracted and appeased. Further south, where appearances and initial reactions may seem more violent, you will need tact, courtesy and patience to allow tranquillity the time to reassert itself.

Avoid seeming to put any local person, especially a minor official, in the wrong. Appeal to his emotions, enlist his magnanimous aid, save his face at all costs. Your own calmness can calm others. If you are delayed or detained, try 'giving up', reading a book, smiling. In the end, things will usually come right. Should you be accused of some misdemeanour, such as 'jumping' a control point, far better to admit your 'mistake' than to be accused of spying—though even this is fairly standard practice in the Third World and shouldn't make you flap unduly.

In much of the Third World, the quicksilver temperament operates, causing attitudes to foreigners to change in a flash. Wherever you go, tones and pitches of voice will vary, 'personal distance' between people conversing may be less than you are used to, attitudes and priorities will differ from your own. Accept people as they are and you can hope that with time and a gentle approach they will accept you also.

'Porter, Sir? Guide?'

by Richard Snailham

Porters, guides, interpreters and their ilk can be a great nuisance—it is difficult to feel at ease at the Pyramids surrounded by droves of importunate Arabs, all called Mahmoud or Ahmed, and all claiming to be the only and indispensable authority on Cheops, Thutmose and co.—but they can just as often be invaluable, as when one is about to venture into wild, badly-mapped country. I would not recommend venturing far from the Pan-American Highway into the remoter corners of the Ecuadorean Andes, for instance, without local guides.

Faced with the first situation, I find it best to appear indifferent, to affect disdain, flourish a guide book and make it clear that I can get myself around without Mahmoud's unwelcome attentions. Generally, some persistent boy will tag along regardless and begin telling me the story whether I like it or not. Not only may he turn out to be a tolerable compromise but might prove as often was the case in the Ethiopian

highlands, an engaging companion whose *pourboire* at the end is thoroughly deserved. It is not difficult to find a guide in these well-visited tourist spots, they just seem to emerge out of the stones.

At some famous sites a guide is obligatory, as we find it is in places closer to home; one would not expect to go round a French château, for example, without a practised, often multi-lingual cicerone.

If a journey is involved, or any form of transport, or any great length of time, it is as well to discover the fee in advance, as one might on engaging a gondola in Venice—if only to allay the shock of the often inordinate sum asked. Some sort of bargaining may be possible, depending on the circumstances: one could haggle with a Tanzanian hire-car driver for a two-day trip to Ngorongoro, but probably not with a Hong Kong Chinese girl courier operating a hotel-to-hotel, all-in bus trip to Repulse Bay and back.

Guides quite often get lost. In open country it is as well to maintain your own navigation by dead reckoning and repeatedly ask the names of prominent features. Never pose questions like "Is it far?' because guides rarely have any communicable concept of distance, nor "Will we get there tonight?" because they tend to wish to please and will give you the answer they know you want to hear.

Porterage has its various aspects too: an often unpleasant but necessary part of travel between airport and hotel; French railway stations with their *prix fixe* per item of baggage (fixed pretty high, too); and at the other extreme there is the uncharted, problem-ridden business of arranging for porters off the beaten track.

Hiring porters amid the clamour of airport or railway station has few rules: follow your gear carefully, have your mind tuned to the local units of currency and the exchange rate, follow the local rules if there are any and don't overtip. I still shudder when I recall watching some of my quite elderly flock walk with their suitcases into a dense mass of grubby bodies outside Karachi airport, the air loud with suppliant screeches and thin, brown fingers clutching urgently at the Gucci handles.

Out in the *ooloo* there are other different considerations to have in mind. Porterage often involves animal transport. Porters and their animals have to be fed and perhaps sheltered. If they are to be away from home for many days some kind of advance payment may be necessary. If you are moving across their country you have to consider how they will get back to their homes.

In all negotiations of this sort it is best to involve a local middleman. If you have an indigenous student or scientist or an attached police officer or soldier with your party he will be the best at overcoming language barriers and agreeing what will seem to him a fair rate for the job. I reckon £1 a day, 'board and lodging' inclusive, is reasonable (in Zaïre in 1975 we paid 1 zaïre 50 *majutas* per man per day, which was just under a pound; in Ecuador in 1976 it was 50 *sucres* or about £1.15). Often porters or guides are best hired through a local chief or landowner. This is essential in most wild parts of Africa or South America where tribal or semi-feudal societies prevail.

Sometimes local men will not wish to come forward as porters whatever the inducements. Even though we could offer marvellous goodies like planks of wood, rope, slightly dented but serviceable jerrycans (to a tribe still using gourds), the proud Shankilla of the lower Blue Nile would not carry for us because it would have shamed them: carrying loads is women's work. Blashford-Snell nearly asked them to send their women along but, since that might have been misunderstood, employed the village boys instead. Other tribesmen, like the Bakongo on the banks of the Zaïre River, where porterage since the advent of Stanley in 1877 has been a minor industry, have no such compunctions. Some of the best porters we have ever had were Angolan FNLA refugees in Bas-Zaïre, who, with their broad smiles and outlandish hats, helped us re-launch our giant inflatables in 1975.

With large numbers of men it is best to deal with a single head-porter and pay him for his pains. But pay each man personally, a portion of his cash at the outset, if you wish, but the bulk not until arrival at the destination. It is an added kindness and a mark of gratitude if there is something in your stores—-plastic bags, wire, spare clothing, cigarettes, even empty tins and other containers—which you can distribute as a bonus.

Porters are generally honest, but it is as well to keep your baggage train in the centre of the column in case the odd load should slip off into the bush. And a porter who is fatigued or taken sick can in this way be relieved by.

I am in favour of taking porters and guides into wild country rather than trying to struggle along without them. They may be difficult, like the Shankilla; they may abandon you precipitately, as ours did on the slopes of Mount Sangay, but they do add a colourful extra dimension to the journey and they can be a form of security in hostile, lawless country. And when you pay them be sure you tell the rest of your party that you have done so, so that they don't, as happened once in Ecuador, get paid twice.

Useful reading—

The Antique Land, Diana Shipton (Hodder & Stoughton, 1950), Chapters 1 and 2 for porterage in th Himalayas.

Dark Companions, Donald Simpson (Elek, 1975), for the historical perspective.

Troublespots and Bureaucratic Hassles—Middle East

The Middle East is a sensitive area politically and militarily, but one of abiding hospitality for the traveller if certain self-imposed rules are observed. A pleasant, good mannered, courteous approach will go a very long way in this area. Shorts and low cut blouses or dresses should not be worn, except perhaps on beaches in Syria and Jordan. Shoes should be off before entering mosques and, in remote areas, permission sought before entering at all. No photography on borders, of police or military personnel and no photography without asking their permission, of ethnic-looking old ladies! In Arab countries, avoid pointing the soles of your feet at your social superiors. And naturally you won't put food into your mouth with your left hand, proffer your left hand in greeting or use it for any other social purpose: in Asia the left hand performs the same function as toilet paper.

Turkey, the first country, has a good and growing network of roads, and is easy to travel in, except in cities, where cars weave in and out of traffic lines and taxis stop frequently to let off passengers. In the remoter eastern areas, stones are sometimes thrown, and travel for lone women may be hazardous. The main border post is in the north, just past Doqubeyazit/Bazagan in Iran. This is the heavy truck route, and a three - four mile queue of trucks is not an unusual sight. But for private vehicles and public transport travellers, delays are directly concerned with the number of people to cross. The normal Turkish customs searches for drugs and antiques are very nominal when travelling eastward, but intense travelling westward. Vehicles may have holes drilled in their sides, and travellers with long hair and 'hippy' clothes may be personally searched.

Turning south, the next border crossing is to the east of Lake Van, at Seroh, reached by an interesting mountain road. The whole area between Lake Van and Lake Rezaiyah is inhabited by Kurdish peoples. Solitary camping by travellers is highly ill-advised. This is a particularly slow crossing point, and even worse if travelling westward.

Still travelling south, the main Turkish/Syrian crossing is between Antakya and Celeppo, *Baba el Hawa* in Arabic. For persons travelling westbound, medical certificates are checked carefully. Public transport travellers are faced with fantastic taxi fares on the Turkish side. Wait patiently until enough travellers gather together. This would bring down the prices. The bus service, once in Antakya, is excellent, eg, 21 hours to Istanbul. On the Syrian side, things are slow but good humoured, a visa can be purchased, currency changed at the bank or at the café. vehicle owners must resign themselves to being patient. Handling of the *carnet de passages* and compulsory insurance details depend on your ability to push. Note: most insurance indemnity *carnets* are not valid for Syria.

Questions may be asked as to your route. Bear in mind that entry will be refused if you state any intention of visiting Israel.

Syria has a good network of major roads, many historical sites and an excellent bus service.

The border between Iraq and Syria was closed but has now been reopened. The Lebanon in its present condition should be avoided by ordinary travellers.

Continuing further south brings you to the only available border crossing, Syria/Jordan at Derae. Customs perform a thorough vehicle and passenger check here. Vehicle owners again have to push through the confusing hubbub. Private vehicles running on diesel are prohibited entry to Jordan for 'environmental' reasons. Visas are obtainable at the border.

Jordan has a good road system and most historical places can be visited easily. Amman, the capital city, has little cheap accommodation, and for drivers parking is particularly difficult and frustrating.

It is possible to visit the 'West Bank' (Israel). Travellers wishing to do so should consult the Ministry of Tourism in Amman.

Generally speaking, tourist visas are not available for Saudi Arabia. North Yemen (the former Yemen) is a fascinating country that normally allows tourists. South Yemen, once Aden Protectorate, is now a totally closed communist country.

So, if the traveller turns east from Jordan, the next country he reaches is Iraq. It is a very long desert drive, even to H4, the Jordan check-point, so all travel documents need to be in order, but exit procedures are straightforward. Once clear of H4, there is another longish drive to the actual border where passports are checked by potentially difficult Iraq police. Without a previously obtained valid visa, entry is not permitted. If your documents are correct, then patient persistence is the correct course.

144kms further along the road is the Iraq customs post, Rutba. Vehicles, luggage and documents are checked slowly and usually thoroughly. Compulsory vehicle insurance has to be taken out. Some currency should be changed here as it may be needed before Baghdad. The importance of 'Room 25'—the office of the Security Police—for clearance must be stressed. Failure to obtain the required piece of paper when travelling from Iraq to Jordan would mean a journey of over 400kms back and forth to the borders. Women travellers also need to show a certain amount of firmness at this border.

Once clear of the border, expect frequent road checks throughout Iraq. But the kindness of the local inhabitants more than compensates for this.

Parts of Iraq, north of Kirkuk, are difficult to visit without permission but everywhere else the historical sites can be seen easily. All the main routes converge on the modern city of Baghdad.

In Iran, following the overthrow of the Shah's régime and the coming to power of the Ayatollah Khomeini, the state of the country is, at the time of writing, uncertain and the future for foreigners there even more so.

Troublespots and Bureaucratic Hassles—Southern Asia

by David Gillespie

As one leaves the smart new Iranian border post at Taiebad and drives the couple of kilometres across the dust-hazed desert landscape towards Afghanistan, one takes another step further into Asia. Previously, officialdom and facilities, although sometimes slow, quaint, perhaps overfriendly, maybe plain obstructionist, were at least recognisable to our European minds. From now on we definitely encounter the different, often devious, logic of the East. On the surface the veneer of western civilisation exists in the form of procedures and institutions that are hangovers from colonial days or a direct result of 'foreign aid' of the superpowers, but underneath the mentality has not changed, as recent political events have proved.

The first indication of this is the

Afghan border post of Islam Qala itself. A group of squat, jaded semi-mud buildings huddle on the plain, where the various officials reflect in their general demeanour that a term of duty here must in some form be a punishment. Sloppy, ill-fitting uniforms and a very ambiguous attitude to working hours give warning that one should no longer expect border formalities just to happen, but that the speed and efficiency of the crossing are often in direct proportion to the amount of work and pleasant talking done by the traveller himself. Beware arriving too late in the day! Offices here are riddled with 'relatives' of the gentleman who runs the seedy border hotel where prices are high and facilities none. Frequently travellers find themselves between customs and immigration just when the offices close, with no alternative but to put up at the hotel for the night. Regular 'customers' such as organised tour operators probably have some special agreement with the authorities, but drivers of private vehicles may encounter a problem. Customs officials are empowered to demand the full (Afghan) value of the vehicle upon entry, issuing a receipt which can be redeemed only as one leaves the country. If you're driving a nice new Mercedes camper you'll need a lot of ready cash to get through!

Afghanistan is still very traditional in its Muslim ways, and attention to 'modest' attire, particularly by females, is desirable. The great days of the hippy trail are over but all contacts with officials can be speeded up by reasonable attention to dress, a pleasant manner, and patience.

The formalities at Islam Qala commence with a visit to the 'health' officer for a check on vaccination certificates. *Two* cholera stamps within the last six months are essential as well as the other normal international vaccinations. Customs and Immigration come next, with completion of efficient-looking customs, immigration and currency forms. This alternates with interminable transcribing of full details of passengers, vehicles and finances in longhand into heavy dusty ledgers. They may count your money, and a vehicle and baggage search will take place, generally cursory and amateurish. For large groups a 'consideration' of a bottle of whisky can be advisable but is not to be encouraged. A visit to the 'bank' and then to the insurance office for compulsory insurance on the vehicle, which will be of no use at all should you be unlucky enough to run down a local a few miles down the road. A final check by a soldier at the barrier when leaving the compound—and you are on your way. An hour and a half is probably the record crossing time—be prepared for longer!

At the other end of the country there is a distinct improvement. Sheltering under the Khyber Pass, nearer the main cities, the Afghanistan border post of Torkham tends to be more lenient. Still arrive in plenty of time to complete both Afghan and Pakistan formalities and get over the Khyber Pass before nightfall to the safety of Peshawar. This border is busier with a lot of traffic between the two countries, and one must be prepared to push slightly for attention as Asians are notoriously bad at queueing. However, always be polite and attentive to official whim. A reverse procedure from those undergone at Islam Qala ensures that Torkham is more efficient and one can be clear in one hour or so unless the border is exceptionally busy. There is an exit tax of 100 Afghanis per person which may or may not be enforced.

Again, through a final checkpost and drive the short distance to the Pakistan post making sure that one obeys the sign to change to driving on the left side of the road—common to the whole Indian subcontinent. Passing through the army-manned checkpost, one reaches the Pakistan border, really a small township sloping up towards the Khyber. A very respectable tourist office, coffee shop, and local shops intersperse with official offices—the police checkpost, Immigration, Khyber Pass toll office, Customs and bank. Most Pakistani officials, particularly the uniformed ones, are smart dressers and expect Europeans to look respectable. They of all the Asians seem to have most successfully assimilated the 'British Raj' system of bureaucracy. Despite the impression that they have all the time in the world to complete formalities, progress is fairly smooth. All paperwork must be correct, though, particularly the *carnet de passages* for the vehicle, as one can see from the forlorn vehicles sitting at this frontier, testimony to several travellers' inattention to correct documentation.

Formalities are the normal, international 'immigration' cards together with the transcription into the record books. The customs search will usually be cursory and there is often a hint that the customs officer would genuinely be grateful for any unwanted paperbacks. Unwanted Afghan money can be easily changed with several touts although it is theoretically illegal to export it from Afghanistan. It is not advisable to stop while crossing the Khyber Pass, as it is militarily a sensitive area.

The border with India is situated between Lahore and Amritsar at Attari Road, and is a conglomeration of checkposts and colonial-style offices distinctively painted in the national colours. Here they combine the worst of the British bureaucratic system with the Indian penchant for petty officialdom. The whole border process is a series of painstakingly thorough checks and multiple entries into record books together with the Immigration and Customs exit and entry forms. It is the only land border open between the two countries and, to give an indication of the time one can expect to spend there, it is usually not possible to begin the process of border crossing after 1.30 - 2 p.m., as they realise that it will be after nightfall before all formalities are completed.

Attention to the detail of correct paperwork is necessary to avoid refusal, delays or perhaps the impounding of vehicles and goods by the customs. The plains Indians have a refreshing tolerance of the eccentricities of some European travellers, so dress and mannerisms are not so important. They do, however, dislike outright rudeness and impatience and scanty, suggestive clothing on women, which can lead to their being unpleasantly accosted. One is handed from one checkpost to another, one office to another, with little that is widely different from other border crossings—just more thorough. The Indian customs can be very painstaking in their searches and no doubt, if there is something to hide, would not be averse to a 'present', but this does not help to speed things up. Pakistani newspapers and magazines will be confiscated, so it is advisable not to carry them; also the import and export of the rupee between the two countries is most definitely not allowed. As usual someone may be found to change them if you do have some, but the exchange rate will be highly unfavourable.

Indian officials can be petty, sometimes vindictive, and it is often with relief that the traveller reaches the border towns of Raxaul and Nautanwa at the foot of the Himalayas. The borders here are far sleepier affairs—the offices often hidden in the bazaars—and it is not unknown for travellers to miss whole chunks of border formalities. Again the entries into the record books are made, accompanied by much social chatter—often covering glaring inaccuracies in visa and other paperwork. Relations are good between India and Nepal and this is reflected in the ease if not the speed of the border crossing. Surprisingly the Nepalese are usually the more efficient at these border points, and a way is usually found to circumvent incorrect visas, for example. The two currencies are mutually negotiable if the banks are closed and many local shopowners will change the one rupee for the other with no problem.

Travellers to India can, before setting out, request a triple-entry visa, so as to be able to visit Nepal and/or Sri Lanka and re-enter India using the same visa.

Finally, one reaches Kathmandu, the only annoyance *en route* from the border being one or two toll and check posts, where although unfailingly polite the officials will require a ridiculous amount of personal and travel information to enter in yet another record book.

Troublespots and Bureaucratic Hassles—South-East Asia

by Tony Wheeler

When I'm asked where in the world I most enjoy travelling, my answer is immediate: no question about it, if I was doomed to do all the rest of my moving around in one region it would have to be South-East Asia. There is more variety in peoples, places, things to see, ways of getting around, food (very, very important) and almost anything else to keep a traveller happy. But just as every clear blue sky has a black cloud lurking somewhere off the edge, South-East Asia has one little drawback—hassles, bureaucracy, rules about short haircuts, even, as they say in Indonesia, *corrupsi*.

One of the commonest annoying pieces of South-East Asian red tape is the infamous 'ticket-out'. Before they will give you a visa to enter country A you first have to show them that you have a ticket to country B—a sort of magic wand to whisk you out of the country when they have had enough of you. Fine for your average 'today-is-day-7-it-must-be-Bangkok' tourist, but what if you are one of those weird and wonderful people who wants to exit Indonesia by the weekly cattle ship (there's no other way to describe it) that runs from Djakarta to Tanjun Pinang near Singapore? Or would like to depart the Philippines by the crazy Swiftair DC-3 from Zamboanga to

Tarakan in Borneo? In both cases you have absolutely no hope of getting those wonderful ticket-outs until you get in. Catch 22, isn't it? The only answer to this idiotic piece of bureaucracy is to buy the cheapest possible ticket-out and get it refunded later on. It's always advisable to get the ticket from a reputable airline—the standard ticket-out of Indonesia, for example, is the short flight from Medan in Sumatra to Penang in Malaysia. But don't get it on the Indonesian airline Merpati who part with refunds like J. Paul Getty did with dollar bills. Get a nice, safe, sensible ticket from MAS (Malaysian Airline Systems). Of course you'll still lose and the airlines will still profit since they inevitably have one exchange rate for selling you a ticket and another, lower one, for refunding it.

More subtle, but equally wasteful of your time and money, is the South-East Asian official's delight in the rubber stamp. Get your pocket calculator out some time and divide the number of pages in your passport by the amount you've paid for the privilege of having it. Then smile every time another beautiful, multi-coloured stamp uses up another precious few square inches of watermarked paper! It's not the cost that really gets me down so much as the sheer inconvenience; a little jaunting around Asia fills up passports with pretty stamps at an incredible pace. Somewhere along the line you then have to settle down and wait while your embassy supplies you with another passport full of nice blank pages to start again on. Americans are a little luckier here than most nationalities since they can get an extra concertina section stuck in their passport when it's full up. The unlucky British just have to pay out for another one. If you're doing much travelling in Asia I'd recommend the cheaper-per-page businessman's passport which is nearly three times the standard size.

Corrupsi is not generally something that affects the everyday traveller but it's worth knowing how to combat the minor price-bending you will meet. Actually it's not really corruption at all, just standard local practice. I certainly felt a lot happier about rip-offs when I realised that there are generally three price levels—one for the immediate family/village/neighbourhood, one for other locals and one for foreigners. Of course somebody from 100 miles away may be considered just as much a foreigner as you are but, whichever way you look at it, you are absolutely not going to get that Number One local price. The best you can do is get the one you are 'entitled' to and the way to do that is to know what it is before you start. Overcharging foreigners is fair game and you'll find that avoidance can be a lot of fun. Find out for certain what the price should be (simply ask a non-involved third party) for your category, insist on your rights and nine times out of ten you'll find your would-be overcharger thinks it is one hell of a good laugh and admires you for seeing through his attempt to double the percentages in life.

One of the most important secrets for smooth travelling almost anywhere in Asia is what I call 'dressing up for borders'. Singapore, with its famous mania for the short back and sides, started this little game and if you want to flow easily through Asia, it's advisable always to look your best when you arrive anywhere. I've almost run experiments in Singapore: once I even came in from Europe wearing a pin-stripe suit just to see if I really could get away with a hair style that would have me branded as a decadent-hippie-national-enemy had I been wearing jeans and a T-shirt. The answer, of course, was yes, I could! It's always worth carrying one otherwise quite unnecessary change of clothes just for those occasions when appearance is all-important. Incidentally, money has a sort of opposite effect to hair length—possession of a very fat wad of travellers' cheques will often serve to balance out hair which is over the length limit.

Money is no problem in most parts of Asia although, like everywhere in the world, they have come to realise in the last year or so that the US dollar is not the only foreign currency worth having. The old Asian black market has more or less disappeared everywhere, but Burma is still a glaring exception. When you've got an official exchange rate which is only a third or a quarter of reality, you've got perfect conditions for dazzling under-the-counter deals. You've also got a perfect situation for being flung behind bars if you get caught dabbling. Fear not, there are other things than money to deal with—the duty free shopsin departure airports in neighbouring countries will quickly inform you which bottle of whisky and brand of cigarettes the well-equipped traveller should take to Burma this week. (It is only in the last 6 or 7 years, by the way, that Burma has allowed tourists in for more than a 24-hour visit to Rangoon.)

In India the black market has virtually dried up but the hassle here is that you're unable to spend rupees for some very useful things—airline tickets in particular. Worse, there is a travel tax on all overseas airline tickets purchased in India, so you're wise to avoid buying them there. Singapore, Malaysia and Brunei once had currencies which were directly interchangeable but they have now gone off in their own directions. You'll find it very easy to spend the more valuable one in the less highly rated countries. Impossible in the other direction. For example, you can spend Hong Kong dollars in Macau (any shop will take them) but you'll have no hope of disposing of Macau *patacas* (which are worth 1-2 per cent less) in Hong Kong.

Airline tickets are very variable in price around Asia. You'll soon discover that it's often much cheaper to fly from B to A than from A to B. In particular the travel agents in Singapore, and even more so in Bangkok, are always ready to talk price. A very sensible way of doing business.

Customs are no problem in Asia so long as you don't try to find out if the profits on dope smuggling really are enormous. If you want to see what being in prison feels like I suggest you do so back home where you'll find the food somewhat better and the firing squad is a very infrequently used instrument of justice. Anyone flying out of Bangkok these days is an object of suspicion and Hong Kong, where they have their own little drug problem, goes over arrivals from Bangkok very carefully.

In Buddhist countries it's wise to be a little respectful to the religion. I still gag every time I look at one well-known overland operator's brochure, which shows a young lady happily sitting on the lap of a Buddha statue. A year or two ago a couple of travellers found themselves trying out a Thai jail after a stunt like that. Ask yourself how the Vatican would like visitors climbing up on the cross for a tourist snapshot?

There's no way the casual visitor is going to master local languages to anything like the standard locals will have of his own—at least the locals who have cause to deal with overseas visitors. English is still very much *the* linking language for many Asians. The old saying about English being the only thing a group of Indians have in common is still quite literally true. Of course you'll soon find that all knowledge of English dries up when Asians don't want to understand. One country where, at least off the beaten track, you may have some communication difficulties, is Thailand—but with an alphabet which has 18 vowels you're unlikely to get too far yourself. There is one language you should definitely have a go at—Indonesian, which is both delightfully simple and a lot of fun to try out. You'll soon pick up what is known locally as *pasar Indonesian*—the 'market' language. A langauge which has a phrase like *jam kerat* ('rubber time'), to describe those buses which are always due to arrive in just another ten minutes, has to be good. Imprecision seems to be something that is built into Asian languages; in Pidgin English-speaking Papua New Guinea any distance is likely to be a *longwe liklik* (a 'little longway') —something between very close and very far. How could anyone not like a region like that?

Troublespots and Bureaucratic Hassles—China

by Richard Snailham

To write a piece about visiting China for inclusion in *The Independent Traveller's Handbook* is currently a contradiction in terms. For the plain fact is that at the time these lines are being written there is no possibility of fully independent travel in China Since late 1977 China has encouraged tourism with the widest of open arms, but all tourists must be members of a group, and are met as such at the point of entry by guides from the China International Travel Service (Lüxingshe), who chaperone them throughout until the moment of exit.

I believe that if you are on diplomatic service in Peking you can get a permit to visit certain other cities in China, but that is about the scale of 'independent' travel in China today. Permits are needed, incidentally, by Chinese or foreigner alike before they may leave the perimeter of any city.

However, within this framework there is scope for a certain amount of independent action. The two Lüxingshe guides who normally accompany a group of 15-20 'Foreign Friends', together with the one or two special guides from each particular centre visited, are not specially leechlike, and indeed exhibit a refreshing degree of flexibility in the choice of places to be visited and the order in which they are taken. In fact, in the case of the Thomas Cook group that I led last April, the guides showed a great capacity for inventiveness and surprise, great good humour, and were eminently teasable.

Visitors who sign up for a tour in advance are asked what they want to see several weeks before departure. Now it is of little use to put down 'The Lop Nor Nuclear Weapon Testing site' or even a humble PLA barracks. But judicious study of a guide book—and the Nagel, even at £20, is worth its considerable weight here—will give an idea of what there is in each locality.

Foreign Friends who feel, during the visit itself, that they are not getting their whack of education, or are missing out on the keenly anticipated look at acupuncture surgery in action, will find that the guides, given a day or two's notice, will arrange something without much difficulty (which is perhaps most of all possible in an essentially unfree, authoritarian society, where no vice-chairman of a hospital or school revolutionary committee dare say no).

I had once imagined that we would be dragged groaning round one ball-bearing factory after another, but my group's three week programme only included one such glimpse of industry—at the remarkable The East is Red tractor plant at Loyang. Everything is very sensibly balanced and due note is taken of individual requests. If you are an ancient music freak, just ask and you will be pointed in the direction of an ancient music shop. No part of the programme is compulsory and personal idiosyncrasies can be indulged at any time. The guides, of course, like to know, but at the same time can be very helpful to the independent. There is no concern shown at guests leaping on buses or taking taxis wherever they wish to go.

If you want to see the Great Wall, the Summer Palace, the Temple of Heaven, and so on, *and* still want to be independent, then you must have stamina. Chinese cities come to life about 5.30 a.m. and there is much to be seen and photographed in the early morning before the 8.0 a.m. breakfast: squads of serious-looking senior citizens under the trees on the Chang'an Avenue in Peking exercising every last muscle in the slow-motion *tai-chi-chuen;* phalanxes of bicycles carrying droves of mao-suited workers weaving their way alongside the big, articulated buses, down the same immensely broad avenue, to the great peril of any jaywalker. Escape from the hotel is also possible after lunch—Chinese guides are great ones for their siesta—or in the evening after supper, if strength will permit: dark streets—entirely safe, milling crowds, some stores still open, excellent ice-cream available.

Much of this private exploration will be done on foot. Needless to say, you cannot yet take your own vehicle into China—and the almost total lack of obvious petrol filling stations would make this a very perilous exercise anyway. The very few private cars in China are, in fact, 'official'.

Entry to China is achieved, as *Time* magazine said recently, 'with a minimum of immigration fuzzbuzz'. What permits are needed? Well, a visa, a cholera inoculation certificate and a smallpox vaccination certificate are about the only prerequisites. If you come from South Africa, Rhodesia, South Korea, Israel or the Vatican City you may have problems—China does not even have postal communications with these states. All visitors, on arrival, have to fill in a Baggage Declaration form for the Customs. On this you state what Chinese and foreign currency you have in bills and cheques, what jewellery or other precious metal or stones you are carrying, and make a list of wristwatches, radio sets, cameras, cinecameras, calculators, recorders and typewriters. A Customs official then marks with a triangle those items which must be taken out of the country on departure, and keeps a duplicate copy of the declaration. At the exit point the visitor may have to produce this form together with the 'triangulated' articles.

Foreign Friends are enjoined to keep all certificates from banks for the exchange of travellers' cheques into *Renminbi* for ultimate examination by currency officials on departure, but in my experience this is a pretty cursory inspection. There is no evidence yet of a black market in China and the official exchange rate is quite realistic (about 31.5 *yuan*=£1). Nor do the Chinese operate the wretched 'minimum expenditure per person' system found in some other developing countries. Once you have paid your tour fee you can be as canny as you like.

Package tours to China are currently run from Britain by half a dozen tour operators, but for those who abhor too much pre-planned package touring it is possible simply to arrive in Hong Kong, arrange for a visa and sign up for a tour. Apply on Wednesday afternoon at the China Travel Service office (77 Queen's Road, Hong Kong. Telephone 5-259121), hand in your passport with three photographs, and you should be all set for a Saturday morning departure by rail from Kowloon to Kwangchow (Canton) a four-day visit last summer cost about £90.

A ten-day tour encompassing Kwangchow and Kweilin departs every Sunday, and there is a one week tour to Peking. The visits to Kwangchow do not operate during the twice yearly Canton Trade Fair (mid-April to mid-May and mid-October to mid-November). It should finally be noted that the Chinese are still wary of journalists and are not happy to accept American Express travellers' cheques.

But everything is subject to rapid change in China just now and these caveats may soon be irrelevant. The Chinese, at present, love Foreign Friends with their funny eyes and bridged noses and hair of varying colours and styles—because we spend plenty of *yuan* and, for the most part, share their apprehensions of the Soviet Union. So go to China, wear what you like, and you will find that within the constraints of the Lüxingshe system you can still be pretty independent.

Troublespots and Bureaucratic Hassles—Japan, Korea and Taiwan

by Ian McQueen

The Chinese in Taiwan are the nicest and friendliest in Asia, a welcome change from the thoroughly unpleasant characters often found in Hong Kong.

Transport facilities continue to improve with new roads and railways being built, and Taipei has a new airport.

On the popular excursion to Ali-san, travel agents block-book the train and would have one believe that there are no individual seats available for reservation. However some travellers have reported that if one turns up at the station there is a good chance of getting a seat. The bus trip down the East Coast is spectacular, if spine chilling, and not to be missed. Taroko Gorge is well worth visiting; overnight accommodation is available from a priest living near the entrance.

Despite widespread dismay at the Americans' full recognition of mainland China, individual Americans are still well treated in Taiwan.

In Tokyo, the *de facto* embassy of Taiwan goes under the name of 'Association of East Asia Relations'. It is located in the Heiwado Bldg. which is a reasonable hike from Kamiyacho subway station. The proper address is Higashi-Azabu 1-chrome 8. The friendly people at the Tourist Information Centre in Yurakucho can give directions on how to get there. The 'phone number is 583-8030, but they probably can't handle inquiries in English very well. This is the office to go to for Taiwan visas.

In Japan and Korea people value courtesy on a person to person basis very highly, so it is unlikely that officials will be gratuitously rude. It is, however, the nature of the Japanese to do everything by the rules—even if it costs more—and they are suspicious of anyone who operates outside normal channels. There is a band of people illegally teaching English in Japan; to renew their visas they go across periodically to Korea on the Shimonoseki Japan)—Pusan (Korea) ferry, which makes the voyage thrice weekly in both directions. Japanese officials on the ferry accordingly look on English-speaking passengers with some suspicion and may refuse the full visa to which the traveller is entitled. Korean officials on the ferry also take nothing for granted and may subject travellers to long scrutiny of documents and drawn-out interviews. *Bona fide* travellers will be well advised to obtain visas for Korea in one of the Japanese cities (e.g. Tokyo, Fukuoka, Osaka), avoiding the linking ferry and its visa trade altogether.

As long as one has reasonable grounds, one can get an extension to stay in Japan longer than the initial two or three months granted by the visa. The Immigration Department is likely to ask to see a return ticket and/or sufficient funds and/or a letter of guarantee. It is a good idea to carry a letter from a bank back home stating that you have arranged to guarantee your repatriation from Japan if this proves necessary.

There is no hassle about long hair such as one gets in Singapore.

In Japan one must carry one's passport or Alien Registration Card with one at all times, otherwise one is liable to arrest if asked for it. The ARC is compulsory for people staying in Japan more than 60 days. There is a hint of racism in these requirements. While individual foreigners are treated with an embarrassing amount of respect and politeness, one gets the feeling that en masse the Japanese still consider themselves *the* or at any rate *a* superior race. However, on balance, the Japanese regulations regarding time that foreigners can spend in the country are probably quite generous; I suspect that our western countries are far less generous in visas for Japanese travellers.

Travellers from Japan to Korea will

find it impossible to buy the Korean currency, *won*, in Japan. They should convert *yen* to US dollars or travellers' cheques either at banks in Shimonoseki (Bank of Tokyo is good) or at the port terminal after going through the check-in door; the rate given is the same as at the banks. In Korea, one then converts the dollars or travellers' cheques into *won*. *Won* can be exchanged back to dollars up to the amount originally changed into *won*—save your bank exchange receipts! Take as many US greenbacks into Korea as you can get your hands on; this is easy in Japan at the two places mentioned above. Unofficially, t is possible to obtain *won* at a rate 7 per cent or so better than the official rate. Conversation with other travellers will reveal where and how you can lo this and whether it is worth the risk nd the trouble.

Korea is beginning to boom and nany people have reasonably good incomes. There is a shortage of quality goods like cameras, tape recorders, adios, etc. (There are Korean-made nes, but the Koreans don't trust hem, preferring Japanese ones). So ny traveller with a good camera will ind people asking him to sell it. Koeans will even buy quite old cameras, ut they prefer new, clean ones. They re not interested in extra lenses. It's oubtful whether they even use the ameras—possibly they just want a staus symbol to hang around their necks. 'he most popular brands are Pentax, linolta, Canon and Nikon. Other rands may be as good, but they are not n', and therefore elicit less interest. ingle lens reflex models are popular. *Von* or, more likely, dollars will be offered and the price for a nearly new one ay be up to double the Hong Kong price. Koreans can pay several hundred dollars for a camera, but don't take the most expensive models. If customs people are the least bit suspicious when you are entering the country, they may record serial numbers of cameras, etc., in your passport and check on the way out that they accompany you. It is useful, therefore, to have an old looking camera, and not to flash it around. On entering there is a customs declaration of valuables including cameras, so it is up to you to decide what to write in. If planning to keep your camera in your baggage while going through Customs, remember that they X-ray all bags at the port of Pusan and probably at the airport. Remove all film from the bags and carry it in by hand!

Slide film is hard to find in Korea and is about double the Japanese price (which is already exorbitant)—buy in Hong Kong if possible. Print film is readily available only in Seoul and Sorak-san; elsewhere it may be impossible to find, and there is a lot of useless stuff being sold as the real thing. Buy only at large shops in Seoul. In Kyongju, the second most important tourist destination, film is not available even at the tourist hotels.

Prices of cameras and electronic equipment are higher than in Hong Kong or Singapore unless you know where to get the best discounts, and even then the best price in Japan is about average for Hong Kong.

It used to be good business to bring three bottles of Johnny Walker Black Label whisky into Japan for sale, but buyers are now harder to find and the price has dropped in recent years. However it may be worth trying to sell the bottles in small bars; if not, they still make great gifts.

Youth Hostels in Japan are plentiful, reasonably priced, and a good place to meet young people. *Minshuku,* roughly equivalent to guest houses, offer the best balance between moderate cost and meeting 'real Japanese'. Next up the list are *ryokan* (inns), which vary in price from the cheap to the fairly expensive. In Korea, the cheapest places to look for are *yogwan:* there will be one or more in most towns. They begin at around £1 and a room can be shared with not too great an increase in price. In Japan on the other hand one pays per person, not per room.

Japan is one of the best places in the world for honesty. You can leave valuables in your room or at the Youth Hostel—but beware mostly of foreign travellers. The ones with the worst reputation in Asia, among the Europeans, are the French.

Korea is quite the opposite to Japan. While honesty is highly regarded by *most* Koreans, thievery is a problem in Korea. Do not let your camera out of your sight for a second! I met an unfortunate fellow who had put his bag on the lawn at the National Palace in Seoul and turned his back on it for two minutes or less—it disappeared, along with a couple of thousand dollars' worth of Nikon lenses. Such stories are common. Another traveller had his baggage rifled while sitting three feet away from it on a crowded bus; he was by the window and an accomplice blocked his view of what was happening. Other Koreans will not say anything if they see something like that happening, and they will not come to your help if you get in a scuffle. While Koreans tend to fight amongst themselves (unlike the Japanese, who are peaceful), they do not look for trouble with foreigners without provocation.

Hitchhiking is the cheapest way to get around in both countries, but there is not much road traffic off the beaten track in Korea. In both countries people are very obliging. In Japan they have been known to go hours out of their way to take someone where he wanted to go. The only way to 'pay' them is to stock up beforehand with rice crackers and sweets (or cigarettes, especially American ones) and give them away. Otherwise they refuse to take money or any other payment.

For English-speaking travellers, language is a distinct problem in all three countries. Despite several years of studying English at high school, the average Japanese cannot speak a word because the emphasis is entirely wrong, stressing the written language (even Shakespeare!) rather than practical spoken forms. Unless the system of education changes, a very doubtful proposition, it will be a long time before a traveller can talk freely with the locals. However, goodwill prevails, and they will be happy if you learn to say 'Thank you' in the local language. Japanese is easier to pronounce; Chinese has to be 'sung'. Korean is similar to Japanese in that way, but is harsher, and it is more difficult to find a source of useful phrases.

There is a plethora of information available at the Tourist Information Centres in Tokyo and Kyoto. Outside those two cities it is difficult to find out anything in English. In Korea, there is a TIC behind Seoul city hall, and they speak English to some extent, but they don't have a fraction of the brochures that the Japanese can offer.

Troublespots and Bureaucratic Hassles—North and West Africa

by Ingrid Cranfield with contributions from John Michael and Tom Sheppard

Countries are a bit like hospitals: your impression of them depends a lot on the treatment you receive there. Who you are, how you behave, whom you meet and the external pressures on those people are the factors contributing to that impression. Generalisations mean little and one traveller's meat is, as always, another traveller's poison.

Yet there are certain things that you would do well to bear in mind when travelling in West and North Africa. You cannot influence whom you meet or national tensions: chance and politics take care of those. But foreknowledge and circumspectness can help to smooth your path.

Specific points of information in the following are drawn from the experiences of two seasoned travellers in North and West Africa, Tom Sheppard and John Michael, and are marked accordingly (T.S. or J.M.).

General

The main attributes for African overlanders are a good sense of humour and a very patient disposition. Many border officials and police suffer from a lack of the experience that breeds maturity and a corresponding excess of experience with bloody-minded Europeans. This tends to make them bossy and loutish. Many are also semi-illiterate. Do not rise to the bait. Keep cool and courteous—though there is no need to fawn—and try to see things from their point of view. Imagine if *you* had been told by your bosses to examine Arabic passports.

In most countries it is unwise to *offer* bribes (T.S.), though there may be occasions when you will be asked for one, e.g. at Maiduguri, Nigeria, where stamped passports were withheld until extra money was handed over (J.M.).

Many African countries are so big —or so rife with unrest—that random checks within the country are made in addition to border controls. Normally the procedure is limited to examining and stamping passports.

Most countries understandably wish to issue their own third party insurance, though comprehensive insurance can sometimes be obtained before departure in the UK for a whole expedition. (This can be very expensive.)

As many countries exact a fee of some kind or another at passport/visa/customs/insurance offices, it is as well to enter the country, if possible, with a little of the local currency, so as to settle accounts without delay.

(i) Algeria

At Nefta, officials were found to be unkempt, bad tempered and rude; at Djanet, thorough, quick and polite (J.M.). As usual, there are good and bad officials, but on the whole helpful and efficient and with an excellent sense of humour if handled well (T.S.).

Until recently, single vehicles were not permitted to travel. This rule seems to have been relaxed lately.

A currency and valuables declaration—the 'Yellow Form'—is filled in at entry and surrendered on departure, at which point a check is made of what you have left. Don't take chances with the Yellow Form, as it will be checked during your travels through the country, by officials with a nose for dishonesty. At Tamanrasset, travellers going south will encounter the customs post before the fuel depot: make it clear to the officials that any Algerian currency you are 'exporting' is intended for fuel purchases to be made while still in Algeria.

UK travellers require no visa.

Carnet is not required. The vehicle is signed in and out of the country on a Blue Card issued to the owner on entry and surrendered on exit. Money exchange and quite cheap third party insurance are obtainable at most border points.

Border with Morocco closed in area of Colomb Beshar, owing to hostilities, and the track is mined! (J.M.)

Distress flares are now classed as firearms and may be confiscated at the border. (J.M.)

Drivers should not honk their car horn in the rhythm of the words *Algérie française*. Arabs who remember the war of independence against the French do not take kindly to being reminded of the slogan.

(ii) **Equatorial Guinea**

For eleven years after independence in 1968, this former Spanish colony according to an International University Exchange Fund report, 'dropped out of the world'. Torture and extermination *were* rife. Industry *had* collapsed. The national bank, the post office, garages and filling stations and most shops *were* closed, hardly any vehicles *were* running and there *were* no buses, taxis or any other public transport. The dictator who presid*ed* over the 'cottage industry Dachau' *was ousted in mid-1979*.

(iii) **Mali**

Officials inherently polite and unfailingly cheerful. (T.S.)

A permit for photography is said to be essential (and a passport photo is needed for the form). This can be obtained overnight in some towns, e.g. Bamako. (T.S.)

A 7-day visa fo UK travellers is obtainable from the Malian Embassy in Belgium. This takes about a week by post. The visa can be renewed without much problem in Mali.

'The Malians were not particularly interested in the *carnet* and I had the impression they stamped only to humour me.' (T.S.)

Insurance is not obtainable at the borders, nor is the traveller usually asked to show proof of it.

Petrol is nearly always in short supply away from the big towns—and even in Tombouctou—and may not be available at all. A 'bond' is sometimes needed. This is a note issued by the local police authorising the petrol attendant to sell you a given quantity (normally as much as you want.) It is a way of making petrol available to tourists where it is not on general sale to the public.

(iv) **Mauritania**

Officials, like the inhabitants at large, consistently pleasant and good humoured.

Movement within the country is now becoming more feasible as a result of the *entente* between the Polisano and the new Mauritanian régime. The Western Sahara to the north remains a problem, as the war between Morocco and the Polisario continues. All the same, southbound travellers usually prefer to route through Mali.

(v) **Morocco**

Officials quite thorough with paperwork. Tourists generally allowed to pass Algerian border unhampered in spite of local tensions.

Vehicle insurance is not obtainable at Figuig/Beni Ounif, but usually at other border points including Oujda. The main tourist exit for northbound travellers is Babsebta (Ceuta). Green Card insurance for Morocco can sometimes be obtained in the UK.

Vehicle searches perfunctory (J.M.)

(iv) **Niger**

For southbound overlanders, the first taste of black Africa. A military-run country, reasonably efficient, with thorough passport and customs controls, including very long delays in processing passports at Dirkou (on border with Algeria). In 1976-77, officials were still overreacting to an attempted coup and prodding bags of sugar, etc., looking for ammunition (J.M.). 100 per cent vehicle searches were common—everything inside and out being removed—and random checks inside the country in operation. One group of travellers was stopped and searched nine times in one stretch of 150 miles (J.M.).

Travellers are required to report to the police at every town—a hassle wasteful both of time and of passport pages, for official stamps are on the large side.

Passport and customs offices are not always located together. In N'Guigmi, the passport office is to the south of the town, while customs control is in the centre.

In times of political unrest, the authorities have deemed it wise not to allow cine-cameras into the country.

For travellers entering Niger from Nigeria: if the Komodugu River is in flood, follow the track to Geidam, where a raft is available to ferry vehicles across.

(vii) **Nigeria**

Officials generally courteous and efficient, although one traveller here encountered his only experience of corruption (see *General* above) and the only instance of incorrect handling of a *carnet* (J.M.). Vehicles not usually searched. Numerous spot checks of passport stamp especially in border areas. One form only to be filled in on entering the country.

Visitors require an Entry Permit, which is granted only on production of a return ticket or the vehicle *carnet*. One recent would-be traveller, wishing to fly out to join an expedition already in Nigeria, was refused a Permit by the High Commission in London.

In order to keep an eye on *carnets* and see that sections are correctly detached and stamped, it is as well to know that for northbound travellers heading up the west side of Lake Chad, the exit certificate should be removed at the customs post at Baga, also known as Kukawa Junction.

(viii) **Tunisia**

Officials courteous, well dressed and efficient.

The main customs searches in the north, for travellers arriving by sea, are concentrated on drugs.

Delays at the vehicle insurance offices are common.

(ix) **Zaïre**

The country at large runs on *matta-biche,* a Zaïrois word for 'tip', 'corruption' or 'graft'. The visitor who understands this will disburse gifts on arrival, departure and at relevant moments in between.

Shaba (formerly Katanga) province is currently troubled by unrest and is no place for foreign travellers.

Troublespots and Bureaucratic Hassles— Southern And East Africa

by Hilary and George Bradt

Perhaps it is because the boot is on the other foot after all those years of colonialism, but some African officials take a little too much pleasure in their authority. We were arrested three times in Africa, so we hope others will learn from our mistakes.

'Ignorance of the law is no defence'—It is illegal to take photographs without a permit in Tanzania. There are no notices to this effect at any border post, and the thousands of tourists who visit game parks and show their slides back home are blissfully unaware that they are technically breaking the law. It is only when you accidentally photograph a politically sensitive subject that the full weight of the law descends on you and informs you that your ignorance is no defence. See *A Traveller's Guide to Photography.* The much photographed Masai of East Africa are well aware of the commercial advantages of camera-wielding tourists, and are not slow to throw stones if you don't pay up.

The political discussion—It seemed innocent enough. A group of Ethiopian students drew us into their discussion of the merits of socialism and the evils of capitalism. We didn't always agree with them, and we didn't notice the army lieutenant sitting alone at a neighbouring table. Another arrest. We were fortunate that his superior was an extremely sophisticated and educated man who let us off with a warning. After that we listened politely when the subject of politics arose, but never allowed ourselves to be drawn into the discussion. The new socialist countries are particularly touchy; so is right-wing Malawi.

Keeping abreast of the news—You may enjoy being oblivious of world events and ignorant of the latest news, but it can also be dangerous. There may be a revolution in the country you are travelling in, or plan to travel in, and you may not find out until you are led off to the police station. We made the mistake of deciding to visit that famous botanical garden in a place called Entebbe, in Uganda. It was a couple of days after the name became famous throughout the world, but we were blissfully unaware of the hijacking and were lucky only to be detained for a few hours.

Puritanism—Many African governments deplore scanty clothing on their own people and tourists, but none take it to the extreme of Malawi. There, a woman may not reveal her knees, nor wear trousers, and a man must be decently dressed and short-haired. A lot of governmental manpower is devoted to enforcing the law, so resign yourself to hiking in a long skirt and getting a haircut. Actually, you are allowed to relax the rule at resorts such as Lake Malawi and in the mountains, but in cities and towns it is rigidly applied.

Tanzania is said to imprison men wearing shorts, although we had no problems of this kind. Strictly, tight and flared trousers and short or clinging skirts are also banned.

It is suicidal to wear anything resembling combat gear in Africa.

The egg roll visa—Few tourists resort to bribery in Africa. It is too risky, since many of the more puritanical government officials are fanatically honest. Or rather, they are fanatical about being seen to be honest. If you have problems with an official, keep your cool, be polite and eventually he will see things your way. Flattery gets you everywhere.

Our only bribe was a Chinese egg roll, transported from the 'all you can eat' Chinese restaurant in Addis Ababa to the Sudanese Embassy to speed up the processing of our visas. It worked.

Crime—Kenya appears to be well in the lead in this field, but it may simply be that there are more tourists to be ripped off in Kenya. Crime there takes a violent form, and you are in real danger from panga-wielding thugs if you walk down dark streets at night, or camp in an unguarded area. You should be on your guard in all countries, particularly in big cities and around the main tourist centres.

The Black Market—Almost all budget travellers exchange their hard currency at black market rates, although this, of course, is illegal. Check with other travellers for current rates and the degree of danger involved. There are certain famous money change areas which you will find out about easily enough through the grapevine. The black market rate is sometimes four times the official one, so it is worth making some effort.

Borders—There are no rules about easy or difficult borders. In our case we had the hardest time entering Rhodesia and the easiest entering Uganda. Hardly to be expected. If you are entering a country known for its border hassles, try an obscure entry point, but make sure your documents are thoroughly in order. In all cases buy your visa in the preceding country so you are aware of the most recent decisions concerning tourist visas. These change frequently. Don't expect to be allowed into war-torn Somalia or the Eritrea province of Ethiopia. At the time of writing, the eastern route trans-Africa, via Cairo and Nairobi, is uncertain because of wars and unrest in Ethiopia and Zimbabwe Rhodesia.

It is important that your vaccination certificate is properly completed. This is often thoroughly scrutinised. If you had your jabs in America make sure the date is written out in full since Africa follows the European date method: day before month.

Make sure you can show 'sufficient funds'. In South Africa you must either show a ticket out of the country, or enough money to buy one to your country of origin.

And finally, don't worry. You will experience delays and frustrations, but nothing more serious if you behave sensibly. Africa's bureaucratic hassles are no better or worse than in other less developed countries.

Troublespots and Bureaucratic Hassles—Central and South America

by Jim Couper

If first impressions were permanent it would be quite possible to travel all the way from Guatemala to Peru with an intense dislike for each of the dozen or so countries passed through.

The entrance formalities at the border invariably leave the visitor angry, harassed, exhausted and wondering why he came to the wretched country in the first place. Fortunately there is time for those first impressions to dissolve as the real country unfolds.

There are two possible approaches for the *gringo* (anyone with pale skin) heading south beyond Mexico. He can retain his puritan upbringing, refuse to pay bribes and fight the lethargic system or he can pay what is asked and flow with the current. The former is cheaper and makes it easier for the next person with the same inclinations. The latter costs more, is easier on the psyche but sets a pattern which makes life more difficult for the next tourist who challenges it.

More travellers in Central and South America use public transport than drive their own cars, owing to the expense of shipping a vehicle between Panama and Colombia and the complications of shipping a vehicle out again on, say, the irregular US-bound services leaving Brazilian ports. You will not be able to sell your car in South America. Almost every road boasts a bus service. Passengers using long-distance buses usually have an easier, if slightly more expensive, time of it at borders than motorists. The passengers are 'processed' together and a special tax levied. Train services in Argentina, Bolivia, Brazil, Chile, Paraguay and Uruguay are good; elsewhere in South America they tend to be limited and the journeys arduous, on archaic rolling stock.

If you arrive by air, most countries require you to show a ticket out to the next country. The best idea is usually to get an airline ticket for the shortest inter-country distance possible and cash it in on arrival—though that process may take six weeks.

Travellers by land or water may also get caught out if they don't possess a ticket out or a substantial amount of money with which they can, in theory, buy one. According to one traveller's tale, a boatload of impecunious ticketless tourists recently took the ferry across Lake Titacaca from Peru to Bolivia. They pooled together all the money they were separately carrying and passed it to the first person in the customs queue. He declared $500 and was allowed in. The money was passed down the line to each subsequent traveller—and with each declaring the same $500 they all made it across the border.

A typical border between *Central American* countries presents a scene like the noisiest, dirtiest bazaar the mind can conjure up. Hordes of people in a state of panic push and shove, money changers refuse to listen to 'no', women sell cakes and biscuits from baskets on their heads, kids beg and guides try to force their services.

To examine the border chaos in

these tropical countries, let's look at timing. Outside business hours, one enters an era known as overtime and a set of charges that depend on the whims of the various issuers of stamps. Since arriving within an hour of the lunchtime break or the official end of the working day will invariably see matters dragged into overtime, there are effectively only about six free hours of business. Bearing in mind that everyone wants to get through during these hours and that borders come in pairs, it is a feat equal to a hole-in-one to clear both borders in a morning or afternoon session without paying overtime. Business hours are usually 8am to 12 noon and 2 to 6pm—but not always. Panama breaks for lunch from 12 noon till 1pm, while Costa Rica has its break between 11am and 12.30pm. Passing from one to the other could mean you're stuck in no-man's land for an hour. To add to the complications there is a time change.

To anyone who has travelled in Europe, or even Mexico, his concept of a border has nothing in common with what lurks between the tropics. There are usually five offices to be visited: customs, immigration, police, car inspection and health. A stamp is to be collected on a piece of paper from each office before the visitor may pass the armed guard. There is a particular order in which the offices must be visited but this is not posted and varies from country to country. There is a hostile, pushing lineup at each, forms to be filled and, during overtime, a charge of about $1 at each. Even during regular hours the tourist will often be asked to pay. The best policy is to question every charge, but if the official insists and can produce an official receipt with a charge printed on it, then comply. If there is a truck driver going through the border procedure at the same time as you, you may take your cues from him—but remember that he too is likely to be conned.

At Central American borders the officials all seem to be into the spraying business. Your tyres, or even your entire vehicle, will be given a spray of what looks like soapy water. This allegedly kills some kind of coffee pest. The charge is about $1.50 and the health inspector stamps your paper. Don't bother to argue that your shoes are more likely to pick up bugs in the countryside than your tyres, which will be humming along at near melting point on the hot road surface: logic counts for nothing. (Speaking of tyres and logic, Costa Rica prohibits the import of tyres that are not on rims. Why? Because mosquitoes could breed in them.) For an extra $1, the interior of your vehicle will be squirted with moth spray. Whether the fellow wielding the can has official duties or is in business for himself remains a mystery.

Visas and exit taxes can be another costly nuisance. Your nationality determines the number of visas needed but exit taxes apply to all. Another hassle is the confiscation of all fruits and vegetables and, if the inspector is choosy, certain grains and cereals. I know one couple who consumed about five pounds of peaches and bananas in front of an official rather than allow him to throw them away.

Continuing south from Central America usually means an ocean journey and entry into a port in Colombia or Ecuador. Colombia is notorious for its uninhibitedly corrupt customs men and for the casual use of knives and guns by the general population. Three in four motorists going there are robbed and all are held up (in both senses of the term) by customs. Paperwork for entering the country takes from two to a record seven days.

Ecuador is considerably less dangerous and somewhat less difficult. At the ports, automobiles go into a compound until the paperwork is completed and owners must rely on public transport to find police, customs and various other officials. The work can be done in a day, but there will be the usual charges for stamps as well as for car storage and the use of dock facilities.

Fortunately, Canadians and Americans can travel in most of *South America* without needing visas and there are no exit taxes. Where there are charges, these are exacted per passport, so it works out cheaper to have children included on a parent's passport.

On entering South America the question of the *carnet de passages* arises. See *Documentation for the International Motorist*. Strictly, Ecuador requires a motorist to present a *carnet* and may refuse him entry without one. A police escort to the Peru border, at the motorist's expense, will be provided. However numerous travellers have motored throughout the whole continent without a *carnet*.

Once into South America proper, the borders become free and easy, as they should be. In Brazil, Chile, Uruguay and Argentina, the officials actually act as though they are glad to see you. Frequent checkpoints in some of the military-run countries are a constant bother, but the tourist encounters no malice. Other than the initial entry into the continent the only borders that are likely to prove tedious are Peru and Bolivia. Paraguay can usually be done in less than an hour, the others in 30 minutes.

Overall, the character of a country usually has nothing in common with that of its frontier. Consider borders as independent little empires and you stand a much better chance of emerging mentally unscathed. If there is an easy way through tropical borders, short of throwing money about, I don't know of it. Neatness of dress and appearance seem to make for no appreciable improvement, although occasionally outrageously long hair on males, which is frowned upon, causes delay while the owner submits to a trim. Where a group of people is going through the system, it is acceptable for one person to show a number of passports; this also leaves someone else free to guard and, if necessary, move the group's vehicle. Allot three hours for leaving one tropical country and entering the next, although once in a while the procedure runs smoothly and is finished in half that time.

Most helpful of all is the ability to speak Spanish. Very little English is known and an understanding of the basics of the language is essential, not only for border crossings but also for getting to know the local people. Although the countries of the Americas, south of Mexico, always seem to be in the midst of a revolution or other political upheaval, the tourist can usually continue on his way since his dollar makes a large contribution to the economy. Although it is difficult to feel comfortable with tanks in the parks and machine guns at the intersections, they can be lived with and the accompanying checkpoints seldom take more than a minute. When soldiers carry out physical checks on bus passengers, women are usually excluded, so it is best for the females of the group to carry such potentially offensive weapons as penknives etc.

Next to border crossings, theft is the tourist's biggest concern. The problem starts at the Mexican border, escalates all the way to Panama, then reaches a peak in Colombia where there is no peace of mind or protection by the police. Ecuador, except for the port of Guayaquil, is reasonably safe as are Paraguay and Bolivia. Uruguay seems to be inhabited by little old couples who sip tea through a communal straw and would never dream of stealing. The energetic folk of Chile and Argentina are both fantastically hospitable and refreshingly honest.

Problems are unlikely in Brazil's southern provinces but in the big cities and the Amazon area possessions can occasionally vanish. Peru, like Colombia, can appear to be a nation of thieves with everyone on the lookout for an unwary tourist. Lima and Arequipa are particularly bad.

The majority of thefts consist of breaking into vehicles, snatching purses, picking pockets and taking anything momentarily left unguarded. Tourists are often victims of watch-snatching; especially sought after are watches with an expanding metal strap. No jewellery of any kind should be worn. But only in Colombia is there a danger of an actual holdup in which the victim will be shot or stabbed.

The question most often asked when travellers talk of places to see is 'How safe is it?'. The most accurate answer is 'It depends on how careful you are'.

The Crow Doesn't Fly Here
A Note about Air Travel in Africa

by Ingrid Cranfield

Air travel between Africa and Europe or North America is usually fairly uncomplicated and reliable. It is flights within Africa that try the patience and the pocket of the traveller.

Often it is impossible, because of international frictions, to fly directly between two capital cities. The two points may be only a couple of miles apart, but if a closed border separates them it is not unknown for an air journey between them to run to thousands of miles and several days. Between Nairobi and Dar es Salaam, owing to tension between Kenya and Tanzania, the air traveller can fly only via the Seychelles, 2,000 miles, or via Lusaka in Zambia, 3,000 miles. Nairobi and Dar es Salaam are some 400 miles apart. Between the Sudan and Ethiopia, Zaïre and Angola, Ethiopia and Somalia, there are no direct connections.

Worst affected are connections between East and West Africa and between the white-ruled countries of southern Africa and their political rivals in black Africa. Flights do cross the continent between west and east but they are slow and inconvenient for they make frequent stops. Seasoned travellers on long hops, eg, Nairobi to Lagos, often choose to go via a European capital, which, strangely enough, costs no more.

Visitors from the north to black Lesotho, Botswana or Swaziland have to change planes at Johannesburg and if they do not have a South African visa they are compelled to remain until their flight out in the very bleak airport hotel where there are no telephones, no television and few creature comforts. Other north-south routes are not quite what they appear on the timetable, and here again passengers sometimes prefer to go to Paris or London to pick up their connections.

On shorter point-to-point flights, the problems are even worse. Some flights only run at all or make scheduled stops if an agreed minimum of passengers is waiting to catch the plane. Punctuality is rare and service erratic. In West Africa especially, airlines seem frequently to issue more boarding passes than there are seats on a flight. Knowing travellers throw caution and courtesy to the winds and make sure of arriving at the departure gate two hours ahead of time, where they claim a position at the head of the queue and guard it until they board the plane. Also in West Africa, particularly at Lagos, a handout tends to make baggage clerks turn a blind eye to overweight baggage—sometimes to the extent of allowing a plane to become grossly overloaded with unrecorded excess weight. Baggage gets lost, possibly even more often than in the channels of western airlines.

Flying in Africa is a test of the independent traveller's ingenuity.

Keeping Track

Travel spin-offs and records

Buying Equipment for Off The Beaten Track Photography

by David Hodgson

You may have heard of Murphy's Law. It states that if something can go wrong it will go wrong—and at the worst possible moment. Even if you've never heard of the law you will have experienced it in action on any expedition. And especially if you have been involved in photography or filming on the trip. Cameras jam, films get lost or stolen. Lenses mist up at the most inconvenient moment or jam on the body when being changed and make you miss that essential shot. Twenty years of magazine photography in some of the least accessible parts of the world have convinced me that Murphy's Law is about as inevitable and universal as the law of gravity.

In this article I have Murphy's Law very much in mind. But I am also making two other assumptions. The first is that photography has some importance to you. That you need, and want, to take first class pictures for serious use rather than a collection of fuzzy snaps for suitable burial in some album. Secondly I assume that funds are limited and that every penny has to be well spent. Incidentally all the money in the world will not protect you from the ravages of Murphy's Law. One of the best financed and most lavishly equipped trips I ever went on started out with £25,000 worth of cameras and finished up being thankful to take their most important pictures on a battered, secondhand Leica.

Choosing Your Equipment

The motto here is—buy tough for travelling rough. The most sophisticated camera in the world is worse than useless if the electronic shutter fails halfway up a mountain. The more things that can go wrong, the more things will go wrong. Built-in meters are very convenient, especially if you can nip down to the local dealer's when a CdS battery fails. But if the battery on one of these meters goes in the wilds you will never replace it and the meter is useless without it. So, if you are buying especially for the trip, why spend money on a built-in meter? Better to put that additional money towards a sturdy shutter and wind-on mechanism and use a separate, selenium cell meter. This is the type where a cell converts the light directly to electrical energy. There is no additional power supply to fail. I would suggest a Weston Meter. You can safely buy them secondhand at very reasonable prices. They are tough and extremely accurate. If you do decide to use a camera with a TTL (through the lens) metering system then take along spare cells and make sure they are protected against excessive damp. But wear braces and a belt. Take a Weston along too.

This brings me to a point about back-up systems. Never put together your equipment on the basis that things will go right, but in the certain knowledge they will go wrong. Have an answer ready when they do. Two camera bodies are a wise investment, especially when you are going to be a long way from repairmen and replacements. You can shoot black and white with one and colour with the other, thus improving your chances of making the widest possible sales when you get home (See *Selling Travel Photography*). But which bodies?

Unless you have some specialist purpose which requires large format photography then 35mm cameras have all the advantages. They are light, easy to use, take 36 pictures on one roll of film, and produce negatives of sufficiently high quality to stand considerable enlargement—provided the exposure and processing have been correctly carried out.

The single lens reflex is probably the most popular camera here. It is a well proven design which allows you to see exactly what you are shooting. This is very valuable when taking close-up pictures, using either a macro-lens or extension tubes, or when using telephoto lenses for capturing distant scenes. If you intend to shoot a lot of extreme close-ups or use lenses longer than 200mm frequently, then I would advise an SLR. Nikon, Pentax, Canon, Leicaflex and Minolta are cameras which have all been tested under professional conditions and come through with flying colours. Nikon, Pentax, Olympus and Canon are probably the most widely used by magazine and newspaper photographers. One of my Nikons was once struck by a jet fighter which was coming in to land. It went on taking pictures. All four of these manufacturers make models which combine great toughness with lightness in a small handy format. The weight factor could make a great difference to transportation problems and should seriously be considered by any expedition photographer who is going to have to carry his or her own gear across difficult terrain.

But do not dismiss that much less popular type of 35mm camera, the rangefinder focus model. Perhaps the best known name here is the Leica. Nikon used to make an excellent rangefinder camera and so did Canon. These were the standard photo-journalist's equipment for decades before the SLR pushed them down into the third division.

Rangefinder cameras have a lot of things going for them as an expedition camera. They are simpler than the SLRs. Less to go wrong can mean fewer problems miles from anywhere. Because they are rather unpopular you can buy them cheaply secondhand. The lack of a mirror makes them quieter to use, which enables candid photography and wild-life studies to be shot more easily. The rangefinder focus is very positive and easy to use under low light levels. They are much less satisfactory, however, when being used with long lenses (above 200mm) or for extreme close-ups. The need for either type of photography really makes an SLR the front runner.

If you are switching from SLR to a rangefinder model then be sure to practise before starting on serious photography. There is a tendency to confuse the overall sharpness of the image as seen through a rangefinder screen with total picture sharpness and to forget to focus as a result! With an SLR, of course, this mistake cannot happen.

Before moving on to the subject of buying cameras secondhand I should mention the Nikonos, a very specialist camera that can be worth its weight in usable pictures on really tough trips. The Nikonos is an underwater camera with a lens fully sealed behind a glass plate and all the controls similarly protected by O-rings. You can use the Nikonos on land as well, in the wettest mud or the worst sand storms and simply wash it out when the muck gets too solid! If there is any chance of your equipment ending up soaked or seriously muddied then the Nikonos could be the answer. There are, of course, other ways of protecting standard gear.

Buying Secondhand Safely

Let us start with the camera body. Check for general signs of wear. Look inside as well as out, although external appearance will tell you a lot about how that camera has been treated. Look at the edges of the leather trim. If these are lifting it may indicate that the camera has been repaired (you have to remove the trim to reach the screws) and perhaps by a none too fastidious mechanic. Internally look at the tops of screws. If they have been turned over and show bright metal they have certainly been got at by a ham-fisted repairer. Look for bare patches on the internal blacking, which could cause flare problems. Run your fingers along the chrome film guides and over the metal rollers which draw the film onto the take-up spool. Any roughness will damage your films, causing what are called 'tram line' scratches along the sensitive emulsion.

Check the shutter blinds visually. If metal, do they look battered? If fabric, do they look worn? Now check the shutter by ear. Run it through *all* the speeds, especially the slow speeds which get used far less frequently and may be sluggish as a result. Does the mechanism sound smooth or a bit unsteady? If you are going to shoot reversal colour film (for transparencies), accurate shutter speeds are essential. Have them checked on a meter. Most well-set-up dealers can provide this service quite cheaply. You will almost certainly find that the speeds are not the same as those indicated by the setting. For example 1/1000th second may be only 1/900th. This does not matter. All shutters show a variation. What is important is that you *know* the exact shutter speed—so that you can set it on the meter—and that you have a constant error.

Check the wind-on by ear and by feel. It should sound as smooth and positive as it is to the touch. If the camera has a built-in meter, check this against a separate meter.

Check the strap connectors. Often these are the weakest part of a camera. If they give, you may lose everything.

Now turn your attention to the lens. There are two ways SLR and rangefinder cameras with interchangeable lenses match body to lens mount. One is by a bayonet lock and the other by screw thread. I prefer the bayonet method because it is so much faster to use in an emergency. You can easily cross the thread when trying to change lenses in a hurry. If you are buying a secondhand camera with a screw thread fastening system then check that the thread has not been damaged by somebody cross-threading the lens in a hurry.

Move the lens from infinity to maximum close-up several times. Does the helical thread movement feel, and sound, smooth? Grit can easily get inside the longer lenses and damage the mounts. Grease, used to pack the threads, can dry and make the focusing stiff or uncertain. The outer condition of the lens mount, again, provides a reliable clue to its previous usage. Look at the surface of the lens, holding it in such a way that any scratches will show up. If you intend to take the lens away on a trip fairly soon after purchase run at least one film through to check the quality—even of a first class lens. The best way of doing this is to set the camera up on a tripod exactly at right angles with a brick wall. Focus on the lines of the bricks and take a picture at each f-stop (changing the shutter speed to correct the exposure of course). After processing examine the frames with a powerful magnifying glass. Notice any distortions (the straight lines of the bricks may appear to bend out or inwards) and see which f-stop gives the best definition. All lenses work best at one or two stops, usually f5.6 or f8.

Additional Equipment

Some photographers set off on trips hung around with more equipment than the average dealer's display window. It is a waste of time, effort and money to take more than you need. Furthermore too many gadgets get between you and the picture-making. Keep life as simple as possible. You may need specialist gear for particular expedition tasks—such as macro-photography for example—but for general shooting here is a basic shopping list.

Lens hoods for all lenses. I prefer the screw on variety to the clip on type. They are less likely to get knocked off and lost. *Ultra violet filters* for all lenses—to cut down haze at high altitudes or when photographing at sea, and to protect the lenses. *Wide camera straps.* These are much more comfortable than the normal, narrow variety. *Light meter.* Separate meter, for reasons already given. *Lenses.* For general purpose 35mm photography my favourite combination is a slightly wide angle lens (a 35mm focal length I have found ideal), together with a slightly long lens. My best buy here would be either a 90mm lens or a 135mm. Both are excellent for candid portraiture, getting details of buildings, statues, etc, and landscape photography. The wide angle lens enables you to work close, frame boldly and operate successfully in a crowd.

What you do not need are as follows: *Tripod:* A tripod solid enough to provide firm support will be too heavy to carry out unless you are fully motorised. Even then it is likely to prove more trouble than it is worth. A much better alternative is a clamp which enables you to fasten the camera to a suitable support. I make my own from a Mole wrench to which a camera locking screw has been attached. Given the flexibility of this type of wrench you can almost always find something steady enough to provide a really stable mount for long exposures. *Flash guns:* I would certainly not rule this out completely. But unless you are intending to shoot a great many flash pictures you will probably find a bulb gun lighter to carry, cheaper to buy and rather more reliable. If you are going a long way off the beaten track bear in mind that an electronic gun will get through batteries far faster than a bulb gun—and provide less light. *Camera cases:* These tend to get in the way when you are working fast. There are better methods of protection.

Protecting the Equipment

My rule here has been to give the equipment maximum protection during transportation and the minimum necessary when in use. Aluminium cases with foam inserts provide the best protection during transportation. They keep out the dust, dirt and damp. You can sit on them on a long trip and stand on them to get a better angle. The locks are more or less useless as a protection. I prefer to rely on a steel cable and a strong padlock of the type used to secure bicycles. Paint a vivid coloured strip around the case as well. This makes it easier to spot when coming off an airport baggage conveyor belt and more conspicuous. Thieves tend to avoid cases and bags which attract a great deal of attention.

On the trip you never know when a shot will present itself. Keep at least one camera ready for use. That means loaded with fresh film with at least half a dozen unexposed frames available, the shutter and f-stops more or less set for the correct exposure. You can protect the chrome finish, if you feel the need to do so, by taping plastic sticky tape over the exposed surfaces. This does preserve them and keeps up the trade-in value of the equipment.

When working in countries where the climate permits, I should wear a waterproofed hunting jacket which had numerous deep pockets. I could keep unexposed film in a right hand pocket and exposed film the other side. Incidentally always wind off *all* your exposed film when using 35mm stock otherwise you can easily reload an already used roll by accident. Other pockets hold the spare lens, light meter and so on. Cameras are fairly sturdy and will survive all but direct immersion in water. But dust can be a real hazard. One way of protecting the camera is to place the whole thing in a plastic bag. Cut a hole for the lens and secure the bag around the mount with a stiff elastic band. A thin, strong, transparent bag enables you to see and operate the controls through the material. But it does slow you down and makes composition and focusing far more difficult. I suggest it only as an emergency measure. If you *know* in advance how tough conditions are going to be then include a Nikonos on the shopping list. Fit an ultra violet filter to *each* lens, not for technical reasons but simply to protect the lens.

Each evening clean up the equipment. There is no need to invest in costly blower brushes; a good quality, small size paint brush is perfectly satisfactory for brushing grit and dust from the camera body. Use chamois leather to clean the UV filter. Do not remove it unless you have to.

If you are using a bulb gun then check the condition of the bulbs (by looking at the colour dot) regularly when working in very damp climates. Before inserting a bulb into the gun, clean and moisten the wire contacts by *gently* rubbing them between your front teeth. This will make a misfire less likely. Remember to use blue coloured bulbs for colour slides.

Specialist Equipment

For close-up work you will need either

a close-up lens, extension tubes or bellows. Some modern standard lenses will focus to within a few inches and you may find this sufficient for all but the most precise record work of very small objects. Otherwise I advise a close-up lens or extension tubes in preference to bellows because they are so much easier to carry around.

Fast focus with long lenses can prove a problem. The best answer is probably the Novoflex follow-focus system where, instead of a helical screw and twist action focus, the lens is focused by pulling in, or releasing, a pistol-like trigger. For very long lenses (400mm and upwards) you will need the shoulder-pod support which is designed especially for use with this lens. Although bulky and rather cumbersome, Novoflex lenses are the best bet for action shots of fast moving subjects, e.g. for bird photography. I paint the barrel of my 400mm Novoflex bright yellow. It makes it look less like a large calibre army weapon and in some areas of the world a mistaken identification could prove most unfortunate for the photographer!

Motorised cameras are increasingly popular. With the exception of the Olympus motor system most are quite bulky and heavy. Take one only if it is going to earn its keep. You can use them for sequence photography or for shooting at a distance using radio-control. But now we are getting into the field of very specialist photography indeed!

Shopping Summary

My own choice, based on extensive personal experience in rough shooting, would be as follows:

Nikon with 35mm and 135mm lenses. Weston Meter. Rox carrying case. Hoods. UV filters. Mole wrench converted to camera clamp. Small screwdriver set (as used by watch makers) for tightening loose screws. Half inch wide paint brush. Bulb gun with spare battery. Spare batteries for CdS meter if carried. Two bodies for choice. Best buy secondhand: the Nikon F.

Alternatively: a secondhand Leica M2 with 35mm and 90mm lens or an Olympus with lenses similar to the Nikon. For a motorised camera: the Olympus on the grounds of lightness.

For very rugged work take a Nikonos with a 35mm lens. But this needs a special flash gun.

I can't guarantee you'll beat Murphy's Law by taking my advice. But you'll be in there with a chance!

A Traveller's Guide to Photography

by Rick Strange, with a contribution by David Hodgson.

Every year hundreds of expeditions set out all over the world, but it is only the larger sponsored ones that can afford to take a professional photographer along with them. And yet financially it is often the photographs, along with a well written report, that can make an expedition a success from the 'after sales' point of view.

Many people feel that they can walk into a camera store, buy a camera and walk out as a fully fledged photographer. Despite the fact that many modern cameras make the whole business of taking pictures remarkably easy, this really applies only to the technical side.

It is still the person who holds the camera, frames the picture and pushes the button who makes a picture good or bad, saleable or non-saleable. After all, any camera, no matter how technically sophisticated, is only a tool, and is only as good as the person who is using it.

On many smaller expeditions that take place every year, the person who is 'in charge of the camera' will undoubtedly have a hundred and one other jobs to do as well, and often under difficult or hazardous conditions. It is therefore even more important that when a couple of seconds can be spared to take a picture, the photograph should be of the highest possible quality. To make the job of picture taking easier for the expedition photographer, the following 'simple rules' should be studied and where applicable practised.

It is essential for the photographer to get to know his photographic equipment. He must practise with it for several weeks before the expedition leaves, so that when it comes to the crunch he will know exactly what to do, instead of having to sit down with the instruction manual while the fantastic shot he was about to take vanishes for ever. Practise all the technical things so that they become second nature: focusing, changing lenses, setting exposure, changing film, following focus, lighting, using flash, closeups, holding the camera correctly, framing, cleaning and the general care of the equipment. As soon as all of this is second nature it is possible to start being creative. It is astounding how picture quality seems to improve 100 per cent overnight, once a few simple rules are followed.

Black and white film is hardier than colour, which needs to be carefully stored, away from heat and humidity and the extremes of cold and dryness. Colour film, if looked after, will survive up to six months before being processed; black and white will usually last a full year. Colour film is often more difficult to find and very expensive in less developed countries than at home, and colour processing less reliable and also very costly. Polaroid film particularly can be very hard to find abroad. The only transparency film that is available nearly everywhere is Agfachrome CT18, though Kodachrome 25 and 64 and Ektachrome ER are usually to be found too. Weigh up these facts against the duration and route of your journey and then decide whether it is best to take with you supplies of all the film you will need or to restock *en route.* On expeditions it is probably best to stick to just one basic general purpose film and to buy all your full stock before leaving. When you think you have all you need, go out and buy another 20 rolls, just in case. After all, it's better to have too much film than too little.

If you are going to very hot or very cold places, the film manufacturers will give you advice on storage and the camera manufacturers will customise your camera equipment for these conditions. Or you can do it yourself. Protect film by storing it in a solid container with polystyrene insulation: for large quantities, an ice chest with a metal exterior would be ideal. In humid conditions put some packets of silica gel into the container to absorb moisture. Remove these in drier areas. Do be a little careful with the new Kodak professional range—these films need careful storage especially in hot countries, and I would strongly advise the use of the ordinary ER type of film.

Keep a record of the pictures you

take. If you don't want the bother of recording every photo—shutter speed, exposure, time, date, etc.—at the very least keep a note of the dates and locations covered by each roll, and number the rolls in the order in which you use them. One photographer's trick is to write the date and location on a piece of paper with a felt-tip pen and make this the subject of the first frame on a roll.

Manufacturer-owned processing laboratories everywhere tend to do a good job. Write to the companies before departure to find out addresses of laboratories in the countries you plan to visit. It may be possible to buy film with prepaid processing in a country where the total cost is cheap and save money by having the processing done elsewhere, where if you were paying on the spot it would be more expensive. Use the special processing envelopes to send film away to be processed or, if using another kind of packaging, mark on the outside 'Film only—not to be X-rayed'. Send film by first class mail and by air if appropriate. Mark on the return address label the number of the roll—to keep your records straight when you get the roll back. At airports, security X-rays, in spite of official denials, do damage about one film in six. To avoid damage, either keep film in a lead-lined bag or insist on handing film around the X-ray check.

When crossing borders, remember that many countries make a sharp distinction between the camera-toting tourist and the professional photographer. Professionals generally have to pay duty on imported equipment, tourists do not; exposed film passes duty-free, unexposed film does not. It is to your advantage to claim to be, and look like, a tourist. You can usually get away with bringing in numbers of rolls of unused films if you remove them from their packages and say that they are exposed.

Before we come to the makings of a good photograph, here are three other guidelines which will improve any picture.

1. Whatever the subject is and wherever a picture is taken, it is most important that camera shake is never permitted to spoil what might otherwise be a good shot. Hold the camera firmly with both hands, brace your arms against the side of your body or chest, and, when actually shooting, squeeze the shutter as if it were the trigger of a rifle. Never jab or stab it. Perform the whole operation in slow motion and keep the camera up to your eye for at least a second after you have pushed the shutter. As well as bracing your arms it is often a very good idea to brace your legs, and even lean against any solid object which may be close at hand, like a tree or a wall. All this in effect turns your body into a tripod and ensures that no body movement will be transmitted to the camera. When shooting in low light conditions which necessitate a slow shutter speed, hold your breath when you push the shutter. This and the bracing positions already mentioned should reduce 'camera shake' to nil.

2. Being prepared for the unexpected is even more important on expeditions than in other conditions because of the very nature of expeditions. You never know exactly what is going to be round the next corner so it is of vital importance that the camera is ready for action all the time. Never put off reloading when the end of the film has been reached. Do it at once and mark the used film so that the pictures on it can be correctly identified. Of course in ad-

verse weather conditions the camera should not be left lying around so that it might be damaged, but it should be ready to go, in its case, which must be near to hand.

3 Seeing—and not just looking—is really a subject all to itself. It might sound ridiculous but it is surprising how few people really see! Without looking tell yourself the colour of the tie the man next to you is wearing. What is the colour of the wallpaper in your office or sitting room? Things that you look at, touch or experience every day are registered by your brain, but only when you have trained yourself to observe are they consciously registered. Things that are so obvious that you do not notice them can ruin a photograph or on the other hand make one that you had not realised was there. Telephone wires running through the middle of a picture ruin it, but most people do not even notice them when they push the shutter, because they have not 'seen' them. When viewing a picture you have taken, how often have you said, 'Oh, I did not notice that was in the picture'? Teach yourself to SEE. Start consciously observing, by talking to yourself about what your eyes are looking at.

Of the three basic rules of photography—focus, exposure and composition—focus is the easiest, but remarkably often neglected. In the heat of the moment the simple tracking in or out of the focus ring on the lens is forgotten. This will soon become second nature if it is practised. Just remember that what you see through the view finder is what the film will reproduce. Always make sure that the subject you are photographing is as sharp as it can be in the view finder.

Exposure is not quite so easy. Many modern SLR cameras have built in exposure meters which read an average light as it falls on the whole subject contained in the view finder. It is very easy to let a bright sky affect the light reading so that the rest of the picture is underexposed. Many countries with a hot climate have white-painted houses. In a village scene, if an average light reading is taken, everything except the white houses will be underexposed. Conversely if a large dark object is part of the picture and the reading is taken off this, then everything else will be overexposed.

An exposure meter is only a simple computer and not a magic instrument for producing perfect pictures. The information it gives must be interpreted with an intelligent appreciation of the situation. The basic rule to follow is that the correct light reading is the one off the main subject of the picture. Thus in a village scene the exposure reading should be taken from the people, not from the houses or from the sky. A local character may be your subject (rather than the white wall he is leaning against): move in close and get your light reading from his face—don't let the brightness of the wall affect the reading.

If the subject is too far away for you to take a reading, use something near to you which is lighted in the same way and has the same tonal values. Several modern makes of cameras have spot reading light meters, which means that the light is read off a very small area in the middle of the frame. This is probably the most accurate type of meter, but again make sure that you take the reading off the subject.

When there are many different light values in a picture (e.g. bright sunlight and dark shadow), the best thing to do is to take a reading off both the light and the dark and then use an exposure which falls halfway between the two. If, however, you find yourself really in trouble because there are so many shades of light and dark and all the various degrees in between, find a nice section of a well lit neutral colour (preferably medium grey) and take your readings from this. The tarmac road ten yards in front of you often gives a very good neutral grey reading.

If you really are not sure what the exposure should be, try taking three pictures of the same subject, one at what you think is the right exposure, one at an f/stop above and another at an f/stop below. This way you can be sure that the picture is in the 'can'. After all, it is much cheaper to use more film and make sure of getting your picture than to have to go all the way back for another shot.

Composition is without a doubt the most difficult of the three basic rules, either to define or to teach. Generally it is something which is inborn or almost instinctive—people with a natural composition ability often become artists of one sort or another—but it can be learnt with practice. Although it is impossible to discuss composition very fully in an article of this length, here are a few practical suggestions which should help the expedition photographer to produce pictures of saleable quality.

The basic rule of composition is known as the Rule of Thirds. All this means is that if you divide any picture into thirds both horizontally and vertically, the points at which the thirds intersect are the optimum subject positions, with the lower right intersection point being the prime position.

Less technical, but just as important, is the policy of keeping your photographs simple. Don't try to say too much with one picture. After you have selected your subject concentrate on it and don't let it become cluttered up with unnecessary objects. You can use the additional lenses you are carrying for selective composition. Don't just use the telephoto lens for distant subjects: use it for close subjects, cutting out the uninteresting foreground. Make sure that there are no trees 'growing' out of the top of people's heads, etc.—remember to see and not just look.

Use your picture to tell a story about the subject. Always start off with an orientation shot which places the subject in its environment—in other words, a long shot. Move in closer to show the subject in more detail, and finally get in really close to show extreme detail.

Always try to fill the view finder of your camera with the subject. Then you won't have to rehearse that all too common story, 'See those two black dots in the distance? Well, that's the Land-Rover and our tent when we were halfway across the Sahara.' It might well be a beautiful shot of the desert, but if the subject is the expedition camp, then fill the frame with the expedition camp.

In many parts of the world you will encounter people who expect to be paid for being photographed (small amounts of money, not gifts, are what they're after) and others who are reluctant to be photographed at all. There are several ways around this. You might use a telephoto lens which can give you a good close-up despite your distance from the subject. You can wait until the subject is occupied and shoot without being noticed. Have the camera prepared, by focusing on something the same distance away as your subject, so that you can shoot without delay if an opportunity occurs. You can pan the camera towards the subject, click the shutter as your subject comes into view and keep panning afterwards, so that it won't be obvious that you've taken the shot. If you're skilled enough, you'll be able to get a good picture while holding the camera at waist-level and looking elsewhere. If all else fails, shoot from a window or a moving vehicle.

Getting out of Trouble

by David Hodgson

I have only been in jail three times in my life and, in each case, the stay was mercifully short! This was just as well since the jails were all in Africa and not amongst the healthiest places to spend a holiday. The cause, I hasten to add, was photographic rather than anything more sinister. A question of pointing a lens in the wrong direction at the wrong time. As a magazine photo-journalist with an editor and office to please I was probably less discreet with my photography than the average explorer or off the beaten track traveller would ever need to be. All the same great difficulties can be created quite unintentionally and with the most innocent of motives.

First of all find out exactly what the restrictions are and then stick to them. In many areas of the world, frontier security problems can turn an innocent border post picture into an excursion into espionage. At best you are likely to find your camera and film confiscated and the worst can be a whole lot worse. Avoid photographing military installations, troop movements, airfields, etc., unless you have a compelling reason for doing so. And I mean one which is worth doing time for! Some countries have a ban on photographing examples of civil engineering, scenes that make the country look primitive—i.e. all the most photogenic places—and industrial plants as well. In Yugoslavia, some years ago, I was arrested for taking shots inside a chemical plant—and this after being given permission to do so. One traveller in Pakistan—which is full of absurd photographic restrictions—was nearly arrested for taking a picture of a river which just happened to have a bridge in the background. Train—and aeroplane-spotters beware: certain Iron Curtain and Third World countries regard the photographic or written record-keeping involved as an offence. A bribe sometimes secures a bending of the rules.

Watch out for religious or cultural prohibitions. These can result in mob violence against you, especially in the remoter parts of the world. If you need or want to take pictures in places where the natives are far from friendly then be careful. Respect their dignity and right of privacy. I should also add that a quite different problem can arise when you are *too* popular as a camera operator. Everybody in the neighbourhood seems to want to get in on the act. This happens mainly when a camera is a rare sight and you are looked on as a piece of street theatre. My advice here is to go through the pretence of taking pictures. If necessary—and you have sufficient film stock—waste a few frames or even a whole roll of film. You never know—some of the pictures may be worthwhile and you will satisfy the crowd's curiosity. When all the fuss

has died down you can carry on with picture taking without arousing much interest. If you are staying in an area for some time and want really candid shots, then let everybody get completely used to seeing you with your camera. Reckon on spending several days simply being seen around. Very quickly your novelty value will disappear.

One good way of persuading reluctant subjects to pose and rewarding them when they do is to carry a Polaroid camera around. Take one shot and let them have it. Then shoot your main pictures. But a word of warning—you can get through a lot of expensive Polaroid film unless you save this tactic for an emergency.

Check List

Camera body (2 if possible) 35mm type with built-in exposure meter and interchangeable lens

Normal macro lens for normal and close-up work

28mm or 35mm wide angle lens

135mm telephoto lens

400mm telephoto lens (for bird and animal photography)

(All lenses fitted with UV filters to protect lens and dust caps for front and back)

Lightweight tripod

Set of jeweller's screwdrivers

Small pair of jeweller's pliers

Lens cleaning tissue and fluid

Antistatic cloth

Spare light meter

Cable release

Electronic flash and spare batteries and cables

Spare batteries for light meters

Strong camera case (aluminium)

Your estimate of film required (plus a reserve)

Hints

Always hold your camera firmly with both hands and brace your arms against the sides of your body.

Squeeze the shutter gently . . . never stab or jab it.

Take exposure readings from the main subject in the picture and don't let the sky or bright or dark surroundings affect the reading.

Frame the subject carefully in the view finder making sure that extraneous objects do not spoil an otherwise good picture.

Fill the view finder with the subject. The amateur generally tends to stand too far back.

Good Luck and Good Shooting!

Filming Your Trip or Expedition

by Adrian Warren

There is no need to emphasise the importance of making a photographic record of an unusual journey, not only for the sake of posterity, but also as proof of claims and achievements. And yet, while millions of tourists every year are shooting millions of feet of film to make home movies as souvenirs or as sentimental keepsakes, many expeditions to remote and seldom-visited parts of the world are ill-equipped photographically and some take no camera with them at all. Priorities have got jumbled somewhere. If you are planning to make your expedition a photographically strong one, the first thing to be emphasised is that there is a world of difference between taking stills and movie film. With stills, if you see something worth taking, you can press the button on the camera and then put the camera away again until the next interesting encounter. Movies, however, cannot consist of a series of unrelated shots; they must be closely enough related to build up into sequences. There are certain golden rules to be followed.

The object being filmed should always be shown in its context, ie in relation to its surroundings. Points of action or interest should be noted and filmed from close-up and from different angles, but *continuity* is important: if you are cutting from a wide shot of a person holding a rope in his right hand, it is no good shooting a close-up of him holding it in his left. Shots which show action detail or human reaction to the main shot are called *cutaways*.

Pans, or camera movement in the horizontal plane, must never be too long and should always have a beginning and an end, with the camera held stationary. Too fast a pan will cause the picture to judder and be uncomfortable to watch; a fast pan is only permissible where a bird or car, for example, is being followed, in which case the background is unimportant. Too slow a pan will bore an audience. Unless you are trying to give an idea of wide open spaces, a pan should always end on a point of interest.

The same basic rules for pans also apply to *tilts,* or camera movement in the vertical plane, but here it is even more important to end the shot on a point of interest. An upward tilt gives the audience a more optimistic feeling than a downward tilt.

Camera movement in the third dimension is called *zooming* or *tracking.* The temptation to use the zoom should be avoided where possible: too many zooms can make the resulting film offensive to watch, although, used in moderation, a zoom can make for very interesting shots. Zooming out from an object to wide angle is often more interesting than the converse movement. Perhaps the zoom lens' greatest attribute is the freedom it offers the cameraman for picture composition prior to shooting, but he must always remember first to adjust the focus on the point of interest, on full zoom, because there is far less latitude in the depth of field with telephoto than with wide angle.

Use a *tripod* wherever possible, but don't always shoot from the same height: if you are filming an animal or a child, adjust to its level—the audience will feel greater intimacy with the subject.

The use of a *wide-angle* lens also brings the audience into closer contact with the subject; its main advantage is the great depth of focus available, so that the subject and its surroundings can be shown in the same shot. Using a wide-angle lens also allows the camera to be hand-held in situations where a tripod is impossible. Movie filming for amateur use would normally be done on the 8mm format, whereas if the film is for professional or commercial use it would be 16mm or 35mm. For normal expeditions, 35mm equipment is far too heavy—and expensive—but the equipment and filmstock is not cheap even with 16mm, while the time consumption, not only during the expedition but after as well, is immense. The professional standards of film and television today are very high; unless the prime object of an expedition is to make a film, the prospect of sticking to stills should be strongly considered.

The 8mm movie camera has been improved to high standards of sophistication in the last few years, and the recording of sound on film is now possible with a single piece of apparatus although it can create problems for the editor. 8mm film may one day be acceptable for television, but—with the possible exception of dramatic newsfilm—not yet. Two basic faults with the small 8mm format are poor definition on the big screen and camera shake. Video tape recorders and cameras are convenient and have the obvious advantage of immediate playback, but image definition is not good.

With the so-called synchronised filming on the 16mm or 35mm format, the sound is recorded on a tape recorder linked to the camera electronically to ensure that both film and magnetic tape pass through the machines at a controlled, steady rate. In this case, the beginning of the shot is marked with the visible and audible clap on a clapperboard. This method of filming is widely used in television and is expected on film submitted from outside film-makers, but it requires at least two experienced operators to handle it—ie two extra mouths to feed and transport.

All films should be at least roughly scripted before shooting to avoid film wastage, but an expedition film is difficult to script because one is never sure what is going to transpire on the journey. It may turn out that the expedition never achieves its aim, in which case one ends up with an expensive film of a story with no conclusion. Here are a few more tips:

Colour negative film, for example 16mm Kodak Eastman colour 7247, is more widely acceptable than colour reversal film, and it is also easier for the laboratories to deal with if they have to make corrections to colour or exposure on the finished film. Also, with negative, copies of the film are of better quality.

While filming, be sure that the camera equipment is always kept scrupulously clean, dry and cool; the working mechanism should be checked between each roll of film and the shutter checked even more frequently to ensure that there are no specks of dust or hairs in the gate.

Always load film in very subdued light, at best in a special black changing bag, otherwise fogging or flare will result.

The filmstock should at all times be kept dry and cool and processed as soon as possible after filming.

A rush print should be obtained from the processing laboratory, at which stage you might consider offering the material for sale to a television company, although it is probably better to edit out the rubbish and the not-so-good shots and cut the rest into some sort of logical order. This is called a rough-cut and is a better stage to start seeking a market. Otherwise the editing and finishing with sound should be done professionally—an expensive undertaking.

If you still feel you can afford it, and dare risk the prospect of no sale, I wish you the very best of luck—film making is most rewarding work!

Selling Travel and Expedition Photography

by John Douglas

A two-man canoe expedition up the Amazon . . . a one-man trek through Afghanistan . . . a full-scale assault on Everest involving a party of sixty . . . a student group studying the fauna and flora of a remote Pacific island.

Question: What two features do these expeditions have in common?
Answer: They will all be short of money and they'll all be taking at least one camera.

The object of this article is to draw attention to the fact that these two features are not unrelated. Too few expeditions, whether they be on the grand scale or simply a student adventure, are aware that the camera can make a substantial contribution to much needed funds. When it is pointed out that a single picture may realise say £100, the hard-pressed expedition treasurer begins to see that he may be neglecting a very substantial source of revenue. While it is true that income from photography may not be received until some considerable time after the expedition has arrived home, it can be used to pay off debts—or perhaps to finance the next excursion.

If photography is to pay, then advance planning is essential. Too often planning is no more than quick decisions regarding types of camera, amount of film to be taken and who is to be responsible for this aspect of the expedition. Of course there *are* essential questions and something might first be said about their relevance to potential markets.

Unless sponsorship and technical assistance are received, a movie camera is certainly not worth taking on the average expedition. The production of a worthwhile expedition film is such an expensive, specialised and time-consuming matter that it is best forgotten. In order to satisfy television and other markets, a film must approach near-professional standards with all that that implies in editing, cutting, dubbing, titling and so on, to say nothing of filming techniques. Of course, if a film unit from, for example, a regional TV network can be persuaded to send along a crew, then some of the profit, as well as a fine record of the expedition's achievements, may accrue to the expedition. But for the average group this is unlikely. By all means allow members to take along a good 8mm movie camera but do not think of this as a source of income.

With still photography, the position is quite different. It *is* worthwhile investing in, or having on loan, a good range of equipment. It will probably be advisable to take perhaps three cameras, two 35mm SLRs and a larger format camera with an interchangeable back. If the latter is not available, then contrary to advice sometimes given, 35mm format is quite satisfactory for most markets except some calendar, postcard and advertising outlets.

A common planning argument is the old black and white/colour controversy. It is *not* true that mono reproduction from colour is unacceptable. Expertly processed, some 70-80 per cent of colour will reproduce satisfactorily in black and white. However conversion is more expensive and difficult than starting in the right medium and there are far more markets for mono than for colour. Although prices paid for black and white will be only some 50-60 per cent of those for colour, it is the larger market that makes it essential to take both sorts of film. A good plan is to take one-third fast black and white film and two-thirds colour reversal fim. For formats larger than 35mm, take colour only. The reason for this imbalance is that it is easier to improve a sub-standard black and white during processing. To all intents and purposes, the quality of a colour picture is fixed once the shutter closes.

It is advisable to keep to one type of film with which you are familiar. Different colour film may reproduce with contrasting colour quality and spoil the effect of, say, an article illustrated with a sequence of colour pictures. Colour prints will not sell.

Little can be said here about the choice of a photographer for an expedition. So often the selection of personnel is determined by such a variety of factors that the individual skills of the group members are subordinate to other considerations. But if the expedition is to sell its photography, it is important that the photographer is experienced and skilled, and that he should be given ample time and scope to practise his art.

Before leaving, the photographer should contact possible outlets for his work. Magazines generally pay well for illustrations, especially if accompanied by an article. Such UK markets as *Expedition*, *Geographical Magazine*, the colour supplements of Sunday newspapers or *Amateur Photographer* can be approached and, although they may not give a firm *yes*, their advice can be helpful. Specialist journals, assuming they are illustrated, may be approached if the work of the expedition is relevant but it should be remembered that the smaller circulation of such journals yields a lower rate of payment. It can be worthwhile to advertise the expedition in the hope of obtaining lucrative photography commissions, but beware of copyright snags if film is provided free.

Overseas magazines such as the American *National Geographic* often pay exceptionally high rates but the market is tight. Much nearer home, local and national newspapers may take some pictures while the expedition is in progress. If the picture editor is approached, he may accept some black and white pictures if they can be sent back through a UK agent. If the expedition is regionally based, local papers will usually be quite enthusiastic, but it is important to agree a reasonable fee beforehand, otherwise a payment of, say, £1 will hardly cover the costs involved. Local papers may also agree to take an illustrated story after the group has returned home but, again, it is important to ensure that adequate payment will be made for pictures published. When an expedition recruits nationally, there is the chance to exploit a number of newspaper outlets with 'local explorer' pictures.

It is not the purpose of this article to discuss techniques of photography but the photographer working with an expedition is well advised to seek guidance, before he leaves home, from others who have worked in the area. There can be problems of climate, customs and the like of which it is as well to be aware before starting out. Photography should be discussed at any planning meeting the expedition has with Embassy or High Commission staff.

Finally, one potentially contentious point *must* be settled before the first picture is taken. This is the matter of copyright ownership and the income received from the sale of photographs. It cannot be left to postexpedition arguments as to who bought which film and whether some photographs are *private* and others are expedition property. In law, copyright is vested in the owner of the film and *not* in the photographer. This can cause headaches if the expedition has no corporate identity in law. It might be best to get a lawyer to sort out an agreement. Certainly something should be in writing at least as far as the official photographer is concerned. One very satisfactory plan is to agree with the photographer that some form of licence shall be granted to the expedition. In any case, the golden rule is to be clear about the arrangements before any argument can start.

Once in the field, the photographer has two professional tasks. The first is to record the work and progress of the expedition. This is quite clearly his major role and in this he will be advised by the expedition leader. Secondly he should seek every opportunity to add to his photographic collections pictures which might have a market. This latter task is often most appropriately carried out prior to and after the expedition is properly under way. Thus the journey to and from the base position will often afford splendid opportunities for photography. Without increasing the overall cost it is sometimes possible for the photographer to make an earlier and more leisurely journey to the starting point of the expedition than the rest of the party.

Not unnaturally the question 'what sells?' will be asked. There is no simple answer except to say that at some time or other almost any technically good photograph may have a market. (Throughout, it is assumed that the photographer is able to produce high quality pictures. There is never a market for the out-of-focus, underexposed disaster!) Statements like, '*the photograph that sells best is the one that no one else has*' may not seem very helpful, yet this is the truth. It is no use building a collection which simply adds to an already saturated market. For example, an expedition passing through Agra will certainly visit the Taj Mahal—and photograph the splendid building. Yet the chances of selling such a photograph on an open market are dismal. It's all been done before, from every angle, in every light and mood. Perhaps a picture of the monument illuminated by a thunderstorm might be unusual enough to find a buyer but the best that can reasonably be hoped for is that the photographer will hit on a new angle or perhaps a human interest picture with the Taj as background.

On the other hand, a picture of village craftsmen at work might sell well, as will anything around which a story can be woven. Landscapes have a limited market but given exceptional conditions of light, then a good scenic picture might reap high rewards in the calendar or advertising markets. The golden rule is to know the markets well enough to foresee the needs. Sometimes the least obvious subjects are suddenly in demand.

Such was the case, for example, in 1976 during the raid on Entebbe Airport by Israeli forces. My own agency, Geoslides,was able to supply television with photographs of the old section of the airport and of the Kampala hospital just when they were needed. Yet who would expect a market for such subjects? Perhaps this is just another reason for carrying plenty of film. My own experience on expeditions is that I am constantly looking around for subjects. Certainly it is no use sitting back waiting for something to appear in the viewfinder. It is wise not to ignore the obvious, everyday scene. As I was preparing this article, Geoslides were asked for a photograph of a hailstorm in our Natal collection. Bad weather photographs sell well, so you should not always wait for brilliant sunshine.

One most important but easily overlooked point is the matter of record keeping. In the conditions experienced by many expeditions, this will not be easy, yet it cannot be emphasised too strongly that meticulous care must be taken to ensure that every picture is fully documented. It is true that certain photographs may be identified at a later date, for example macro-photography of plants, but no shot should be taken without some recording of at least its subject and location. It is usually best to number the films in advance and to have an identification tag on the camera which will indicate the film being exposed. A notebook can also be prepared in the days before the expedition leaves.

With the advertising market in mind, it is helpful to make sure that good photographs are taken which include the expedition's equipment. This may have already been agreed with suppliers. Less obviously, there is a market for photographs of proprietary brands of food, magazines, newspapers, items of clothing and equipment and so on in exotic or unusual settings.

If the expedition is to be away for a long time, it can be important to get some of the exposed film back home. There are dangers in this procedure because of the uncertainty of postal services but, provided some care is taken—perhaps with arrangements made through embassies—then there are advantages. Apart from the obvious problems involved in keeping exposed film in sub-optimum conditions, some preparatory work can be carried out by the expedition's agent. Of course if the film is sent home, it is essential that labelling and recording are foolproof.

Once the expedition has returned home, the serious business of selling begins. Topicality is a selling point, so there is no excuse for taking even a few days off, no matter how exhausted you may feel. Processing the film is clearly the first task, followed by cataloguing and the production of sample black and white enlargements. No one is going to buy if the goods are badly presented, so it is worth making sure that a portfolio of high quality mono enlargements and colour transparencies is prepared with a really professional appearance.

The first market to tackle will be local newspapers. Following up the advances made before the expedition set out is very important, no matter how lukewarm was the original response. It will be necessary for the photographer to work closely with whoever has the responsibility for the expedition's literary output. It often *looks* more professional if there is both a writer and a photographer to produce a magazine article, but it should be made clear to editors that a separate fee is expected for text and illustrations. This is invariably better than a lump sum or space-payment.

A direct source of income from photography can be slide shows for which the audience is charged. These are relatively easy to organise but must be prepared with slides of maps and perhaps an accompanying tape or live commentary. Incidentally, do not mix vertical and horizontal frames. It gives an untidy appearance to the show even when the screen actually accommodates the verticals. The bigger the screen the better. If these shows are to have a wide audience, it may be necessary to put the organisation into the hands of an agent.

A photographic exhibition can provide helpful publicity but it will probably raise little or no income in itself. Branch librarians are usually helpful in accommodating exhibitions and if these showings precede some other post-expedition event like a lecture or slide show, they can be indirect money spinners. For an exhibition, great care should be taken in making the display as professional as possible. Again, the bigger the enlargements the better. As far as photography is concerned, 'big is beautiful!' It is worth investing in a few really giant blow-ups.

Depending on the standing of the expedition, it can be a good plan to show some prints to the publicity department of the camera company or franchise agent whose equipment has been used, expecially if you have made exclusive use of one company's products. The same may apply to the makers of the film that has been used.

If the expedition has not been too far off the beaten track throughout its travels, then the firms may take photographs with which to illustrate brochures and posters. However, as with the calendar and postcard market, it must be pointed out that this is a specialist field, requiring not only particular sorts of photographs but pictures of a very high technical quality. This also applies to photographs used for advertising, although the suggestions made earlier regarding pictures of proprietary brands and expedition equipment leave this door slightly wider open than usual.

Whenever an original transparency or a negative is sent to or left with a publisher or agent, a signature must be obtained for it, a value placed on it should it be lost or damaged (anything between £50 and £100 per original), and a record kept of its location.

Lastly, when the cataloguing is complete, the expedition will wish to put the whole of its saleable photograph collection on the market. Now a decision must be reached on the thorny issue of whether or not to use an agency. Of course direct sale by the expedition would mean an almost 100 per cent profit, while agency sales will probably net only 50 per cent of the reproduction rights fees. But, as so often, it is the enlargement of the market, the professional expertise and marketing facilities of the agency which are attractive. It is worth making enquiries of a number of agencies (see *The Writers' and Artists' Yearbook*) and finding a company which offers the sort of terms and assistance that satisfy the expedition's requirements. It is usually preferable to deal with a company which does not expect to hold the collection but simply calls for pictures when needed. This allows much greater freedom to the copyright owner as well as being a check on what is happening in the market. Some agencies, such as Geoslides, offer additional services to associate photographers in the way of help with the placing of literary as well as photographic material, and in the organisation of lecture services.

It may well be better to contact an

agency before the expedition leaves. For a small consultancy fee, a good agency may be able to advise on the sort of pictures which sell well and on the level of reproduction fees which should be charged. There is nothing more annoying than selling rights for £5 and then finding that the market would have stood £25. Many amateurs sell their pictures for too low a fee and others assume that there is a set price irrespective of the use to which photographic material is put. In fact the market for photographic reproduction rights is something of a jungle and it may be better to gain professional advice rather than get lost. The same applies to locating markets. It is almost impossible for the inexperienced amateur to identify the likely markets for his work. There are thousands of possible outlets and a small fortune could be lost in trying to locate a buyer for a particular picture, no matter how high its quality.

An expedition of modest size and aims should expect to make a substantial profit from its photography, providing an effort is made along the lines indicated. In the case of a specialised and well-publicised expedition, it is not unknown for the whole of the cost of mounting the venture to be recouped from the sale of pictures. There are some simple points to remember. Don't treat the camera as a toy. Don't give the job of photography to a non-specialist. Don't put all those transparencies and negatives in the back of a drawer when you get home. As a money spinner, the camera may be the most important piece of equipment an expedition carries.

Freelance Travel Writing

by Richard Harrington

This chapter will not tell you how to write a travel article. That is something that you either have an aptitude for or you don't. What I shall try to tell you about is how to make it easier for yourself if you do decide to set out on the difficult road to becoming a professional freelance travel writer, or simply a good amateur with enough talent and determination to write occasionally for newspapers and magazines to bring in a little extra income, if only to cover the cost of the trip.

Let me begin with a warning. Many people aspire to writing but, as with photography, very few people do it well. You may feel you have real talent. Any editor will tell you, however, that reality usually (but not always) lies a long way from most people's opinion of their own ability. I've been on the receiving end of enough manuscripts in the last few years to know that most articles submitted by amateurs inevitably fail to make the grade.

Don't be discouraged, however. You may have done a trip so unusual that a good editor will work away with you at all costs to produce publishable material just because the trip was so exciting and unique, even if you've never put pen to paper in your life before and your writing talent is minimal. This is the reason why a lot of good modern expedition material gets published. A growing number of explorers in the last few years have turned their hand to writing, photography, film-making and lecturing in order to cover the cost of their expeditions. Some even make a living at it.

For the straight travel writer, the situation is different. There are very few successful travel writers who are totally freelance, either in Europe or in North America. A lot of travel writing is done by staff writers working for the big magazines and newspapers. Some of these may be general writers, others may be specialists in travel only. The travel editor will usually be a specialist, and will do some of the trips and some of the writing himself, but most of the in-house material will normally be delegated. Other articles will be commissioned from freelances or bought in from syndicates on a regular basis. A syndicate is a group of writers who between them supply material to periodicals. Because the buyers do not obtain exclusive rights, they pay reduced rates for the material, but the syndicate profits by selling the same material to a large number of publications simultaneously. Smaller magazines and more particularly newspapers have come to rely heavily on syndicate material over the years, and this has led to the development of specialist travel syndicates in North America. The bigger news syndicates also tend to have specialised travel departments.

A few hints on syndicates are appropriate at this point. To join a reputable syndicate you will usually be required to have a track record of published material. Or you could consider starting your own syndicate, but this is not a course I would advise unless you are an incredible salesman as well as writer, or intend to go on your own having proved yourself successfully with an existing recognised syndicate. Bear in mind that my remarks on syndicate writing are more appropriate in North America, and that, in the UK at least, there is not much scope for freelance travel writers hoping to get published through one of the big news syndicates. The lesser syndicates in the UK are often little more than picture agencies operating as agents for photographers who can write a little as well as take good photographs.

Being a freelance writer has never been easy, and in the travel field it is especially difficult. Somehow there seems to be the notion that if you're getting a lot of 'expenses-paid' trips, you can't expect to get paid as well for writing about your vacations. So as you may already have guessed, the competition is strong. Here as elsewhere, one solution to the problem may be to specialise. I've already touched briefly on the possibilities for the expedition writer. But the scope in this field is relatively limited.

One kind of travel writing I find distasteful is the traditional airline inflight magazine type, which seems to whisk you by jet from one poolside to another—only the Martinis and the Rum Punches seem to change. Regrettably this seems to be the predominant type of present-day travel writing, a sure sign that the hotels, airlines and national tourist offices are out to ensure that the hen that lays the golden egg gets its just reward. In other words, there isn't much objective travel writing around in the general tourism field. A few publications like the *New York Times* insist that all services should have been paid for by the writer as a measure aimed at something approaching objectivity. However few travel writers (understandably) and few publications are willing to pay for what they could have for nothing. Even if you prefer to pay your own way and tell the awful truth about the hassles you endured somewhere deep in Central America, recognise the fact that the editor to whom you submit your piece isn't interested in telling his readers about the countries they should avoid —only about the great places he feels they ought to visit. And if he doesn't feel that way, then the advertising sales department will soon remind him, via the managing editor, that their regular advertisers do happen to include a number of airlines, resort hotels and national tourist offices.

Let's come back to areas of specialisation, now that we've briefly mentioned the kind of tourism that this handbook is not about. One area in particular is Third World countries, especially those that still fortunately do not have a tourism ministry of any size and have thus far avoided the Hilton complex. As more people travel to more places, countries that were once unvisited by tourists are gradually becoming known. A classic example would be the Yemen (not to be confused with Soviet-controlled South Yemen.) Here is a country that really is little changed by western ways, and where very few tourists are seen. The same could be said of mainland China until about five years ago, but already that is changing.

I believe that a writer could do worse than specialise in far-flung lands where mass tourism is still a long way off. Or he could specialise in countries that are politically risky, as so many African countries now are, producing stories with strong political overtones. You could write about tourism in communist countries, the more easily if you carried the right card and political ideology around with you. Or you could take an economic viewpoint and examine the way that tourism is being encouraged by a Third World country as a foreign exchange earner or means of reducing unemployment. You could look at the way countries begin to receive foreign visitors again after a dead period resulting from war, guerilla incursions or natural disasters. These ideas may not quickly enrich the novice freelance writer, but at least they may salve his conscience in the face of so much present day 'bought' travel writing. It's certainly time for more writers to tell the truth in contrast with the sham and humbug that most editors choose to print. However I cannot counsel this as its the surest way to make a living.

Some travel pieces are commissioned and some are submitted speculatively. Obviously the former stand a better chance of getting published —and without a series of rejections. Few editors, however, will commission a piece from a freelance writer unless he has an excellent history of publication before. Inevitably the word 'freelance' comes to imply 'free' only in a certain sense. Most freelances return time and again to the same papers and magazines, and some may even have a more or less formal arrangement to supply a set number of pieces each year. Other writers may know that they are always the first to be called if a particular editor wants an article on say game parks or desert islands. Frequently, airlines, national tourist offices, hotels and resorts will retain particular writers through an agreement stipulating a certain amount of coverage each year. Such arrangements may be extremely lucrative for the hireling writer bribed in this way. However, the piper who calls the tune is unlikely to sign up the novice. Instead he will need to see a portfolio proving that the writer has good editorial connections and can virtually guarantee publication—not just anywhere, but in the right newspapers and magazines, the same ones that probably appear on his advertising schedule. On the subject of editorial freedom from advertising pressure I leave you to draw your own conclusions, except to say that connections between editorial and advertising interests are strongest in controlled circulation publications and in publications that are distinctly of the second rank in their field. I would add however that editorial integrity is higher in the UK than it is in North America.

Sad to say, the only way you might get started as a travel writer is by accepting 'freebies' from airlines, hotels and so on. Some (not all) will make you put your intentions in writing as to the coverage you'll give them, and may even demand a fixed minimum of column inches and the right to review the article before submission. If the airline/hotel/national tourist office/resort/tour operator is desperate or you have a lot of muscle with editors (proved by your bulging portfolio), it may grant you the necessary services *gratis* without a full assignment from a publication. More likely, however, there will be a demand to see a letter on the publication's notepaper showing more than just a passing interest in the story you may have proposed to them, or indeed a hard written commitment to publish. Just as many advertisers take a hard look at publications before they take paid advertising in them, so too they scrutinise readership demographics—socio-economic breakdowns, psychographic profiles—before committing themselves to granting free services to a travel writer. In the process, there will be meetings and phone calls. Nothing is undertaken lightly. Too many promoters of tourist services have had their fingers burnt in the past.

So who makes the first approach? That's something to which there is no set pattern. It might be you, the budding writer, or it may be an editor, or the promoter of a tourist service, or their public relations agency. One thing is sure. If you're willing to take the venal road and accept a retainer from say a national tourist office for so many pieces published per year, you could end up getting doubly paid for the same work—the retainer for the plugs you gave their banana republic and the fee you were paid by the publication. Better still, if your piece is syndicated and bought by say twenty newspapers, you could do very well out of that one article, and get a suntan and free vacation into the bargain.

If it sounds easy, it isn't. In the UK, I can think of no solely freelance travel writers making a really good income out of it. (Here I count syndicate members as freelance.) In North America, where there are far more full-time travel writers (freelance and staff writers), the situation is a little healthier. However I would still say there were no more than twenty real freelances making a good living at travel writing in North America.

The Guild of Travel Writers in the UK and the Association of American Travel Writers in the US comprise a healthy number of generally less successful travel writers, most of them part-timers with other sources of income. Conspicuously absent from both lists are the staff travel writers of the main magazines and newspapers, as well as the few really successful freelance writers. The reason is not hard to find. The travel writers' associations don't have the muscle to improve the lot of their members and top-notch writers do better negotiating contracts and terms on their own, without the help of a back-up organisation. A recent study showed that, in dollar terms, travel writers were being paid about the same as ten years ago. In real terms, that means they're getting half as much, and there is nothing their professional societies can do to change that situation. Increasingly publications are moving away from freeance writers and using in-house staff where they possibly can. One reason is to keep costs down by keeping it 'in the family'; the other is that it's actually a very good way of providing staff with vacations they could not otherwise afford. That may even have money-saving implications on the salaries front.

Some publications pay set rates for so many words, while others may be willing to negotiate. Some very large circulation magazines and papers take advantage of their size to pay less on the assumption that most writers will value the prestige of being in their publication and getting the cutting into their portfolio. In the end it's all a question of how much that editor needs you and how much you need him. Frequently reputations and supply and demand all come into play and overturn the set rates normally paid by the publication. The bigger a name you are, the higher the fee you can command. It follows from this fact of life that a commissioned article will normally earn more than one presented 'on spec'.

The terms of sale will also have a bearing on how and where you sell. Few publications are so powerful that they can demand exclusive rights worldwide in perpetuity. Indeed few would try. Normally if an article is commissioned by a publication (especially if an advance is paid), it will demand first publication rights and stipulate how long must elapse before the article is resold either abroad or in the same country, and in the same form or rewritten. With a speculatively submitted article, the writer is much more likely to be able to sell it simultaneously elsewhere. But bear in mind that he is also likely to be paid a good deal less in this situation. Whatever arrangement he comes to for publication, he should bear in mind that everything should be in writing and legally watertight, including such things as reprint rights, the publication's right to resell the article and so on. Under the new copyright laws in the US, copyright always remains with the author. In Europe, on the other hand, a writer may sell his copyright outright if he chooses to.

Always ensure that you write with a particular publication and even a particular editor in mind. A lot of editors won't even see you if you're unknown, but you should at least make the effort. That way, you'll begin to get a feel for what they want and be able to produce material accordingly. If a publication can supply you with written copy requirements and guidelines for articles (*Expedition* magazine and the *National Geographic* do this), get a copy and read it carefully. A major reason why articles are rejected is the writer's failure to study the publication's requirements. The very best articles need little or no editing after submission.

One subject that cannot be omitted from an article like this is photography. Even hometown newspapers like at least one picture to go with a travel article. The message is brief and to the point. Either take a good photographer with you on your trips or make sure you can take good pictures yourself. Taking good quality photographs is an ability that some are born with but few can learn. If friends don't get ecstatic about your pictures, and if editors keep describing your favourite shots as 'technically competent', then don't even persevere. Give up, admit you don't have it and get acquainted with a photographer who can take the kind of pictures you need.

Whereas good freelance travel writers generally work independently or through a syndicate (you'd probably be wasting your time to have an agent), most photographers work through agents. This may reflect their more artistic temperaments, their other-worldliness. It may also have something to do with the fact that while photographers do tend to specialise, there are many who do a lot of good travel work but cannot be classified as travel photographers since they also produce first rate pictures in other fields.

Photographers can make a living in travel and tourism in a number of areas. Just as many travel writers on the way up get involved in writing advertising and brochure copy for tour services, so a lot of photographers find more than enough regular work providing the images for travel ads. and promotional literature. Some editors will take a good story and try to find pictures to match it. Others work the other way round and, having a good set of photographs, will call up a writer they feel can write a good story around them. However the average editor is only human and really prefers to have his work done for him. Try to do yourself a favour, therefore, and think in terms of submitting article and photographs together. A lot of publications will give consideration to your piece only if both elements are there—copy and pictures. In fact many may review the whole submission initially on the strength of the pictures alone and, if these are not up to standard, may not even waste their time with reading the article. An editor's time is always short. Make sure that transparencies are all in clear plastic folders, and that black and white prints are large format and printed to professional standard. In fact presentation has a lot to do with your chances of success—not just how your manuscript is typed, but also whether or not you take the photographer along to meet the editor, whether you meet him formally in an office or outside it for a drink. You'll have to take your cue from the editor but a lot will be up to you: if you have the option, go for the formal approach in the USA, the casual in the UK. Remember that the freelance travel writer has always got to be a first class salesman.

There aren't any secrets of the trade when it comes to making it as a travel writer. One way up the ladder is first to become a staff travel writer on a major newspaper or magazine. The thinking behind this is quite simple. As the travel editor of a major magazine—let's say a New York-based general interest man's magazine with a circulation just over one million—your name comes to prominence to a degree that far outstrips your writing and editing talent. You appear on the list of the magazine's editorial staff, you get the chance to do some of the trips and articles yourself and you meet every freelance writer in town—from the right side of the desk. Within a year of taking up the title, you find you can go where you like. As a freelance or syndicate writer, the hard work has been done, and diplomacy rather than talent made it all so easy. There's only one snag to this approach and that is that very few travel editors of major publications quit this post to become freelances. Financially, they think it just isn't worth it, and they're probably right. What they tend to do instead is work out a deal with their employers that will allow them to do some freelance (including syndicate) work, while still filling the rôle of travel editor. On this basis, their publisher may even be able to afford to pay a slightly reduced salary.

You may wonder at this point how it is that one walks into a major magazine and becomes a travel editor. In the US today, it's not all that difficult to make it to this position at a relatively young age, provided that you have a degree in journalism, a little talent, an ability to hustle and the determination to work hard and make it to the top of your field. Most New York-based major magazines have a travel editor, and unless it's specifically a travel magazine, the travel editor will rate fairly low in the hierarchy of editors. It's relatively easy to reach such a position of power by starting as an editorial assistant in virtually any department and steering your way in the general direction of the travel editor's office. In due course, with some luck, you'll go from being an assistant on say the good food column, to being the travel editor, when the travel editor gets promoted to editor in charge of all services columns.

I've written these broad hints for the would-be freelance with the US mainly in mind, since freelance travel writing in Europe (with the possible exception of Germany) still has a long way to go. New York is undoubtedly the world centre of professional travel writing, much of it nevertheless of an extremely low standard. In the US today, there is a large market awaiting the talented freelance travel writer willing to take on the struggle. In Europe it's a lot tougher, but the market is growing in size and quality. If you really feel you could make it, go right ahead and try. I wish you luck.

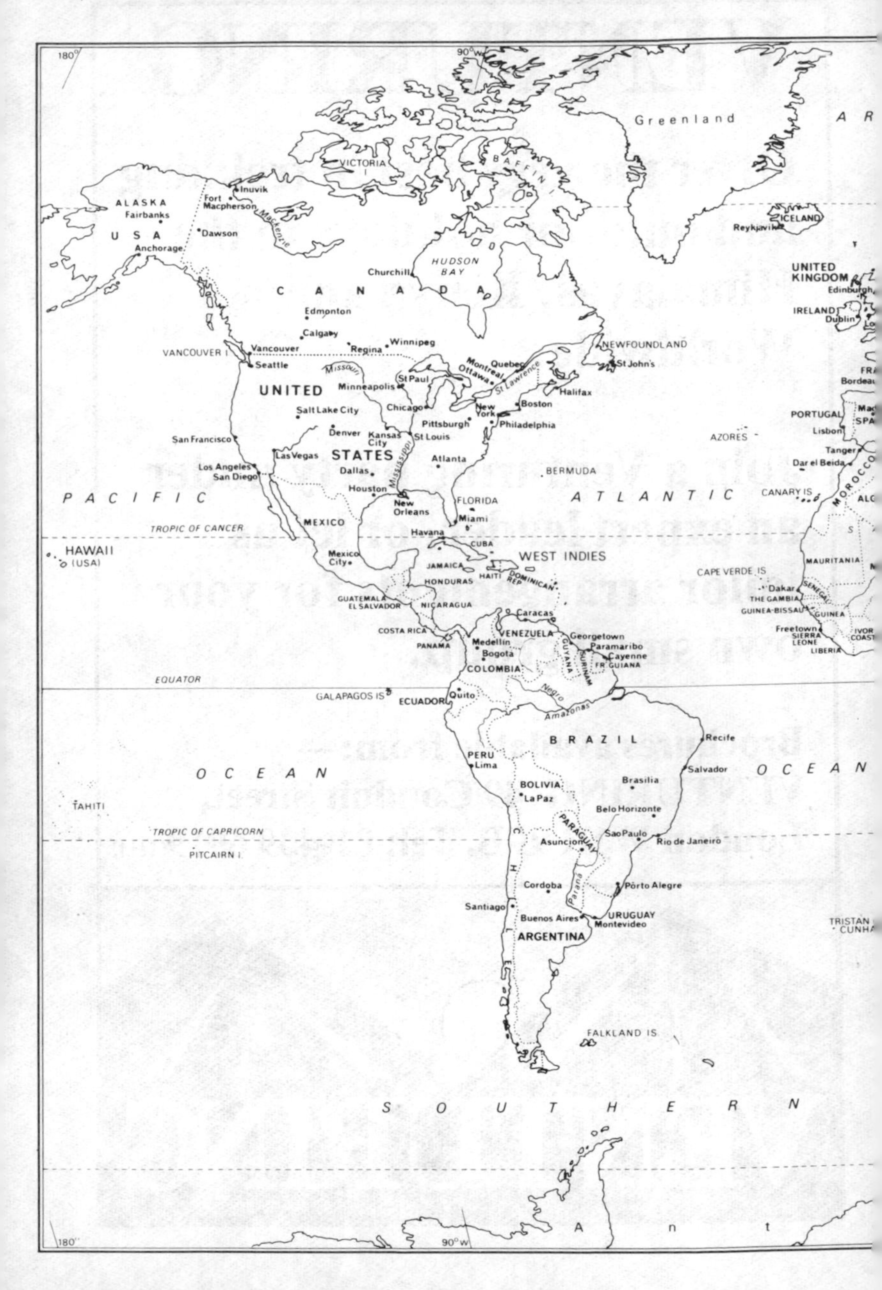
180°
90°W
Greenland
VICTORIA I
BAFFIN I
ALASKA
Fairbanks
USA
Anchorage
Inuvik
Fort Macpherson
Dawson
Mackenzie
ICELAND
Reykjavik
HUDSON BAY
Churchill
CANADA
Edmonton
Calgary
Regina
Winnipeg
VANCOUVER I
Vancouver
Seattle
Missouri
UNITED STATES
Minneapolis
St Paul
Chicago
Salt Lake City
Denver
Kansas City
St Louis
Pittsburgh
San Francisco
Las Vegas
Los Angeles
San Diego
Dallas
Houston
Mississippi
Atlanta
New Orleans
Montreal
Ottawa
Quebec
St Lawrence
New York
Philadelphia
Boston
Halifax
NEWFOUNDLAND
St John's
UNITED KINGDOM
Edinburgh
IRELAND
Dublin
PORTUGAL
Lisbon
AZORES
Tanger
Dar el Beida
MOROCCO
CANARY IS
BERMUDA
FLORIDA
Miami
PACIFIC
ATLANTIC
TROPIC OF CANCER
MEXICO
Havana
CUBA
HAWAII (USA)
Mexico City
WEST INDIES
JAMAICA
HAITI
DOMINICAN REP
HONDURAS
GUATEMALA
EL SALVADOR
NICARAGUA
COSTA RICA
PANAMA
Caracas
VENEZUELA
Medellin
Bogota
COLOMBIA
Georgetown
Paramaribo
Cayenne
GUYANA
SURINAM
FR GUIANA
MAURITANIA
CAPE VERDE IS
Dakar
SENEGAL
THE GAMBIA
GUINEA-BISSAU
GUINEA
Freetown
SIERRA LEONE
LIBERIA
EQUATOR
GALAPAGOS IS
ECUADOR
Quito
Negro
Amazonas
BRAZIL
Recife
PERU
Lima
Salvador
OCEAN
OCEAN
BOLIVIA
La Paz
Brasilia
TAHITI
Belo Horizonte
PARAGUAY
TROPIC OF CAPRICORN
Asuncion
Sao Paulo
Rio de Janeiro
PITCAIRN I
CHILE
Parana
Cordoba
Pôrto Alegre
Santiago
Buenos Aires
URUGUAY
Montevideo
ARGENTINA
TRISTAN DA CUNHA
FALKLAND IS
SOUTHERN
Ant
180°
90°W

90°E
180°
OCEAN
Siberia
Yenisey
Ob
Lena
Yakutsk
USSR
FINLAND
SWEDEN
Helsinki
Leningrad
Oslo
Stockholm
Copenhagen
Volga
Sverdlovsk
Novosibirsk
Irtysh
Irkutsk
Amur
Moscow
Berlin
Warsaw
POLAND
Kuybyshev
Kiev
CZECH
Danube
Vienna
AUST
HUN
ROMANIA
Belgrade
YUGOSLAVIA
BULGARIA
BLACK SEA
Istanbul
Rome
Naples
Volgograd
Tbilisi
CASPIAN SEA
Alma Ata
Tashkent
SINKIANG
MONGOLIA
Haerhpin
Vladivostok
Shenyang
Peking
CHINA
Hwang Ho
Seoul
KOREA
JAPAN
Tokyo
Ankara
TURKEY
Athens
GREECE
Tripoli
Tehran
IRAN
Kabul
AFGHANISTAN
Indus
IRAQ
SYRIA
ISRAEL
JORDAN
Jerusalem
Baghdad
TIBET
Sian
Wuhan
Shanghai
Yangtze
Chungking
Kathmandu
NEPAL
Delhi
PAKISTAN
Cairo
LIBYA
EGYPT
SAUDI ARABIA
Riyadh
INDIA
Dacca
Kwangchow
Taipei
TAIWAN
Hanoi
Hong Kong
BURMA
Nile
OMAN
DEM REP OF YEMEN
YEMEN
Bombay
ARABIAN SEA
BANGLADESH
Rangoon
THAILAND
LAOS
VIETNAM
CHINA SEA
Bangkok
Manila
PHILIPPINES
MARIANA IS
CHAD
SUDAN
Madras
ANDAMAN IS
KAMPUCHEA
Saigon
Addis Ababa
ETHIOPIA
SRI LANKA
Colombo
NICOBAR IS
CENTRAL AFRICAN REP
CAMEROUN
SOMALIA
MALAYSIA
Kuala Lumpur
SUMATRA
CAROLINE IS
KENYA
UGANDA
Mogadiscio
CONGO
ZAIRE
Nairobi
Singapore
NEW GUINEA
SOLOMON IS
Kinshasa
SEYCHELLES
TANZANIA
Dar es Salaam
Djakarta
INDONESIA
JAVA
Luanda
ANGOLA
COMORO IS
MALAWI
Darwin
ZAMBIA
Lusaka
Zambezi
Mocambique
INDIAN
SW AFRICA
Salisbury
RHODESIA
MALAGASY REPUBLIC
MAURITIUS
REUNION
Alice Springs
BOTSWANA
(NAMIBIA)
Pretoria
Maputo
AUSTRALIA
Brisbane
SOUTH AFRICA
Durban
LESOTHO
OCEAN
Perth
Kalgoorlie
Canberra
Sydney
Melbourne
TASMANIA
Hobart
Auckland
Wellington
Christchurch
NEW ZEALAND
OCEAN
90°E
180

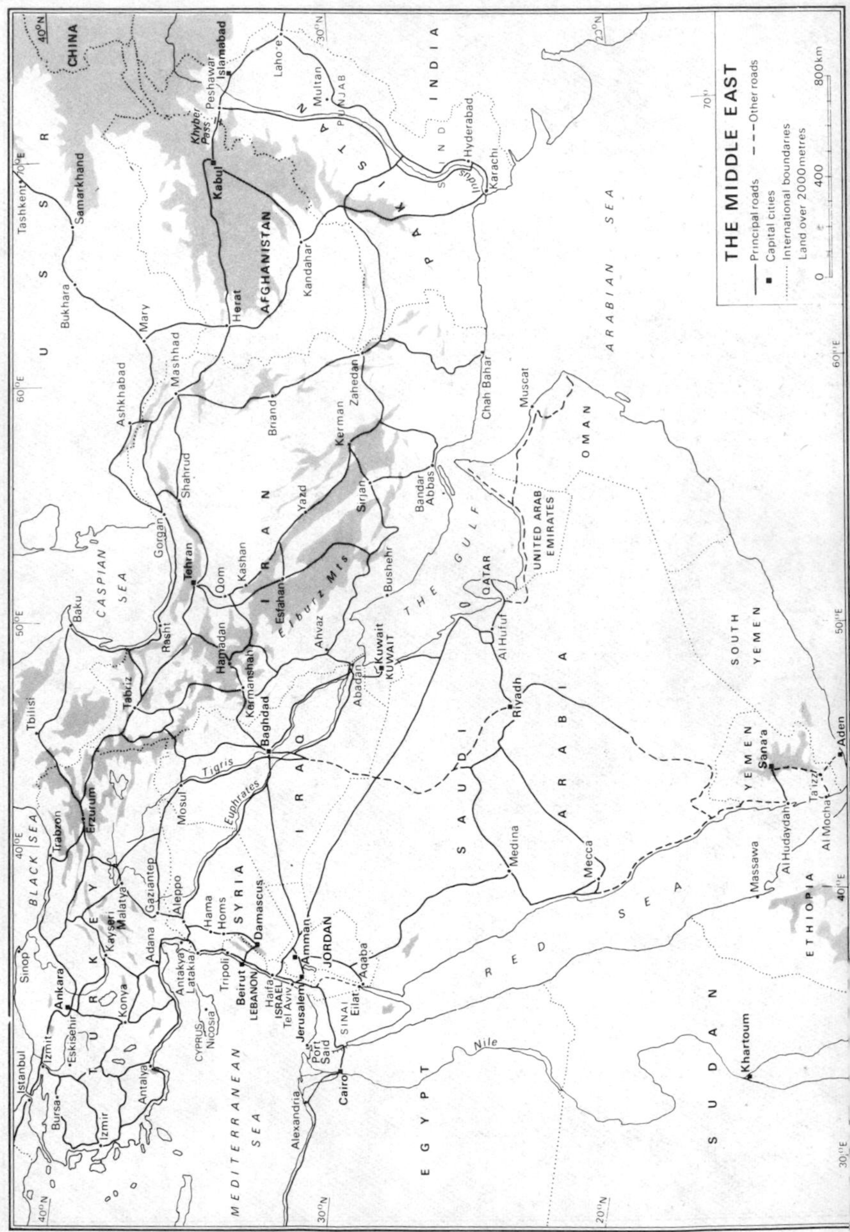
THE MIDDLE EAST
Principal roads
Other roads
Capital cities
International boundaries
Land over 2000 metres
0 400 800km
U S S R
CHINA
AFGHANISTAN
P A K I S T A N
I N D I A
I R A N
I R A Q
S Y R I A
T U R K E Y
JORDAN
LEBANON
ISRAEL
KUWAIT
QATAR
UNITED ARAB EMIRATES
OMAN
SAUDI ARABIA
YEMEN
SOUTH YEMEN
ETHIOPIA
S U D A N
E G Y P T
CYPRUS
SINAI
PUNJAB
S I N D
BLACK SEA
CASPIAN SEA
MEDITERRANEAN SEA
RED SEA
THE GULF
ARABIAN SEA
Elburz Mts
Tigris
Euphrates
Nile
Indus
Khyber Pass
Tashkent
Samarkhand
Bukhara
Mary
Ashkhabad
Mashhad
Herat
Kabul
Kandahar
Peshawar
Islamabad
Lahore
Multan
Hyderabad
Karachi
Zahedan
Birjand
Kerman
Yazd
Sirjan
Bandar Abbas
Chah Bahar
Muscat
Shahrud
Gorgan
Tehran
Qom
Kashan
Esfahan
Bushehr
Ahvaz
Abadan
Kuwait
Hamadan
Kermanshah
Rasht
Baku
Tabriz
Tbilisi
Baghdad
Mosul
Erzurum
Trabzon
Malatya
Gaziantep
Aleppo
Hama
Homs
Damascus
Amman
Aqaba
Eilat
Jerusalem
Tel Aviv
Haifa
Beirut
Tripoli
Latakia
Antakya
Adana
Kayseri
Konya
Ankara
Sinop
Istanbul
Izmit
Eskisehir
Bursa
Izmir
Antalya
Nicosia
Port Said
Alexandria
Cairo
Riyadh
Al Hufuf
Medina
Mecca
Sana'a
Ta'izz
Aden
Al Hudaydah
Al Mocha
Massawa
Khartoum
40°N
30°N
20°N
40°E
50°E
60°E
70°E

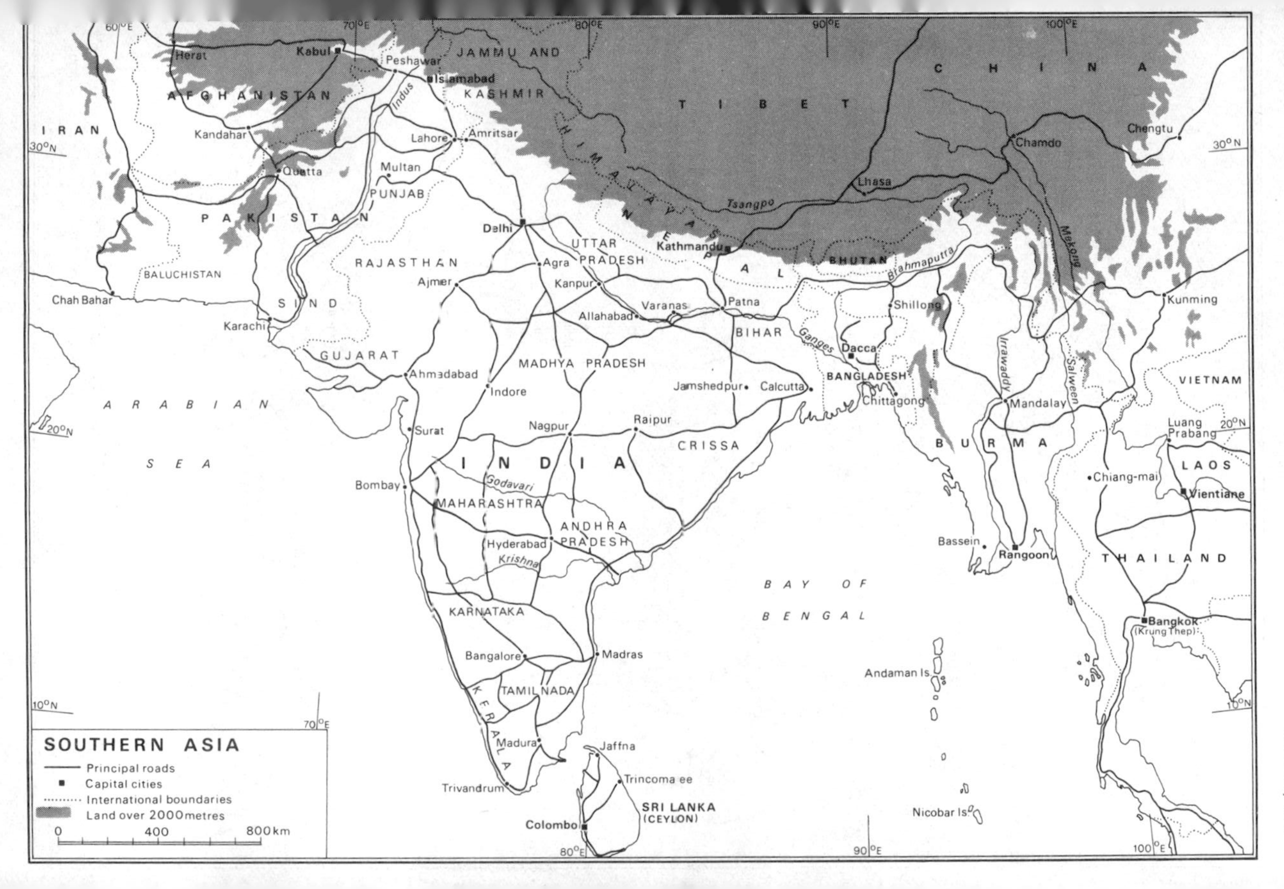
SOUTHERN ASIA
Principal roads
Capital cities
International boundaries
Land over 2000 metres
0
400
800 km
60°E
70°E
80°E
90°E
100°E
30°N
20°N
10°N
IRAN
AFGHANISTAN
Herat
Kabul
Kandahar
Peshawar
Islamabad
JAMMU AND
KASHMIR
Indus
Lahore
Amritsar
Multan
PUNJAB
Quetta
PAKISTAN
BALUCHISTAN
Chah Bahar
SIND
Karachi
GUJARAT
Ahmadabad
RAJASTHAN
Ajmer
Delhi
Agra
UTTAR
PRADESH
Kanpur
Allahabad
Varanas
Patna
BIHAR
HIMALAYAS
NEPAL
Kathmandu
TIBET
Tsangpo
Lhasa
BHUTAN
Brahmaputra
CHINA
Chamdo
Chengtu
Mekong
Kunming
Shillong
Ganges
Dacca
BANGLADESH
Chittagong
MADHYA PRADESH
Indore
Jamshedpur
Calcutta
Raipur
CRISSA
Nagpur
Surat
INDIA
Bombay
Godavari
MAHARASHTRA
ANDHRA
PRADESH
Hyderabad
Krishna
KARNATAKA
Bangalore
Madras
TAMIL NADA
KERALA
Madura
Trivandrum
Jaffna
Trincoma ee
SRI LANKA
(CEYLON)
Colombo
ARABIAN
SEA
BAY OF
BENGAL
Andaman Is.
Nicobar Is.
Irrawaddy
Mandalay
Salween
BURMA
Bassein
Rangoon
VIETNAM
Luang
Prabang
LAOS
Chiang-mai
Vientiane
THAILAND
Bangkok
(Krung Thep)

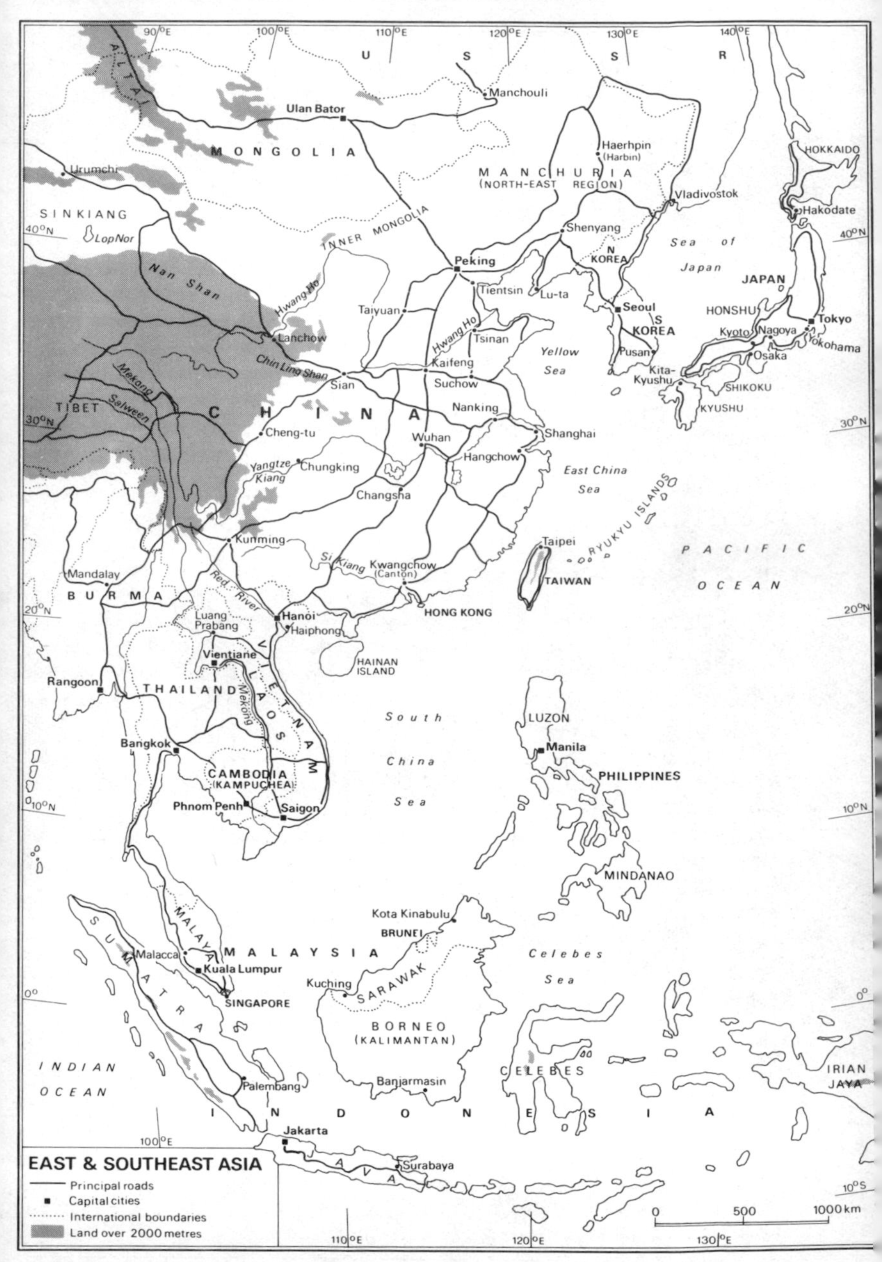
EAST & SOUTHEAST ASIA
Principal roads
Capital cities
International boundaries
Land over 2000 metres
U S S R
MONGOLIA
Ulan Bator
Manchouli
ALTAI
Urumchi
SINKIANG
Lop Nor
INNER MONGOLIA
MANCHURIA
(NORTH-EAST REGION)
Haerhpin
(Harbin)
Vladivostok
Shenyang
N KOREA
S KOREA
Seoul
Pusan
Peking
Tientsin
Lu-ta
Nan Shan
Hwang Ho
Taiyuan
Tsinan
Lanchow
Chin Ling Shan
Sian
Kaifeng
Suchow
Nanking
Shanghai
Hangchow
Wuhan
Cheng-tu
Chungking
Yangtze Kiang
Changsha
TIBET
Mekong
Salween
CHINA
Yellow Sea
East China Sea
Sea of Japan
JAPAN
HOKKAIDO
Hakodate
HONSHU
Tokyo
Kyoto
Nagoya
Yokohama
Osaka
Kita-Kyushu
SHIKOKU
KYUSHU
RYUKYU ISLANDS
PACIFIC OCEAN
Taipei
TAIWAN
Kunming
Si Kiang
Kwangchow
(Canton)
HONG KONG
Red River
Mandalay
BURMA
Hanoi
Haiphong
Luang Prabang
Vientiane
HAINAN ISLAND
Rangoon
THAILAND
LAOS
VIETNAM
Mekong
Bangkok
CAMBODIA
(KAMPUCHEA)
Phnom Penh
Saigon
South China Sea
LUZON
Manila
PHILIPPINES
MINDANAO
Kota Kinabulu
BRUNEI
MALAYA
MALAYSIA
Malacca
Kuala Lumpur
SINGAPORE
SARAWAK
Kuching
BORNEO
(KALIMANTAN)
Celebes Sea
SUMATRA
Palembang
Banjarmasin
CELEBES
INDIAN OCEAN
INDONESIA
IRIAN JAYA
Jakarta
JAVA
Surabaya
90°E
100°E
110°E
120°E
130°E
140°E
40°N
30°N
20°N
10°N
0°
10°S
0
500
1000 km

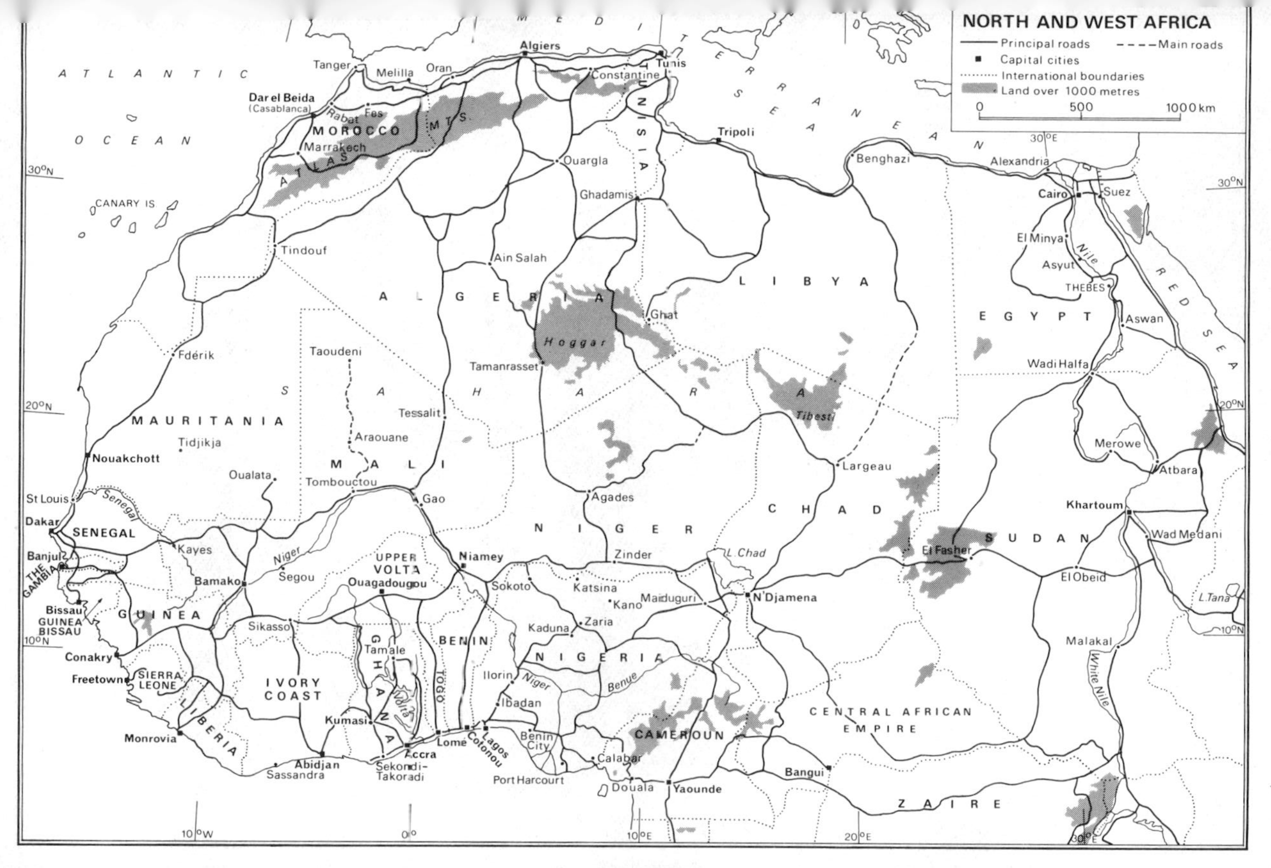

NORTH AND WEST AFRICA
Principal roads
Main roads
Capital cities
International boundaries
Land over 1000 metres
0
500
1000 km
ATLANTIC OCEAN
MEDITERRANEAN SEA
RED SEA
CANARY IS
MOROCCO
ATLAS MTS
ALGERIA
TUNISIA
LIBYA
EGYPT
SUDAN
CHAD
NIGER
MALI
MAURITANIA
SAHARA
SENEGAL
THE GAMBIA
GUINEA BISSAU
GUINEA
SIERRA LEONE
LIBERIA
IVORY COAST
UPPER VOLTA
GHANA
TOGO
BENIN
NIGERIA
CAMEROUN
CENTRAL AFRICAN EMPIRE
ZAIRE
Hoggar
Tibesti
Tanger
Dar el Beida (Casablanca)
Rabat
Fes
Marrakech
Melilla
Oran
Algiers
Constantine
Tunis
Tripoli
Benghazi
Alexandria
Cairo
Suez
El Minya
Asyut
Nile
THEBES
Aswan
Wadi Halfa
Merowe
Atbara
Khartoum
Wad Medani
L. Tana
El Obeid
Malakal
White Nile
El Fasher
Largeau
N'Djamena
L. Chad
Maiduguri
Kano
Zinder
Agades
Ghat
Ghadamis
Ouargla
Ain Salah
Tamanrasset
Tindouf
Fdérik
Taoudeni
Tessalit
Araouane
Tidjikja
Nouakchott
Oualata
Tombouctou
Gao
St Louis
Senegal
Dakar
Kayes
Niger
Segou
Niamey
Banjul
Bamako
Ouagadougou
Sokoto
Katsina
Bissau
Sikasso
Kaduna
Zaria
Conakry
Tamale
Freetown
Ilorin
Ibadan
Benue
Volta
Kumasi
Monrovia
Lome
Cotonou
Lagos
Accra
Abidjan
Sassandra
Sekondi-Takoradi
Benin City
Port Harcourt
Calabar
Douala
Yaounde
Bangui
30°N
20°N
10°N
10°W
0°
10°E
20°E
30°E

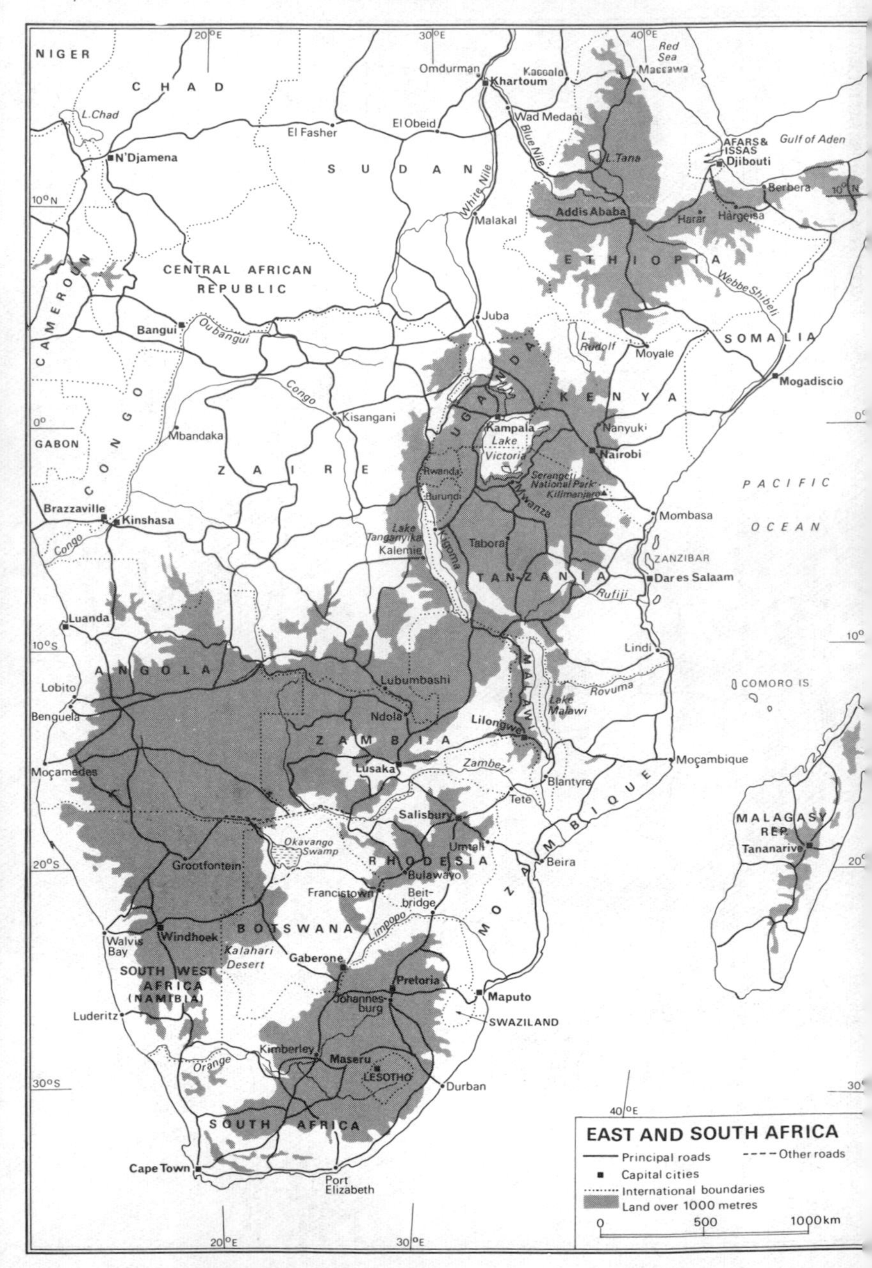
EAST AND SOUTH AFRICA
Principal roads
Other roads
Capital cities
International boundaries
Land over 1000 metres
0
500
1000km
NIGER
CHAD
SUDAN
ETHIOPIA
SOMALIA
KENYA
UGANDA
TANZANIA
ZAIRE
CENTRAL AFRICAN REPUBLIC
CAMEROUN
CONGO
GABON
ANGOLA
ZAMBIA
MALAWI
MOZAMBIQUE
RHODESIA
BOTSWANA
SOUTH WEST AFRICA (NAMIBIA)
SOUTH AFRICA
LESOTHO
SWAZILAND
MALAGASY REP.
AFARS & ISSAS
Red Sea
Gulf of Aden
PACIFIC OCEAN
ZANZIBAR
COMORO IS.
Khartoum
Omdurman
Wad Medani
El Obeid
El Fasher
N'Djamena
Addis Ababa
Djibouti
Berbera
Hargeisa
Harar
Malakal
Juba
Mogadiscio
Moyale
Nanyuki
Nairobi
Kampala
Mombasa
Dar es Salaam
Tabora
Kigoma
Kisangani
Mbandaka
Bangui
Brazzaville
Kinshasa
Luanda
Lobito
Benguela
Moçamedes
Lubumbashi
Ndola
Lusaka
Lilongwe
Blantyre
Tete
Moçambique
Lindi
Salisbury
Umtali
Beira
Bulawayo
Francistown
Beitbridge
Grootfontein
Windhoek
Walvis Bay
Luderitz
Gaberone
Pretoria
Johannesburg
Maputo
Kimberley
Maseru
Durban
Cape Town
Port Elizabeth
Tananarive
L. Chad
L. Tana
L. Rudolf
Lake Victoria
Lake Tanganyika
Lake Malawi
Serengeti National Park
Kilimanjaro
Okavango Swamp
Kalahari Desert
Congo
Oubangui
White Nile
Blue Nile
Webbe Shibeli
Zambezi
Limpopo
Orange
Rovuma
Rufiji
Rwanda
Burundi
Kalemie
Mwanza
20°E
30°E
40°E
10°N
0°
10°S
20°S
30°S

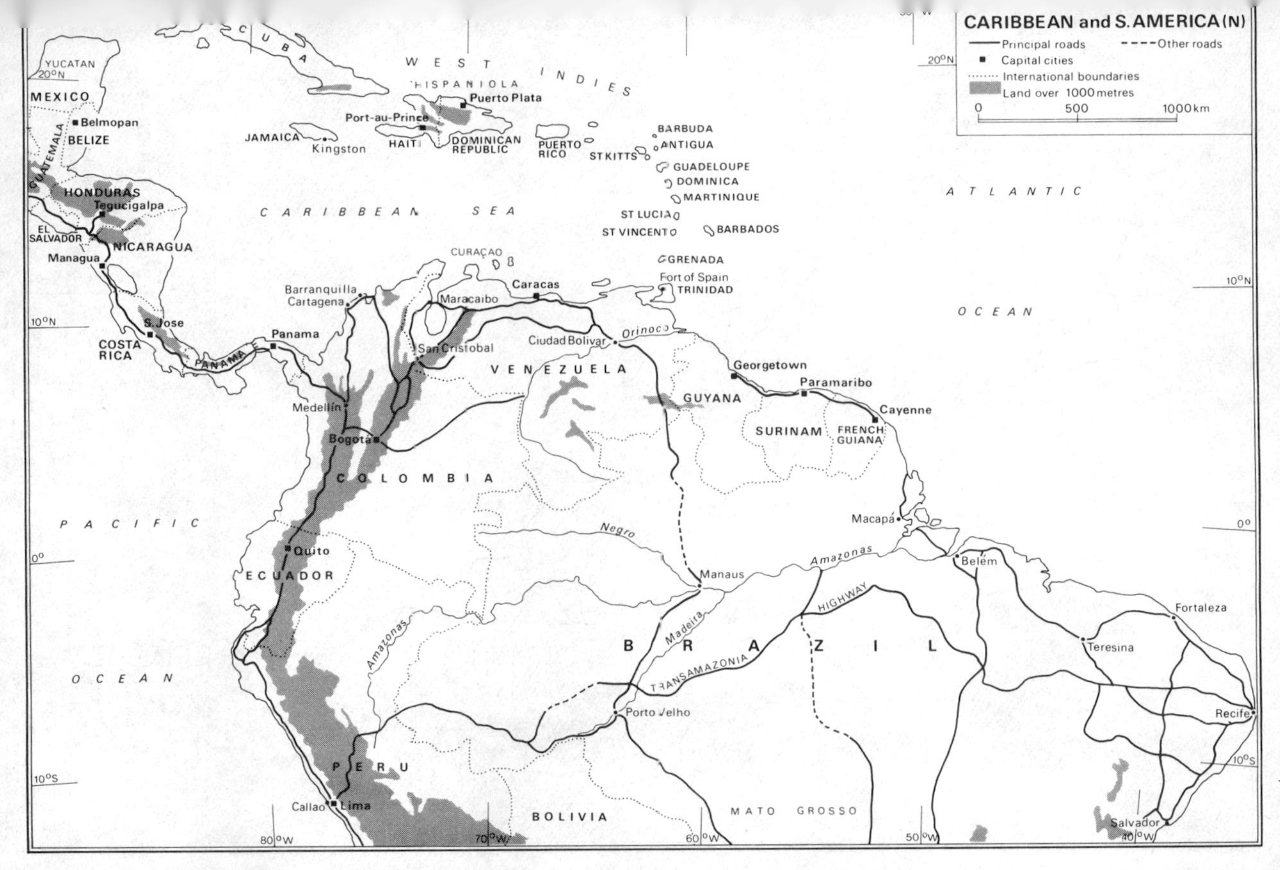
CARIBBEAN and S. AMERICA (N)
Principal roads
Other roads
Capital cities
International boundaries
Land over 1000 metres
0 500 1000 km
CUBA
WEST INDIES
HISPANIOLA
Puerto Plata
Port-au-Prince
HAITI
DOMINICAN REPUBLIC
JAMAICA
Kingston
PUERTO RICO
ST KITTS
BARBUDA
ANTIGUA
GUADELOUPE
DOMINICA
MARTINIQUE
ST LUCIA
ST VINCENT
BARBADOS
GRENADA
Fort of Spain
TRINIDAD
ATLANTIC
OCEAN
CARIBBEAN SEA
CURAÇAO
YUCATAN
MEXICO
Belmopan
BELIZE
GUATEMALA
HONDURAS
Tegucigalpa
EL SALVADOR
NICARAGUA
Managua
S. Jose
COSTA RICA
PANAMA
Panama
Barranquilla
Cartagena
Maracaibo
Caracas
San Cristobal
Ciudad Bolivar
Orinoco
VENEZUELA
Medellín
Bogotá
COLOMBIA
GUYANA
Georgetown
SURINAM
Paramaribo
FRENCH GUIANA
Cayenne
Macapá
Belém
Negro
Manaus
Amazonas
HIGHWAY
TRANSAMAZONIA
Madeira
BRAZIL
Fortaleza
Teresina
Recife
Salvador
Porto Velho
MATO GROSSO
BOLIVIA
PACIFIC
OCEAN
Quito
ECUADOR
PERU
Callao
Lima
20°N
10°N
0°
10°S
80°W
70°W
60°W
50°W
40°W

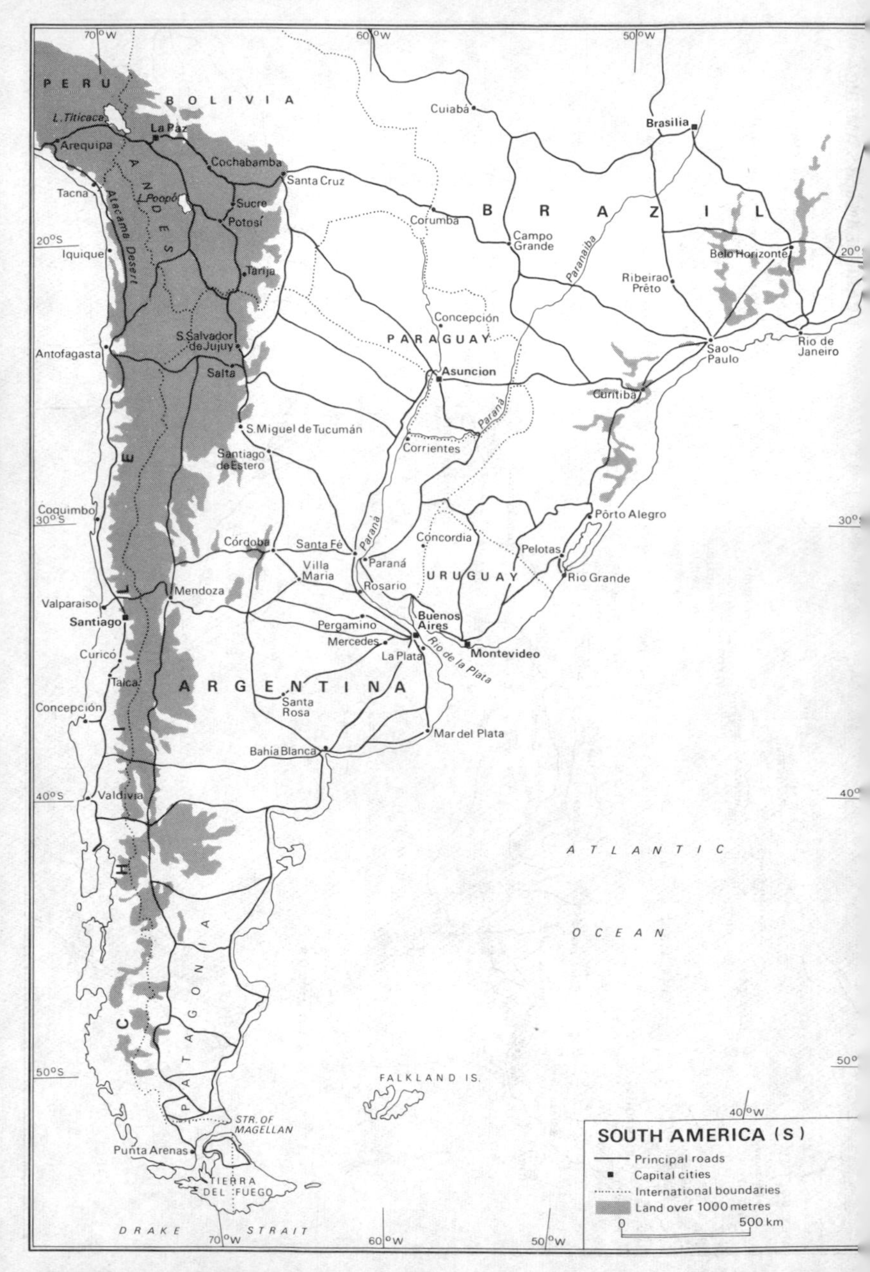
SOUTH AMERICA (S)
Principal roads
Capital cities
International boundaries
Land over 1000 metres
0
500 km
PERU
BOLIVIA
BRAZIL
PARAGUAY
URUGUAY
ARGENTINA
CHILE
ANDES
PATAGONIA
Atacama Desert
ATLANTIC
OCEAN
L. Titicaca
L. Poopó
La Paz
Arequipa
Cochabamba
Santa Cruz
Tacna
Sucre
Potosí
Iquique
Tarija
Antofagasta
S.Salvador de Jujuy
Salta
S. Miguel de Tucumán
Santiago de Estero
Coquimbo
Córdoba
Santa Fé
Villa Maria
Paraná
Rosario
Mendoza
Valparaiso
Santiago
Curicó
Talca
Concepción
Pergamino
Mercedes
Buenos Aires
La Plata
Rio de la Plata
Montevideo
Santa Rosa
Mar del Plata
Bahía Blanca
Valdivia
Punta Arenas
STR. OF MAGELLAN
TIERRA DEL FUEGO
DRAKE STRAIT
FALKLAND IS.
Cuiabá
Brasilia
Corumba
Campo Grande
Paranaiba
Belo Horizonte
Ribeirao Prêto
Concepción
Asuncion
Sao Paulo
Rio de Janeiro
Curitiba
Corrientes
Paraná
Pôrto Alegro
Concordia
Pelotas
Rio Grande
70° W
60° W
50° W
40° W
20° S
30° S
40° S
50° S

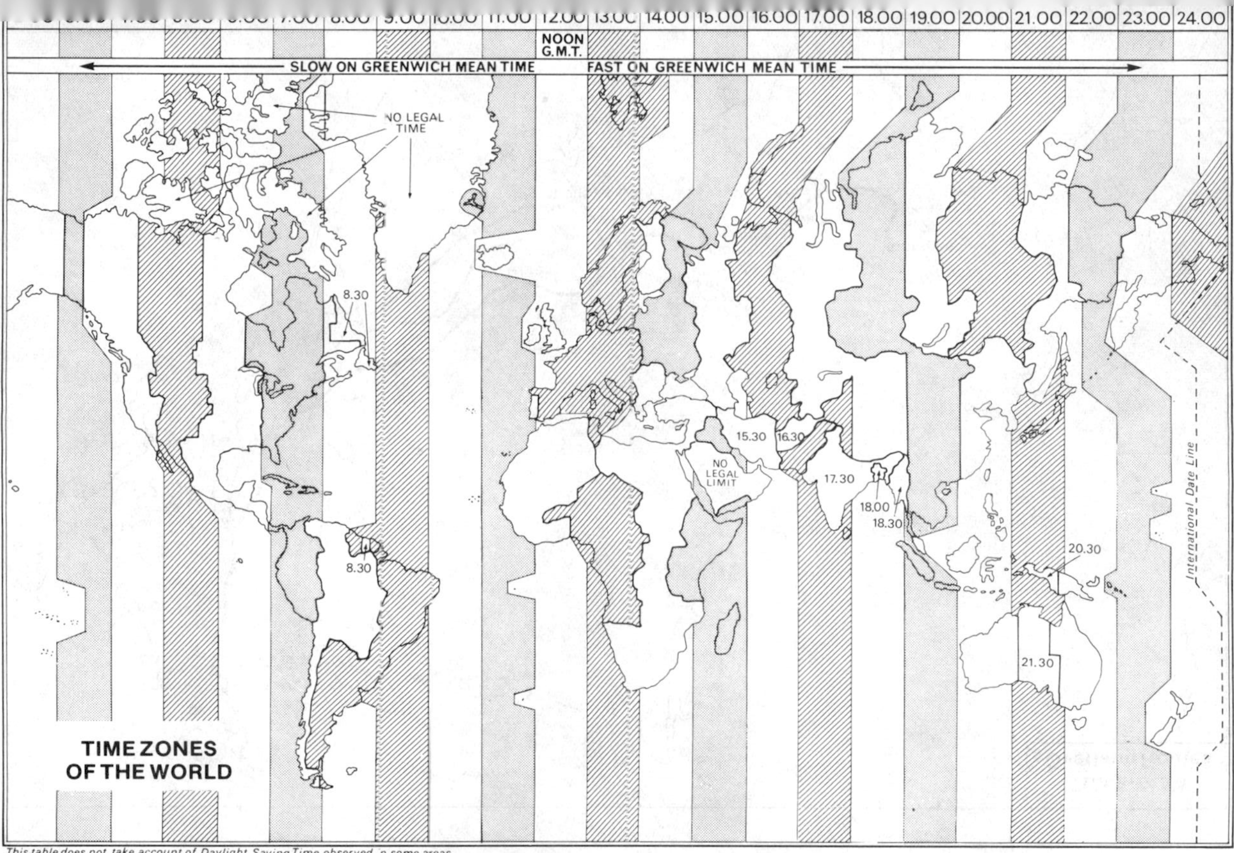

9.00
10.00
11.00
12.00
13.00
14.00
15.00
16.00
17.00
18.00
19.00
20.00
21.00
22.00
23.00
24.00
NOON G.M.T.
SLOW ON GREENWICH MEAN TIME
FAST ON GREENWICH MEAN TIME
NO LEGAL TIME
8.30
8.30
15.30
16.30
NO LEGAL LIMIT
17.30
18.00
18.30
20.30
21.30
International Date Line
TIME ZONES OF THE WORLD
This table does not take account of Daylight Saving Time observed in some areas

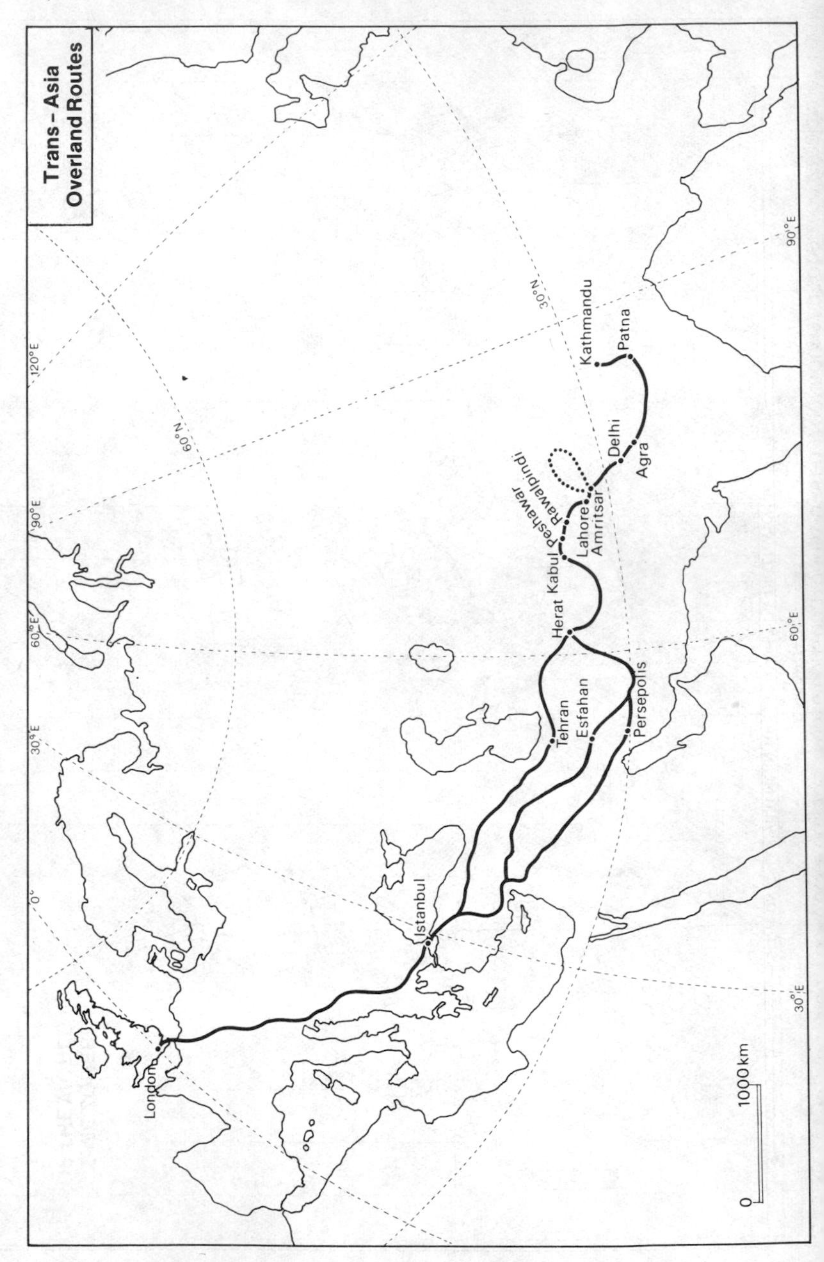
Trans - Asia
Overland Routes
London
Istanbul
Tehran
Esfahan
Persepolis
Herat
Kabul
Peshawar
Rawalpindi
Lahore
Amritsar
Delhi
Agra
Patna
Kathmandu
0
1000 km
0°
30°E
60°E
90°E
120°E
30°N
60°N

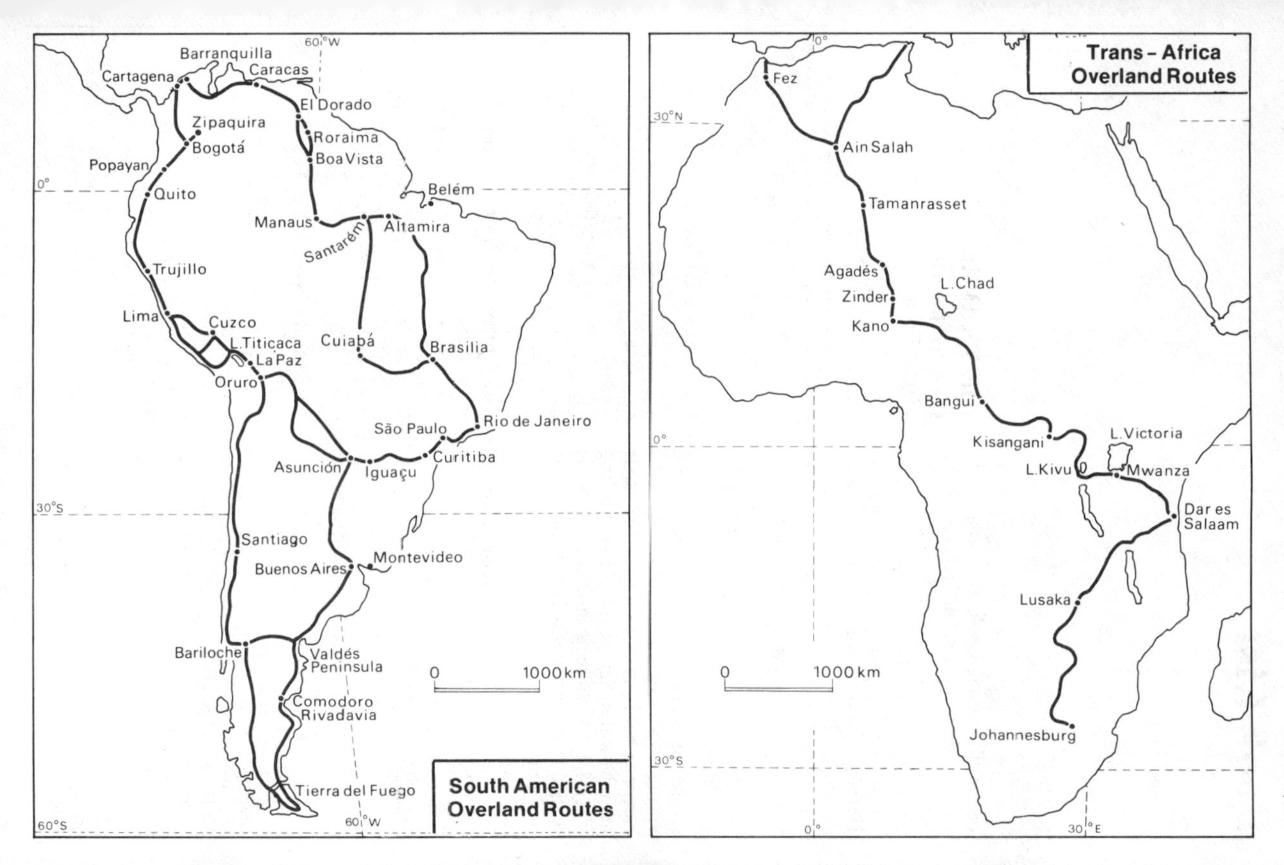

South American
Overland Routes
Barranquilla
Cartagena
Caracas
El Dorado
Zipaquira
Roraima
Bogotá
Boa Vista
Popayan
Quito
Belém
Manaus
Altamira
Santarém
Trujillo
Lima
Cuzco
L. Titicaca
La Paz
Oruro
Cuiabá
Brasilia
São Paulo
Rio de Janeiro
Asunción
Iguaçu
Curitiba
Santiago
Buenos Aires
Montevideo
Bariloche
Valdés
Peninsula
Comodoro
Rivadavia
Tierra del Fuego
0
1000 km
60°W
0°
30°S
60°S
Trans - Africa
Overland Routes
Fez
Ain Salah
Tamanrasset
Agadés
Zinder
Kano
L. Chad
Bangui
Kisangani
L. Kivu
L. Victoria
Mwanza
Dar es
Salaam
Lusaka
Johannesburg
0
1000 km
0°
30°N
30°S
30°E

Book List

For well known book and map publishers, distributors, or retailers, the addresses are not given. Addresses are given for lesser known book sources, except for a recurring few such as Harian Publications, Trail Finders (Services) Ltd., Lonely Planet Publications, Roger Lascelles. The addresses of these sources are to be found in the Directory of Services, under Book and map distributors, publishers or retailers, pp. 0000.

The categories in which books are arranged are as below. Within each category books are to be found in their alphabetical order of title (disregarding *The, A,* etc.) where no author or editor is named or in alphabetical order of author or editor.

Air Travel
Camping, backpacking, bikepacking, hostelling and motor caravanning
Desert travel
Equipment guides
Expedition planning
Guidebooks 1. Australia
.. 2. Asia
.. 3. Africa
.. 4. Central and South America
.. 5. Arctic
.. 6. General
Guidebook series
Guides to adventure trips
Language learning
Low cost travel
Medicine and first aid
Navigation
Overland guides
Photography
Railway travel
Ship travel
Survival
Travel tips
Vehicles

AIR TRAVEL

Care in the Air
Flight Plan

Pub. Airline Users' Committee
PO Box 41, Cheltenham
Gloucester
Free booklets. The first has advice specifically for handicapped travellers. The second covers all aspects of travelling by air, from choosing a flight, packing, checking in, safety on board and buying duty free to advice on passengers' rights and emergency drill.

St. James's Guide to ABC Flights

Pub. St. James Press
3 Percy Street
London W1
England
Lists all Advance Booking Charter Flights, airlines and their destinations, details of bookings, stopovers, length of stay requirements.

The Complete Skytraveller

by David Beaty

Pub. Magnum (Methuen Paperbacks Ltd.), London, 1979
How to choose an airline, best seasons for travel, best seats on a plane. Aircraft, crew, airports, safety, security, comfort. 240 pp.

Hickmans World Air Travel Guide 1979

Eds. R. H. and M. E. Hickman

Pub. Elm Tree Books Ltd.
90 Great Russell Street
London WC1
from: W. H. Smith, Menzies, Boots, airports shops and other large booksellers.
560 pages of information on 173 countries, 443 airports, 192 airlines and miscellaneous other aspects of air travel, such as booking, checking in, in-flight regulations, advice on visas, baggage, etc.

£3.95

Airport International

by Brian Moynahan

Pub. Pan, in association with Macmillan, London, 1978

The inside story on airports and commercial aircraft operations. Revealing.

Paperback 80p

The Flier's Handbook

Ed. Helen Varley

Pub. Marshall Editions and Pan Books, London, 1978

Covers flight planning, economics, security, airport vehicles, runways, guide to airports and guide to air travel, including how to read an airline ticket.

CAMPING, BACKPACKING, BIKEPACKING, HOSTELLING, MOTOR CARAVANNING

Backpacking and Hiking with Children

Pub. Windfall Press
Box 67, Yosemite, CA. 95389
USA

Concise, common sense guide covering trail techniques, equipment, planning, first aid, recipes.

$2.50 incl. p. & p.

Backpacking, Tenting & Trailering

Pub. Argus Books, Los Angeles

$4.95

The Complete Book of Practical Camping

Pub. Argus Books, Los Angeles

$5.95

Handbook of Camping and Caravanning

Pub. The Camping Club of Great Britain, London

Outdoor Cook's Bible

by Joseph D. Bates, Jr.
New York, Doubleday, 1964
Illustrated.

Paperback $2.50

Two-Wheel Touring and Camping

by Cliff Boswell and George Hays

Pub. 1970
Bagnall Publishing Co.
Box 507, Lake Arrowhead
CA 92352, USA

A 'how-to' book covering selection of a bike, planning a trip, gear and packing and information on specific travel areas.

$3.50

Backpackers' Africa

17 Walks off the Cape to Cairo Route

Backpacking in Mexico and Central America
Backpacking in Peru and Bolivia

A Guide to the Ancient Ways and Inca Roads
2nd edn., revised

Backpacking South America

by Hilary and George Bradt

Pub. 1977-79
Bradt Enterprises
Chalfont St. Peter
Bucks., UK

from: Overmead
Monument Lane
Chalfont St. Peter
Bucks. SL9 0HY

in USA from:
S. N. Bradt
409 Beacon Street
Boston
Mass. 02115

in Australia from:
G. Mortimer
c/o Yugilbar
Rankine's Road
St. Andrews
Vic. 3761

Firsthand information about regulations, food, accommodation, transport, wildlife, parks, security, culture; with drawings, photos and entertaining asides and anecdotes.

£1.50 to £2.60

The Complete Guide to Hiking and Backpacking

Ed. Andrew J. Carra

Pub. Winchester Press, New York, 1978
from: A. & C. Black
35 Bedford Row
London WC1R 4JH

A 'how-to' manual with a wealth of information from which the hiker will choose what he needs for his own specific purpose and which the experienced hiker will use to jog his memory and add to his knowledge.

£5.00

Camper's & Backpacker's Bible

Ed. C. B. Colby

Pub. Stoeger Publishing Co., South Hackensack, NJ, 1977

34 articles by experts on different aspects of camping and backpacking, with advice on tents, campfires, clothing, meals and attendant sports and hobbies such as canoeing and fishing; 10 extracts from the writings of authors from the past, such as R. L. Stevenson and Mark Twain, on hiking, sleeping out and so on; and a vast reference section detailing trails, map sources, books & periodicals, organisations, manufacturers & suppliers; a section giving equipment specifications under 18 headings, with selected illustrations; and a similar section giving specifications for accessories, under 28 headings. 480 pp. Exceptional comprehensiveness and value–except for the one-page index, which is not much use.

Soft cover $7.95

Camping

by Department of Education and Science

Pub. Her Majesty's Stationery Office, London, 1970

The Complete Light-Pack Camping and Trail-Foods Cookbook

Edwin P. Drew

McGraw-Hill, New York, 1977

Contains home-drying and packaging instructions, recipes, food value charts for both bought and home-prepared foods, and a test report on stoves.

$3.95

Walking, Hiking and Backpacking

by Anthony Greenbank

Pub. Constable, London, 1977

The Complete Book of Practical Camping

John Jobson

Pub. Stoeger Publishing Co., South Hackensack, NJ, 1977

Deals with camping in all types of weather and on every type of terrain. Author has tested equipment professionally for many years.

Softbound $5.95

Travel Light Handbook

by Judy Keene

Pub. Contemporary Books, Chicago, 1979

Planning, equipment selection and packing advice for the wilderness backpacker.

$3.95

Motor Caravanning

A Complete Guide
by Henry Myhill

Pub. Ward Lock, London, 1976

Guide to acquiring and utilising a motorised home, with much practical and also some delightful anecdotal material.

£3.95

Wilderness Camping's Wilderness Adventure

Ed. Harry N. Roberts

Pub. Wilderness Camping
1597 Union Street
Schenectady, NY 12309, USA, 1978

Annual for backpackers and hikers, with articles on tents, boots, knives, backpacks, physiology and the philosophy of wilderness travel.

$2.00 plus 75c postage

Camping Complete

by Sqn. Ldr. P. F. Williams, RAF

Pub. Pelham Books, London, 1972

Sections on equipment, techniques, cooking, planning a camping holiday for family camping; and on various aspects of adventure camping, including trekking, making fires, loading a bike, canoe camping.

£2.50

DESERT TRAVEL

Common Sense in Desert Travel

How to Enjoy Boondocking without Screaming for the Rescue Squad
by Carl and Barbara Austin

Maturango Museum
China Lake
CA, USA
from: Cepek

Deals mainly with survival, but also discusses off road driving.

$1.49

Desert Travel and Research

J. L. Cloudsley-Thompson

Pub. Institute of Biology, London, 1975

Cruising the Sahara

by Gerard Morgan-Grenville

Pub. David and Charles, Devon, 1974
Detailed reference book on every aspect of desert planning and travel.

Algeria and the Sahara

A Handbook for Desert Travellers
Valerie and Jon Stevens

Pub. Constable, London, 1978
Guidebook including a detailed survey of the principal motor routes, a review of attitudes to be found towards tourists amongst Algerians, and many travel tips.

£4.95

EQUIPMENT GUIDES

Backpacking Equipment Guide

Pub. 1977
Field & Trek (Equipment) Ltd.
23-25 Kings Road
Brentwood
Essex CM14 4ER
Aims to help purchasers of equipment to compare technical details of various types before buying.

25p

Backpacking Equipment Buyer's Guide

William Kemsley and the editors of *Backpacker* magazine

Pub. Collier Books, New York, 1977
Gives advice on evaluating buyer's own needs and on what to look for when choosing equipment to correspond with those needs, with extra information such as how to pack a load and how boots are made.

$8.95

Movin' Out

Equipment and Technique for Hikers
by Harry Roberts

Pub. 1977
Stone Wall Press, Inc.
5 Byron Street
Boston
MA 02108, USA
144 pages. 100 illus. Excellent, down-to-earth book on backpacking information, by the editor of Wilderness Camping. *Covers clothing, protection, shelter, food, the effects of cold on the body, etc.*

$4.95

The Backpacker's Guide

Kate Spencer and Peter Lumley

Pub. Kate Spencer Agency Ltd.
1 Warwick Avenue
Whickham
Newcastle upon Tyne
2nd annual edn., 1978
Mostly lists, with commentary, of equipment, suppliers, etc.

£1.25

The Budget Backpacker

How to Select or Make, Maintain and Repair Your Own Lightweight Backpacking and Camping Equipment
L. A. Zakreski

Pub. 1977
Winchester Press
205 East 42nd Street
New York
NY, 10017, USA
Distributed in the UK by A. & C. Black
Clear and detailed handbook written by an experienced backpacker, a Canadian forest ranger in Alberta's Jasper National Park

$9.95
£5.50

EXPEDITION PLANNING

Expeditions: The Experts' Way

Eds. John Blashford-Snell and Alastair Ballantine

Pub. Faber & Faber, London, 1977
Includes: Expedition Planning and Logistics (Kelvin Kent), Expedition Health and Medicine (Peter Steele), Jungle Exploration (John Blashford-Snell), Desert and Bush Exploration (David Hall), The Polar Regions (Wally Herbert), Underwater Exploration (Christopher Roads), Ocean Sailing (Chay Blyth), River and White Water Exploration (Roger Chapman), Caving (Russell Gurnee), Mountains (Malcolm Slesser), Kit & Equipment (Kelvin Kent); with bibliography and biographical notes on the editors and contributors.

The Expedition Organiser's Guide

by John Blashford-Snell and Richard Snailham

Pub. The Daily Telegraph, London
from:
Bob Johnson
Marketing Department
The Daily Telegraph
135 Fleet Street
London EC4
Revised edn., 1978

Includes sections on planning, information, research and organisation; personnel; finance; stores and vehicles; leadership; winding up an expedition.

The Expedition Handbook

Ed. Tony Land
For the Brathay Exploration Group
Pub. Butterworths in association with The Geographical Magazine, London and Bristol, 1978
With a foreword by David Attenborough. Initial considerations, planning details, expedition field studies, appendices including addresses. Tips on fund-raising activities, getting reduced air fares, packing, insurance, etc.

The Young Explorers' Trust Expedition Equipment Guide
and
Expedition Food and Rations Planning Manual

both from:
Young Explorers' Trust
c/o Royal Geographical Society
1 Kensington Gore
London SW7 2AR

Equipment Guide:
Covers personal equipment, footwear, sleeping bags, carrying gear, tentage, cooking equipment, home-made equipment, sources of general, specialist and material suppliers, with information on costs and quantities, manufacture and maintenance. Very suitable for any novice expeditioner, as also for the ordinary camper, hiker, walker or even do-it-yourself enthusiast.

Food and Rations Planning Manual:
Covers basic nutrition, practical hints on food preparations, packing, suggested lists, emergency rations. Written for organisers of expeditions abroad composed of young people, but invaluable for many other expedition or even holiday organisers.
Limited number of copies of both on general sale.

GUIDEBOOKS
1. AUSTRALASIA

Indonesia Handbook

by Bill Dalton
Pub. Moon Publications
PO Box 732
Rutland
VT 05701
USA
2nd edn., 1979

£4.50

Indonesia and Papua New Guinea

A Traveler's Notes
by Bill Dalton
Pub. Moon Publications
PO Box 88
South Yarra
Victoria 3141
Australia
from:
Roger Lascelles
7th edn., 1978
Poorly written but informative guide to all the major islands.

Papua New Guinea Handbook

by July Tudor
Pub. 1978
Pacific Publications
Box 3408
Sydney
NSW 2001
Australia
from: Edward Stanford Ltd.
12-14 Long Acre
London WC2E 9LP
Description of the country, area by area.

£6.50

Papua New Guinea

A Travel Survival Kit
by Tony Wheeler
Pub. Lonely Planet Publications, South Yarra, 1979

2. ASIA

All-Asia Guide

Pub. Far East Economic Review
Tong Cheng Street
PO Box 47
Hong Kong
from:
Roger Lascelles, and Trail Finders, UK
or:
Charles E. Tuttle Co., Inc.
26-30 Main Street
Rutland
VT 05701
USA
10th annual edn., 1979
Classical middle depth guide. 584 pages of travel advice for businessmen and travellers, from Afghanistan to Japan.

£4.95
(+55p p. & p.)

Des Alpes à l'Himalaya I, II

from:
Librairie de la Maison des Voyages
28 rue du Pont-Louis-Philippe
75004 Paris
France
Two volumes covering vehicle preparation, itineraries from Geneva to Pakistan and detours via the Middle East.

Nagel's China

Pub. Nagel Publishers
Geneva, Switzerland, 1976
from: Godfrey Cave Associates Ltd.
42 Bloomsbury Street
London
WC1B 3QJ
Said to be the most complete travel guide to mainland China ever assembled.
$35

Nepal Trekking Guidebooks

Pub. Peter Purna Books
Avalok
PO Box 1465
Kathmandu, Nepal
from: YHA Services, London
Series of small guidebooks on treks in Nepal.
£2.55 each

101 Books on Asia to Read Before Going

Pub. Passport Publications
20 North Whacker Drive
Chicago, IL 60606, USA
20 page booklet.
$3.00

Travellers' Guide to the Middle East

Pub. International Communications
63 Long Acre
London WC2E 9JH
2nd edn., 1979
Guide for the businessman or tourist including country by country information on how to get there, accommodation, local customs, business hours, public holidays, climate, transport, culture, etc.
£10.00

Travellers' Survival Kit to the East

Pub. Vacation Work
9 Park End Street
Oxford
£1.95

Trekking in Nepal
Nepal Visitors' Guide

from:
The Royal Nepalese Embassy
2131 Leroy Pl., NW,
Washington, DC 20008
USA
Trekking in Nepal, *published in Kathmandu in 1976, offers information on planning a trek and describes 8 treks in Nepal.*
Nepal Visitors' Guide *gives a brief general introduction and details on accommodation, points of interest, etc.*

An Introduction to Nepal

by Stan Armington
from: Trail Finders (Services) Ltd.
Written for overland travellers with a week or two to spend in Nepal. What to do, where to stay.

Trekking in the Himalayas

by Stan Armington
Pub. Lonely Planet Publications
from:
Roger Lascelles
2nd edn., 1979

A Guide to Trekking in Nepal

by Stephen Bezruchka
Pub. 1974
Sahayogi Press
Tripureshwar
Kathmandu
Nepal
Detailed descriptions of many routes and advice on organising a trek.

Overland to India

by Douglas Brown
Pub. Outerbridge & Dienstfrey
New York, 1971
'Underground' book with much attention to drug regulations. Readable, but not a comprehensive guide.

The Asian Highway

The Complete Overland Guide from Europe to Australia
by Jack Jackson and Ellen Crampton
Pub. Angus & Robertson, London, 1979
A useful manual for any overlander, giving the benefit of Jack Jackson's experience of camping, customs, driving, preparations and bureaucracy.

Student Guide to Asia

Ed. David Jenkins for Australian Union of Students

Pub. E. P. Dutton, New York
in UK from: Roger Lascelles
Useful low-cost guide for students and other budget travellers visiting any of 24 countries in southern and eastern Asia. Visa requirements, health requirements, currency, transport, accommodation, food and sights.

Into India

by John Keay

Pub. John Murray, London, 1973
An area by area coverage of modern India and Indian life, to introduce the country to the intending traveller or to help the experienced traveller to interpret what he has seen.

The Gulf Handbook

A Guide for Businessmen and Visitors
Eds. Peter Kilner and Jonathan Wallace

Pub. Trade and Travel Publications Ltd.
The Mendip Press
Parsonage Lane
Bath BA1 1EN
and
The Middle East Economic Digest Ltd.
84-86 Chancery Lane
London WC2A 1DL
Annual guide.

Hardback £5

India and the Sub-Continent

An Overland Odyssey
by Tom King
Pub. 1979
Michael Boot
23 Koornay Road
Carnegie
Vic. 3163
Australia
from: Roger Lascelles
A photographic account, with some full page colour, of an overland journey undertaken with Capricorn Travel & Tours Ltd.

£6.95

Trans-Siberia by Rail and a Month in Japan

by Barbara Lamplugh

Pub. Roger Lascelles, London, 1979

Trans-Asia Motoring

by Colin McElduff

Pub. Wilton House Gentry,
London, 1976
For travellers between Istanbul and Bangladesh. Main sections cover preparation, individual countries, conditions and facilities along the main routes. Information also on ferries and ships, campsites, entry and customs regulations.

£5.95

Exploring Japan

A Guide to Travel and Understanding
by Ian McQueen

Pub. Lonely Planet Publications,
South Yarra, 1979

$A4.95
£2.95
$5.95

Korea

A Travel Survival kit
Ian McQueen

Pub. Lonely Planet Publications,
South Yarra, 1979

$A2.95
$2.95
£1.75

Malaysia and Singapore

Gladys Nicol

Pub. Batsford, London, 1978
Factual information and a loving account of the ways of life in the two countries.

£5.50

The Road to India

Guide to the Overland Routes to the East
by John Prendergast

Pub. John Murray, London, 1977
By a traveller who has made 8 return trips to India by road since 1960. The book falls into two sections, the first covering preparation, of both vehicle and driver, and maintenance, eg servicing, spares, clothing, medical advice, etc. The second part deals with the route itself: distances and times, stopping places, petrol stations, gradients, surfaces, side trips and bypasses, etc.

£5.95

Kathmandu and the Kingdom of Nepal

by Prakash A. Raj

Pub. Lonely Planet Publications
from:
Roger Lascelles
2nd edn., 1978

History, culture, festivals, things to do and see, places to stay and eat.

Asia for the Hitchhiker

By Mik Schultz
Pub. 1972
Faxebugtens Bogtrykkeri
Faxe
Denmark
from:
Roger Lascelles, UK
and:
Information Exchange
TEJ 22
West Monroe Street
Chicago
IL 60603
and CIEE, USA
Detailed advice for budget travellers. Translated from Danish original.

Asia Overland

A Practical Economy-Minded Guide to the Exotic Wonders of the East
Dan Spitzer and Marzi Schorin
Pub. Stonehill Publishing Co.
New York, 1978
Covers preparation and general advice, food and health and the following individual countries: Turkey, Iran, Afghanistan, Pakistan, India, Sri Lanka, Nepal, Burma, Thailand, Laos, Malaysia, Singapore and Indonesia. Includes notes on the drug scene and regulations, tips for women, advice on visas, customs, local laws, and reports on accommodation, transport, shopping, festivals, food.
$5.95

Island Hopping through the Indonesian Archipelago

by Maurice Taylor with Trail Finders
Pub. Wilton House Gentry
London, 1976
A guide to travel at leisure and on the cheap through Thailand, Malaysia, Indonesia, including Sumatra, Java and Bali, Timor and on to Darwin. With town plans and very detailed recommendations on where to find bargains in lodging or food in many tiny, out-of-the-way places.
£5.40

Nepal für Globetrotter

Ludmilla Tüting
Pub. the author
from:
Hassleyerstrasse 4a
D-5800 Hagen
West Germany
3rd edn., 1978
Contains trekking guide to 35 treks.

The On-Your-Own Guide to Asia

by Volunteers in Asia
Pub. Charles E. Tuttle Co., Inc.
26-30 Main Street, Rutland
VT 05701, USA
Covers East and South-east Asia, giving practical tourist information as well as cultural and historical background.
$3.95

Tips für Tips

by F. von Engel
Pub. 1973
from:
31 Celle
Stauffenbergstrasse 9
West Germany
Booklet on overland journey by vehicle from Germany to India and beyond; with facts, figures and costs.

Guide to Western Asia—Overland Route

by E. M. Wheeler
Pub. Study Centre Publications
Burton Chambers
24 Hamilton Road
Felixstowe
Suffolk
from: YHA Services, London

Across Asia on the Cheap

A complete guide to making the overland trip with minimum cost and hassles
by Tony Wheeler
Pub. Lonely Planet Publications
from: Roger Lascelles
3rd edn., 1978
Pocket guide for travellers between Australia or New Zealand and Europe, with advice on bargains, what to avoid, and some background information on the countries concerned.

Burma

A Travel Survival Kit
by Tony Wheeler
Pub. Lonely Planet Publications, South Yarra, 1979
From Rangoon to Mandalay with sidetrips to Pagan and the Inle Lakes.
$A2.95
$2.95
£1.75

South-East Asia on a Shoestring

by Tony Wheeler

Pub. Lonely Planet Publications
from: Roger Lascelles, UK,
and: Bookpeople, USA
Revised edn., 1977
Guide for the budget traveller, with practical, and some background, information, on 8 south-east Asian countries.

Sri Lanka

A Travel Survival Kit
by Tony Wheeler

Pub. Lonely Planet Publications, South Yarra, 1979

$A2.95
$2.95
£1.75

A Handbook for Travellers in India, Pakistan, Nepal, Sri Lanka and Bangladesh

by Prof. L. F. Rushbrook Williams

Pub. John Murray, London, 1975
Periodically revised and virtually the only guide book available on Pakistan, Nepal and Bangladesh. Book first published in 1859.

£12.50

3. AFRICA

Durch Africa

Pub. Därr's Expedition Service/Tourist Club of Switzerland, Munich
from:
Kirchheimerstrasse 2
D-8016 Heimstetten
Munich
West Germany
Description, in German, of roads and tracks in Africa

Pre-Departure Handbook, Trans-Africa

from: Trail Finders (Services) Ltd
Deals with vehicle type and modification, tools and spares, camping and cooking equipment, medical, immigration and health requirements, visas, carnets de passages, *insurance, estimate of total budget, photography.*

Transafrique

from:
Librairie de la Maison des Voyages
28 rue du Pont-Louis-Philippe
75004 Paris, France
Routes and practical information. 400 pages.

Traveller's Guide to Africa

Pub. 1979
International Communications
63 Long Acre
London WC2E 9JH
3rd edn., 1979
Guide for the businessman or tourist including country by country information on how to get there, accommodation, local customs, business hours, public holidays, climate, transport, culture, etc.

$10.00

The Traveler's Africa

A Guide to the Entire Continent
by Philip M. Allen and Aaron Segal

Pub. 1973
Hopkinson & Blake
185 Madison Avenue, New York
NY 10016, USA
Country by country guide to Africa, with information on history, geography, weather, social attitudes and more conventional tourist information. Thoroughly researched and superbly organised. Nearly 1000 pp.

$14.95

Africa for the Hitchhiker

by Fin Bjering-Sorensen and Torben Jorgensen

Pub. Bramsen & Hjort, Copenhagen, 1974
from: Roger Lascelles, UK
and: Information Exchange
TEJ 22
West Monroe Street
Chicago
IL 60603
and: CIEE, USA
Detailed advice for budget travellers, though some areas (eg Malagasy Republic) receive shallow treatment. Translated from Danish original.

Bright Continent

A Shoestring Guide to Sub-Saharan Africa
by Susan Blumenthal

Pub. Doubleday (Anchor Books), New York, 1974
from: CIEE
550-page thorough guide to low-cost travel in Africa, written with feeling for the local people. Combines narrative with information.

Published price $5.95
CIEE price $2.00

Africa on Wheels

A Scrounger's Guide to Motoring in Africa
by John J. Byrne

Pub. 1973
Patrick Stephens and Haessner Publications
Newfoundland
New Jersey
USA

Practical information and anecdotes on security, documents, food, etc, based on author's journey from Casablanca to Cape Town in eight months.

Travellers' Guide to East Africa

(includes Ethiopia)

Travellers' Guide to Southern Africa

(includes Namibia)
by Thornton Cox

Pub. Hastings House Publishers. New York, 1974
from: Trail Finders (Services) Ltd
Game parks, sights to see, places to stay and eat. Background on history, culture, society. For short-term holidaymakers seeking average diversions and comforts.

Africa on the Cheap

by Geoff Crowther

Pub. Lonely Planet Publications, South Yarra, 1978
from: Roger Lascelles, UK
and: Bookpeople, USA
Covers 51 countries. Updated edition of the BIT Information and Help Services guide, Overland Overland through Africa.

$A4.95
$4.95
£2.50

Trans Sahara

by Klaus and Erika Därr
Pub. Därr's Expedition Service, Munich
Written by a German couple who are experts on vehicle crossing of the Sahara.

With Love . . . . Siri and Ebba

by SirpFraser and Ebba Pederson

Pub. 1973
from:
Nicholas Saunders
2 Neals Yard
London WC2
and:
BIT Information and Help Services
97a Talbot Road
London W11
Letters home from two young girls travelling in Africa

Crossing the Sahara by Car

by Eric Jackson
from:
the author
Spring Grove House, Chapel Hill
Clayton West
Huddersfield, Yorks.
Excellent 26-page booklet based on author's considerable experience.

Morocco by Car

by Leslie Keating

Pub. 1972
from:
Kenneth Mason
13 Homewell
Havant
Hampshire
Routes, useful tips for planning.

Cairo to Nairobi

A Handbook for the Independent Traveller
by Trevor Kenworthy

Pub. Wilton House Gentry, London, 1977
Details on transport and accommodation, etc, for the traveller with limited time at his disposal.

£5.95

The Nile Journey

by Trevor Kenworthy

from: Trail Finders (Services) Ltd
Information on the journey down the Nile by public transport, with details of train and steamer fares, costings, equipment, connecting air travel, visas. Also advice on health, food, drink, sightseeing, souvenirs. Sketch maps of major towns en route.

The Travellers' Guide to Morocco

by Christopher Kininmonth

Pub. Jonathan Cape, London, 1972

Trans-Africa Motoring

by Colin McElduff

Pub. Wilton House Gentry, London, 1975
Equipment, visas health regulations and other details in Part I. All major routes of Contient, with condition, facilities, maps, etc, in Part II. Author an experienced traveller himself, Head of the Foreign Routes Department of the RAC.

£5.95

Travellers' Guide to Africa

Eds. Richard Synge and Colin Legum

Pub. 1978
IC Publications Ltd
63 Long Acre
London WC2E 9JH
Rather 'glossy', but good background on each country.

£4.95

4. CENTRAL AND SOUTH AMERICA

Central America Travel Guide

Pub. 1973
from:
Cliff Cross
PO Box 301
North Palm Springs
California
USA

Guia Latino Americana de Omnibus

from:
Entre Rios 1137
Piso 80, depto. 5
Rosario, Argentina
Annually revised publication giving bus timetables from Caracas right down to Rio Gallegos, and details of hotels and excursions.

The South American Handbook

Pub. Trade & Travel Publications Ltd
The Mendip Press
Parsonage Lane
Bath BA1 1EN
and
Rand McNally & Co.
PO Box 7600
Chicago
IL. 60680, USA
57th edn., 1980
Annually revised. Cost of living, climate, clothing, health, language, sights, accommodation, restaurants, motoring, currency, the economy, transport. Covers Mexico, the West Indies, Bermuda and Latin America as well as South America. Called by J. S. Gordon, author of Overlanding, *'the best guidebook to anywhere by anyone'.*

Mexico and Central America

A Handbook for the Independent Traveller by Frank Bellamy

Pub. Wilton House Gentry, London, and New York, Two Continents, 1977
Information on transport, health, climate, places to see; with simplified maps and plans. Some timetables, notes on fares, and background information on each country.

£5.95

Student Guide to Latin America

by Marjorie A. Cohen

Pub. E. P. Dutton, New York
from: CIEE and other distributors
Written for the Council on International Educational Exchange. 150 pages.

$2.95

South America on a Shoestring

by Geoff Crowther

Pub. Lonely Planet Publications, South Yarra, 1979

$A5.95
$6.95
£2.95

Guide to Peru

Handbook for Travellers
by G. de Reparaz

Pub. Lima
from:
Mountain Travel, Inc.
1398 Solano Avenue
Albany
CA 94706
USA
Essential guide for the visitor. 480 pp.

Paperback $6.00

Along the Gringo Trail

by Jack Epstein

Pub. And/Or Press
PO Box 2246
Berkeley
CA 94702, USA
from: Roger Lascelles
A guide to Latin America for the budget traveller. The author describes an actual route, taken by himself in 1974 and by many other young travellers, and compares prices, services, schedules in the various countries, as well as adding information on transport schedules, visas, currency, climate, etc.

Latin American Motoring

by Jonathan Hewat

Pub. Wilton House Gentry, London, 1977

Detailed route instructions, information on how to get to Latin America from Europe, Australia and North America, maps and town plans, hints on accommodation.

£5.95

Latin American Travel Guide including the Pan American Highway Guide

Ed. Ernst A. Jahn

Pub. Compsco Publishing Co.
663 Fifth Avenue
New York
NY 10022
USA
6th edn., 1979
Prepared in cooperation with the Pan American Union and the American Automobile Association. Covers Mexico, Central and South America, Alaska, the Yukon, British Columbia and the western US. Includes profile chart of the Pan American Highway, 14 page mileage table. Facts and figures, not a guide to tourist attractions.

South Continent Down

by Dr. David Powell

from:
the author
WNMU
Silver City
New Mex 88061
USA
Book on the 'dangers, discomforts, rip-offs, confusions and comic insanities involved in Latin American land travel'.

$6.00

Latin America for the Hitchhiker

by Mik Schultz

Pub. Multivers, Copenhagen, 1973
from:
Roger Lascelles UK
and:
Information Exchange
TEJ 22
West Monroe Street
Chicago
IL 60603, USA
and:
CIEE
USA
Detailed advice for budget travellers. Translated from Danish original.

South American Survival

A Handbook for the Independent Traveller
by Maurice Taylor

Pub. Trail Finbers Ltd., 1975
Pub. Wilton House Gentry, London.
Expanbed edn., 1977
Full details on routes (including Amazonas), regional differences, statistics, sketch maps of countries, regions and towns.

£6.95

Von Alaska bis Feuerland

by LudmillwTüting

Pub. the author
from:
D-5800 Hagen
Hassleyerstrasse 4a
West Germany
New edn., 1978
Guide based on journey by Volkswagen bus through South, Central and North America.

5. ARCTIC

The Arctic Highway

by JohmB ouglas

from:
Gcoslides (Books Div.)
4 Christian Fields
London SW16
A full description of north Norway's famous road to the Arctic, with details of the northern road system and a discussion of the environment and history.

6. GENERAL

Travelers' Picture Dictionary

Pub. AJS International
6101 Palo Alto Drive
Huntington Beach
CA 92647
USA
Pocket-sized, indexed, picture dictionary with sections on personal needs, drinks, food, clothes, medical aid, services.

$2.00

Whole World Handbook

A Student Guide to Work, Study and Travel Abroad
by Marjorie A. Cohen and Mari aret E. Sherman

Pub. Council on International Educational Exchange and the Frommer/Pasmantier Publishing Corporation, New York
5th edn., 1978-79
Excellent compendium on student travel, including books to read, how to get there, the 'bare essentials', study programmes and reports from students on any useful aspect of travel or study abroad.

$3.95

Worldwide What & Where

Geographic Glossary and Traveler's Guide
by Ralph DeSola

Pub. Avon Books, New York, 1975
Geographical gazetteer, glossary and lexicon; with lists of international capitals, ferry routes around the world, festivals and museums.

Paperback $5.95

Le Guide du Routard

by Michel Duval anb Philippe Gloaguen

Europe
Afrique
Amérique du Nord & Centrale
Amérique du Sud
Moyen-Orient . . . Indes
Asie du Sud-Est
Pub, Librarie Hachette, Paris
Various 1970s editions of each 'hip' practical guide, with information on how to get there, country by country guides including notes on places to stay and to see, methods of travel, advice to hitchhikers, etc.

Le Manuel du Routard

by Michel Duval and Philippz Gloaguen

Pub. Librairie Hachette, Paris, 1977
Excellent, fact-filled book of tips, addresses, advice, lists, aimed at the budget traveller and written in breezy, informal and readable style.

Travel Guidebooks in Review

Ed. Jon O. Heise
Pub. University of Michigan International Center
603 East Madison Street
Ann Arbor, MI 48104
USA
3rd edn, 1979. Formerly called **Suit Your Spirit**
Guide to about 100 books themselves covering travel, written to provide 'a comprehensive overview' of this field.

New Horizons: Living Abroad

Pan American's Guide to Living Conditions in 88 Countries
by E. P. Pierce

Pub. Pan American
PO Box 1790
New York
NY 10017
USA

GUIDEBOOK SERIES

A to Z Guides

by Robert S. Kane
Pub. Doubleday, New York

Bazak Guides

Pub. Harper & Row, New York
Popular travel guides.

Blue Guide series

Pub. Ernest Benn Ltd
Tonbridge
Kent
Distributed in USA by Rand McNally
Full-sized guides printed in very small type.

Explorer Guides

Explorer I: North America, Central America, South America, Bahamas, Bermuda, Caribbean
Explorer II: India, Pakistan, Sri Lanka, Far East, South East Asia, Australia, New Zealand, Pacific islands
Pub. British Airways, London
Paperbacks. Give information on hotels, restaurants, local transport, climate, etc.

Fielding's Guides

Pub. William Morrow and Co, Inc, New York

Fodor's Guides

Caribbean
India
Israel
Japan
Mexico
Morocco
S America
SE Asia
Soviet Union
Pub. David McKay, New York, for Fodor's Modern Guides
750 Third Avenue
New York
NY 10017
USA
and Hodder & Stoughton, London

£4.95 to £6.95 each

Get it and Go

Pub. Houghton Mifflin, Boston
First title, Mexico, *published 1979*

Michelin Guides

Pub. Michelin Tyre Co., France
The famous green and red guides giving background and tourist information and recommendations on places to visit, to stay and to buy meals.

Nagel Encyclopaedia Guides

39 titles
Pub. Nagel Publishers
Geneva
Switzerland
in UK from:
Godfrey Cave Associates Ltd
42 Bloomsbury Street
London WC1B 3QJ

$10-a-Day Guides

Israel on $15 a Day
Mexico and Guatemala on $15 a Day
South America on $10 a Day
Spain and Morocco (plus the Canary Is.) on $10 & $15 a Day
Turkey on $10 & $15 a Day
Pub. Arthur Frommer, Inc
380 Madison Avenue
New York
NY 10017, USA
in UK from: Roger Lascelles

£3.20 to £3.45
$4.50 to $4.95

GUIDES TO ADVENTURE TRIPS

Adventure TravelGuide

Ed. Bob Citron

Pub. American Adventurers Association, Seattle, Wash., 1979

Describes 2000 worldwide adventure trips and expeditions open to public participation. Compiled for members of the Association.

$9.95

A Guide to Unusual Vacations

by Wilbur Cross and Farrell Cross

Pub. Hart Publishing Co, New York, 1973

Descriptions of over 150 selected trips, giving itinerary, main attractions, notes on gear and accommodation, cost and name and address of trip organiser.

$3.95

Sport and Adventure Holidays

Ed. Hilary Sewell

Pub. Central Bureau for Educational Visits and Exchanges, London, 1976

Details on activity holidays in Britain and abroad, ranging from go-karting and canal cruising to trekking, mountaineering and overland expeditions in Asia, Australia, Africa or the Americas. Bibliography.

Adventure Holiday Guide Abroad 1978

Ed. David Stevens

Pub. 1978
Vacation Work
9 Park End Street
Oxford
from: W. H. Smith booksellers

Guides giving tour operators by activity offered, eg. canoeing, hiking, overland, and then by geographical area; and an alphabetical list of tour operators with details of experience, number of years in operation and staff.

£1.50 each

LANGUAGE LEARNING

Phrasebooks

Berlitz
20 titles in the languages series.

95p each

Collins
10 titles

75p each

Penguin
7 titles.

from 75p each

Teach Yourself Series

Over 40 titles in the languages series.

Pub. English Universities Press, London
Distributed in the USA by
David McKay Co, Inc
750 Third Avenue
New York
NY 10017, USA

How to Get What You Want (in Eight Languages)

by Lixi

Pub. Julian Friedman

Sensibly straightforward illustrated phrasebook.

95p

Disabled Traveller's International Phrasebook

by Ian McNeil

Pub. 1979
Disability Press Ltd
1 Farthing Grove
Netherfield
Milton Keynes MK6 4JP

200 words and 50 phrases given in English and several European languages, including eg crutches, calipers, wheelchair, artificial leg/arm. Volumes covering non-European languages in progress.

70p

LOW COST TRAVEL

Hints to Businessmen

Pub. Department of Trade & Industry
Export Series Division
Hillgate House
35 Old Bailey
London EC4
Useful information about individual countries: bank/shop hours, import/export regulations, local customs, recommended dress, etc.

Survival Kit

Pub. CIEE, NY, USA
250 pp. Gives details of intra-European student charter flights, 230 hotels in 33 countries where bed and breakfast may be cheaply obtained.

The Good, The Bad, and The Bargains in Travel

Pub. Joyer
8401 Connecticut Ave.
Washington, DC, 20015
USA
184 pp. Contains letters from readers of the Joyer Travel Report Newsletter, *with information from their personal travel experiences around the world.*
$4.95 incl. post

How to Travel Without Being Rich

by Norman D. Ford, with the assistance of Richard Leavitt and Raymond P. Quin

Pub. Harian Publications, USA 1972
Where to go on a budget, and the best ways of travelling, by ship, rail, bus or aeroplane.

Charter Flight Directory

by Jens Jurgen

Pub. Travel Information Bureau
Kings Park
NY, USA
Paperback, revised annually. Gives tips on special air fares, ways of obtaining stopovers, of getting airlines to pay for accommodation, of transferring tickets, or joining charter flights, etc. Also includes many addresses of airlines, operators, etc.

Air Travel Bargains

Ed. Jim Woodman

from:
PO Box 987 Coconut Grove
Miami
Florida 33133
USA
Revised annually.

MEDICINE AND FIRST AID

A.M.A. First Aid Manual

Pub. American Medical Association
535 No. Dearborn Street
Chicago
IL, USA

American Red Cross First Aid Textbook

Pub. 1973
The Country Life Press
Garden City
NY, USA

First Aid

Pub. St. John Ambulance, St. Andrew's Ambulance Association, and British Red Cross Society.

3rd edn., 1972

A Foreign Language Guide to Health Care

Pub. Blue Cross Association
Blue Cross/Blue Shield
622 Third Avenue
New York
NY 10017, USA
Free booklet giving translations of phrases concerning health and medicine into French, German, Italian and Spanish.

Guide Médical des Voyages

from: Librairie de la Maison des Voyages
28 rue bu Pont-Louis-Philippe
75004 Paris, France

Preservaton of Personal Health in Warm Climates

Pub. Ross Institute of Tropical Hygiene
London School of Hygiene and
Tropical Medicine
Keppel Street
Gower Street
London WC1E 7HT
7th edn., 1971

The Ship Captain's Medical Guide

Pub. HMSO, London
20th edn., 1967
Excellent handbook for anyone responsible for the health of others and reliant on his own resources.

U.S. Certified Doctors Abroad

Pub. American Express Co, USA
List of over 3,500 English-speaking doctors who practise outside North America, including names, addresses, qualifications and medical background of each. The listing covers 120 foreign countries.

A Traveller's Guide to Health

A guide to prevention, diagnosis and cure for travellers and explorers
by Lt.-Col. James M. Adam

Pub. Sphere Books, London
from: Royal Geographical Society
1 Kensington Gore
London SW7 2AR
Reprint of 1966 guide which is much cherished by explorers for its basic advice though some of its details are outdated.

62p from RGS

Being Your Own Wilderness Doctor

by Bradford Angier

Pub. Stackpole, Harrisburg, PA, 1975

Paperback $4.95

Medical Advice for the Traveler

Dg. Kevin M. Cahill

Pub. Popular Library, New York, 1972

$0.95

Mountaineering Medicine

A Wilderness Medical Guide
by Fred T. Darvil, Jr.

Pub. Darvil Outdoor, USA
7th edn., 1975

Paperback $1.00

The Year-Round Travelers' Health Guide

by Patrick J. Doyle and James E. Banta

Pub. Acropolis Books, Washington, DC, 1978
Lightweight guide to preparation and precautions, with chapters on special circumstances, eg travel by air, with children, for the handicapped, and useful check-lists and examples of medical forms.

Exploration Medicine

Being a Practical Guide for Those Going on Expeditions
Eds. O. G. Edholm and A. L. Bacharach

Pub. John Wright & Sons Ltd, Bristol, 1965

New Advanced First Aid

by A. W. Gardner and P. J. Roylance

Pub. John Wright & Sons Ltd, Bristol
2nd edn., 1977

Pye's Surgical Handicraft

Ed. J. Kyle

Pub. John Wright & Sons Ltd, Bristol
20th edn., 1977
Includes reprint of most of H. S. M. Crabb's 1969 article 'Emergency Dental Treatment'.

Health Hints for the Tropics

Ed. Harry Most, M.D.

Pub. American Society of Tropical Medicine and Hygiene, 1964

Foreign Travel Immunization Guide

by Hans H. Neumann, M.D.

Pub. Medical Economics Co
Book Division
Oradell
NJ 07649, USA
6th edn., 1977
Guide to the immunisations necessary for each area and a discussion of some common travellers' ills.

$2.95

The Outdoorsman's Medical Guide

Common Sense Advice & Essential Health Care for Campers, Hikers & Backpackers
by Alan E. Nourse, M.D.

Pub. Harper & Row, New York, 1974

$3.95

The Emergency Book: You Can Save a Life!

by Bradley Smith and Gus Stevens

Pub. Simon & Schuster, New York, 1979
Readable, up-to-date book on first aid, covering two of the major breakthroughs of recent years: cardio-pulmonary resuscitation as treatment for cardiac arrest, and abdominal thrusts to clear an obstructed airway. 160 pp.

Hardback $8.95

Dr. Taylor's Self-Help Medical Guide

by Robert Taylor

from:
Joyer Travel Report
8401 Connecticut Avenue
Washington, DC 20015
USA

Gives advice on what to do for yourself, when you do not need a doctor and when you do.
$10.45 incl. p. & p.

The Traveller's Health Guide

by Dr Anthony C. Turner
from: Roger Laschelles
2nd edn., 1977
Paperback by the Senior Overseas Medical Officer of British Airways Medical Services and Hon. Associate Physician and Lecturer at the Hospital for Tropical Diseases, London. Invaluable for visitors to Africa, Asia or Latin America.

Health Information for International Travel

by U.S. Department of Health, Education anb Welfare
Pub. U.S. Government Printing Office, Washington, DC
from:
Public Information Service
Centre for Disease Control
Atlanta, GA. 30301
USA
Free leaflet, revised annually.

Emergency Dentistry

by Dr David Watt
Pub. 1975
Clausen Publications
115 St. Mary's Road
Weybridge KT13 9QA
Surrey
Useful especially for larger expeditions. 'Intended for those who must treat the occasional dental patient'.

Medicine for Mountaineering

Ed. James A. Wilkerson
Pub. 1972
Vail Ballou, Inc.
Seattle
Washington
USA
Recommended for any group travelling off the beaten track.

How to Stay Healthy While Traveling

by Bob Young, M.D.
from:
PO Box 567
Santa Barbara
CA 93102
USA
Dr Young writes a regular column for International Travel News. *Aimed at the budget traveller who may have unexpected medical problems and will expect to deal with them with improvised amateur methods.*
$2.00 incl. p. & p.

NAVIGATION

A Simple Method of Navigating in Deserts

by D. N. Hall
Pub. in Geographical Journal, Vol. 133, Pt. 2, June 1967

Be Expert with Map & Compass

The Orienteering Handbook
by Björn Kjellstrom
Pub. Charles Scribner's Sons, New York, 1972
Paperback $3.95

OVERLAND GUIDES

Overland through Central and South America
Overland through Africa
Overland to India and Australia

Pub. BIT Information and Help Services
97a Talbot Road
London W11
'Underground' guides intended to counter-balance glossy literature on the overland routes. With information on accommodation, visas, contacts, how to find work, special hazards (including health), police using 'heavy' methods, transport, tour operators, firms offering medical insurance, sources of vaccinations. Recommended reading.

Around the World on Wheels

by Harry Coleman and Peggy Larson
Pub. 1978
from:
the authors
9444 Forbes Avenue
Sepulveda
CA 91343
USA
A 'how-to' book, with the emphasis on overlanding. 1,000 page account of travels in a VW Camper.

Overland

by Peter Fraenkel

Pub. David & Charles, UK, 1975
Indispensable, money-saving guide to planning, preparation, equipment, vehicle modification, etc, (but not routes) by author with 100,000 miles of overland experience in Africa, Asia and Europe.

Overlanding

How to Explore the World on Four Wheels
by John Steele Gordon

Pub. Harper & Row, New York, 1975
A wealth of practical information on overland travel, from planning an itinerary to border procedures, selecting companions to buying maps. Well written and readable.

Paperback $4.95

Overland and Beyond

Advice for Overland Travellers
by Theresa and Jonathan Hewat

Pub. Gulliver Press
25 Lloyd Baker Street
London WC1
from:
106 West Street
Corfe Castle
Dorset BH20 5HE
and other distributors.
Planning, equipment and the other essentials of overland travel, based on authors' 3-year trip around the world in a Volkswagen Campervan.

Wanderlust

Overland through Asia and Africa
by Dan Spitzer

Pub. 1979
Richard Marek Publishers, Inc
200 Madison Avenue
New York
NY 10016, USA
276 pp.

World Understanding on Two Wheels

by Paul R. Pratt

Pub. 1976
Alemar-Phoenix Publishing House, Inc
927 Quezon Boulevard Extension
Quezon City, Philippines
Motorcycle journeys in Canada, USA, Mexico and Central America 1961-3, then, between 1966 and 1975, the longest continuous journey in motorcycle history, through Mexico, Central America, the USA, the Far East, USSR and South East Asia. By a Yorkshire-born electronics engineer who has lived both in Britain and the USA. With tips on costs, climate, altitude, crossing frontiers, etc.

PHOTOGRAPHY

What You Must Know When You Travel With a Camera

Pub. Harian Publications, NY

The Photographer's Handbook

by John Hedgecoe

Pub. Ebury Press, London, 1978
Use of equipment, procedures, glossary, special projects.

£8.95

How to Earn £5,000 a Year with Your Camera

by David Hodgson
from:
Airfairs Ltd
21-23 Balham Hill
London SW12 9DY
For the freelance photographer and photo-journalist.

Photography on Expeditions

by D. H. O. John

Pub. Focal Press, London, 1965

Wilderness Photography

by Boyd Norton

Pub. Reader's Digest Press, New York, 1978
Geared towards serious outdoor photographers using sophisticated cameras and equipment, but good for the less accomplished cameraman too.

$14.95

Travel Photography

Pub. Time/Life Library of Photography, 1973
Inquiries to:
Time-Life Books
c/o Time and Life Building
New Bond Street
London W1Y 0AA

RAILWAY TRAVEL

Fodor's Railways of the World

Ed. Robert C. Fisher
Foreword by Paul Theroux

Pub. David McKay Co., New York, 1977
Descriptions of railways, including itineraries, accommodation, meals, times, frequencies, types of train, of railways worldwide; and a chapter on train travel in general.

$9.95

SHIP TRAVEL

Freighter Days

How to Travel by Freighter

Pub. Harian Publications, USA
Life aboard a passenger-carrying freighter.

Freighter Travel Guidebook

Pub. Freighter Travel Service
201 East 77th Street, New York
NY 10021
USA

Trip Log Guide

Pub. Air and Marine Travel Service
501 Madison Avenue
New York
NY 10022
USA
Gives dates, departures and itineraries of freighters sailing out of the USA to ports worldwide.

$2.00

The Freighter Travel Manual

by Bradford Angier

Pub. 1974
The Chilton Book Co
Radnor, PA
USA
Destinations, routes, times and advantages of sailings by freighter.

The Total Traveler by Ship

by Ethel Blum

from:
Travel Publications, Inc
One Lincoln Road
Suite 214
Miami
FA 33139
at $6.50 incl. postage to subscribers to the Joyer Travel Report.
1978-79 edn
Comprehensive book about cruising, both on passenger ships and on freighters, with tips on packing, choosing a ship, picking a cabin, etc, and profiles of some 50 ships.

$7.95

Ford's Freighter Travel Guide

Ed. Merrian Clark

Pub. Ford's Travel Guides
PO Box 505
22151 Clarendon Street
Woodland Hills
CA 91365
USA
Published twice annually since 1952. Includes listings of travel agents, ports, freighter and cargo-passenger services, steamship lines.

$4.50 post paid

Today's Outstanding Buys in Freighter Travel

by Norman Ford

Pub. Harian Publications, USA
A selection of the world's outstanding passenger-carrying freighters.

Travel Routes Around the World

The Traveler's Directory to Passenger-Carrying Freighters and Liners.
Eds. Fredric E. Tyarks, Norman D. Ford and M. Drouet

Pub. Harian Publications, USA
700 freighter lines sailing from almost every port in the world, their destinations, prices, accommodation, tips on what to take.

SURVIVAL

Hip Pocket Emergency Survival Handbook

Pub. American Outdoor Safety League
13256 Northrup Way, Suite 8
Bellevue
WA 98005, USA
and
Cordée
249 Knighton Church Road
Leicester
UK

95p

Jungle Survival
Desert Survival
Sea Survival
Snow Survival
Survival against the Elements

Pub. Ministry of Defence, London
Booklets not usually available to the general public, but readers may be able to obtain copies if they have any contacts in the Services. All have useful information on first aid, possible sources of injury, illness or danger.

Survival

Pub. Department of the Air Force
Washington, DC, 1959
Printed by US Government Printing Office, Washington, DC, Air Force Manual 64-5.

Survival Cards

Pub. Survival Cards
Box 805, Bloomington
IN 47401
USA
Ten 3" x 5" plastic pages giving 300 techniques-solar still, snares, signals, snakebite, hypothermia, amputation, etc–for temperate, desert, artic and tropical conditions.

$2.50

The Outdoorsman's Emergency Manual

by Anthony J. Acerrano
Pub. Winchester Press, New York, 1976
Well-organised handbook describing dangers likely to be encountered outdoors and methods of dealing with them. Illustrated.

$10.00

Survival with Style

by Bradford Angier
Pub. Stackpole, Harrisburg, PA, and Vintage Press, New York, 1972

$2.45

The Complete Wilderness Almanac

by Berndt Berglund
Pub. John Wiley & Sons, Canada, Ltd
22 Worcester Rd
Rexdale, Ontario M9W 1L1
Canada
The outdoor manual on wilderness survival, wilderness living, and wilderness cooking.

The Spur Book of Survival and Rescue

by Terry Brown and Rob Hunter
Pub. Spurbooks Ltd, Bucks, 1977

Survival in Cold Water

The Physiology and Treatment of Immersion Hypothermia and of Drowning
by W. R. Keatinge
Pub. Blackwell Scientific Publications, Oxford, 1969

Stay Alive in the Desert

by Dr K. E. M. Melville
Pub. Roger Lascelles
Also available from the author:
Laws House
nr. Duns
Berwickshire
or from: Trail Finders (Services) Ltd
New edn., 1977
Deals with the hazards of desert driving, including advice on equipping the vehicle, procedure if stuck or stranded, precautions when driving, against ill-health or accidents. Also covers health and hygiene. By author with seven years' experience as a doctor in Libya.

Outdoor Survival Handbook

by David Platten
Pub. David & Charles, Devon, 1979
160 pages. For trekkers and campers.

£3.95

TRAVEL TIPS

All in One Travel Information

from: North Bergen Travelinfo Publishers
PO Box 9097
North Bergen
NJ 07047
USA
Worldwide climate, currency, tipping, car rental, addresses of tourist offices, entry requirements of all countries.

$3.00 incl. p. & p.

The Best and Worst of Travel

Pub. The Travel Advisor, Washington, DC, 1978
Results of a survey of over 2,000 world travellers, rating travel companies and giving customer comments.

Paperback $4.95

The Care of Babies and Young Children in the Tropics

from:
National Association for Maternal and Child Welfare
Tavistock House North
Tavistock Square
London WC1
Free booklet. Mostly about medical problems, their prevention and treatment.

Guide de l'Aventure et du Voyage

Pub. Guilde Européenne du Raid
15 quai de Conti
75006 Paris
France
Comprehensive alphabetical reference work giving advice, information and addresses covering the range of adventure and travel in France. Items include: insurance, hitchhiking services, green cards, map sources, grants, adventure schools, libraries, books and guides, air transport, international driving licences, photography, health, vehicle servicing.

Hints to Travellers

Pub. Royal Geographical Society, London, 1930; reprinted 1947
An impressive fund of information, some of it outdated but much still relevant. Especially useful on selecting personal equipment.

Holiday Insurance

Pub. British Insurance Association
Aldermary House
Queen St.
London EC4P 4JD
Free information leaflet. (Enclose s.a.e.)

International Youth Hostel Federation Handbook

Vol 1: Europe and the Mediterranean;
Vol 2: Africa, America, Asia, Australasia
Pub. 1973
IYHF
Midland Bank Chambers
Howardsgate
Welwyn Garden City
Herts.
also from: YHA Services
Guide to youth hostelling principles and regulations, and lists of all youth hostels in the areas named with notes on their amenities, location, costs, etc.
£1.25 each, plus 20p p. & p.

List of Guidebooks for Handicapped Travelers

Pub. The President's Committee on Employment of the Handicapped
1111 20th Street NW
Washington, DC 20036
USA
Free pamphlet.

Notice to Travellers

Pub. Bank of England, London
Indicates countries with special currency restrictions, sources of best rates of exchange, etc.

Travelfree

Pub. 1978
916 Helen Avenue
Yuba City
CA 95991
USA
Booklet on free travel, including tips on leading tours, taking groups, organisations to write to.
$3.90 plus tax

The Traveller's World Guide to Public Holidays

Pub. Export Times
60 Fleet Street
London EC4
UK
Leaflet.

The Weather Handbook

Pub. Conway Research
Atlanta
Georgia, USA

YMCA Directory
YWCA Directory

Pub. YMCA
640 Forest Road
Walthamstow
London E17 3DZ
and
YWCA of Great Britain
Hampden House
2 Weymouth Street
London W1
List own addresses in over 70 countries, plus other useful addresses of doctors, lawyers, consulates, priests, etc.

The Complete Traveller

Everything you need to know about travel at home and abroad
by Joan Bakewell

Pub. Hamlyn Paperbacks, Feltham, Middlesex, 1979
Extensive book covering travel at home and abroad, full of snippets of information, written in informal, personal style, and aimed mainly at the traveller who is conventional in terms of means of travel, destination, funds available, attitude.

Paperback £1.50

Consumer Handbook for Travelers

by Hal Gieseking

Pub. Travel Advisor/High Street Press
Washington, DC, USA

$5.95

1001 Sources for Free Travel Information

Valuable Freebies You Should Know About
by Jens Jurgen

Pub. 1978
Travel Information Bureau
PO Box 105
King's Park
NY 11754
USA
Lists of embassies, consulates, tourist offices, airline information centres for 200 countries and all states of the USA; and chapters on stop-overs, planning trips, moving and retiring, and notes on visas and passports, youth hostels, travel for the handicapped.

$3.95

World Traveler's Almanac

Eds. Nancy and Larry Meyer

Pub. Rand McNally, Chicago, 1975
Bumper book on travel, with chapters on travel agents, flying shopping, cruising, business travel, brief country by country summaries, list of guide books. Emphasis on European travel by Americans.

$6.95

Explorers Source Book

Ed. Al Perrin

Pub. 1977
Explorers Ltd
2107 Grant Avenue
Wilmington
Delaware 19809
USA
Advice, addresses, etc on training, equipment, books and maps, governing bodies; by type of activity eg canoeing, climbing, backpacking. Comprehensive and extremely useful.

How to Beat the High Cost of Travel

by Betty Ross
Pub. 1977
US News & World Report Books
Suite J, 2300 N Street
Washington,
DC 20037 USA
Information on rail passes, charter flights, home exchanges.

The Climate Advisor

by Gilbert Schwartz

Pub. Climate Guide Publications
Box 323A, Station C
Flushing
NY 11367
USA
Non-technical reference guide to climate and weather in the United States, Canada, Mexico and the Caribbean, with climate charts for over 350 locations and accompanying descriptions of the climate at each.

$7.90 post paid
from publishers

MsAdventures

Worldwide Travelguide for Independent Women
by Gail Rubin Sereny

Pub. Chronicle Books, San Francisco, 1978
Light reading, hard facts, a good guide for the lone woman traveller.

Paperback $5.95

Area Handbooks

Compiled by US State Department

from: Superintendent of Documents
US Government Printing Office
Washington, DC 20402, USA
Detailed information on history, politics, economics, population, health of very many areas. Request by country.

$5 to $10

Background Notes on Countries of the World

Compiled by US State Department

from: Superintendent of Documents
US Government Printing Office
Washington, DC 20402, USA
Synopses of countries' history, geography, climate, economy, current exchange rates, addresses of consulates in the US. Approx. 75 countries are updated each year.

35c to 70c

Travel on the Cheap

by Jonathan Walters

Pub. Penguin Books (paperback), Kestrel Books (hardback), 1978

Advice from a young but seasoned traveller.

Around the World in Eighty Rules

by Jack Womeldorf

Pub. 1974
Welfare and Recreation Association Travel
Club of the Library of Congress
142 E St.
Washington, DC 20003 USA

Pamphlet aimed at those who have not yet travelled extensively, dealing with planning and travel bargains.

$1.25 postpaid

VEHICLES

Autobooks
Autocare Manuals
Handybooks

Pub. Autobooks Ltd
Golden Lane
Brighton
Sussex BN1 2QJ

Comprehensive range of owners' do-it-yourself manuals and vehicle reference books.

Chilton Guides

Blazer/Jimmy
Jeep Universal
International Scout
Jeep Wagoneer, Commando & Cherokee Bronco
Toyota Land Cruiser

Pub. Chilton Book Co.
Radnor
PA
from:
Four Wheeler Books
PO Box 547
Chatsworth
CA 91311
USA
or from: Argus or Autobooks, USA

Repair and tune-up guides for the novice or semi-professional mechanic wishing to maintain his own offroad vehicle. Annually updated.

Cloth $8.95
Paperback $6.95

A Close Look at 4-Wheel Driving

Pub. Ford Motor Company
Public Relations—Truck and Recreational Products Operations
PO Box 2053
Dearborn
MI 48121
USA

Free 15 page pamphlet covering principles of four-wheel driving, driving in problem conditions, four-wheel-drive and how it works. Straightforward and well done.

Clymer Maintenance Manuals

Pub. Clymer Publications
222 N Virgil
Los Angeles
CA 90004, USA
from: Argus
Autobooks
Cepek

A series of vehicle maintenance books including volumes on all major US 4WD vehicles. Annually updated.

$6.00 or $7.00 each

A Guide to Land-Rover Expeditions

Pub. 1972
Jaguar Rover Triumph Ltd
Solihull
Warwickshire
In USA from:
British Leyland Motors
600 Willow Tree Road
Leonia
NJ 07605

Free booklet. Information on the vehicle, hints on cross-country driving and some general information on travel, eg visas, maps, insurance. 15 pages.

How to Keep your VW Alive

Pub. John Muir Publications
Box 613
Santa Fe
New Mexico 87501
USA

Light Truck Buying Made Easier

from:
Ford Motor Co.
Box 1978
The American Road
Dearborn
MI 48121
USA

Free. Guide to light trucking including four-wheel-drive. Emphasis on Ford vehicles.

Fitness for the Motorist

by Dennis Chambers

Pub. Letts Guides, 1976

Four Wheel Drive Handbook

by James T. Crow and Cameron A. Warren

Pub. CBS Publications, New York
from:
Autobooks
Cepek
5th edn., 1976
Covers choice and preparation of vehicle. equipment, driving tips. Chapters on troubleshooting and correction of carburettor and ignition malfunctions.

$3.95

All About Pickup Campers, Van Conversions and Motor Homes

by John Gartner
Pub. 1969
Trail-R-Club of America
Beverly Hills
CA
USA

$4.50

Pickup Camper Manual

by Clinton Hull
Pub. Trail-R-Club of America
Beverly Hills
CA
USA

$5.95

Petersen's Complete Book of Four-Wheel Drive

Ed. Spence Murray

Pub. Petersen Publishing Co.
8490 Sunset Blvd.
Los Angeles
CA 90069, USA
from:
Autobooks Cepek
2nd edn., 1976
Combines history of the military Jeep with practical tips for modern four-wheel drives.
Includes report on sixteen 4WD vehicles which the editor and his staff submitted to comparison tests.

$3.95

The Do-It-Yourself Custom Van Book

by Franklynn Peterson and Judi R. Kesselman

from: Argus
Covers fittings, accessories.

$6.95

Riding the Dirt

by Bob Sanford

Pub. Bond-Parkhurst Publications
1499 Monrovia Avenue
Newport Beach
CA 92663
USA
One of the best of the recent books on trail and off road driving and motorcycling. Techniques, design, maintenance, clothing, physical fitness. Very readable and informative.

$6.95

To Hell on Wheels

The Illustrated Manual of Desert Survival
by Alan H. Siebert

Pub. 1976
Brown Burro Press
Pasadena
CA
from: Cepek
Divided into two sections: vehicle survival and personal survival. Includes advice on the emergency use of Citizens Band radio.

$2.95

Two Wheel Travel: Motorcycle Camping and Touring Bicycle Camping and Touring

by Peter Tobey

Pub. Tobey Publishing Co
Box 428
New Canaan
CT 06840
USA
Two useful books.

Motorcycle $3.00
Bicycle $4.95

Off Road Handbook with Back Country Travel Tips

by Bob Waar

Pub. 1975
HP Books
Box 50640
Tucson
Ariz. 85703
USA
from:
Argus
Autobooks
Cepek
Covers vehicle preparation, driving techniques and maintenance. Good sections on winches and tyres.

$4.95

Periodicals

Numbers given at right indicate number of issues per year.

ABC AIR TRAVEL ATLAS **2**
ABC GUIDE TO INTERNATIONAL TRAVEL **4**
ABC RAIL GUIDE **12**
ABC WORLD AIRWAYS GUIDE **12**
ABC SHIPPING GUIDE **12**

ABC Travel Guides Ltd
Oldhill
London Road
Dunstable
Beds. LU6 3EB
Subscriptions
30-40 Bowling Green Lane
London EC1R 0NE

Guide to International Travel
Ed: J. Frank Holt
Consulates, journey times, baggage charges, taxes, temperature and humidity charts, WHO maps of spread of diseases, etc.

Shipping Guide
Details of fares, sailing dates, company addresses, for passenger shipping lines all over the world.

Air Travel Atlas
Maps showing scheduled air routes, trunk and regional routes; also lists of city and country codes and airline designators.

World Airways Guide
Timetables for all airlines, details of through flights and main transfer connections.

Rail Guide
Covers routes and fares from London, with timetables to main stations. Other timetables and details of shipping services.

ACTIONS **3**

Ed. Hughes Toussaint
Guide Européenne du Raid
15 quai de Conti
75006 Paris
France
Reports on grants made by the Guilde, forthcoming events, organised tours, classified advertisements.

ADVENTURE TRAVEL

Ed. Anthony Peagam
Adventure Travel Magazine Ltd
Chamberhouse Mill
Thatcham
Berks. RG13 4NU
Launched autumn 1979.

ADVENTURE TRAVEL **6**

Adventure Travel Publications, Inc
A subsidiary of the American Adventurers Association, Inc
Suite 301
444 NE Ravenna Blvd
Seattle
WA 98115
USA
Feature articles, news, reviews, opinion, equipment surveys, etc.

ADVENTURE TRAVEL NEWSLETTER **12**

American Adventurers Association, Inc
Suite 301
444 NE Ravenna Blvd
Seattle
WA 98115
USA
News, lists of adventure trip operators, books, associations, courses.

AFRICA GUIDE **1**

Africa Guide Co
21 Gold Street
Saffron Walden
Essex
England
Useful, comprehensive guide to some 25 countries.

AFRICA SOUTH OF THE SAHARA **1**
FAR EAST AND AUSTRALASIA **1**
MIDDLE EAST AND NORTH AFRICA **1**
Europa Publications Ltd
18 Bedford Square
London WC1B 3JN
England
Background information, especially on political and economic aspects.

AVVENTURE NEL MONDO **4-5**
Via Vitellia 81
00152 Rome
Italy
Magazine of the club of the same name, giving details of adventure trips.

BACKPACKER **6**
BACKPACKER FOOTNOTES **6**
Backpacker Magazine
65 Adams Street
Bedford Hills
NY 10507
USA
Periodicals published in alternate months. Notes and features on hiking, backpacking and skiing. Magazine has equipment evaluations, in-depth articles, interviews. Footnotes has tips, news, recipes, written mainly by readers.

BACKPACKING JOURNAL **2**
Davis Publishing Co
380 Lexington Avenue
New York
NY 10017
USA
Interviews, news, equipment tests, articles on training, survival, conservation.

BUSHDRIVER
Rick Williams & Associates
2/2 Stokes Street
Lane Cove
NSW 2066
Australia
Off roading the Australian way.

BUSINESS TRAVELLER **4**
The insider's guide to cheaper travel
Export Times Publishing Ltd
60 Fleet Street
London EC4Y 1LA
Features, news, travellers' forum, notes on fares and flights, 'air fares minimiser section'–guide to principal air fares and cut-price business travel packages from London to Europe.

CAMPING **12**
Caravan Publications Ltd
Link House
Dingwall Avenue
Croydon
Surrey CR9 2TA
England

CAMPING **8**
Galloway Publications
5 Mountain Avenue
North Plainfield
NJ 07060
USA
Established 1926.

CAMPING & CARAVANNING **12**
Magazine of the Camping Club of Great Britain and Ireland
Camping Club of Great Britain and Ireland Ltd
11 Lower Grosvenor Place
London SW1W 0EY
England

CAMPING JOURNAL **8**
Davis Publishing Co
380 Lexington Avenue
New York
NY 10017
USA
Established 1962.

CAMPING JOURNAL **8**
The magazine of family outdoor recreation
Camping Journal
Box 2620
Greenwich
CT 06835
USA
Articles on destinations, kit, vehicles, accessories, etc.

CAMPING MAGAZINE **8**
American Camping Association
Bradford Woods
Martinsville
Indiana 46151
USA
Established 1926

CYCLETOURING 6

Magazine of the Cyclists' Touring Club
Cotterell House
69 Meadrow
Godalming
Surrey, England
Free to members, available to others.

ECONOMY TRAVELER 12

The international travel consumer's resource
Publisher: Ed Perkins
Box 353
Menlo Park
CA 94025
USA
Founded 1977 as Economy Traveletter, name changed after four issues. Stresses comprehensive examination of major travel questions, with company-by-company dollars-and-cents comparisons of competitive travel services based on 'own original, independent, professional research'. Feature-length articles on places, issues.

EXPEDITION 4

The Magazine for the Independent Traveller
WEXAS International
45 Brompton Road
Knightsbridge
London SW3 1DE
England
Established 1970. Photofeatures on travels and expeditions. Letters, editorials classified and display ads, latest news on travel, expeditions and overland, book reviews. For members only.

EXPLORERS JOURNAL 4

Official quarterly of the Explorers Club
The Explorers Club
46 East 70th Street
New York
NY 10021
USA
Established 1902.
Articles on scientific discoveries, expeditions, ornithology, personalities and many other branches of exploration. Reviews. Embraces space exploration.

FINANCIAL TIMES WORLD HOTEL DIRECTORY 1

Financial Times Ltd
Bracken House
Cannon Street
London EC4
England
Mainly for businessmen. Covers availability of conference rooms, secretarial help, telex facilities, parking, etc, at some 3,500, mostly top class, hotels in over 150 countries.

4x4s AND OFF-ROAD VEHICLES 12

E-GO Enterprises, Inc
PO Box 5755
Sherman Oaks
CA 91403;
15825 Stagg Street
Van Nuys
CA 91406
USA

FOUR WHEEL AND OFF-ROAD 4

Petersen Publishing Co
8490 Sunset Blvd.
Los Angeles
CA 90069
USA

FOUR WHEELER 12

Four Wheeler Publishing
Editorial office:
PO Box 2547
Chatsworth
CA 91311
USA
Subscriptions:
PO Box 243
Mt Morris
IL 61054
Covers racing and touring, with emphasis on 4WD.

FREIGHTER TRAVEL NEWS 26

PO Box 504
Newport
Oregon 97365
USA
Fortnightly newsletter on freighter routes and bargains worldwide.

GEO 12

Gruner & Jahr AG & Co
Editorial office:
Warburgstrasse 45
2000 Hamburg 36
West Germany
Subscriptions:
Postfach 111629
2000 Hamburg 11
West Germany
Travel and places (but not academic issues) in the style of the National Geographic Magazine.

THE GEOGRAPHICAL MAGAZINE 12

1 Kensington Gore
London SW7 2AR
England
Articles, notes, news, reviews, classified advertisements.

GLOBE **6**

Newsletter of the Globetrotters Club
The Globetrotters Club
BCM/Roving
London WC1V 6XX
England
Travel information. Articles on individual experiences, tips, news of 'members on the move', mutual aid column for members.

GO **10**

The Adventure Newsletter for the Go People
W. J. Hunt
Box 571
Barrington
IL 60010
USA
Up-to-date travel/adventure news and holiday ideas and a feature article.

GOOD DEALS **6**

. . . on Low Cost Charter Flights and Tours
Good Deals, Inc
1116 Summer Street
Stamford
CT 06905, USA
Information on fares and flights, insurance, prices and tours (including dates) to all parts of the world.

GREAT EXPEDITIONS **6**

Great Expeditions, Inc
Box 46499
Vancouver
BC V6R 4G7
Canada
Started March/April 1978. Offers a free classified ads service to subscribers and non-subscribers alike, features on great explorers, news of expeditions and field studies, travel notes, etc. Concerned mainly with non-commercial and educational expeditions.

HOLIDAY WHICH? **4**

Consumers Association
14 Buckingham Street London WC2
England
Available to subscribers only. Advice on tours, packages, insurance, exchange rates, etc.

HOSTELING **12**

American Youth Hostels
Metropolitan New York Council
132 Spring Street
New York
NY 10012
USA
Established 1972.

INTERNATIONAL TRAVEL NEWS **12**

Martin Publications, Inc
2120 28th Street
Sacramento
CA 95818
USA
News source for the business and/or pleasure traveller who often goes abroad. Contributions mostly from readers.

JOYER TRAVEL REPORT **12**

'The newsletter that brings you the bargains in travel—in the USA and around the world'.
Phillips Publishing, Inc
Editorial Office:
8401 Connecticut Avenue
Washington DC 20015
USA
Subscriptions:
Box 707
Corona del Mar
CA 92625
USA
Established about 1968. Includes 'travelers' exchange' for subscribers who wish to swap ideas, homes, services, experiences. News and information; 'feedback', readers' letters.

THE LAZY WAY TO BOOK YOUR CAR FERRIES **1**

Car Ferry Enquiries Ltd
9A Spur Road
Isleworth
Middlesex TW7 5BD
Ferry booking advice.

MARIAH OUTSIDE **6**

Mariah Publications
Corporation
3401 West Division Street
Chicago
IL 60651
USA
Devoted 'to the spirit of any person who is sensitive to the beauty and freedom nature offers'. A merger between Mariah, *established 1976, and* Outside, *the magazine 'for doers' with equipment tests, expert advice, etc.*

NATIONAL GEOGRAPHIC MAGAZINE **12**

National Geographic Society
Washington DC
USA

This long-established magazine is familiar to many people. Noted for the quality of its photography, articles cover travel and expeditions and many other fields.

NOMAD **irregular**

BCM-Nomad
London WC1V 6XX
or
Box 4137
Grand Central Post Office
New York
NY 10017, USA
Newsletter aimed at people on the move and written by peripatetic publisher, with many readers' reports. 'Current issues are sent to any address on receipt of $1.00, and for $2.00 a press card is also included.' Press card is for people to use 'for what they can get away with'.

OAG WORLDWIDE TOUR GUIDE **3**

Reuben H. Donnelley Corp., USA
450-500 page reference guide containing nearly 100 pages devoted to special interest tours in over 70 categories, and several hundred pages grouping tours by areas.

OFF BEAT **4**

1250 Vallejo Street
San Francisco
CA 94109
USA
Newsletter offering practical advice on visiting unusual travel destinations.

OFF-ROAD **12**

Combined with OFF-ROAD VEHICLES and OFF-ROAD VEHICLES & ADVENTURE
Ed. Tom Madigan
Argus Publishers Corporation
12301 Wilshire Blvd.
Los Angeles
CA 90025
USA
Subscriptions:
Box 49659
Los Angeles
CA 90049
Vehicle test reports, technical and feature articles.

OFF-ROAD ADVERTISER **12**

California Off-Road Vehicle Association
PO Box 1327
Arcadia
CA 91006
USA
Subscriptions:
9466 Cambridge Street
Cypress
CA 90630
Magazine containing information on equipment, parts, local events, legislation and services; and a buyers' and sellers' guide to services. Also classified ads.

OFF-ROAD AUSTRALIA **12**

ORA Publisning Co Pty Ltd
3rd Floor
Ryrie House
15 Boundary Street
Rushcutters Bay
NSW 2001
Australia

ON THE ROAD **irregular**

Pub. Jurgen Bischoff
Bernd Vahle
Oskar-Hoffmanstrasse 46
4630 Bochum 1
West Germany
Practical information for hitchhikers and other budget travellers.

O.R.V.

The Magazine for Off-Roading
7950 Deering Avenue
Canoga Park
CA 91304
USA

OVERLANDER **6**

PM Publications Pty Ltd
19A Boundary Street
Rushcutters Bay
NSW 2011
Australia
Started 1976. Articles about off-beat journeys, product-testing, news, tips, etc; letters, even fiction.

PARTIR **11**

32 rue du Pont-Louis-Philippe
75004 Paris
France.
Articles on places and expeditions; travel tips, including how to reach places mentioned in text. Designed mainly for independent travellers overlanding on a tight budget.

PASSPORT **12**

20 North Wacker Drive
Chicago
IL 60606
USA
Newsletter on sightseeing, places to stay, readers' tips, warnings about hotels, overbookings, etc.

PETERSEN'S 4 WHEEL & OFF ROAD **12**

Petersen Publishing Co
8490 Sunset Boulevard
Los Angeles
CA 90069
USA

PICKUP, VAN & 4WD **12**

CBS Publications
Editorial office:
1499 Monrovia Ave
Newport Beach
CA 92663
USA
Subscriptions:
555 Crowe Road
Marion
OH 43302
One of the newest and best of the off-road periodicals, with technical information, road tests, etc.

POPULAR OFF-ROADING **12**

Challenge Publications, Inc
7950 Deering Avenue
Canoga Park
CA 91304
USA

PRACTICAL CAMPER **12**

Haymarket Publishing Ltd
Gillow House
5 Winsley Street
Oxford Circus
London W1A 2HG
Subscriptions:
34 Foubert's Place,
London W1
England
Articles on sites, attractions, lightweight equipment, practical tips, buyer's guide, advertisements.

SAFARIPOSTEN **1**

Topas Globetrotterklub
Safarihuset
Lounsvej 29
Louns
DK-9640 Farsø
Denmark
Annual expedition publication in Danish.

SIERRA **10**

The Sierra Club Bulletin
Sierra Club
530 Bush Street
San Francisco
California 94108
USA
Articles on geography, conservation, urban development, history, etc. List of Sierra Club books (guides, handbooks, 'exhibit format' books). Advertisements (suppliers, adventure holiday organisers).

THE SOUTH AMERICAN EXPLORER **4**

Official journal of the South American Explorers Club
South American Explorers Club
Casilla 3714
Lima 1
Peru
Sections include area field guides, news, club activities, book reviews, an equipment directory, classified advertisements, letters. Plus full-length articles, not by 'professional adventurers'.

TAKING OFF **2**

Educational Cooperative
176 W. Adams
Suite 2121
Chicago
IL 60603
USA
First issue Spring/Summer 1977. General travel tips, information on inexpensive/student/youth transport, adventure trips, charter flights, reference books. The Cooperative is a non-profit making corporation.

THOMAS COOK INTERNATIONAL TIMETABLE **12**

Thomas Cook & Son
PO Box 36
Peterborough England
Comprehensive timetable of all trains known to run in the civilised world, and of local shipping services.

TIME OUT 52

Tower House
Southampton Street
London WC2
Excellent classified ads section useful for overland travel, cheap flights, travelling companions.

TRAIL 6

For campers, caravanners—and all who love the great outdoors
The Automobile Association
Fanum House
Basingstoke
Hampshire RG21 2EA
Started 1978. Contains tests of vehicles and equipment, site reports and feature articles.

TRAILFINDER 3

Trail Finders Ltd
46-48 Earls Court Road
London W8 6EJ
Illustrated newspaper with news, features and advertisements on budget flights and overland travel.

THE TRAIL VOICE 4

The International Backpackers Association
PO Box 85
Lincoln Center
Maine 04458
USA
Articles on conservation, backpacking areas and expeditions, helpful hints, book and bulletin sales service.

TRAVEL DIRECTORY 1
TRAVEL PHONE GUIDE 1

St James Press
3 Percy Street
London W1
England
List tour operators, travel agents, hotel groups, foreign hotels and their UK representatives, vehicle hire firms. The **Phone Guide** *lists companies by telephone number only.*

TRAVELERS DIRECTORY 1

Compiled by Tom Linn
6244 Baynton Street
Philadelphia
PA 19144
USA
For the 'underground' traveller, an international register of people willing to offer a home or hospitality to travellers. The Directory *is available only to people listed in it. Listings cost $10 per year; this fee includes a quarterly* Newsletter *containing 'underground' travel information.*

TRAVEL INFORMATION MANUAL (TIM) 12

Compiled by IATA
TIM
PO Box 7627
1118zj
Schiphol Airport
Netherlands
Gives full information on most recent visa, health, customs, currency regulations. Aimed mainly at the travel industry.

TRAVEL NEWS TIPS 12

A monthly newsletter for the discriminating traveller
Compsco Publishing Co.
663 Fifth Avenue
New York
NY 10022
USA
Gives 'new developments from all four corners of the traveller's world'.

TRAVEL SMART 12

Communications House
40 Beechdale Road
Dobbs Ferry
NY 10522
USA
Newletter for the budget traveller, with charter listings, aiming to help readers save money on travel.

TRAVEL TRADE GAZETTE 1

Morgan-Grampian (Publishers) Ltd
Calderwood Street
London SE18
Lists agents, operators, airlines, vehicle hire firms, insurance companies, hotels and hotel representatives, in the UK and Irish Republic.

TRAVELTIPS FREIGHTER TRAVEL ASSOCIATION BULLETIN 6

40-21 Bell Blvd.
Bayside
NY 11361
USA
First-person accounts of freighter travel to all parts of the world.

TRAVELWORLD 4

O. Hayward
38 University Road
Colliers Wood
London SW19
Established June 1979. Feature articles on conservation, archaeology, communications, weather and items of human interest. Plus a 32-page supplement from time to time on off the beaten tack places, how to get there, tour operators.

TREADS

Murray Publishers Pty Ltd
142 Clarence Street
Sydney
NSW 2000
Australia
Off-road magazine, with features on different kinds of vehicle and of motoring.

TROTTER 5-6

Deutsche Zentrale für Globetrotter e.V.
Hassleyerstrasse 4a
D-5800 Hagen
West Germany
News and articles from outside Europe only; tips, readers' reports, classified ads; for club members only, though non-members may advertise or obtain information on payment of a fee.

VAGABONDING 6

PO Box 20095
Oklahoma City
OK 73120
USA
Magazine for 'liberated international vagabonds'.

WESTERN TRAVEL ADVENTURES

Sunset Magazines
80 Willow Road
Menlo Park
CA 94025
USA

WILDERNESS CAMPING 6

Official magazine of the American Youth Hostels and US Canoe Association
1597 Union Street
Shenectady
New York
NY 12309
USA

WORLD CRUISE NEWS 12

SCCH
(Sail Crew Clearing House)
Box 1976
Orlando
FL 32802
USA
Monthly newsletter which is the only regularly published source of berth and employment information on private sailing yachts.

WORLDWIDE YACHT CHARTER AND BOAT RENTAL DIRECTORY 1

18226 Mack
Grosse Pointe
Michigan 48236
USA
Annual, listing over 10,000 charters available worldwide, from power boats to sailing vessels.

Worldwide Directory of Services

To locate the firm or association you want, first consult this 'yellow pages' list of the categories to be found in the Directory of Services. You will save time by knowing which category or categories will fit your needs.

Accommodation
Aircraft
Awards and grants
Book and map distributors
Book and map publishers
Book and map retailers
Book clubs
Camping and hostelling associations
Communications
Courses
Currency information
Customs
Equipment: appliances
Equipment: backpacking
Equipment: camping, outdoor, expedition and travel
Equipment: clothing and footwear
Equipment: cold weather
Equipment: instruments
Equipment: optical
Equipment: photographic
Equipment: secondhand
Equipment: survival
Equipment: tents
Equipment: water purifiers
Expedition consultancies
Expedition food
Expedition reports
Exploration, travel and adventure associations
Ferry services
Film hire
Freight
Freighter advisory services
Hitchhiking
Inflatable boat suppliers and outfitters
Insurance
Language
Libraries
Low cost travel
Medical care
Medical supplies
Picture libraries
Route planning
Specialist associations
Travel finance/currency
Travellers' aid centres/clubs
Travellers' aids suppliers
Vaccination centres
Vaccination information
Vehicles: accessories, spares, outfitting
Vehicles: purchase, hire and conversion
Vehicles: shipment
Visas
Weather information

With a few exceptions, information is given alphabetically within the following subsections: UK, USA, Canada, Europe, Australasia, South America, Asia, Africa. Within subsections covering whole continents, individual countries are also arranged alphabetically.

Accommodation International

Servas

Is an international cooperative system of hosts and travellers, established to promote peace and goodwill amongst nations. It was first conceived by an international group of youths studying in Denmark in 1948. Servas hosts worldwide are willing to accept travelling members as guests for a few days in their homes. To become approved as a Servas traveller, contact the national coordinator in native country. The British coordinator is at 15 Valerie Grove, Birmingham 22A and the American at US Servas Committee, PO Box 790, Old Chelsea Station PO, New York, NY 10011, USA.

Aircraft UK

Air Associate Ltd.

40 St. Peters Road
Hammersmith
London W6 9BH
Tel. (01) 748 0222

Offer consultancy services to expeditions where air support is required; and obtain aircraft for expeditions to purchase, lease or charter, whether for one occasion or in the longer term: a group may purchase an aircraft so that it is then available where and when required for expeditions.

Awards and grants Europe

Touring-Club Royal de Belgique

44 rue de la Loi
1040 Brussels
Belgium
Tel. (02) 513 8240

Makes grants known as Les Bourses de Voyage-Jeunesse to Belgians aged between 17 and 35, for extensive travel. As well as the sizeable grant, successful applicants also receive cameras, vehicle accessories, maps, travel tickets and various coupons.

Dotation Kodak

Service des relations extérieures
8—14 rue Villiot
75012 Paris
Tel. 347 90 00

Makes a fixed number of gifts of film per year to 18–25 year-old travellers who are required to submit a report of their trip.

Dotation Nationale de l'Aventure

15 quai de Conti
75006 Paris
France
Tel. 033 52 53

Makes grants to French travellers aged between 18 and 35 years. A brochure is available on request.

Dotation Renault

Régie Renault
Service des relations extérieures
53 avenue des Champs-Elysées
75098 Paris
France
Tel. 225 18 57

Makes an award of a dozen 4Ls each year to young travellers, between 18 and 35, who are making a cultural or scientific vehicle trip or expedition. Candidates must submit an expedition plan with details of organisation, itinerary, aims, etc. Recipients of the awards receive also petrol coupons, tools, spares, insurance and other assistance.

Dotation 3M

Service des relations publiques
Blvd. de l'Oise
95000 Cergy-Pontoise
France
Tel. 031 61 61 and 031 75 36

Makes awards to 18–30 year olds travelling as a team that includes a photographer. The awards consist of film, magnetic tape and mini-cassettes.

Fédération Française des Automobiles Clubs

61—67 rue Haxo
75020 Paris
France
Tel. 797 43 29

Makes awards called Dotation 'Jeunes Grands Voyageurs' to young French travellers aged between 18 and 30 who are members of an Automobile Club affiliated with the FFCA and who are undertaking a journey with a specific purpose, by vehicle, covering at least 5000 km.

Book and map distributors UK

BAS Overseas Publications

45 Sheen Lane
London SW14 8LP
Tel. (01) 876 2131

General sales agents for a wide range of British, American and European timetables, hotel directories, etc.

Cordee

249 Knighton Church Road
Leicester LE2 3JQ

Are distributors of mountaineering, adventure and travel books.

Michelin Tyre Co.

81 Fulham Road
London SW3
Tel. (01) 589 1460

Sell Michelin maps and English editions of the Michelin Red and Green Guides.

USA

Bookpeople
2940 Seventh Street
Berkeley
Ca 94710

Distribute Lonely Planet Publications (Australia) books on Africa, New Zealand and Kathmandu.

Bradt Enterprises
409 Beacon Street
Boston
Mass 02115

Is the US for Bradt Enterprises of Chalfont St. Peter, Bucks., UK publishers of backpacking books.

David & Charles
North Pomfret.
Vt.

Are distributors for David & Charles, UK, publishers of overland and travel books.

French and European Publications
610 Fifth Avenue
New York
NY 10020

Import Michelin maps into the USA.

Michelin Guides and Maps
PO Box 188
Roslyn Heights
NY 11577

Distribute Michelin road maps, Red Guides and Green Guides.

Two Continents
30 East 42nd Street
New York
NY 10017

Distribute Lonely Planet Publications (Australia) books on Australia, South East Asia, Hong Kong, Papua New Guinea and Trekking in the Himalayas.

Writer's Digest
9933 Alliance Road
Cincinatti
Ohio 45242

Is the US distributor for Vacation Work Publications (UK), publishers of books on employment and budget travel in foreign countries.

Canada

Dominie Press
55 Nuggett Avenue
Unit J
Agincourt
Ontario

Are the Canadian distributors of Lonely Planet Publications (Australia) books.

Australasia

Caveman Press
Box 1458
Dunedin
New Zealand

Distribute Lonely Planet Publications (Australia) books in New Zealand.

Book and map publishers

UK

Autobooks Ltd
Golden Lane
Brighton
Sussex BN1 2QJ
Tel. 0273 721721

Publish a very comprehensive range of owners' do-it-yourself manuals and reference books intended for the practical layman. Autobooks and Handybooks are the two principal lines of books. They are easy to follow, being somewhat less technical than factory manuals.

John Bartholomew & Son Ltd
12 Duncan Street
Edinburgh EH9 1TA
Tel. (031) 667 9341

Publish tourist, road, topographic/general maps and atlases. Free catalogue from Marketing Department.

BIT Information and Help Service
97a Talbot Road
London W11
Tel. (01) 229 8219

Publishes 'alternative' overland guides, compiled and regularly updated with contributions from travellers, and sells these by mail or from office.

Bradt Enterprises
Overmead
Monument Lane
Chalfont St. Peter
Bucks SL9 0HY
Tel. 02407 3865

Are publishers of the Backpacking Guide Series *of books by Hilary and George Bradt.*

David & Charles
Brunel House
Forde Road
Newton Abbot
Devon
Tel. 0626 61121
Bookshop: 33 Baker Street
London W1
Tel. (01) 486 6959

Publish overland and travel books and a hardback series on islands with their geography and historical background.

Directorate of Overseas Surveys
Kingston Road
Tolworth
Surbiton
Surrey KT5 9N3
Tel. (01) 937 8661

Are the official UK publishers of maps of former (and current) British possessions. Edward Stanford Ltd. (q.v.) are the agents for the sale of D.O.S. maps.

Geographia Ltd
63 Fleet Street
London EC4
Tel. (01) 353 2701
93 St. Peter's Street
St. Albans
Herts
Tel. 56 30121

Publish a wide range of maps and atlases.

Hodder & Stoughton Ltd
St. Paul's House
Warwick Lane
London EC4P 4AH
Tel. (01) 248 5797

Are publishers in the UK of Fodor's Guides.

Hydrographic Department
MOD (Navy)
Taunton
Somerset TA1 2DN
Tel. 0823 87900

Publishes Admiralty Charts of the world.

Roger Lascelles
3 Holland Park Mansions
16 Holland Park Gardens
London W14 8DY.
Tel. (01) 603 8489

Is a travel and cartographic publisher and sells maps, town plans, guides and narrative books on travel, especially Asia, Africa and Europe. Please send sae with enquiries.

Ordnance Survey
Romsey Road
Maybush
Southampton SO9 4DH
Tel. 0703 75555

Is the official civilian mapping agency for the UK.

George Philip & Son Ltd
12 Long Acre
London WC2
Tel. (01) 836 1915
Trade Counter: 27a Floral Street
London WC2
Tel. (01) 836 1321
Works and Cartographic Department:
Victoria Road
London NW10
Tel. (01) 965 7431

Publish a wide range of topographical & thematic maps, globes, atlases and charts.

Royal Geographical Society
Publications Department
1 Kensington Gore
London SW7 2AR
Tel. (01) 589 5466

Sells maps originally published in the Geographical Journal *and maps published separately by the Society, and expedition pamphlets,* G.J. *reprints and other papers on geography, expeditions and related subjects. List available on request.*

Vacation Work Publications
9 Park End Street
Oxford
Oxon OX1 1HJ
Tel. 0865 41978

Publishes books on employment and budget travel abroad.

USA

Department of Defense Mapping Agency
Hydrographic Center
Washington
DC 20390

Publishes charts of oceans and coasts of all areas of the world; and pilot charts.

Department of Defense, Mapping Agency
Topographic Center
Washington
DC 20315

Supplies maps or photocopies of maps on request provided that the exact area is specified.

Fielding's Guides. see Wm. Morrow & Co., Inc

Arthur Frommer Inc
380 Madison Avenue
New York
NY 10017

Publish the Frommer Guides.

Harian Publications
203 Thomas Street
Greenlawn
New York
NY 11740

Specialise in publishing travel, cruise and retirement guide books.

Manuals
Four-wheel-drive shop manuals may be obtained as follows:

Toyota:
Allied Graphics Inc. PO Box 4339, Whittier, Ca 90607.

Dodge:
Chrysler Motors Corp. Dodge Division Service Dept., PO Box 857, Detroit, Mi 48231.

Ford:
Ford Service Publications. PO Box 7750, Detroit, Mi 48231.

Chevrolet:
Helm Inc. Chevrolet Manual Distribution Dept, PO Box 7706, Detroit, Mi 48207

International:
International Harvester Company.
401 N. Michigan Avenue, Chicago, Ill 60611.

Jeep:
Jeep Corp. Owner Relations, 14250 Plymouth Road, Detroit, Mi 48207.

David McKay Co. Inc.
750 Third Avenue
New York
NY 10017

Are the publishers in the USA of Fodor's Guides.

Rand McNally & Co.
PO Box 2600
Chicago
Ill 60680
Retail: 10 East 53rd Street
New York
NY.
Publish maps, atlases, guides and globes.

Wm. Morrow & Co., Inc
105 Madison Avenue
New York
NY 10016

Are the publishers of Fielding's Guides.

National Geographic Society
17th and M Streets
Washington
DC 20036

Publishes mainly topographical maps to accompany the National Geographic Magazine, *also sells wall, relief and archaeological maps, atlases and globes.*

National Ocean Survey
Distribution Division
C-44
Washington
DC 20235

Publishes and sells Operational Navigation Charts (I:1,000,000).

Passport Publications
20 N. Wacker Drive
Chicago
Ill 60606
Tel. (312) 332 3571

Publish Passport *newsletter and other travel booklets and guides.*

North Bergen Travelinfo Publishers
PO Box 9097
North Bergen
NJ 07047

Are specialist travel publishers.

Petersen Publishing Co.
6725 Sunset Blvd
Los Angeles
Ca 90028

Publish automotive and motorcycle maintenance books.

Travel Information Bureau
PO Box 105
Kings Park
NY 11754

Are specialist travel publishers.

Travel Publications, Inc.
1 Lincoln Road
Suite 214
Miami
Fl 33139

Are specialist travel publishers.

US Army Topographical Command
Corps of Engineers
Washington
DC 20315

Publishes and sells USATC road maps, political and topographical maps.

Europe

Freytag-Berndt u. Artaria KG
Kartographische Anstalt
Schottenfeldgasse 62
Vienna VII
Austria

Are large general map publishers.

Le Guide du Routard
5 rue de l'Arrivée
92190 Meudon
France

Compile the six Guides du Routard *and* Le Manuel du Routard, *guides for the budget traveller.*

Institut Géographique National
Direction Général
136 bis, rue de Grenelle
Paris VII
Retail: 107 rue la Boëtie
75008 Paris
Mail order sales: Bureau de Vente des Cartes par Correspondance (BVCC)
2 avenue Pasteur
94160 Saint-Mande
Tel. 808 2918/6336

Publish and sell maps of France and very many of the former French possessions.

Manuals
Sources of vehicle maintenance manuals in France:

Citroën
13 quai André Citroën
75447 Paris
Tel. 578 61 61

Publishes a free leaflet about Citroën vehicles and advises long-distance Citroën motorists.

Librarie de l'Automobile
83 rue de Rennes
75006 Paris

Specialises in books of all kinds about vehicles.

Renault
53 avenue des Champs-Elysées
75098 Paris
Tel. 225 18 57

Publishes Renault maintenance manuals.

Revue technique Automobile
22 rue de la Saussière
92 Boulogne-sur-Seine

Publishes a technical guide of the same name which describes vehicle maintenance, parts and spares.

Volkswagen
02600 Villers-Cotterets
Tel. (23) 53 19 03

Publishes a free brochure called Tropiques.

Michelin Guides
46 avenue de Broteuil
75341 Paris

Publish maps, and the famous Red and Green Guides.

Kiepert KG
Hardenbergstrasse 4-5
1000 Berlin 12
Germany, Federal Rep.

Is a large map publisher.

Ludmilla Tüting
D-5800 Hagen
Hassleyerstrasse 4a
Germany, Federal Rep.

Publishes and sells handbooks for globetrotters and overlanders.

Berlitz Guides
1 avenue des Jordils
Lausanne 6
Switzerland

Publish handy pocket guides (published in the UK by Cassell and in the USA by Macmillan).

Kümmerly und Frey
Hallerstrasse 6-8
Berne
Switzerland

Publishes charts and political, topographic, road and other maps.

Australia

Lonely Planet Publications
PO Box 88
South Yarra
Vic 3141
Tel. (03) 429 5268

Publish low-cost travel guides to Australasia, Africa and Asia.

Book and map retailers

UK

Compendium Bookshop
234 Camden High Street
London NW1
Tel. (01) 485 8944
267 1525

A useful source of travel books that are difficult to obtain elsewhere.

Cook, Hammond & Kell Ltd
22 Caxton Street
London SW1
Tel. (01) 222 2466

Are agents for Ordnance Survey maps and retailers of other maps.

Foyles Travel Department
First Floor
North Building
121 Charing Cross Road
London WC2

The Map House
54 Beauchamp Place
London SW3
Tel. (01) 589 4325

Sell a wide selection of maps and travel books.

The Map Shop
A.T. Atkinson & Partner
4 Court Street
Upton-on-Severn
Worcestershire
Tel. 06846 3146

Sell Ordnance Survey, Michelin, Geographia, Philip, A to Z and HMSO publications and guides. Send for catalogue.

Dick Phillips
Whitehall House
Nenthead
Alston
Cumbria CA9 3PS

Specialises in books and maps on Iceland, including Check-List of Principal Books on Iceland and Faroe, *a paperback compiled by Dick Phillips himself, which is a bibliography of over 240 books in English on Iceland, excluding saga translation and specialised scientific publications. He stocks maps at scales of between 1: 750,000 and 1:25,000 (general maps) and thematic or specialised maps held in stock or usually obtainable by special order.*

J.D. Potter Ltd
145 The Minories
London EC3
Tel. (01) 709 9076

Is one of the best sources of Admiralty Charts and other Hydrographic Department publications. List available.

RAC Touring Services
PO Box 92
Croydon
Surrey CR9 6HN

Sells a number of guides, handbooks, phrase books and maps for travellers in the UK and on the Continent; reduced prices to members.

Rallymaps of West Wellow
PO Box 11
Romsey
Hampshire SO5 8XX

Are Ordnance Survey postal specialists and sell reproduction maps, map gadgets, Michelin maps and guides, geological maps and books, etc. Send for catalogue with prices and laminated map sample.

Sherratt and Hughes
17 St. Ann's Square
Manchester 2

Are leading map retailers.

Snowden Smith Books
41 Godfrey Street
London SW3 3SX
Tel. (01) 352 6756

Sells books on travel and anthropology, especially first editions, and issues a catalogue.

Edward Stanford Ltd
12-14 Long Acre
London WC2
Tel. (01) 836 1321

Is the largest mapseller in London, carrying a wide range of maps, globes, charts and atlases, including Ordnance Survey and Directorate of Overseas Surveys maps. For anyone within reach of London plans to buy specific maps, Stanford's should be the first port of call.

Trail Finders (Services) Ltd
46-48 Earls Court Road
London W8 6EJ
Tel. (01) 937 9631

Stock quite a number of overland and budget guides.

Capt. O.M. Watts Ltd
45 Abermarle Street
London W1
Tel. (01) 493 4633

Sells a complete line of Admiralty Charts, tide tables and navigational instruments.

YHA Services
14 Southampton Street
London WC2E 7HY
Tel. (01) 836 8541

Sell books, maps and guides for campers, hostellers, adventure sportsmen and budget travellers.

USA

Adventure Travel Literature Service
401 N.E. 45th Street
Seattle
Wa 98105
Tel. (206) 634 3453

Is a mail order service selling guidebooks, handbooks and descriptive books on adventure, sport and travel; and topographical and Michelin maps.

Argus Books
PO Box 49659
12301 Wilshire Blvd
Los Angeles
Ca 90049

Are specialists in books on motoring, especially 4WD. Free catalogue.

Autobooks
2900 West Magnolia Blvd
Burbank
Ca 91503
Tel. (213) 849 1294

Are retailers of books and manuals 'for motor enthusiasts'. No connection with the British firm of the same name.

Backpacker Books
Bellows Falls
Vt 15101
and Main Street

Orwell
Vt 05760
Tel. (802) 948 2770

Sells trail guides, books on backpacking, mountaineering, mapping, cycling, caving, skiing, canoeing, survival and walking; and aim to stock every trail guide of any importance.

Bikecentennial
PO Box 8308
Missoula
Mt 59807

Sells books and guides for the cycle-tourist.

Books-on-File
Union City
Nj 07087

Will locate virtually any book, from its title alone, no matter how old or long out of print, fiction or non-fiction, all authors and subjects.

Dick Cepek
9201 California Avenue
South Gate
Ca 90280
Tel. (213) 569 1675

Sells books on off-road motoring.

Council on International Educational Exchange (C.I.E.E.)
777 United Nations Plaza
New York
NY 10017

Publish and sell a number of students and budget travel guides and sourcebooks.

Dawson's Book Shop
535 North Larchmont Blvd
Los Angeles
Ca 90004

Handles rare and unusual books at moderate prices, on a range of subjects.

Hammond Map Store, Inc
10 East 41st Street
New York
NY 10017

Sell world, road and travel maps, political maps and weather charts.

Hippocrene Books
171 Madison Avenue
New York
NY 10016

Are specialists in travel literature.

Humble Touring Service
15 W. 51st St
New York
NY 10019

Supplies free maps of almost every country in Latin America and also of Mexico and the Caribbean.

Information Exchange
T.E.J. 22 West Monroe Street
Chicago
Ill 60603

Sells travel and guide books.

Joyer Travel Report Bookshelf
8401 Connecticut Avenue
Washington
DC 20015

Sells by mail order a selection of travel books, guides and maps, especially of the Americas, the Pacific and Caribbean, and Europe.

E.S. Schechter & Co
181 Glen Avenue
Sea Cliff
NY 11579

Offer a complete range of service manuals for all the popular makes of 4WD vehicles and vans, as well as books and manuals on repair and maintenance, engine construction and servicing, parts, etc.

Trail-R-Club of America
Box 1376
Beverly Hills
Ca 90213

Sells books on recreational vehicles. Free catalogue.

Travel Library
PO Box 249
La Canada
Ca 91011

Sells travel books, guides and maps.

Travel Library
PO Box 2975
Shawnee Mission
Kansas 66201

Sells maps and travel books.

The Trip
509 Main
Glen Ellyn
Ill 60137

Sells Kümmerly & Frey maps and world and national atlases, by mail order.

Wide World Bookshop
401 N.E. 45th
Seattle
Wa 98105
Tel. (206) 634 3453
Proprietors: Royce S and Garth M Wilson

Sell a selection of unusual travel books, many of them imported, maps and globes, antique books, rare prints and 'other trivia'.

Wilderness Book Service
PO Box 979
Port Hueneme
Ca 93041
Tel. (805) 487 2322

Are suppliers of books and maps on wilderness and wilderness activities. Also offer personalised expedition research service. Free catalogue.

Canada

Oxbow Books
Box 244 Clarkson
Mississauga
Ont L5J 3Y1

Sell out of print books on travel, exploration, geography, mountaineering, caving.

Travel Tips Ltd
PO Box 1051
1040 Speers Road
Oakville
Ont L6J 5E9
Tel. (416) 844 7273

Publishes own booklets on Europe, distributes road maps, guidebooks and pamphlets covering all parts of the world. Catalogue.

Europe

Librairie du Voyage
Galerie Defré
Avenue Defré 82
1180 Brussels
Belgium
Tel. 374 29 65

Is a recently-opened travel bookshop.

L'Asiathèque
6 rue Christine
75006 Paris
France
Tel. 325 34 57

Offers books on the Far East and South East Asia.

L'Astrolabe
46 rue de Provence
75009 Paris
France
Tel. 285 42 95

Sells travel books, old and new, and maps.

La Grande Porte
4 rue Dialpozzo
06000 Nice
France

Specialises in books on long distance travel, with works also on seafaring and mountaineering.

Librairie de la Maison des Voyages
28 rue du Pont-Louis-Philippe
75004 Paris
France
Tel. 277 63 55 and 277 30 75

Sells travel books, guides.

Librairie Gilbert
First floor
5 pl Saint-Michel
75005 Paris
France

Has a wide choice of route maps and tourist guides.

Librairie Ulysse
35 rue Saint-Louis-en-l'Ile
75004 Paris
France
Tel. 325 17 35

Sells geographical, travel and guide books, and maps.

Le Tour du Monde
9 rue de la Pompe
75016 Paris
France
Tel. 267 52 13

Sells travel books and guides, old and new, in or out of print and on all aspects of foreign countries.

Geocenter
7 Stuttgart 80
Vaihingen
Postfach 80 08 30
Germany, Federal Rep.
Tel. (07 11) 73 50 31

8 Munich 22
Liebherrstrasse 5
Germany, Federal Rep.
Tel. (08 11) 22 62 45

Is one of the largest and best retail outlets for maps in Europe.

Travel Bookshop
Gisela Treichler
Seilergraben 11
8025 Zurich
Switzerland

Australia

Gregory's Map Centre
Ground Floor
Magazine House

142 Clarence Street
Sydney
NSW 2000
Tel. 29 3761

Sells road maps and guides, overseas maps, street directories, government maps.

Rex Map Centre
413 Pacific Highway
Artarmon
NSW
Tel. 428 3566

Is the best retail outlet for maps in New South Wales. There is another branch called:

All Maps
Arbitration Street
Circular Quay
Sydney
NSW

Book Clubs

UK

The Travel Book Club
119 Charing Cross Road
London WC2

Is one of the Foyles book clubs.

USA

Explorers Book Club
Riverside
NJ 08075

Offers 3 books for $1.00 each to members on joining and reduced prices on additional books. The Book Club News *is published and sent to members 15 times a year, describing selections.*

The Travel Book Club, Inc
Attention: Ms Shirley Jarris
Suite 910
23 East 26th Street
New York
NY 10010

Camping & hostelling associations

UK

International Youth Hostel Federation
Midland Bank Chambers
Howardsgate
Welwyn Garden City
Herts

USA

American Youth Hostels
National Headquarters
Delaplane
Va 22025
Tel. (703) 592 3271

Maintains 130 hostels in the United States, sponsors inexpensive educational and recreational outdoor travel opportunities and sports such as skiing, canoeing and cycling.

Canada

Canadian Camping Association
102 Eglinton Avenue E.
Suite 203
Toronto
Ontario M4P 1E1
Tel. (416) 488 7345

Canadian Youth Hostels Association
333 River Road
Vanier City
Ontario K1L 8B9
Tel. (416) 746 0060

Publishes the CYHA Handbook.

Australia

Australian Youth Hostels Association Incorporated
383 George Street
Sydney
NSW 2000

Is encouraged by government grants. It has 92 hostels, usually within a comfortable day's drive of each other. In capital cities, hostels sleep an average of 15-25 persons. The Association has 78,000 members, who are encouraged to use their own vehicles in travelling between hostels. A handbook is available to members.

Communications

USA

America Calling
Three Hamburg Turnpike
Pompton Lakes
New Jersey 07442

Are originators of a quick and economical way for travellers to communicate with those at home. Subscribers are supplied with two identical code books, one for the traveller and one for his contacts at home, and the traveller is assigned an identity code based on his initials and digits. The code books list various messages in code form, e.g. 'call home immediately, there is a serious illness in the family' = AA, and these together with the traveller's identity code can be inserted in the columns of the International Herald Tribune, a daily newspaper that circulates throughout Europe. Messages appear in the newspaper for two consecutive days. Since the alphabet yields 676 two-letter combinations, a wide range of prearranged messages is available.

Local telephone company

Can issue travellers with a credit card, free of charge, which permits the user to make calls on credit to the USA, and can supply useful booklets on overseas telephone rates and on a method for avoiding the high surcharges imposed by some foreign hotels, known as the 'Teleplan'. Off the beaten track countries that honour the credit card for calls to the US include: Australia, Brazil, Chile, Hong Kong, Israel, Japan, Korea, Mexico, New Zealand, Peru, Philippines, Singapore, Turkey, USSR, Venezuela.

Courses

UK

Belair School
5 Denmark Street
London WC2
Tel. (01) 836 1316

Offers evening classes in airline fare construction, 20 lessons from £70.

Jaguar Rover Triumph Ltd Public Relations Department
Solihull
Warwickshire

Publishers of an invaluable 'Guide to Land-Rover Expeditions'. The company runs a three-day course on maintenance.

The Royal Photographic Society
14 South Audley Street
London W1Y 5DP
Tel. (01) 493 3967

Offers a series of weekend 'teach-ins' for both amateurs and professionals, aiming to provide background knowledge as well as specialist information. Of special interest to travellers may be the courses on Photography and the Imagery of Nature and The Scope and Technique of Adventure Photography.

Currency information

UK

Bank of England
(Foreign Department)
Threadneedle Street
London EC2
Tel. (01) 601 4444

Customs

UK

HM Customs
Kings Beam House
Mark Lane
London EC3
Tel. (01) 283 8911

USA

US Customs Service
1301 Constitution Avenue, N.W.
Washington
DC 20229

Is the head office. There are also regional offices.

Europe

Bureau de Douane
11 rue Léon Jouhaud
75019 Paris
France
Tel. 202 61 10

Is the central French passport office. Offices in other regions of the country may be located through the:

Centre de Renseignements Douaniers
8 rue de la Tour-des-Dames
75009 Paris
Tel. 280 13 26

Equipment: appliances

UK

Brexton Ltd
Pensnett Trading Estate
Brierly Hill
Staffs DY6 7NW

Are suppliers of the Coleman No. 639 paraffin pressure lantern which is particularly good for detailed work in the evening on an expedition, for example.

Camping Gaz (GB) Ltd
126-130 St Leonards Road
Windsor
Berks
Tel: 55011/7

Is the best source of information on the availability of Camping Gaz throughout the world.

Camping Gaz International
PTC-Langdon Ltd
Dorking
Surrey

Are makers of gas containers, heaters, lanterns and blowtorches, and of the Globe Trotter portable stove, which is very small and light.

Dudley Hill Engineering Co Ltd
300 Goswell Road
London EC1V 7LU
Tel. (01) 837 2828

Manufacture Jet Gaz appliances for cooking, lighting, heating, including table stove, express percolator; also containers. Spare parts are available in over 30 countries from Europe to the Far East.

Electrolux Ltd
Luton
Bedfordshire

Make Camping Box refrigerators; portable, top-opening: can be operated by 220-240 volt mains electricity or by a 12 volt car battery.

Europleasure Gas Ltd
Curtis Road
Dorking
Surrey RH4 1XA

Manufacture EPI gas butane-fuelled leisure appliances which operate from almost any available butane fuel and which may be obtained from good camping, leisure and hardware stores. They include a picnic stove, 'backpacker' stove, lantern, heater, blowlamp and converter units.

Ever Ready Co Ltd
Ever Ready House
High Road
London N20
Tel. (01) 446 1313

Are makers of batteries.

Firemaster Ltd
Friendly Place
Lewisham Road
London SE13
Tel. (01) 692 6231

Supply vehicle fire extinguishers.

Joy & King Ltd
15 Alperton Lane
Perivale
Greenford
Middx
Tel. (01) 997 5653

Are makers of 'Jaykay' brand camping accessories including heaters, lights, tyre inflaters, ovens and cookware. Catalogue available from distributors or from address above, on receipt of 20p.

A.B. Optimus Ltd
Mill Road
Sharnbrook
Beds
Tel. 0234 781924

Are makers of the well known Optimus stoves, and of lanterns, heaters, ovens, grills and fire/security systems.

Yachts International Ltd
Marine Centre
Beckett's Wharf
Lower Teddington Road
Kingston-upon-Thames
Surrey KT1 4ER
Tel. (01) 943 0161

Fridges supplied and fitted for boats and vehicles.

USA

The Anite Co
PO Box 375
Pinole
Ca 94564

Makes the Nightwatch, an effective method of detecting anyone or anything around an area. It has 4,000 feet of tripwire which the user loops around the campsite in two circuits. The device both beeps and flashes and gives an indication first that there is an intruder and second of his direction.

Zebco Consumer Div., Brunswick Corp
PO Box 270
Tulsa
Okl 74101

Supply the Traveler 4000 flameless catalytic heater using disposable cartridges. This can be adapted for bottle-fed gas, if you write for a spare 'pad' unit.

Europe

Camping Gaz International
ADG 15
75 rue de Châteaubriand
Paris 8
France

Butane gas camping equipment.

Rexor
56 avenue Victor-Hugo
75016 Paris
France
Tel. 704 96 80

Sell an anti-glare, heat-reflecting windscreen shield which sticks onto the inside of the windscreen.

Equipment: backpacking

UK

Brown, Best
102 Old Kent Road
London SE1
Tel. (01) 703 8111

Are rucksack specialists.

USA

Alpine Designs

PO Box 3561
Boulder
Co 80303

Manufacture rucksacks and packs, which are sold through dealers.

Don Gleason's Campers Supply, Inc

Pearl Street
Northampton
Mass 01060
Tel. (413) 584 4895

Sell camping and backpacking equipment. Comprehensive catalogue.

Living Things Inc

980 Chapel Street
New Haven
Conn 06510
Tel. (203) 624 3017

Is a wilderness club, has an international programme, mounts research projects and expeditions and has an environmental college. It also sells mountaineering and backpacking equipment retail and by mail order. Profits go to environmental education and research.

Wilderness Experience

20120 Plummer
Chatsworth
Ca 91311

Design and make packs.

The Yak Works

PO Box 70256-K
Seattle
Wa 98107

Are makers of the Yakpak, a frameless pack which can be folded but not bent or broken.

Pacific/Ascente

PO Box 2028
Fresno
Ca 93718

Make waterproof backpacking gear.

Europe

Camp Trails International Ltd

Waterford Industrial Estate
Waterford
Eire
Tel. Waterford 32469

Supply all-weather pack, bags and frames of tough nylon with waterproof urethane coating, 'designed by experts' and field tested. Free catalogue and guide to European Long-Distance Foot-trails.

Equipment: camping, outdoor, expedition and travel

UK

Berghaus

34 Dean Street
Newcastle-upon-Tyne NE1 1PG
Tel. 0632 23561

Sell boots, equipment and clothing for the outdoor enthusiast.

Blacks of Greenock

22-24 Gray's Inn Road
London WC1
Tel. (01) 405 4426

Port Glasgow
Renfrewshire PA14 5XN
Scotland

Ruxley Corner
Sidcup
Kent DA14 5AQ

132 St Vincent Street
Glasgow
Scotland
Tel. (041) 221 4007

Have tents and camping equipment for hire from centres in Aberdeen, Birmingham, Bristol, Edinburgh, Fareham, Hull, Leicester, Leeds, Liverpool, London (3 centres), Nottingham, Stoke on Trent, Swansea and Wolverhampton. They supply lightweight, patrol, frame, mountain and touring tents, camp furniture, kitchen kits, stoves and lamps, clothing and accessories, convertible, specialist and summer-weight sleeping bags.

Blacks Outdoor Centre

202-4 Deansgate
Manchester
Tel. (061) 833 0340

21-2 Grand Arcade
Leeds
Tel. (0532) 458634

93-117 Princes Street
Dundee
Scotland
Tel. (0382) 43766

34 Edgbaston Shopping Centre
Hagley Road
Birmingham
Tel. (021) 454 8771

Camping & Outdoor Centres (Scout Shops, The Scout Association Ltd)

Churchill Industrial Estate
Lancing
Sussex BN15 8UG
Tel. 09063 64231/5352

Offer a free catalogue. There are eleven Camping & Outdoor Centres in Britain, stocking all kinds of camping, outdoor and survival equipment, as well as Scouting uniforms, souvenirs, trophies, badges, patches and flags; and books on Scouting, survival, first aid, outdoor activities and camping.

Field & Trek (Equipment) Ltd

23-25 Kings Road
Brentwood
Essex CM14 4ER
Tel. 0277 221259/219418/210913
0277 224647 (Contract Dept)

Field & Trek Ltd offers WEXAS members discounts of 15% off the list price plus VAT on most leading makes of expedition equipment including tents, rucksacks, boots, waterproof clothing, sleeping bags and mountaineering gear. Large mail order department. Expedition bulk orders entitled to additional discount. *Ask for special quotation. Send for free comprehensive price list.*

The Hampton Works (Stampings) Ltd

Twyning Road
Stirchley
Birmingham B30 2XZ

Also at London, Manchester, Bristol and Glasgow. Are manufacturers of the Hampton range of camping accessories.

Howards (Bolton) Ltd

235 Blackburn Road
Bolton
Lancs BL1 8HB
Tel. Bolton 25729/34379

Have a permanent camping exhibition featuring many leading makes of tent, camping equipment and furniture. There is a mail order service. The firm arranges part exchange of tents and trailer tents and operates a tent and trailer hire service.

Karrimor International Ltd

Avenue Parade
Accrington
Lancs

Manufacture mountaineering and outdoor pursuits equipment, including sacs, frames, tent, overboots, gaiters, Karrimats. They have supplied equipment to the International Everest Expedition of 1971, the 1972 (British) Everest South West Face Expedition, the Japanese Ladies Everest Expedition of 1975, the Japanese South West Face Everest Expedition and the Army Mountaineering Association Everest Expedition 1976.

Pindi Contracts

373-5 Uxbridge Road
Acton
London W3
Tel. (01) 992 6641

The contract division of Pindisports, suppliers of gear to bona fide bulk buyers such as Education Authorities and expeditions˜ but not *to individuals.*

Pindisports

14 Holborn
London EC1
Tel. (01) 242 3278

1098 Whitgift Centre
Croydon
Surrey
Tel. (01) 688 2667

373 Uxbridge Road
London W3
Tel. (01) 992 6641/2
(this branch handles all mail orders)

27 Martineau Square
Birmingham
Tel. 021 236 9383

5 Welsh Back
Bristol
Avon
Tel. 0272 211577

Mitcham Corner
34 Chesterton Road
Cambridge
Tel. 0223 56207

(in Hornes)
The Butts Centre
1 Oxford Road
Reading
Berks
Tel. 0734 584157

Provided some equipment for the 1972 Everest South West Face Expedition and the 1970 Annapurna South Face Expedition and are suppliers to the John Ridgway Adventure School at Ardmore. They supply equipment for hill-walking, rockclimbing, big-wall climbing, alpinism, expeditions, shelter and survival; also guide books and magazines.

Raclet Ltd and The Camping Centre

24 Lonsdale Road
Kilburn
London NW6
Tel. (01) 328 2166

Supply tents and camping accessories, including the Campcase: an Antler suitcase with essentials for camping packed into it – including a Raclet tent with awning, airbeds and sleeping bags, cooker, billies, plates, pump, wash bowl, water carrier, mugs, cutlery, salt and pepper, clothes hanger clip, stools, night light – altogether weighing 43½lb.

YHA Services Ltd

14 Southampton Street
London WC2E 7HY
Tel. (01) 836 8541

35 Cannon Street
Birmingham B2 5EE
Tel. (021) 643 5180

166 Deansgate
Manchester M3 5FE
Tel. (061) 834 7119

Supply lightweight tents and equipment for overland travel, mountaineering and caving, also weather-proof clothing and a good selection of boots.

USA

Arcata Transit Authority
10th Street
Arcata
Ca 95521
Tel. (707) 822 2204

Make and sell Blue Puma lightweight and innovative outdoor equipment; shelters, sleeping bags and garments. Free catalogue.

L.L. Bean, Inc
Freeport
Maine 04033
Tel. (207) 865 3111

Operate a mail order service and have a salesroom which is open 24 hours a day, 365 days a year. Firm sells outdoor garments and accessories, boots and other footwear, canoes, compasses, axes, knives, binoculars, thermometers, stoves, tents, sleeping bags, packs and frames, skis and snowshoes, camp ware, travel bags, lamps, blankets.

Early Winters Ltd
110-MF Prefontaine Pl. S.
Seattle
Wa 98104
Tel. (206) 622 5203

Are makers of backpacking and outdoor gear, including the Thousand Mile Socks, which are guaranteed not to wear out before 1 year or 1000 miles of walking (whichever comes last). Free colour catalogue.

Eddie Bauer
417 E. Pine Street
Seattle
Wa 98124
Tel. (406) 622 2766

Free catalogue.

Famous Department Store
530 S. Main
Los Angeles
Ca 90013

Free catalogue.

Herter's, Inc
R.F.D. 2 Interstate 90
Mitchell
S Dakota 57301
Tel. (605) 996 6621

Highway 13 South
Waseca
Minn 56093
Tel. (507) 835 4011

2425 Marvin Road
Olympia
Wa 98506
Tel. (206) 491 5033

Highway 151 South
Beaver Dam
Wisc 53916
Tel. (414) 885 9330

Highway 55 South
Glenwood
Minn 56334
Tel. (612) 634 4577

Sell and make all kinds of sporting and outdoor equipment, for fishermen, hunters and trappers, guides, gunsmiths, tacklemakers, forest rangers, explorers, expeditions and backpackers. Free catalogue.

Kelty's
1801 Victory Blvd
Glendale
Ca 91201

Free catalogue.

Recreational Equipment, Inc
PO Box 22090
Seattle
Wa 98122
Tel. (800) 562 4894 or
(toll free, except from Hawaii or Alaska)
(800) 426 4840

Was established in 1938 and operates as a co-operative, with members receiving annual dividends on purchases. Shops also in Portland, Ore., Berkeley and Los Angeles, Cal. Outdoor and expedition equipment.

Canada

ABC Recreational Equipment
555 Richards Street
Vancouver
BC V6B 2Z5

Sell hiking and climbing gear. Free catalogue.

Margesson's
17 Adelaide Street East
1 block north of King East of Yonge
Toronto
Ontario M5C 1H4
Tel. 366 2741

Established 1934, are lightweight camping specialists, selling canoes, tents, packs, sleeping bags, down-filled ski jackets, hiking boots and so on. Staff comprises active canoers and hikers experienced in camping problems.

Europe

Au Vieux Campeur

48-50 rue des Ecoles
1-3 rue de Latran
75005 Paris
France
Tel. 329 12 32

Are major French outdoor equipment specialists.

Daerr's Expeditionsservice

Kirchheimerstrasse 2
D-8016 Heimstetten
Munich
Germany, Federal Rep.
Tel. (089) 903 1519

Claims to sell everything for travel in the Third World, including expedition equipment and books. Free catalogue.

Gerhard Lauche and Wolfgang Maas

Alte Allee 28
8 Munich 60 (Pasing)
Germany, Federal Rep.
Tel. (089) 880705

Are expedition outfitters.

Equipment: clothing and footwear

UK

Damart Thermawear Ltd

Bowling Green Mills
Bingley
Yorks BD16 3BR

Are makers of thermal underwear.

Europleasure Ltd

PO Box 80
Duke Street
Liverpool

Make Man Alive clothing: shirts, anoraks, some sweaters and accessories; and distribute an impressive range of tents, including Maréchal.

Ultimate Equipment Ltd

Willowburn Trading Estate
Alnwick
Northumbria NE66 2PF
Tel. 0665 3621/2/3

Sell shell clothing, climbers' helmets, lightweight tents, waterproof jackets, and other climbing and climbing equipment.

USA

W.L. Gore & Associates

PO Box 1220
Elkton
Md 21921

Make Gore-Tex Laminates.

Neptune Mountaineering

2020 30th Street
Boulder
Co 80301
Tel. (303) 442 3551

Repair boots in 'one week of shop time'. Write for information.

Ultimate Equipment Ltd

c/o Royal Robbins
Mountain Paraphernalia
906 Durant Street
Modesto
Ca

Are the US counterparts of the UK firm (q.v.)

Equipment: cold weather

UK

Polywarm Products Ltd

Quay Road
Rutherglen
Glasgow
Scotland G73 1RS

Electronic House
Churchfield Road
Weybridge
Surrey KT13 8DB
Tel. Weybridge 53377

Manufacture a selection of lightweight and compact specialist sleeping bags suitable for the mountaineer and hiker, including conventional and mummy-shaped bags, made of washable man-made fibres and stitched by a special process that prevents the filling from moving. The sleeping bags are all guaranteed for 12 months. The firm also manufactures the Insul Vest thermal waistcoat.

Vista (Thermal Products) Ltd

Is a manufacturing company founded in 1977 and heavily backed by the parent company,
Mountain Equipment Ltd
of George Street
Glossop
Derbyshire

Heading its range of products is the:
Eskimo Bag, made from synthetic pile, which is body-shaped and is claimed to maintain comfortable body heat in air temperatures down to -5°C with minimal clothing and -15°C in normal clothing; the bag is said to eliminate cold spots, maximise insulation and give excellent wet/cold performance.

Additional products from Vista are the: Shearwater, a lightweight pile bag that may be used as a liner for the Eskimo or any down or synthetic bag or as a summer bag on its own; Nylon outer cover for the Eskimo or Shearwater; Nomad, a bivouac shell made of PTFE/nylon laminate, that is waterproof but allows moisture vapour to pass through from the inside, thus reducing condensation.

USA

Alpine Products
1309 Windward Circle
West Sacramento
Ca 95691

Manufacture PolarGuard bags and jackets. Catalogue.

Canada

Thomas Black & Son
225 Strathcona Avenue
Ottawa
Ont K1S 1X7

Make excellent down gear. Catalogue.

Australasia

Antarctic Products Co Ltd
PO Box 223
Nelson
New Zealand

Make sleeping bags, down clothing, down blankets and woollen jumpers. Money-back guarantee if not satisfied. Free booklet.

Equipment: instruments

UK

Chilecrest Ltd
155-7 Pitshanger Lane
Ealing
London W5

Manufacture the Milograph: a thermometer-type route mileage calculator, suitable for use with maps ranging in scale from 0.1 to 40 miles to the inch. It is supplied with a plastic slipcase, instruction leaflet and mile/kilometre conversion table.

Navigemus (IoM) Ltd
Great Downs
Tollesbury
Maldon
Essex CM9 8RD
Tel. 062 183 280

Are makers of the Cruiserfix, a new device which will fix the user's position within a few miles anywhere in the world, whether used on land, at sea or in the air. It is light, compact and inexpensive; and easy to use, provided that the user has some means of determining True North and South and Greenwich Mean Time. Used between 0900 hrs and 1500 hrs, the Cruiserfix gives accurate readings for longitude and latitude. The device costs £5.18 including VAT.

Survey & General Instrument Co Ltd
Fircroft Way
Edenbridge
Kent
Tel. 0732 71 2434

Sell the Thommen 'Everest' Altimeter, an accurate pocket instrument giving readings up to 15,000 or 18,000 feet.

B.J. Ward Ltd
130 Westminster Bridge Road
London SE1

Sell 'Silva' compasses of the protractor type, suitable for expedition work, with a discount to educational bodies.

USA

Silva, Inc
A division of Johnson Diversified, Inc
Highway 39 North
La Porte
Indiana 46350

Make Silva compasses.

Equipment: optical

UK

Heron Optical Co
23-25 Kings Road
Brentwood
Essex CM14 4ER
Tel. 0277 221259/219418

Stock all leading makes of binoculars, telescopes and rifle scopes. Associate company of Field & Trek (Equipment) Ltd (q.v.)

Equipment: photographic

UK

Agfa Gevaert Ltd
20 Piccadilly
London W1
Tel. (01) 734 4854

Bulk film supplies.

Kodak Ltd

Kodak Ltd
PO Box 66
Kodak House
Station Road
Hemel Hempstead

Hertfordshire HP1 1JU
Tel. 0442 61122
Enquiries to: PR Division, ext 221

Supply a limited number of films on trade terms to expeditions having the support of the Royal Geographical Society or a similar authority, provided that purchases are made in bulk, that one order is placed at a minimum value of £50 and that delivery at an address in the UK (excluding docks and airports) is accepted.

Photomarket
147 Earls Court Road
London SW5
Tel. (01) 370 5979

High Street Kensington Tube Station
London W8
Tel (01) 937 3173

Buy and sell used equipment, also offer a same-day colour photo processing service.

USA

TAMRAC
12351 Hartsook Street
PO Box 4690
No. Hollywood
Ca 91607

Make the Camera Pak 35, a foam padded, weatherproof camera case to hold any 35mm camera (with neck strap) with lens up to 3½in, and still provide instant access.

Europe

Photo Poste (Odéon-Photo)
110 blvd St-Germain
75006 Paris
France
Tel. 329 40 50

Is a photographic developing and printing service that will process photos or films sent to the firm from anywhere in the world and send the results on anywhere; will undertake a variety of processes; will give advice on film handling and photographic technique; will retain negatives safely until your journey is over; and charges reasonable prices for these services.

Equipment: secondhand

UK

Nimbus
121 Uxbridge Road
Shepherds Bush
London W12
Tel. (01) 743 0487

Buy and sell used camping equipment, including Camping Gaz equipment.

Europe

CIDJ
101 quai Branly
75015 Paris
France
Tel. 566 40 44

Has a notice-board on which people may place advertisements free. This is a good source of secondhand camping gear, packs, sleeping bags, tents and so on and a good place to sell as well.

Equipment: survival

UK

Major P.H. Clayton
Defence Equipment Manager
Schermuly Ltd
Newdigate
Dorking
Surrey

May be contacted by expeditions for information and leaflets about distress rockets and flares.

Primarc Ltd
Busgrove Lane
Stoke Row
Henley-on-Thames
Oxon

Supply the Emer Lite, an emergency flashing light for detection, a compact, hand-held, battery-operated unit that is waterproof and weighs only 6 ozs.

Ski International
32 White Cross Road
York
N. Yorks
Tel. 0904 28566

Sell the Pieps avalanche victim detector, a transceiver operating on the agreed frequency of the UIAA.

USA

Buckhead Outfitters, Inc
3130 Maple Drive
Atlanta
Ga 30305

Make the 'päk-kit', a compact, lightweight emergency kit containing 24 essential items including means of providing shelter, fire, basic first aid, life sustaining food, and notes on survival. Free catalogue listing other products.

Colt Industries
Hartford
Conn 06102

Make the Colt Aerial Flare Kit.

Crossroads of Sport
5 East 47th Street
New York
NY 10017

Market the ResQ Locating Balloon *which is a kit for inflating a 16-in diameter heavy duty rubber orange ball with 200 sq ins of surface area, that is visible for miles in good weather. It has a reflective metal strip that shows up distinctly on an aircraft's radar screen. It will stay aloft for 16-24 hours at the end of its 600 feet of monofilament. The pack weighs only 12 ozs and comes in a handy small case.*

Lee Nading
PO Box 805
Bloomington
In 47401

Supplies 'survival cards', measuring 3" x 5" and made from plastic, which are crammed with survival information, including edible plant classification, emergency shelter construction, first aid, Morse Code, climbing techniques and knot tying.

Naval Undersea Research and Development Center
San Diego
Ca 92132

Have developed a large bag of thin, strong, very light material with inflatable collars at the top which will completely conceal a 'castaway' at sea and serve as a shark screen. The bag can also be used as a sleeping bag, pup tent, lean-to, stretcher or solar still.

Survival Research Laboratories
17 Maryland Road
Colorado Springs
Co 80906

Make snake bite kits.

Therma-Pak Co
Somerville
Mass

Manufacture a heating device in the form of a pouch filled with a chemical formula, activated by pouring two tablespoons of water through a resealable flap. Within five minutes the pouch begins to give off moist heat that lasts up to eight hours. The device can be used outdoors in place of hand warmers, indoors instead of electric heating pads or hot water bottles or whenever warmth is needed in an emergency to prevent a casualty from going into shock. The heat pad is safe for any part of the body and can be reused as often as nine times.

Canada

Motronic Electronics Canada Ltd
Mike Wiegele
Box 1824
Banff
Alberta TOL OCO
Tel. (403) 762 4171

Manufacture an avalanche electronic detector device which emits and receives a 'peep' signal, which may help a rescuer to locate a buried person.

Equipment: tents

UK

Advance Tents
Aireworth Road
Keighley
West Yorkshire BD21 4DN

Make the Roamer one-pole, lightweight tent with a tunnelled guy-rope system which ensures that all pressure from tightening and loading is transferred directly to the structure and not exerted on the canvas; adjustments for tension and loading may be made from inside by adjustment of just one rope. The tent is both well ventilated and protected from the elements and there are optional extras such as a table which fits onto the central pole. Floor space 102 square feet (9.48 m^2).

Bradimports
Wigan
Lancs WN5 8BR

Are distributors of André Jamet tents. Colour catalogues are available from 'Freepost' at the above address.

Edward R. Buck & Sons Ltd
Bukta House
Brinksway
Stockport
Cheshire
Tel. (061) 480 9721

Make the Hermit, *a tent and rucksack combined in one.*

H. Burden Ltd
Pytchley Lodge Road
Industrial Estate
Kettering
Northants

Make OBI frame tents, which are sold by 28 dealers throughout Britain.

Campari Ltd
Priestley Way
London NW2
Tel. (01) 450 6681

Makers of Campari tents and airbeds.

Carawagon Coachbuilders Ltd
Thames Street
Sunbury on Thames
Middlesex TW16 5QR
Tel. 76 85205

Have designed and market themselves a tent, called the Caranex, which fits onto the back of almost any car.

Continental Wholesale (Fareham)
137 Gosport Road
Fareham
Hants
Tel. 0329 235303

Are manufacturers and distributors of ridge tents and specialised outdoor camping equipment. Makers of 'Conquest' range of ridge tents, suppliers to H.M. Armed Forces for the Adventurous Training and Resource and Initiative Training Programmes.

Conway Trailers
Appley Bridge
nr Wigan
Lancs
Tel. Appley Bridge 2424

Are manufacturers of tents and trailers.

R. Garside (Liverpool) Ltd
35-43 Pitt Street
Liverpool L1 5BY

Are tent manufacturers.

Graham Tent Service
90 Station Road
London N11
Tel. (01) 368 1515

Carry out repairs, reproofing and extensions or modernisations to tents.

Overland Camping Services
256-263 Arches
Grosvenor Terrace
London SE5
Tel. (01) 701 6607

Are specialists in supplying tents and ancillary equipment to expeditions, safari and camping tour operators and in manufacturing and renovating tents and camping equipment.

J Rands & Jeckell Ltd
Foundation Street
Ipswich
Suffolk
Tel. 0473 52266/7

Are manufacturers of Aquatite tents and equipment.

Rentatent Ltd
Twitch Hill
Horbury
Wakefield
Yorks
Tel. 0924 5131/2

36 London Road
Wembley
Middx
Tel. (01) 903 3473

40 Ducie Street
Piccadilly
Manchester 1
Tel. (061) 273 3626

Have tents and camping equipment for hire and sale.

Roy Ryder Ltd
St Oswalds Road
Gloucester
Tel. 0452 33116

Can supply the new Stormryder Tent System, tested on Everest in 1976 by the Army Mountaineering Association team.

Robert Saunders (Chigwell) Ltd
Five Oaks Lane
Chigwell
Essex

Make Top Tents, 9 models of tent for mountain, moorland and meadow.

Freddie Slazenger Ltd
4 Spring Street
London W2 3RA
Tel. (01) 262 9058

Are manufacturers of Gale Force tents.

Tarpaulin & Tent Manufacturing Co
Head Office
101-3 Brixton Hill
London SW2
Tel. (01) 674 0121/3
(Catalogue available from Head Office)

137 Clapham High Street
London SW4
Tel. (01) 720 5451

19 Wimbledon Bridge
London SW19
Tel. (01) 946 9600
(accessories only)

Suppliers to commercial and private expeditions of all kinds of tent from the marquee to the 2-man, of camp beds, air beds, stoves, mosquito nets, sand ladders, jerry cans (water and petrol), wire ropes and slings, towing chains, all types of ropes and tarpaulins. They also repair tents.

Vango (Scotland) Ltd
47 Colvend Street
Glasgow G40 4DH
Tel. 041 556 7621

Have stockists in other parts of the UK and are manufacturers of Vango Force Ten Tents (supplied to the 1975 Everest South West Face Expedition) and of the recently developed Force Ten Micro-weights. Also suppliers of rucksacks, the Vango Stove/Lamp and other climbing and camping equipment.

Equipment: tents

USA

Bishop's Ultimate Outdoor Equipment
2938 Chain Bridge Road
Oakton
Va 22124

Manufacture The Ultimate tents and Bishop Packlite tents, all of them superbly designed and made.

Clear Creek
14361 Catalina Street
San Leandro
Ca 94577

Make the Ogive, a tent with a Dynaflex suspension system, that stays up without pegs or ropes.

Tent Works Ltd
Camden
Me

Opened in 1975 to manufacture the unusual tent designs of Bill Moss, which are light, easily assembled, leak-proof and sturdy, deriving much of their strength from curved surfaces which replace the traditional poles and staked-in ropes. Among his innovations are: the Eave Two-Man and Three-Man tent, rounded on top, assembled in two minutes, with a waterproof 'eave' that overhangs a porous cloth roof through which the tent 'breathes'; the Trillium, a bulbous tent consisting of three lobes each with its own entrance, joined in a central 'common room', freestanding, weighing 13 lbs and sleeping 6; the Poly-Dome, an instant plastic cottage, 110 square feet, which emerges from a package 9 ft by 2 ft and is erected by unfolding, pushing out curved panels and zipping up the roof.

Europe

Campri Sport Handels GmbH
Dieselstrasse 22
8046 Garching
Munich
Germany, Federal Rep.
Tel. (089) 329 1071

Are makers of Campari tents.

Campri (L.D.) BV
Flevolaan 12-16
Weesp
Netherlands
Tel. 02940 15171

Are makers of Campari tents.

Campri Fritid AB
Flojelbergsgatan 12
431/37 Molndal
Sweden
Tel. (031) 272150

Are makers of Campari tents.

Equipment: water purifiers

UK

S.H. Johnson & Co Ltd
Carpenters Road
Stratford
London E15

Make Millbank Bags for filtering water.

Kirby Pharmaceuticals
Mildenhall
Bury St Edmunds
Suffolk

Make Puritabs (in two sizes) which are effervescent water purifying tablets.

Portasyl Ltd
Cannon Lane
Tonbridge
Kent

Make a 1¼ gallon pump filter for purifying water before drinking, whether it has been fouled by effluent or contains mica or rock flour in continuous suspension (e.g. glacier water).

Safari (Water Treatments) Ltd
28 The Spain
Petersfield
Hampshire
Tel. 0730 4452

Manufacture a wide range of lightweight water purifiers which guarantee to remove all harmful and offensive elements from water from any non-salt water source. One model is the portable, battery operated Safari Filtron.

USA

Davco, Inc
Suite 601 Wolverine Tower
3001 South State Street
Ann Arbor
Mi 48104
Tel. (313) 662 1717

Make the new lightweight Walbron Water Purifier, which is said to destroy instantly, on contact, E. coli and most other types of bacteria and viruses. It has been approved by the Environmental Protection Agency and selected for use on all future NASA space flights.

Europe

Beam Trading NV
Langstraat 52 Gel A
2200 Borgerhout
Belgium

Agents for portable water purifying filters manufactured by Pollution Control Products, Inc, USA.

Laboratoires Helmer
12 rue d'Ingersheim
BP 329
68006 Colmar
France

Manufacture the Filtron Sachs, which is an electronic water purifier, that works in one minute and is the size of two cigarette packets. It is sold at chemists and specialist shops.

Fichtel & Sachs AG
Ernst-Sachs-Strasse 62
D-8720 Schweinfurt
Germany, Federal Rep.

Are manufacturers of the Sachs Filtron CAMP 3000 purification plant which electronically purifies and sterilises surface water from rivers, lakes and pools, with the exception of that which is chemically polluted; and the Sachs Filtron, a small portable battery operated water purification unit. British distributor is: Sachs Motor-Service, Scotlands Industrial Estate, London Road, Coalville, Leics LE6 2JJ. Tel. 0530 36106/39511.

Elektro-Industri A/S
PO Box 34
1701 Sarpsborg
Norway

Make a drinking water cooler with two filters for water treatment – a mechanical filter for removing particles and a coal filter for reducing unwelcome tastes and/or smells. The cooler keeps water at an average temperature of 10°C. A carafe filler is an optional extra.

Katadyn Products Ltd
Industriestrasse 27
CH 8309
Wallisellen
Switzerland
Tel. (01) 830 36 77

Sharp Associates
PO Box 1618
Deming
New Mexico 88030
(USA representatives)

39 avenue des Piliers
94210 La Varenne
France
Tel. 283 67 87

Chez Alpino
129 chaussée de Leuze
9600 Renaix
Belgium

Deutsche Katadyn GmbH
Schaufeleinstrasse 20
8 Munich 21
Germany, Federal Rep.
Tel. (089) 57 56 73

Make portable water filters which clear and decontaminate water.

Expedition consultancies

UK

Christopher Portway
22 Tower Road
Brighton
Sussex BW2 2GF
Tel. 0273 682783

This travel consultancy aims to assist the individual or group with the planning of a mainly overland holiday or journey, outside the scope of a package tour, to suit individual whims and requirements. It specialises in giving advice on surface transport and particularly railway journeys of the world, arranging regimental association battlefield tours, helping with individual expeditions and offering, from personal experience, hints and suggestions that will make a dream travel project cheaper and more attainable than you think. Send SAE for more details.

Quest 4 Ltd
Ashton Wold
Peterborough
Cambridgeshire PE8 5LD
Tel. (083) 22 2614/3676

Expedition consultants who organise and advise expeditions to remote places, including in the past numerous expeditions to North Africa with such varied objectives as collecting for museums, ballooning, observing a total eclipse of the sun, filming and anthropological and wildlife research. Quest 4 undertake planning and preparation in all aspects of an expedition: vehicles, route notes, equipment, etc. They can supply Land-Rovers modified to suit the needs of any expedition and can service 4WD vehicles.

Trail Finders (Services) Ltd
46-48 Earls Court Road
London W8 6EJ
Tel. (01) 937 9631

Provides information for the private overlander and a booking service for those making reservations on commercial overland services. Covers Asian overland routes in particular, but also Africa and the Americas. Sells maps, books, insurance.

Young Explorers' Trust
c/o Royal Geographical Society
1 Kensington Gore
London SW7 2AR
Tel. (01) 589 9724

Can provide to members information on particular expedition areas: Svalbard, South America, Labrador-Ungava, Baffin Island, Algeria, Morocco, Iceland.

USA

Expedition Research, Inc
PO Box 467
Annapolis
Md 21404

159 Conduit Street
Annapolis
Md 21401
Tel. (301) 268 3322
Director: James P. Stout

Provides, to registered members only, a consultancy service on scientific and exploratory expeditions all over the world. It acts as an information pool on current expeditions and on the current situation regarding such things as permits and visas, as a marketing service and as a means of putting expedition organisers and hopeful team members in touch with each other. The firm publishes a quarterly newsletter and an annual journal, circulated to members only.

Expedition food

UK

L & I Equipment
14 Bromborough Road
Bebington
Merseyside
Tel. (051) 645 7577

Supply to the trade only Turblokken emergency rations for survival kits.

Andrew Lusk Ltd
4 Thames Road
Barking
Essex
Tel. (01) 594 3822

Specialise in the acquisition and packing of expedition rations. The service is said to be excellent, though not cheap.

J.L. Priestley & Co Ltd
93 High Street
Heckington
Sleaford
Lincs

Sell bulk packs of various foods, by mail order only.

Ranch House Meals
Direct Foods Ltd
20 Lavant Street
Petersfield
Hampshire GU32 3LT
Tel. 0730 4911

Supply lightweight packets of meals which take under 10 minutes to prepare.

Raven Leisure Products Ltd
23 Oxendon Road
Arthingworth
Market Harborough
Leics LE16 8LA
Tel. 085 886 349

Manufacture complete lightweight meals for campers and backpackers. The meals only need 5 minutes' cooking.

Springlow Sales Ltd
Marsland Industrial Estate
Werneth
Oldham
Lancs
Tel. (061) 624 7904

Supply one man, one pan meals and rucksack weekend packs of food and will give advice to people planning menus. Their specialised lightweight foods include quick-cooking meals for children, vegetables, dried egg, meats and fruits, ideal for trekkers and overlanders. Products are available at camping, general outdoor shops, caravan and caving shops. Send for details of nearest stockist or mail order catalogue.

Stow Away Foods
2 Blithfield Street
London W8

Produce canned dehydrated foods which may suit expeditions. The food cans are packed in cases, can be stored almost indefinitely and are designed to occupy the minimum space and weight. The foods are unaffected by temperature.

Swel Foods Ltd
Wharf Road
Crowle
nr Scunthorpe
South Humberside DN17 4JR
Tel. 0724 710421

Sell, through camping shops, a range of dehydrated foods which keep well in dry store for several months. Also mail order.

USA and Canada

Manufacturers and retailers of lightweight food in the USA and Canada:

Chuck Wagon Foods
Nucro Drive
Woburn
Mass 01801
USA

Sells retail and by mail order. Catalogue.

Dri-Lite Foods
11333 Atlantic
Lynwood
Ca 90262
USA

Make tasty foods. Brochure available.

Freeze-Dry Foods Ltd
579 Speers Road
Oarville
Ont
Canada

201 Savings Bank Bldg
Ithaca
NY 14850
USA

Make and deliver foods throughout North America. Catalogue.

Indiana Camp Supply, Inc
PO Box 344 W
Pittsboro
Ind 46167
USA
Tel. (317) 547 1134

Sell freeze-dried and dehydrated foods; also quality backpacking equipment.

Kamp Pack Foods, Select Food Products Ltd
120 Sunrise Avenue
Toronto 16
Ont
Canada

Catalogue.

National Packaged Trail Foods — Seidel
632 East 185th Street
Cleveland
Oh 44119
USA

Sell by mail order and offer a free sample selection. Foods are suitable for preparation at lower altitudes.

Oregon Freeze Dry Foods, Inc
PO Box 1048
770 West 29th Avenue
Albany
Ore 97321
USA
Tel. (503) 926 6001

Make the widely recommended Mountain House brand and also the Tea Kettle brand of freeze dried foods, both in the form of regular foods packed in foil pouches and in the form of compressed foods, called 'Space Savors'. The firm makes meats, vegetables, main courses, entire meals for 4 people, emergency packs of 6 cans (designed to serve one person for 8 days and weighing only 11.4 lbs), survival packs, fruits, beverages, soups and snacks, even including ice cream.

Permapak
40 East 2430 South
Salt Lake City
Utah 84155
USA

Sell retail or by mail order. Catalogue.

Rich-Moor Corporation
PO Box 2728
Van Nuys
Ca 91404
USA

Manufacture high quality lightweight low moisture food and sell through outfitters or by mail order. Catalogue and price list. Branches in all states of the USA.

Stow-A-Way
166 Cushing Highway
Cohasset
Mass 02025
USA

Catalogue.

Trip-Lite
S. Gumpert Co of Canada
31 Brock Avenue
Toronto 3
Ont
Canada

S. Gumpert & Co
812 Jersey Avenue
Jersey City
NJ 07302
USA

Manufacture easily prepared food designed especially for children. They sell direct, in case lots only.

200

Expedition reports

UK

Royal Geographical Society
1 Kensington Gore
London SW7 2AR
Tel. (01) 589 5466

Scientific Exploration Society
Home Farm
Mildenhall
nr Marlborough
Wilts
Tel. 098 02 2075

Young Explorers' Trust

c/o Royal Geographical Society
address as above
Tel. (01) 589 9724

Exploration, travel and adventure associations

UK

British Mountaineering Council

Crawford House
Precinct Centre
Manchester University
Booth Street East
Manchester M13 9RZ
Tel. (061) 273 5839

Is the national representative body of British mountaineers. It offers members a comprehensive insurance scheme, information on access problems which it is constantly tackling and the use of an equipment pool for parties supported by the Council. It also gives financial aid, through the Sports Council, to expeditions that meet the criteria of the Mount Everest Foundation. The BMC's services include a monthly magazine, Climber and Rambler, *and an information service of books, brochures, leaflets and films dealing with specialised items of equipment and safety. Advice is given on all aspects of mountaineering, including climbing possibilities, planning, finance, etc.*

Globetrotters Club

BCM Roving
London WC1V 6XX

A small informal association of travellers from all over the world, linked by an interest in low-cost travel and the desire to study the cultures of other lands at first hand. Members share their personal experiences and detailed knowledge of local conditions. The club is small and personal and concentrates on attracting as new members only those 'non-tourists' with a genuine empathy for the people in other lands. The club's newsletter Globe *has been coming out regularly for thirty years. It is full of travel information, articles and diary notes. Members may advertise in it for information, travelling companions, etc. Globetrotters groups meet regularly for films and talks in London and southern California.*

Iceland Unit

86 Dovedale Crescent
Buxton
Derbyshire SK17 9BQ

Established by the Young Explorers' Trust, the Unit exists to improve communication between past and present expeditions to Iceland. With a central working group and a body of consultants specialising in different fields of study, the Unit is building up a record of information useful to expeditions. (Inquiries should be accompanied by SAE).

Royal Geographical Society

1 Kensington Gore
London SW7 2AR
Tel. (01) 589 5466

Is a focal point for geographers and explorers. It directly organises and finances its own scientific expeditions and gives financial support, the loan of instruments, approval and advice to some 30-40 expeditions each year. The Society honours outstanding geographers and explorers with a series of annual medals and awards.
The RGS maintains the largest private map collection in Europe, with over half a million sheets. It has a drawing office with a staff of expert cartographers and a library with over 100,000 books and periodicals on geography, travel and exploration. There is also an archive of historical records and expedition reports.
There are regular lectures, children's lectures, discussions, symposia and academic meetings in the Society's 760-seat lecture hall. Most of the leading names in exploration, mountaineering and geography have addressed the Society. The RGS publishes the Geographical Journal *three times a year and is closely associated with the* Geographical Magazine. *There is now a Young Members' Committee.*
Anyone with a geographical interest can apply for a Fellowship of the RGS. An applicant must be proposed and seconded by existing Fellows.

Royal Scottish Geographical Society

10 Randolph Crescent
off Queensferry Street
Edinburgh EH3 7TU
Tel. (031) 225 3330

Also has branches in Aberdeen, Dundee and Dunfermline. It offers the following classes of Membership: Ordinary, Life, Student Associate, Junior, School Corporate, Country Areas and Overseas. The Society houses a library, a map collection numbering over 50,000 sheets and a collection of over 200 periodicals. It arranges tours, excursions and lectures, and sells map reproductions and publications.

Scientific Exploration Society

Home Farm
Mildenhall
nr Marlborough
Wiltshire

Was formed in 1969 by a group of explorers, many of whom had been together on expeditions, with the aim of making their association more permanent so that personnel and useful equipment would not be dispersed but instead kept together for future undertakings. The Society exists to organise expeditions and to help others – universities, schools, services and individuals – to organise their own. It maintains close links with commerce, industry, educational establishments, the services and other kindred scientific and exploration organisations.

The Society has 250 members, many of them expert explorers. All are eligible to take part in expeditions. Other classes of membership are available for non-participant supporters. Fully sponsored expeditions generally appoint their Leader, Secretary and Treasurer and many of their personnel from among the Society's membership. Other expeditions can be given the approval and support of the SES by the Council and may then borrow equipment, receive advice and use the SES name in their publicity. Though the Society 'approves and supports' expeditions, it rarely gives cash to any project.
Members have to be proposed and seconded by existing members, and then elected by the Council.

WEXAS

45 Brompton Road
Knightsbridge
London SW3 1DE
Tel. (01) 589 0500/3315

WEXAS (for World Expeditionary Association) was founded in London in 1970 to provide an information and travel service for expeditions. Membership has since become open to anyone, and WEXAS currently has many thousands of members, spread over 88 countries, for the majority of whom WEXAS is predominantly a travel club. In this context, WEXAS' appeal derives from its worldwide programme of low cost flights, and from its Discoverers programme of two, three and four week adventure holidays (at special low rates). The Awards Committee makes financial grants to promising UK-based expeditions each year. Members of WEXAS receive Expedition *magazine, quarterly, and are eligible to receive other WEXAS publications.*

Young Explorers' Trust

(The Association of British Youth Exploration Societies)
c/o Royal Geographical Society
1 Kensington Gore
London SW7 2AR
Tel. (01) 589 9724

Exists to promote youth exploration and to provide a forum within which societies and individuals can exchange information and act together for their mutual benefit. It does not organise its own expeditions, place individuals on expeditions or make travel bookings. The Trust is a registered charity. Membership is open to groups or societies wishing to take part in the Trust's activities and to contribute to the Trust's aims. Present members include all the major national and regional bodies active in the field of youth expeditions, as well as school and university groups.
Information is available on a wide range of topics and on a variety of foreign locations.
A Bulletin, twice yearly, and Yetmag *carry news, papers and speeches, preliminary expedition reports and other information. An expedition index enables members with particular problems to be rapidly put in touch with others with relevant advice to offer.*

USA

Airguide Travel Club

92 21st Avenue
San Mateo
Ca 94403

Was founded in 1978 to represent private pilots in the USA. Members receive books, maps, discounts and assistance in planning trips.

American Adventurers' Association

Suite 301
444 N.E. Ravenna Blvd
Seattle
Washington 98115
Tel. (206) 527 1621

Founded in 1977 to become a clearing house for information on adventure travel opportunities available to the general public and evaluate these opportunities, to establish an adventure trip booking service, to publish a monthly magazine Adventure Travel *and to publish the* Worldwide Adventure TravelGuide, *a 600-page directory listing the numbers, types, geographical areas and brief descriptions of the thousands of adventure trips offered throughout the world. Three different types of membership are offered: for organisations, individuals and students. The AAA's publications are not related to* Adventure Travel USA *published by Adventure Guides, Inc.*

American Geographical Society

Broadway at 156th Street
New York
NY 10032
Tel. (212) 234 8100

Is devoted to research in geography and the dissemination of geographical knowledge. It produces maps, engages in environmental research, sponsors lectures, presents medals and awards and maintains a large geographical library numbering a quarter of a million books, nearly 350,000 maps and over 5,000 atlases, as well as many photographs covering all aspects of geography and related sciences. The Society also publishes a number of periodicals.

Circumnavigators Club

271 Madison Avenue
New York
NY 10016

Was founded in 1902 as an association of world travellers by most means and for most purposes. There are about 800 members. The Club publishes a bi-monthly magazine called The Log.

The Explorers' Club

46 East 70th Street
New York
NY 10021

Was founded in 1904 as an institution of serious

purpose designed for and dedicated to the search for new knowledge of the earth and outer space. It serves as a focal point and catalyst in the identification and stimulation of institutional exploration, independent investigators and students. The Club has over 1700 members who continue to contribute actively in the constructive rôle of the explorer. The classes of membership are: Member, Fellow, Student, Corporate Associate Fellow, Corporate Associate Sponsor, Corporate Associate Founder; each class being divided into Resident (living within 50 miles of the Headquarters) and Non-Resident. The Club has financed over 140 expeditions and awarded its flag to over 300 expeditions.

The James L. Ford Memorial Library *contains over 25,000 items, including maps, charts, archives and photographs, and is probably the largest collection in North America wholly devoted to exploration. The Club publishes the quarterly* Explorers' Journal, *a students'* Newsletter *and other literature, including* Told at the Explorers' Club, *and historical/educational films on exploration. Lectures, seminars and special events and an Annual Dinner are held. There is an annual presentation of honours and awards. The Club offers members opportunities to participate in unusual travel programmes.*

Formally constituted groups exist in Washington, DC, Philadelphia, Los Angeles, Seattle and Juneau. The New England Chapter *of the Explorers' Club shares an address with the* Expedition Training Institute *(PO Box 171, Prudential Center Station, Boston, Mass 02199. Tel. (617) 894 7674), which is the youth activity arm of the Club. Its Outlook programme encourages the scientific endeavours of high school and college people, assisting their preparation for, and rèplacement on, expeditions (generally within the Americas).*

The International Backpackers Association

PO Box 85
Lincoln Center
Me 04458
Tel (207) 794 6062

Is a non-profit making body promoting the protection and development of non-motorised trails.

National Geographic Society

17th and M Streets N.W.
Washington
DC 20036

The Society's aim is to pursue and promulgate geographical knowledge and to promote research and exploration. The Society publishes the National Geographic Magazine, *and occasionally sponsors significant expeditions. Articles may be sold to the magazine but the standards required (particularly for photographs) are very high.*

Canada

Travel Wise Club

444 Robson Street
Vancouver
British Columbia

A club for low-budget travellers, with aims similar to those of the Globetrotters Club.

Europe

Club 'Antipodes'

66 avenue des Paradisiers
1160 Brussels
Belgium
Tel. (02) 673 1291

Is a non-profitmaking club which brings together lovers of travel and serves as a clearing house for information about travel. It holds meetings and conferences and publishes a monthly newsletter.

Topas Globetrotterklub

Safarihuset
Lounsvej 29
Louns
DK-9640 Farsø
Denmark
Tel. (08) 63 83 85

Was established in 1973 and organises overland expeditions in Africa, Asia and northern Scandinavia, has 700 members, publishes a monthly newsletter and an annual expedition publication, Safariposten, *in Danish.*

La Centrale du Voyageur

28 rue du Pont-Louis-Philippe
75004 Paris
France
Tel. 277 63 55 and 277 30 75

Is a meeting place for travellers which holds lecture evenings, maintains a map library and acts as a ticket agent. There are branches at Bordeaux, Brest, Grasse, Marseilles, Mont du Château and Rennes.

Guilde Européenne du Raid

15 quai de Conti
75006 Paris
France
Tel. 033 52 53

Promotes adventure in remote parts of the world. It organises meetings, conferences, lectures, film evenings and permanent exhibitions, and a small number of organised tours/treks. Since 1971 it has made annual grants under the heading Dotation Nationale de l'Aventure to responsible but adventurous expeditions having a definite goal, usually those mounted by individuals or small groups. Other grants and donations in money or kind are available through the Guilde, *as are also the advice of experts, access to documents, discounts and benefits from associated bodies, and help with the safe-keeping and sale of expedition reports and photos.*

Deutsche Zentrale für Globetrotter e.v.

Hassleyerstrasse 42
D-5800 Hagen
Germany, Federal Rep.

Is a globetrotters' club with 700 members, mostly in Germany but many foreign members as well. Globetrotters exchange information via the club and a newsletter, Trotter. *The club operates from a private address, which also publishes various travel handbooks. People can join the club if they can prove that they have travelled on their own in countries outside Europe for at least three months. The moderately high entrance fee and annual subscription are meant to discourage all but the keenest.*

Australasia

Overland Club

Third Floor
Queen's Arcade
Queen Street
Auckland 1
New Zealand
Secretary: Miss Marie R. Dornauf
Flat 5
62 Church Street
Northcote
Auckland 9
New Zealand
Tel. 484 484

Holds meetings twice a month.

South America

South American Explorers' Club

Casilla 3714
Lima 1
Peru

Exists to promote travel and sporting aspects of exploration; and to record, coordinate and publicise academic research on a wide variety of natural and social sciences. Membership is open to all. It publishes a magazine The South American Explorer. *Its reference library reading rooms (and rooftop garden) are at Avenida Portugal 146, Breña distr., Lima, Peru. Tel. 31 4480.*

Ferry Services

Africa

Limadet-Ferry

13 rue Prince Moulay Abdellah
Tangier
Morocco
Tel. 33625

Compañia Trasmediterranea SA

Méjico 39
Tangier
Morocco
Tel. 34101

Run a car ferry service between Tangier and Algeciras or Malaga. Both companies also have offices in Malaga, Algeciras and Paris.

Film hire

UK

Explorer Films

3 Fairway
Merrow
Guildford
Surrey
Tel. 0483 35468

Expedition films, for hire or outright purchase, some of them made by the Brathay Exploration Group.

Stewart Film Distributors Ltd

107 Long Acre
London WC2
Tel. (01) 240 5148

Supplies expedition films, including 'Discovering Iceland' (16mm, colour).

Young Explorers' Trust

c/o Maurice Dybeck
The Village College
Sawtry
Huntingdon
Cambs

Maintain a directory of films of young people's expeditions, their sources and their availability for hire to interested YET members. The list is being constantly augmented and updated and information is therefore being sought by its co-ordinator, Maurice Dybeck, from whom a questionnaire may be obtained.

Freight

UK

Aerotrans Services Ltd

48-52 Brighton Road
Salfords
Surrey
Tel. 02934 72432

Provide a special freighting service for the advance movement of baggage, expedition equipment, etc, with savings of up to 50% on normal airline and shipping rates. Members can also take advantage of consolidated air charter rates for bulk freight loads in excess of 100kg, to an increasing number of destinations. The company undertakes collection, packing and insurance and provides a free export advisory service on customs procedures, particularly important for travellers who intend to bring exported items back into the country.

Freighter advisory services

USA

Freighter Travel Club
PO Box 12693
Salem
Ore 97309

Publishes a monthly newsletter on freighter travel, ideas and tips.

Pearl's Freighter Tips
175 Great Neck Road
Suite 306
Great Neck
NY 11021
Tel. (516) 487 8351 and (212) 895 7646

This organisation help plan and book worldwide freighter trips. No charge for their services.

TravLtips Freighter Travel Association
40-21 Bell Blvd
Bayside
NY 11361
Tel. (212) 428 4646

Gives personalised help to members with planning, reservations and advice, handles bookings on freighters, and publishes a 32-page illustrated booklet full of members' reports, bi-monthly.

Hitchhiking: organisations which put prospective drivers and passengers in touch with each other

UK

Lift Exchange Centre
14 Broadway
London SW1
Tel. (01) 834 9225

Europe or beyond.

USA

People's Trans-Share
PO Box 40303
Portland
Ore 97207
Tel. (800) 547 0933

Is a national non-profit making service for passengers and motorists or private aeroplane pilots. Walk-in registration centres in many US cities.

The Trail Voice
c/o The International Backpackers Association
PO Box 85-M
Lincoln Center
Me 04458

Is a quarterly newsletter which features a section 'Hiking Partners Wanted' in which members may advertise free.

Europe

Taxi-Stop
6 place des Brabançons
13400 Louvain-la-Neuve
Belgium
Tel. 41 81 99

86 rue des Templiers
1301 Bierges (Wawre)
Belgium
Tel. 653 88 65 during working hours, 653 19 42 answering service.

Allostop (France)
Is the collective name for the associations Allauto, Provoya and Stop-Voyages. The offices are at:

65 passage Brady
75010 Paris
Tel. 246 00 66

Au CIJA
5 rue Duffour-Dubergier
33000 Bordeaux
Tel. 48 55 50

3 rue Mérentié
13005 Marseilles
Tel. 42 68 80

Chez Sound-Music
18 rue d'Espagne
64000 Bayonne
Tel. 25 25 83

A L'Office du Tourisme
Palais Rihour
59800 Lille
Tel. 52 96 69

28 rue des 4 Eglises
54000 Nancy
Tel. 36 79 08

Feu Vert pour l'Aventure
75 rue Delambre
Paris 14
France

Téléstop
7 place des Hauts Murats
31000 Toulouse
France
Tel. 52 21 37

Mitfahrdienst
Arnulfstrasse (behind main railway station)
Munich
Germany, Federal Rep.
and other offices in West Germany.

Australasia

Travel Mate Hostels
496 Newcastle Street
West Perth
WA 6005
Australia
Tel. 328 6685

For travellers from Perth for the eastern and northern states. Registration fee. Also provides hostel and shared private accommodation for overseas and interstate visitors.

Inflatable boat suppliers and outfitters

UK

Avon Inflatables Ltd
Dafen
Llanelli
Dyfed SA14 8NA
Wales
Tel. 055 42 59171

Pioneered the use of inflatables in both Britain and the USA. They make dinghies, sports boats, life rafts, river runners and accessories such as life jackets.

E.P. Barrus Ltd
12-16 Brunel Road
London W3 7UY
Tel. (01) 743 3425

Are sole concessionaires for the UK and Eire of the Zodiac inflatable boat, which is light, manoeuvrable, rugged, economical and with a high payload.

International Diving and Watersports
1 New Road
Shoreham-by-Sea
Sussex
Tel. (07917) 5892

Manufacture the Bombard 'Commando', an inflatable with a hard wood internal keel, made in four models and lengths from 11'10" to 19'8".

Liferaft Servicing Co
14 Chapel Road
Tiptree
Essex CO5 0RA
Tel. (0621) 815549

Service and repair inflatables; sell and hire liferafts; and are authorised agents for Zodiac, Avon, etc. Vans collect and deliver in most parts of the country.

USA

Avon Rubber Co Ltd
Seagull Marine Sales
(Western US distributor)
1851 McGaw Avenue
Irvine
Ca 92705

Inland Marine Co
(Midwest and Central Eastern US distributor)
79 East Jackson Street
Wilkes-Barre
Pa 18701

Sell Avon heavy-duty nylon rafts, 11 models ranging in length from 9 to 17 feet.

Zodiac of North America, Inc
11 Lee Street
Annapolis
Maryland 21401
Tel. (301) 268 2009

Manufacturers of Zodiac Inflatable Boats.

Insurance

UK

Aaron Assurance Consultants
61 Whitefield Road
New Duston
Northampton
Tel. 0604 54878

Life assurance for expedition members.

Alexander Howden Ltd
22 Billiter Street
London EC3M 2SA
Tel. (01) 488 0808

Provide efficient carnet *indemnity and double indemnity.*

Antony Gibbs, Sage Ltd
International Insurance Brokers and Consultants
Standard House
Bonhill Street
London EC2A 4RZ
Tel. (01) 588 4111

Can arrange cover for adventure trips or any other hazardous pursuit. Send full details of planned trip for further information.

Automobile Association
Fanum House
Leicester Square
London W1
Tel. (01) 954 7373

Has a reasonably priced scheme to cover overland travel abroad.

Baggot, Evans & Co Ltd

99 Church Road
Upper Norwood
London SE19 2PR
Tel. (01) 771 9691

Can arrange insurance on motor vehicles of most types in the UK, throughout the whole of Europe including the USSR, in the Middle East as far as the border between Iran and Afghanistan and over the whole of North Africa down to 20°N; also partial cover on vehicles from the border of Afghanistan as far as Singapore. At Lloyds they can obtain cover for sea transits of vehicles to countries other than those mentioned and for goods and equipment to all parts of the world. They also offer personal accident, sickness and baggage insurance.

Campbell Irvine Ltd

48 Earls Court Road
London W8 6EJ
Tel. (01) 937 9903/6981

Specialise in unusual insurance and can offer travellers insurance against medical expenses, repatriation, personal accident, cancellation and curtailment and personal liability, also baggage and money cover subject to certain restrictions. Vehicle insurance can be arranged and usually takes the form of Third Party insurance (for countries where British insurers have adequate representation) or accidental damage, fire and theft insurance (worldwide, including sea transit risks). Carnet *indemnity insurance, which travellers need to obtain the* carnet de passages, *can also be arranged.*

Davison & Co

1 Devonshire Square
London EC2
Tel. (01) 247 8366

Offer carnet *indemnity insurance for travellers in Asia and elsewhere.*

W.E. Found & Co Ltd

Boundary House
7 Jewry Street
London EC3 2HH
Tel. (01) 709 9051/9508

Can arrange insurance cover for carnets *to travel overland by vehicle.*

Harold Yates, Burgess & Co Ltd

Victory House
Leicester Square
London WC2H 7NB
Tel. (01) 734 4177

74 King Street
Manchester M2 7AX
Tel. (061) 832 7531

Are insurance brokers who offer cover for all aspects of insurance to expeditions with the approval of the Young Explorers' Trust and some other suitable expeditions, providing proper declarations are made. Specialists in high risk business.

Jardine d'Ambrumenil International Ltd

56 Artillery Lane
London E1 7LS
Tel. (01) 377 9266

Will 'consider sympathetically all enquiries from expedition organisers'.

Slugocki, Norman & Co Ltd

48 Earls Court Road
London W8 6EJ
Tel (01) 937 6981

Insurance brokers who can arrange vehicle insurance for long-distance travel abroad, including insurance on carnets.

Travellers' Insurance Association Ltd

82 Pall Mall
London SW1Y 5HQ
Tel. (01) 930 5276

Insurance for overland expeditions.

Willis, Faber & Dumas Ltd

10 Trinity Square
London EC3
Tel. (01) 488 8111

Is a brokerage which has extensive experience of insuring mountaineering expeditions to the highest ranges of the world. It arranges cover for the Scout Association.

USA

Assist-Card

745 Fifth Avenue
New York
NY 10022
Tel. (800) 221 4564

Is an organisation designed to help with travel crises such as loss of passport, illness, theft, legal trouble. Card holders may telephone toll free numbers in 28 European, 4 Middle Eastern, 5 North African and most South American countries, where multilingual staff are on call 24 hours a day. Assist-Card then mobilises assistance from a wide range of contracted services, including use of a private jet. The cost of the card varies with the length of validity – 10, 16, 30, 45, 60 or 90 days.

Kemper Insurance

Kemper Insurance

Long Grave
Ill 60049

Offer a 12 month travel accident policy which gives the same cover and at the same premium as the insurance offered at airport terminals for only 21 days' cover. The policy, which must be ordered a week in advance, covers approved charter flights.

Europe

Touring Assistance

44 rue de la Loi
1040 Brussels
Belgium
Tel. 513 92 40

Offers travellers' insurance covering Europe, Turkey and North Africa, either for motorists or for travellers by train, 'plane, coach, boat or other public transport.

Centre de Documentation et d'Information de l'Assurance

2 rue de la Chaussée-d'Antin
75009 Paris
France
Tel. 824 96 12

15 place des Terreaux
69 Lyon ler
France

Will give advice to travellers on insurance problems.

International Assistance

92-98 blvd Victor Hugo
92115 Clichy
France
Tel. 739 33 11

Offers assistance to travellers: repatriation, payment of fines or medical fees, the service of doctors or chauffeurs, care of domestic pets, care of children or animals left behind, accident assistance and various other services. Travellers wishing to subscribe to the service should write for the Demande d'abonnement *form.*

Pinon Assureur

43 rue La Fayette
75009 Paris
France
Tel. 878 02 98

Is one of the rare insurance companies that will insure cameras and photographic equipment. Premium amounts to about 3% of the value of the items insured and the firm will insure for a minimum premium of 200 Francs.

Language

USA

Contemporary Marketing, Inc

790 Maple Lane
Bensenville
Ill 60106

Sell the Lexicon LK-3000 Language Translator, a hand-held language conversion computer, available with cartridges for translating from English into Spanish, French, Italian, German or Portuguese, or vice versa. There is also a 'Person to Person' cartridge, featuring basic social conversation in 6 languages. 8 more language cartridges are in preparation.

J S & A Group, Inc

One J S & A Plaza
Northbrook
Ill 60062
Tel. (312) 564 7000, or toll free
(800) 323 6400

Market the Craig Translator, made by the Craig Corporation, a hand-held computer that stores up to 7,000 words and translates them from one language to another. It features indexing and learning systems. Languages available are French, German, Italian, Spanish and Japanese. Other language, and more advanced, cartridges are in preparation. J S & A also market a miniature alarm clock, the J S & A Mini Travel Alarm.

Libraries

UK

The Royal Geographical Society

1 Kensington Gore
London SW7 2AR
Tel. (01) 589 5466

Encouraged by the success of the first in 1976, now holds annually in mid-November, jointly with the Young Explorers' Trust, a one-day seminar at its headquarters on 'Planning a Small Expedition'. *Speakers cover personnel, logistics, finance and budgeting, public relations, equipment, supplies, transport; and panels deal with specialised functions such as surveying, photography and collecting and with different types of terrain such as jungle, desert and river. Available to participants at the seminar is* a list of research sources in the UK., *prepared by the Director of the RGS, which includes most of the major general and specialised libraries, institutions, museums, map libraries and government bodies to which expedition planners may need to refer.*

Low cost travel

UK

WEXAS

International Office
45 Brompton Road
Knightsbridge
London SW3 1DE
Tel. (01) 589 0500/3315

WEXAS, Europe's largest travel club (members in 88 countries), offers reliable low cost air travel to 118 worldwide destinations. Most flights are on the scheduled services of large international airlines. Savings through WEXAS may be as high as 45 per cent.

Europe

Nord Norge Bussen

Kongensgate 66
Narvik
Norway

A Norwegian-based bus service operating a service along the Arctic Highway.

Australasia

Trans-Australia Airlines and Ansett Airlines

Both offer a selection of discount fares on routes around Australia.

Aussiepass

Allows unlimited travel over the Pioneer Express coach network in Australia for 35 days or 2 months. Details from Thomas Cook.

Eaglepass

Allows unlimited travel on the Greyhound system in Australia for 30 days.

Austrailpass

Permits unlimited first class travel on 27,000 miles of railways throughout Australia. Different rates are available for 14 and 21 days, 1, 2 and 3 months. Obtainable from any office of Thomas Cook and from many travel agents.

New Zealand Rail Pass

Gives 14 days of unlimited travel on all railways and bus routes and on the ferry between Wellington and Picton, between June 1 and November 30. Extensions may be purchased at a daily rate, up to a maximum of 28 days.

South America

Amerailpass

Allows unlimited travel over 30,000 miles of railways during a limited time period – one, two or three months – in and between Argentina, Bolivia, Brazil, Chile, Paraguay and Uruguay.
Available from:

Asociacion Latino-Americana de Ferrocarriles

ALAF
Florida 783
Buenos Aires
Argentina

Amerbuspass

Allows first-class bus travel throughout South America for 180 days. Available from:

TISA

Bernado de Irigoyen 1370
Buenos Aires
Argentina

Asia

Indrail Pass

Is a rail pass available to visitors and Indians who live outside their country, valid for 7, 21, 30, 60 or 90 days, and permitting unlimited travel on Indian railways.

Medical care (see also Insurance)

UK

Europ Assistance Ltd

269/273 High Street
Croydon
Surrey

Provides medical advisers, casualty surgery specialists, air and road ambulances, all on call in Western and Eastern Europe, Turkey, Morocco, Tunisia, the Mediterranean islands, the Canary Islands and Madeira. A medical expenses insurance is included in the package offered. The firm maintains a 24-hour 7-day-per-week 'phone service (holidays included) with a multilingual staff.

The Flying Doctors' Society of Africa

27 Dover Street
London W1X 3PA
Tel. (01) 629 7137

Exists to provide financial backing for the East African Flying Doctor Services (part of the African and Medical Research Foundation). Members are guaranteed free air transport if injured or taken seriously ill while on safari in East Africa. Temporary membership is available for tourists or visitors to East Africa, and confers the same benefits. There is also an annual lottery based on membership card numbers, for which the prize is a round trip by air to Africa or from Africa to Europe. The Society depends upon subscriptions and donations.

The Ross Institute of Tropical Hygiene

Keppel Street
London WC1E 7HT

Performs tests on returning travellers in cases where disease is suspected.

St John Ambulance Headquarters

1 Grosvenor Crescent
London SW1X 7EF
Tel. (01) 235 5231 during office hours,
(01) 235 5238 at other times.

Provides doctors and nurses and lay members with specialised aeromedical training to escort sick and injured persons by air to or from any part of the world. It owns special ambulances equipped to travel to and from the Continent with patients when air travel is not possible.

Trans-Care International Ltd

Group House
Woodlands Avenue
London W3
Tel. (01) 992 5077/8/9

Is the only British private organisation specialising in the repatriation of the sick and injured by air, land and sea from any country in the world. It also sends out medical teams and nursing escorts and its London Control Centre is continuously manned. For an annual fee members can avail themselves of this service and contact the company direct in the event of need. Certain insurance companies incorporate the services of Trans-Care in their policies for holidaymakers, etc.

The Young Explorers' Trust

c/o Royal Geographical Society
1 Kensington Gore
London SW7 2AR
Tel. (01) 589 9724

YET has a card index of medically qualified people, some with, some without, expedition experience, who may be available to offer their services to expeditions. Those wishing to be placed on the register may write to the Secretary with details of their experience and preferences. Expedition organisers wishing to consult the list may do so through membership of the YET and by application to the Secretary. The list is not for general distribution.

USA and Canada

International Association for Medical Assistance to Travellers (IAMAT)

Suite 5620
350 Fifth Avenue
New York
NY 10001
USA
and
1268 St Clair Avenue W.
Toronto
Ontario
and

Intermedic

777 Third Avenue
New York
NY 10017
USA
Tel. (212) 486 8974

Both issue a directory of their own offices in major cities abroad and will recommend English-speaking doctors to member travellers. The doctors agree to attend patients according to a fixed scale of fees. Both organisations also provide each member with a chart on which to enter his personal medical history. Intermedic *levies a small membership fee; IAMAT does not, but welcome donations.*

Both issue a directory of their own offices in major cities abroad and will recommend English-speaking doctors to member travellers. The doctors agree to attend patients according to a fixed scale of fees. Both organisations also provide each member with a chart on which to enter his personal medical history. Intermedic *levies a small membership fee; IAMAT does not, but welcomes donations.*

Public Information Service Center for Disease Control

Atlanta
Ga 30301
USA

Gives free health information to international travellers.

Europe

Caisse Primaire Centrale de la Sécurité Sociale

Division des relations internationales
84 rue Charles Michel
93200 St-Denis
France
Tel. 820 61 05

Is the office to which French travellers should apply with any queries or claims about reimbursement of medical expenses incurred abroad.

The three most reliable French companies which offer assistance to international travellers in times of difficulty, as opposed to offering reimbursement of expenses afterwards, are:

Elvia

15 rue de Berri
75008 Paris
Tel. 359 55 09

Europ-Assistance

23-25 rue Chaptal
75009 Paris
Tel. 285 85 85

Which provides a daily 24-hour service and a network of representatives in Europe, North Africa and the USA, enabling the company to supply immediately a doctor, mechanic, medically-equipped aircraft or other assistance as needed.

France Secours-International
36 rue Tronchet
75009 Paris
Tel. 260 39 39

Elvia *and* France Secours-International *do not maintain reps. abroad, but will still make every effort to assist the traveller in need.*

Medical supplies

UK

Simonsen & Weel Ltd
Hatherley House
Hatherley Road
Sidcup
Kent DA14 4BR
Tel. (01) 300 1128

Make the portable/Tauranga Thomas Rescue Splint (for fractures of the femur).

Smith & Nephew Ltd
Welwyn Garden City
Herts AL7 1HF
Tel. 96 25151

Make a plaster of Paris 'Emergency Splint Pack', and

Vickers Ltd Medical Engineering
Priestly Road
Basingstoke
Hants RG24 9NP
Tel. (0256 29141

Make inflatable splints;

All are makers and suppliers of splints.

Slaney's
40A Church Road
Richmond-upon-Thames
Surrey TW10 6LN
Tel. (01) 940 8634

Suppliers of first-aid essentials, especially for expeditions, including snake sera, tropical and marine supplies, foil blankets.

Chemists in London which specialise in making up expedition supplies are:

John Bell and Croyden
50 Wigmore Street
London W1
Tel. (01) 935 5555

Stanleys
243 Regent Street
London W1
Tel. (01) 437 7926

USA

The Medic Alert Foundation
Turlock
Ca 95380

Provides a necklace or bracelet on which is recorded the most important items of the wearer's medical condition and history, e.g. blood group, a heart condition, an allergy to antibiotics.

Stat-Pack
252B Concord Road
Aston
Pa

Are makers of the Stat-Pak, which is a waterproof card the size of a credit card on which is stored the traveller's medical history and other data, including his allergies, illnesses, current conditions, electro-cardiogram and blood analysis chart, his immunisations and a list of emergency contacts.

Wyeth
PO Box 8299
Philadelphia
Pa 19101

Manufacture a wide range of antivenins against many kinds of snake bite. The serum is sold in a freeze-dried condition, making it ideally suited for expeditions (no need for refrigeration), and in small quantities.

Picture libraries

UK

Alan Hutchinson Library
2 Logan Mews
London W8 6QP
Tel. (01) 373 4830

Collect first-class transparencies and expedition photos, specialising in Third World subjects; sell single shots and selections of photos, both in the UK and abroad.

Bruce Coleman Colour Picture Library
16a/17a Windsor Street
Uxbridge
Middlesex UB8 1AB

Established in 1959, represents a small number of well-known photographers including Chris Bonington, specialises in wildlife, geography, anthropology, archaeology, science. Literature available on the agency's requirements and terms of business on receipt of SAE.

Fotofass Travel Pictures
59 Shaftesbury Avenue
London W1
Tel. (01) 437 5951

Specialises in travel pictures.

Geoslides Photographic Library
4 Christian Fields
London SW16

Collect photos taken by travellers, especially in Africa, Asia, the Arctic, sub-Arctic and Antarctica, sell reproduction rights to publishers in a wide field, especially educational, and publish free leaflets (send SAE): 'Services for Freelance Photographers', 'Services for Teachers, Lecturers and Resource Centres' and 'Services for Publishers'.

Nigel Press Associates
167 Coldharbour Lane
London SE5 9PA
Tel. (01) 274 1705

Offer a free service to bona fide *expeditions for the acquisition and interpretation of satellite photography, of which they have a large holding covering most parts of the world. Satellite photographs can be used to produce accurate maps at scales between 1:1 million and 1:200,000 or even larger.*

Robert Harding Associates Picture Library
5 Botts Mews
Chepstow Road
London W2
Tel (01) 229 2234

Represent many expedition photographers, collect 35mm or 2¼ in square transparencies of exceptional quality on geographical, anthropological, botanical or zoological subjects and sell reproduction rights in photos of remote areas.

USA

Black Star, Inc
450 Park Avenue South
New York
NY 10016

Bruce Coleman Colour Picture Library
15 East 36th Street
New York
NY 10016

Is the American office of the UK firm (q.v.)

ROLOC
Box 1715 KO
Washington
DC 20013

Carries a stock of captioned colour slides of the sights of North America, Europe, the USSR, the Far and Middle East, Latin America and the Caribbean, Africa and the Pacific. Travellers can purchase these to complement or complete their own collection. Catalogues are available – for countries other than the USA these cost 10 cents each.

Europe

AFIP
8 rue de l'Ecole Polytechnique
75005 Paris
France
Tel. 326 80 13

Is an agency that is prepared to accept photographs from amateurs, though, like most agencies and picture libraries, it pays the photographer only when a picture is placed.

Rapho Agency
8 rue d'Alger
75001 Paris
France

Route planning
International road information may be obtained from:

Automobile Association
Overseas Route Planning Department
Basingstoke
Hampshire
Tel. 0256 20123

The A.A. Route Planning Department plans routes through Africa and Asia to order for members. Six weeks should be allowed for preparation of route plans.

Royal Automobile Club
Foreign Touring Department
PO Box 100
Croydon
Surrey CR9 6HN
Tel. (01) 686 2525

The RAC draws up routes for members on an individual basis.

Specialist associations

UK

Automobile Association
Fanum House
Leicester Square
London W1
Tel. (01) 954 7373

Information on visas, inoculation and en route accommodation is available to members.

Royal Automobile Club
RAC Touring Services
RAC House
Lansdowne Road
Croydon CR9 9DH

PO Box 100
Croydon CR9 6HN
Tel. (01) 686 2525

The RAC advises motorists and produces a number of touring publications useful to the motorist travelling overseas.

USA

International Backpackers Association, Inc
PO Box 85
Lincoln Center
Maine 04458
Tel. (207) 794 6062

Aims to promote trail care, trail education, wilderness care and positive legislation to protect and build more hiking trails, as well as protect wildlife, wildlands and wilderness.

National 4 Wheel Drive Association (N4WDA)
PO Box 386
Rosemead
Ca 91770

Is a national body providing information on 4WD vehicles and promoting the interests of 4WD vehicle owners. It publishes a monthly newsletter, sent to all members, and a conservation bulletin.

Travel finance/currency

UK

Travel Finance Ltd
56 Newhall Street
Birmingham B3 3RJ
Tel. (021) 236 7252

Offers easy payment terms for all holiday and travel arrangements. Application form and details on request.

Universal Air Travellers Plan (U.A.T.P.)
Provides for the issue by British Airways of an unlimited number of charge cards to a company, on receipt of deposit of $425, for use by its employees in paying for flights on British Airways and some 155 other airlines.

USA

Deak-Perera, Inc
677 S. Figueroa Street
Los Angeles
Ca 90017
Tel. (213) 624 4221
or toll free, in Southern California
(800) 252 9324, outside California
(800) 421 8391
182 Geary Street
San Francisco
Ca 94108
Tel. (415) 362 3452
or toll free, in Southern California
(800) 792 0825, outside California
(800) 227 4076
630 Fifth Avenue
New York
NY

Offers a currency exchange service, direct or through local travel agents, for travellers needing to purchase foreign currency before leaving the USA.

Travellers' aid centres and clubs

USA

USC Student Travel Service
Student Union 301
University of Southern California
Los Angeles
Ca 90007
Tel. (213) 741 7580

Acts as agent for tour operators, contractors and other travel services; issues a newsletter-type advertising brochure which gives information on tours, treks, expeditions and adventures, cruises, individual travel destinations, accommodation and flights.

Europe

Automobile Club de France
8 place Vendôme
75001 Paris
France; and

Touring Club de France
65 avenue de la Grande-Armée
75116 Paris
France
Tel. 553 39 59

Will both issue carnets de passages en douane *to their own members only.*

French travellers in difficulty should note that

Conventions administration et Affaires consulaires
21 bis rue La Pérouse
75016 Paris
France
Tel. 553 52 00

Is a government office responsible to the ministry of foreign affairs which handles aid to French citizens in distress abroad, repatriation, legal aid, protection of the interests of French citizens resident abroad and so on.

Foyer International d'Accueil de Paris
30 rue Cabanis
75014 Paris
France
Tel. 707 25 69

Holds discussion 'forums' for the exchange of information between travellers.

Touring Club Suisse (TCS)
9 rue Pierre Fatio
1211 Geneva 3
Switzerland
Tel. (022) 35 76 11

Issues international driving licences to members and also carnets de passages en douane.

South America

Bolivia 6000
Casilla 8412
La Paz
Bolivia

Is a company offering two kinds of service for those interested in exploring the country: equipment for the extremely varied terrain, including camping gear, specially adapted vehicles, cartographic and photographic data; and information, advice, guides and transport for the various kinds of expedition to which Bolivia lends itself, from alpine climbing to jungle trekking.

South American Travel Organisation
El Rosario 240
ler piso
Miraflores
Lima
Peru

Will provide on request specific information, e.g. on bus services in Peru or any other South American country, to overland travellers.

Travellers' aids suppliers (see also Equipment)

UK

Copper Wood Travel Comforts
Southwater Industrial Estate
Southwater
Horsham
Sussex
Tel. 0403 731124

Specialises in products and kits designed to overcome emergencies and problems abroad, e.g. multi-purpose knife, luggage straps, crisis kit.

USA

Dana
PO Box 161723
Sacramento
Ca 95816

Sell globes, alarm clocks, padlocks, wallets, magnifying glasses, books and atlases and other accessories.

Franzus Co
352 Park Avenue South
New York
NY 10010; and

Parks Products
3611 Cahuenga
Hollywood
Ca 90068
Tel. (213) 876 5454

Make voltage converters and adapter plugs to fit electronic/portable appliances anywhere in the world.

The Forearmed Traveler
227 Scenic Avenue
Piedmont
Ca 94611

Provides maps, language tapes, guidebooks and other goods and services for foreign travel. Catalogue available.

The Trip
509 Main Street
Glen Ellyn
Ill 60137

Sells luggage, maps and a wide range of other items for the traveller.

Traveler's Checklist
Cornwall Bridge Road
Sharon
Conn 06069
Tel. (203) 364 0144

Sells electrical devices, luggage holdalls, vehicle accessories, travel gifts and accessories and travel books.

Travelers' Service
681 Ellis Street
San Franscisco
Ca 94109

Sell water filters, money bags, 'talking cards' with translations of useful phrases into three foreign languages.

Travel-Pac
Box 1213
Thousand Oaks
Ca 91360

Make travel belts of urethane-coated lightweight nylon, for safeguarding travellers' cheques, money, passports, etc.

Trip-Along
138-B Ardmore Avenue
Hermosa Beach
Ca 90254

Supplies field tested travel aids by mail order, including immersion heaters, money belts. Free brochure.

Wahlstrom Travel Aids
9737 White Oak Avenue
Northridge
Ca 91325

Sell a range of travel items, including the Mini II Travel Water Purifier which provides micro-biologically pure water and weighs only 4 ozs.

Canada

Travel Tips Ltd
PO Box 1051
1040 Speers Road
Oakville
Ont L6J 5E9
Tel. (416) 844 7273

Publish and distribute maps, guidebooks and pamphlets and stock a wide range of travel aids and accessories. Catalogue.

Europe

La Bagagerie
Galerie marchande de la Tour Montparnasse
Paris
France
Tel. 538 65 53

Pépin
Galerie marchande du palais de Congrès
Paris
France
Tel. 758 23 06

Sports et Climat
223 blvd St-Germain
75007 Paris
France
Tel. 548 30 21

To-Day
51 blvd St-Michel
75005 Paris
France
Tel. 033 35 35

All for insect repellents, money belts, etc.

Librairie de la Maison des Voyages
28 rue du Pont-Louis-Philippe
75004 Paris
France

Sell travel/camping accessories, including food in tubes, storm matches, jerry cans, lamps, torches, plugs.

Vaccination centres

UK

British Airways Immunisation Centre
Victoria Terminal
London SW1
Tel. (01) 834 2323

British Airways Medical Centre
Speedbird House
London Heathrow Airport
Hounslow
Middlesex
Tel. (01) 759 5511

Central Public Health Laboratories
Colindale Avenue
London NW9
Tel. (01) 205 7041

Gives gamma globulin immunisation against hepatitis.

The Hospital for Tropical Diseases
4 St Pancras Way
London NW1
Tel. (01) 387 4411

West London Designated Vaccinating Centre
53 Great Cumberland Place
London W1
Tel. (01) 262 6456

Europe

In **Belgium** vaccinations may be given by a GP, in which case the yellow card must be countersigned by an authorised centre, or by one of the following vaccination centres, when no further endorsement of the record is required:

Ministère de la Santé Publique et de la Famille
Cité Administrative de l'Etat
Quartier de l'Esplanade
1000 Brussels;

Centre Médical du Ministère des Affaires Etrangères
9 rue Bréderode
1000 Brussels

In **France** vaccinations may be performed by a GP, in which case his signature must be endorsed by the

Direction départementale d'Action sanitaire et sociale
57 blvd de Sébastopol
75001 Paris
Tel. 508 96 90

or by local offices; or they may be performed at vaccination centres such as those of the:

Institut Pasteur
25 rue du Docteur-Roux
75015 Paris
Tel. 566 58 00

or the

Cité Universitaire
40 blvd Jourdan
75014 Paris

in which case no further endorsement of the record is needed.

The International Certificate of Vaccination, or 'yellow card', is obtainable in France from chemists, travel agencies or vaccination centres.
There are many vaccination centres in France that offer free immunisation to travellers, though as usual the yellow fever vaccination is harder to come by and is available from the Institut Pasteur only.

In **Switzerland** vaccinations are performed by GPs, at clinics or at vaccination centres, with exception of the yellow fever vaccination which can be obtained only at

L'Institut d'Hygiène
2 quai du Cheval Blanc
1200 Geneva
Tel. (022) 43 80 75

Asia

Municipal Health Centre
Parliament Street
New Delhi
India

Vaccination information

UK

Health Control Unit
Terminal 3 Arrivals
Heathrow Airport
Hounslow
Middlesex TW6 1NB
Tel. (01) 759 7208

Can give at any time up-to-date information on compulsory and recommended immunisations for different countries.

In the UK, vaccination against diseases other than yellow fever can be carried out by the traveller's own doctor or, exceptionally and by arrangement, at a hospital. A list of yellow fever vaccination centres, together with details of health precautions for the traveller, is given in the booklet **Notice to travellers: Health protection,** prepared by the Health Departments of England, Wales and Northern Ireland and the Central Office of Information. This is available from the Department of Health and Social Security in London, the Welsh Office in Cardiff, the Department of Health and Social Services in Belfast, the Scottish Home and Health Department in Edinburgh, or from local offices of these departments.

Vehicles: accessories, spares, outfitting

UK

Brownchurch (Components) Ltd
308-310 Hare Row
off Cambridge Heath Road
London W2
Tel. (01) 729 3606

Cover all Land-Rover needs for trips anywhere, including the fitting of jerry cans and holders, sand ladders, sump and light guards, crash bars, winches, water purifying plants, suspension seats, roof racks (custom-made if necessary), overdrive units. They also supply new vehicles and offer a service, maintenance and spares service for Land and Range Rovers.

Dunlop Group of Companies
Dunlop House
Ryder Street
London SW1
Tel. (01) 930 6700

Information on Dunlop special purpose tyres.

Edmonton Car Parts
10 Aberdeen Parade
Angel Road (North Circular)
Edmonton
London N18
Tel. (01) 807 0491

Sell cheap tyres.

Fairey Winches
South Station Yard
Whitchurch Road
Tavistock PL19 9DR
Tel. 0822 4101

Supply winches suitable for use with expedition/overland vehicles.

FLM Pandcraft
32-35 The Arches
Broughton Street
London SW8
Tel. (01) 662 2080

K.G. Hubbard, Hubbard Engineering Co
Hill View
Otley
nr Ipswich
Suffolk IP6 9NP
Tel. 047 339 522
Unit F
Tunnel Avenue Trading Estate
279-287 Tunnel Avenue
London SE10
Tel. (01) 858 6475

Fit air conditioning units to expedition vehicles, to the owner's specifications, and advise on servicing and maintenance abroad. The firm's work is of a high standard and they are willing to consider 'one-off' jobs as well as the easier fittings.

Michelin Tyre Co Ltd
81 Fulham Road
London SW3
Tel. (01) 589 1460

Special purpose tyres, especially desert tyres.

Monorep Ltd
Poplar Avenue
Norwood Green
Southall
Middlesex UB2 4PN
Tel. (01) 574 0722 and 571 1550

Vehicle air conditioning suppliers and fitters.

Oxley Coachcraft
Craven Street
Holderness Road (6)
Hull
Yorks
Tel. 0482 20072

Sell kits for fitting an elevated roof to almost any make of van. Also sell motor caravans.

RAC Touring Services
PO Box 92
Croydon CR9 6HN

Sell a range of touring aids, including headlamp beam deflector sets, emergency windscreens, holdalls and even Linguaphone records, at reduced prices to members.

J.C. Salt & Co
27 Speedwell Road
Hay Mill
Birmingham 25
Tel. 021 772 7675

Stevenson Motors
Cannon Lane
Tunbridge Wells
Kent
Tel. 0892 64444

Do Land-Rover and Range Rover conversion work.

TVC Sales Ltd
13 Sussex Mansions
Old Brompton Road
London SW7 3JZ
Tel. (01) 584 2938

Quality roof tents (for 2) and baggage trailers. A long wheelbase Land-Rover can accommodate 2 roof tents (4 people). Also camping tents, roof racks and closed lockable plastic roof rack containers.

USA

Advance Adaptors
Box 206
Bell
Ca 90201

Adapters for engine conversions with complete instructions and parts list.

Advanced Four Wheel Drive
2354 E. Huntington Drive
Duarte
Ca 91010
Tel. (213) 357 3094

Repairs, swaps and accessories.

Air Lift Company
Box 449
Lansing
Mi 48902

Air springs.

Alaskan Campers Northeast-Lanheim, Inc
401 West End Avenue
Manheim
Pa 17545

Make hydraulic telescopic camper tops.

Alondra, Inc
826 W. Hyde Park
Inglewood
Ca 90302

Fuel pressure regulators and filters. Catalogue.

American Speed Equipment Supply Corp
300 Bethpage Spagnoli Road
Melville
NY 11746

Large range of 4WD equipment. Catalogue.

Archer Bros
19745 Meekland Avenue
Hayward
Ca 94541

Hubs, tops, accessories, wheels, rollbars, etc., for Jeeps.

Auto-Haus
PO Box 428
Buena Park
Ca 90621
Tel. (714) 521 5120

Baja 500 Off Road Mart
3536 E. Colorado Blvd
Pasadena
Ca 91107
Tel. (213) 796 4580

Berens Associates, Inc
6046 Claremont Avenue
Oakland
Ca 94618

Wide selection of winches, wheels, tops and all types of 4WD accessories. Ships anywhere.

Bestop
Box 318
Boulder
Ca 80302

All types canvas top for all 4WDs.

Big Wheels
12865 Main Street
Garden Grove
Ca 92640

Off road tyres and wheels to 10in wide. Free catalogue.

Bill Stroppe & Associates
Box 1891
Long Beach
Ca 90801

Everything for Bronco and Ford trucks.

Dick Cepek
9201 California Avenue
South Gate
Ca 90280
Tel. (213) 569 1675

The leading 4WD tyre specialists. Also has warehouse full of camping and off road accessories. Free catalogue.

Chuchua 4-Wheel Drive Center
Box 301
Fullerton
Ca 92632

Jeep parts, conversion kit for installing automatic transmission.

Con-Ferr, Inc
123 S. Front Street
Burbank
Ca 91502
Tel. 849 1800;

Con-Ferr Jeep Center
300 N. Victory Blvd
Burbank
Ca 91502
Tel. 849 3554

Has over 800 off road accessories. Emphasis on ruggedness and Jeep parts.

Custom Gas Tanks
11719 McBean Drive
El Monte
Ca 91732

Will build what they don't have in stock. Free literature.

DataMotive, Inc
14041 Mt Bismark
Reno
Nv 89506

Independent automotive and tyre test lab for manufacturers. Write for capabilities.

Desert Dynamics
13720 E. Rosecrans Avenue
Santa Fe Springs
Ca 90670
Tel. (213) 921 0421

Grille guards, rollbars and other accessories. Free catalogue for Ford and Chevy units.

Desert Vehicles
1598 Fayette Street
El Cajon
Ca 92020

Makers of suspension kits. Makers of Rough Country shocks.

Don Gleason's Campers Supply Inc
Pearl Street
Northampton
Mq 01060

Camper top and other camping equipment. Catalogue.

Downey Toyota
8851 Lakewood Blvd
Downey
Ca 90241

Everything for Toyotas. Catalogue.

Drive-Gard Products
PO Boy 4190
Downey
Ca 90241

Dualmatic Manufacturing Co
PO Box 1119
Longmont
Ca 80501

Full line in 4WD hardware. Free catalogue.

Duffy's Enterprises
16611 Normandie Avenue
Gardena
Ca 90247

Suspension accessories. Catalogue.

Edelbrock Equipment Co
411 Coral Circle
El Segundo
Ca 90245

Intake manifolds. Catalogue.

Edgewood National, Inc
6603 North Meridian
Puyallup
Wa 98371

Automotive and recreational vehicle parts and accessories specialising in 4WD. Catalogue.

Fairway 4x4 & RV Center
1350 Yorba Linda Blvd
Placentia
Ca 92670
Tel. (213) 620 9387

Wide range of 4WD accessories. Catalogue.

Fiber-Kraft Manufacturing
1051 West 48th Avenue
Wheatridge
Ca 80033

Makers of a pop-up camper for 4WD vehicles.

Fisher Industries, Inc
PO Box 245
Marlette
Mi 48453

Heavy-duty steel grids called Easy Outs, which will slip in front of tyres and allow an immobilised vehicle to ride on top and get back on the road. A telescoping version of a sand ladder.

Four Wheel
2890 West 62nd Avenue
Denver
Co 80221

Make campers for 4WD vehicles.

Four Wheel Drive Center of San Francisco
2865 San Bruno Avenue
San Francisco
Ca 94134
Tel (415) 468 1250

Four Wheel Parts Wholesalers
Box 54572
Terminal Annex
Los Angeles
Ca 90054

A discount house for Jeep parts and for many accessories. Catalogue.

Fredson RV and Van Supply
815 N. Harbor Blvd
Santa Ana
Ca 92703

Huge range of camper top supplies. Catalogue.

General Technology
Box 1134
Scottsdale
Az 85252

Swingout tyre carriers, side or rear carrying position, for Blazer, Scout, Bronco and pickups.

Good-Time Recreational Supply
PO Box 615
Madison
Wi 53701

Enormous range of recreational vehicle supplies. Catalogue.

Hanford 4 Wheel Drive
950 E. Lacey Blvd
Hanford
Ca 93230
Tel. (209) 582 1204

Harrah Mfg Co
46 W. Spring Street
Bloomfield
Ind 47424

'Handyman' jacks, Loc-Racs and other accessories.

Hellwig Products
16237 Avenue 296
Visalia
Ca 93277

Torsion sway bars, accessory springs and other suspension devices.

Herter's, Inc
Waseca
Minn 56093

Hickey Enterprises, Inc
1645 Callens Road
Ventura
Ca 93003
Tel. (805) 644 5571

The well-known sidewinder winch and a heavy line of accessories for Blazer, Scout and Bronco. Speciality is Blazer/Jimmy.

Hone Manufacturing Company
11748 E. Washington Blvd
Santa Fe Springs
Ca 90670

Overdrives for all units.

Hooker Headers
1032 Brook Street West
Ontario
Ca 91762

Headers for nearly all 4WDs. Also muffler and tail pipe packages.

Hoosier Machine Products
314 S.E. 6th Street
Pendleton
Or 97801

Bronco, Jeep and Scout engine and transmission hardware. Free catalogue.

Husky Products Company
Box 824
Longmont
Co 80501

Overdrives, hubs, tops, tyre carriers, seats, etc.

Indiana Camp Supply, Inc
PO Box 344
Pittsboro
Ind 46167

Camper top and other camping equipment. Catalogue.

K C Hilites, Inc
PO Box 155
Williams
Az 86046

Go-ahead manufacturer of a wide range of accessory lights for off roading. Free catalogue.

K & N Engineering, Inc
PO Box 1329
Riverside
Ca 92502

A wide range of air filters for off road vehicles. Free catalogue. Makers of K & N Performance Products.

Kelly Manufacturing Co
5611 Raritan Road
Denver
Co 80221

Jeep accessories.

Koenig Iron Works, Inc
Box 7726
Houston
Tx 77007

King winches. Free literature.

Laacke and Joys
1432 N. Water Street
Milwaukee
Wi 53202

Camper top and other camping equipment including own tents. Catalogue.

Lakewood Industries
4800 Briar Road
Cleveland
Oh 44135

Come-A-Longs, rollbars, metallic brake linings. Catalogue.

Lubens Industries
233 S. 7th Avenue
City of Industry
Ca 91744

Off road equipment, especially for Fords.

Man-A-Fre Co
18736 Parthenia Avenue
Northridge
Ca 91321

Specialist in Toyota Land Cruiser parts and accessories. Also headers and carb conversions for all 4WDs.

Master Mechanic Manufacturing Co
PO Box A
Burlington
Wi 53105

Winches.

McCain Hub Winch Company
345 Mill Street
Eugene
Or 97401

Winches. Free literature.

Midwest Four Wheel Drive Center
5511 No. Lindbergh
St Louis
Mo 63042

Range of off road accessories. Catalogue.

Mile Post One
PO Box 5799
Commerce Park
Athens
Ga 30601

Complete range of basic 4WD equipment. Catalogue.

Mister Tire 4 Wheel Drive Center
Warehouse No 5
740 Brockton Avenue
Abington
Ma 02351

Complete range of off roading accessories. Catalogue.

Monroe Auto Equipment Company
1426 E. First Street
Monroe
Mi 48161

Air adjustable shocks plus conventional shocks.

Mountain Man Industries
PO Box 1147
One Delta Way
Grand Junction
Co 81501

4WD accessories, mainly carriers, racks, etc. Free catalogue.

Mountain Roads Four Wheel Drive
185 Sebastopol Road
Santa Rosa
Ca 95401
Tel. (707) 525 9393

National Four Wheel Drive Association
PO Box 386
Rosemead
Ca 91770

National Surplus Center
240 East Cass Street
Joliet
Ill 60432

Extensive range of government surplus equipment at low prices. Catalogue.

North Country 4WD
Box 558
Saxtons River
Vt 05154

Off pavement equipment for 4WDs.

Novak Enterprises
Box 1324
Whittier
Ca 90609

Engine conversion hardware. Conversion instructions. Free catalogue.

Nova Products
15934 S. Figueroa Street
Gardena
Ca 90248

Manufacturers of Shur-Lift winches, shock absorbers and hydraulic accessories. Free catalogue.

Offenhauser
5300 Alhambra Avenue
Los Angeles
Ca 90032

Covers the 4WD field with manifolds for all vehicles. Free 4WD application list. Catalogue.

Pathfinder Equipment Co, Inc
PO Box 86
San Gabriel
Ca 91778
Tel. (213) 285 9747

Make 4WD van conversions.

Perfectune
PO Box 26848
Tempe
Az 85282

Quadrajet and 2G Rochester carb re-jet kits for altitude, mileage and performance. Free details.

Per-Lux
804 East Edna Place
Covina
Ca 91722

Off road lights.

Pick-up Equipment Co, Inc
311 Hewitt Drive
PO Drawer C
Hewitt
Tx 76643

Racer Brown
9270 Borden Avenue
Sun Valley
Ca 91352

Performance and economy camshafts. Catalogue

F. & W. Rallye Engineering
5N775 Campton Ridge
St Charles
Ill 60174

Acrylic plastic covers for headlights and auxiliary lights which are inexpensive, very durable and do not hinder or distort the light pattern.

Ramsey Winch Company
Box 15829
Tulsa
Ok 74115

A complete list of PTO and electric winches for 4WD vehicles. Free winch booklet.

Rancho Jeep Supply
6309 Paramount Blvd
Long Beach
Ca 90805

Largest supply of Jeep parts in the world. Also has Rancho overdrive, oversize brake kits and rollbars.

Rancho Off-Road Center
7028 Reseda Blvd
Reseda
Ca 91335

All the good accessories for Chevys.

Rapid Cool
306 So. Center Street
Santa Ana
Ca 92703

Engine oil and transmission fluid coolers with lifetime guarantee.

Recreational Vehicle Accessories
2111 S. Leyden Street
Denver
Co 80222

Four wheel drive accessories. Catalogue.

R.J.'s Off Road Center
945 Yonkers Avenue
Yonkers
NY 10704

Rough Country, Inc
1080 N. Marshall Avenue
El Cajon
Ca 92020

Sand Tires Unlimited
1401 West 178th Street
Gardena
Ca 90247
Sand tyres. Free literature.

Sears, Roebuck & Co
4640 Roosevelt Blvd
Philadelphia
Pa 19124

Tools and general maintenance supplies.

Shur-lift
See Nova Products.

Simpson's On and Off Road Accessory Center
8456 Firestone Blvd
Downey
Ca 90241
Tel. (213) 869 1528

Extensive range of 4WD accessories. Catalogue.

Smith Jeep, Inc
Rt 32,
No. Franklin
Ct 06254

Complete 4WD accessory line. Free catalogue.

Smittybilt
2124 N. Lee Avenue
S. El Monte
Ca 91733

Off road vehicle rollbars, tow bars. Free catalogue.

Southern California RV Products
2907 N. San Fernando Blvd
Burbank
Ca

Auxiliary or replacement tanks, skid plates, rollbars and other off road heavy accessories. Free catalogue.

Subaru Town Motors, Inc
427 E. 90th Street
New York
NY 10028

Range of accessories for Subaru 4 WD vehicles.

Superwinch Inc
Pomfret
Ct 06258

The famous 'Miniwinch' in three sizes.

Suzuki Fun Center
515 N. Victory Blvd
Burbank
Ca 91502

Motorcycle gear and accessories, repair manuals. Catalogue.

The North Face
PO Box 2399
Station A
Berkeley
Ca 94702

Camper top and other camping and outdoor equipment. Catalogue.

Trailblazer
2217 N. 36th Street
Boise
Id 83703

Jeep, Wagoneer, Commando, Cherokee gas tanks, skid plates, tyre mounts. Free price list.

Trail City
8600 E. Main Street
Reynoldsburg
Oh 43068

Four-wheel-drive supermarket. Catalogue.

Travel Accessories, Inc
Box 248
Orland
Ca
Three Rivers
Mi and in Calgary, Canada

Jumbo travel tanks for 4WD pickups, Blazer and Suburban. Features an electric tank selector valve and associated gas gauge.

US General Supply Corp
100 General Place
Jericho
NY 11753

Wide selection of tools.

Viking Camper Supply, Inc
99 Glenwood Avenue
Minneapolis
Minn 55403

Complete range of recreational vehicle accessories. Catalogue.

Warehouse Sales
PO Box 36
Rossford
Oh 43460

Winch manufacturers.

Warn Industries
19450 68th Avenue

Warn Industries
19450 68th Avenue South
Kent
Wa 98031
Tel. 854 5350

Winches, mini-winches, post-traction, locking hubs, utility hoists. Free literature.

Weiand Automotive Industries
2316 San Fernando Road
Los Angeles
Ca 90065

Intake manifolds. Catalogue.

Western Auto
2107 Grand Avenue
Kansas City
Mo 64108

Large mail order catalogue including numerous items for camping and camper tops.

Western Off-Road Wholesalers
16612 Beach Blvd
Huntington Beach
Ca 92647
Tel. (714) 847 5555

Very large selection of 4WD accessories. Catalogue.

Western Tire Center
3645 S. Palo Verde
Tucson
Az 85713

Wheels, tyres and a variety of 4WD accessories. Catalogue.

Whitco, White Automotive Corporation
Box 1209
Colorado Springs
Co 80901

Soft tops for most 4WD vehicles. Free literature.

J.C. Whitney & Co
1917-19 Archer Avenue
PO Box 8410
Chicago
Ill 60680

One of the world's largest vehicle accessories suppliers, selling mainly by mail order. Instruction books and vehicle manuals available. Catalogue.

Canada

Four Wheel Drive Center of Ontario

50 Howden Road
Scarborough
Ontario

Suppliers of general 4WD equipment and accessories. Catalogue.

The Offroader

RR No 12
Dawson Road
Thunder Bay
Ontario

Suppliers of off road accessories. Catalogue.

Travel Accessories, Inc

Box 5722
St A
Calgary

Jumbo travel tanks for 4WD pick-ups, Blazer and Suburban. Features an electric tank selector valve and associated gas gauge. Branches in Orland, California, and Three Rivers, Michigan, USA.

Europe

France

A.B. Accessoire

237 rue de Bellevue
92700 Colombes

Wire mesh headlight screen.

Air Camping

Roux Carrossier
30 route de Bagnols
30130 Pont St-Esprit
BP 27
Tel. (66) 890233

Tents.

Aqua Marine

73 avenue Jean Jaurès
93300 Aubervilliers
Tel. 833 58 44

Portable freezers.

Bineau moteur

87 blvd Saint-Denis
92400 Courbevoie
Tel. 333 7314

Desert driving aids.

P. Bordas

28 rue Armonville
92200 Neuilly
Tel. 637 1400

Mann air filters.

Le Caillebotis Industriel

103 rue Lafayette
75010 Paris
Tel. 285 28 47

Aluminium sand ladders.

Calvet

17 quai de l'orme Sully
78230 Le Pecq
Tel. 976 62 96

Galvanised steel jerry cans.

Ceuci

8 rue du Bitche
92400 Courbevoie
Tel. 333 25 36

Off road tyres.

Société De Carbon

8 avenue Foch
92250 La Garenne Colombes.
Tel. 780 7171 and 242 4319

Shock absorbers.

Delhomme

47 rue Jules Guesdes
92300 Levallois
Tel. 737 1400 and 737 0531

Water tanks.

ENAC

6 route des Carrières
95870 Bezons
Tel. 982 67 75

Case for spare wheel.

Lacroix

8 rue Malheur
75004 Paris

Distress flares and rockets.

Mod Plastia

24 rue Auger
93500 Pantin
Tel. 844 64 63

Sand mats.

Super Flexit

45 rue des Minimes
92400 Courbevoie
Tel. 333 75 30

Pneumatic jack for foot pump.

Tractel

85-87 rue Jean Lolive
93170 Bagnolet
Tel. 858 05 40

Towing gear.

Switzerland

Maloja Tyres
Ave Appia 1211
Geneva
Tel. (022) 34 60 61

Special purpose tyres.

Australia

Arthur Garthon Motors
492 Forest Road
Penshurst
NSW 2222
Tel. (02) 57 6520

Accessories for Land-Rover and Range Rover.

Australian Winch & Machinery Co Pty Ltd
PO Box 18
South Yarra
Vic 3141
Tel. (03) 336 7575 or 26 4264; and
Port Wakefield
Virginia
SA 5120
Tel. (08) 380 9177

Ramsey electric winches.

Bennett and Wood Ltd
Joynton Avenue
Zetland
NSW 2017

Power winches.

Bull Bag Australia
26 Harriett Street
Marrickville
NSW 2204
Tel. 55 5699

Air compressors and bull bag jacks.

Bush Ranger 4WD
203 Springvale Road
Springvale
Vic 3171
Tel. (03) 547 6969

General off road accessories.

Chatswood Motors
Pacific Highway
Chatswood
NSW
Tel. 419 2498

Specialists in Toyota Landcruiser conversions.

Discount Tyre Service
Barrier Road
Fyshwick
Canberra
ACT
Tel. 95 6304

Off road wheels and tyres.

DP Off-Road Accessories Pty Ltd
39 Barry Street
Bayswater
Qld
Tel. 720 1108

Off road accessories.

Duncan's Tyre Service
831 Nepean Highway
Moorabin
NSW
Tel. 972 881

Off road tyres.

4 x 4 Australia Pty Ltd
601 Forest Road
Bexley
Sydney
NSW 2207
Tel. 502 1244

Vehicles, parts and accessories; also houseboats, caravans, canoes and other outdoor equipment.

Four Wheel Drive Equipment and Service
210 Palmer Street
East Sydney
NSW 2010
Tel. (02) 31 3710

Lightweight bull bars.

Four Wheel Drives
304 Middleborough Road
Blackburn South
Vic 3130
Tel. (03) 89 0509

Manual winches and Land-Rover spares.

Four-Wheel-Drive Service Centre
221 Harris Street
Pyrmont
Sydney
NSW
Tel. (02) 660 2386 or 467 1475 (emergency calls)

Emergency, field or workshop repairs on all types of 4WD vehicle.

Four-Wheeler Centre
2/698 Pittwater Road
Brookvale
NSW 2100
Off road equipment.

Frank Murphy 4 Wheel Drive Accessories
592 Burwood Road
Hawthorn
Melbourne
Vic 3122
Tel. (03) 82 2295
Wide range of 4WD accessories.

Gilbert Lodge & Co Ltd
13 Parramatta Road
Lidcombe
NSW 2141
Tel. 648 0331
Manual winches.

Goulds
516 High Street
Northcote
Vic
Tel. 48 6456
Also at
High Street
Thomastown
Vic
Tel. 465 2977
Toyota Landcruiser specialists.

The Jeep Factory
1367 Botany Road
Botany
NSW 2019
Tel. (02) 666 5639/40
All Jeep accessories.

Keep on Truckin' Pty Ltd
121 Majors Bay Road
Concord
NSW 2137
Tel. 73 4424
General 4WD accessories.

Kevin Donnellan Performance Tyres
1-3 Errol Street
North Melbourne
Vic 3051
Tel. 329 5107
Off road tyres.

Land Rover Spare Parts
See Four Wheel Drives.

Land Vehicle Spares
10 Fitzpatrick Street
Revesby
NSW 2212
Tel. (02) 772 1566
Land-Rover parts and accessories.

M.S. McLeod Ltd
276 Parramatta Road
Auburn
NSW
Tel. 648 0497
and at the other McLeod outlets.
Wide range of off road tyres.

The Motor Sport Shop
286 Pennant Hills Road
Thornleigh
NSW 2120
Tel. (02) 848 0211
Off road accessories.

The Motor Sport Workshop
32 Carlingford Street
Regents Park
NSW 2143
Tel. (02) 645 2920
Off road accessories.

Newcastle 4 Wheel Drive
42-44 Park Avenue
Adamstown
NSW 2289
Tel. (049) 575 961

Off Road Equipment (Sydney)
10 Erskine Road
Taren Point
NSW 2229
Tel. 525 0758

Off-Road Equipment
176 Rocky Point Road
Kogarah
NSW 2217
Tel. (02) 529 4854
Recovery equipment and a large range of accessories.

Off-Road Equipment
40 Sarich Court
Osborne Park
WA 6107
Tel. 446 8686
Largest 4WD accessory stockists in Western Australia.

Overland Accessories Centre
3 Kitson Road
Maddington
WA 6109
Tel. 459 5282

Rallyspeed
rear 264 Pacific Highway
Crows Nest
NSW
Tel. (02) 438 1195

Rollbars and bull bars.

Robertson & Marchant Pty Ltd
PO Box 147
Concord
NSW 2137

Battery isolators.

Shop 7
Cedar Village
18-22 Bridge Street
Eltham
Vic
Tel. 439 2876

Wide range of off road accessories.

Shute-Upton Engineering Pty Ltd
Evans Road
Salisbury
North Brisbane
Qld
Tel. 277 5822

Winches and free-wheeling hubs.

Sunraysia RV Equipment
72 Stacey Street
Bankstown
NSW 2200
Tel. (02) 708 8821

Wide range of off road and RV accessories.

Terrain Tyres WA
Unit 4
210 Walter Road
Morley
WA 6062
Tel. (09) 276 5237

Off road accessories.

TJM Products
Morrisby Street
Geebung
Brisbane
Qld
Tel. (07) 350 2794/350 2877

General 4WD accessories.

Tom Fry's Suzukiland
86 Princes Highway
Sylvania
Tel. 522 7646/7534

Everything for Suzuki 4WD.

Town and Country Bull Bars
57 Rookwood Road
Yagoona
NSW
Tel. 709 5600

Bull bars and general off road accessories.

Tyre Services Pty Ltd
971 Ipswich Road
Moorooka
Brisbane
Qld 4105
Tel. (07) 48 3180; and
PO Box 4
Moorvale
Qld 4105
Tel. (07) 48 3157

Queensland's largest off road equipment specialists.

ULR Pty Ltd
1339-1347 High Street
Malvern
Vic 3144
Tel. 20 2130/1911

Specialists in Land-Rover and Range Rover.

Victoria's Off-Road Specialists
203 Springvale Road
Springvale
Vic 3171
Tel. (03) 547 6969

4WD accessories.

Winch Industries Pty Ltd
Coin Street
Moorooka
Qld 4105
Tel. (07) 277 5166

Free wheeling hubs and electric winches.

Vehicles: purchase, hire and conversion

UK

Carawagon Coachbuilders Ltd
Thames Street
Sunbury on Thames
Middlesex TW16 5QR
Tel. 76 85205

Will convert Land-Rovers and Range Rovers for safari and expedition use.

Dunsfold Land-Rovers Ltd
Common House Road
Dunsfold
Surrey
Tel. (048 649) 567

Offers free advice to those contemplating overland travel; expedition hardware, air conditioning, lefthand drive conversions, comprehensive stores, rebuilding to owner's specifications; and sales of new and secondhand Land-Rovers.

Four by Four Hire Ltd
Twickenham Road
Hanworth
Middlesex TW13 6JF
Tel. (01) 894 1211

Glasselfield
Lockerbie
Dumfriesshire
Tel. 05762 3334

Have Land-Rovers for hire (over 200 at any time), mostly equipped for camping.

Fram Expeditions Ltd
John Hamilton
69 Ledi Drive
Bearsden
Glasgow G61 4JN
Tel. 041 942 6254, evenings

A small but reliable organisation which hires out diesel Land-Rovers for expeditions. Founded in 1953 by Stan Whitehead, Fram now has four 12-seater station wagons available, three in Birmingham and one in Glasgow. The service is intended to be neither a charity nor profitmaking, but merely self-supporting. Fram does not organise expeditions, but hires out vehicles to ready-made expeditions.

Guerba Expeditions Ltd
Stokehill Farm
Earlestoke
Devizes
Wilts SN10 5UB
Tel. 0380 83 476

Offers a range of 2- or 4-wheel-drive expedition vehicles – motor homes, safari buses, expedition trucks and Land-Rovers – suitable for private parties, for sale or hire, complete with spares and equipment; and a guide service for Africa, especially for Saharan trips. Also stocks books and maps on overland travel, and offers comprehensive planning advice and crew training as part of each purchase.

Harvey Hudson
Woodford
London E18 1AS
Tel. (01) 989 6644

Land-Rover specialists, suppliers of new and used vehicles to expeditions.

Jaguar Rover Triumph Ltd
Solihull
Warwickshire
Tel. 021 743 4242

Manufacturers of Land-Rovers and Range Rovers. Purchase must be through authorised dealers, however. No secondhand vehicles or vehicles for hire are available from the Rover Company.

Land-Roving Ltd
30 Sidbury
Worcester
Tel. 0905 26988

Land-Rovers for sale and hire.

Power Torque Special Products
Pickford Brook
Allesley
Coventry CV5 9AN
Tel. 020 334 3181

Are distributors of Stonefield vehicles.

Roverhire
12 Seagrave Road
London SW6
Tel. (01) 385 5291

Land-Rovers for sale and hire.

Stonefield Vehicles Ltd
Cumnock
Ayrshire
Scotland KA18 1SH
Tel. 0290 21822

Are manufacturers of Stonefield 4WD trucks.

Torcars of Devon
High Bullen
Torrington
Devon
Tel. 080 52 2461

Make motor caravans and do conversions. They deal direct rather than through distribution agents.

Europe

Topaz Explorers
Safarihuset
Lounsvej 29
DK-9640 Farsø
Denmark
Tel. (08) 63 83 85

Complete conversion and expedition outfitting of 4WD trucks, Land-Rovers and Volkswagens. Expedition research and planning. Hire of 4WD vehicles and expedition crew.

Bineau Moteur
87 blvd Saint-Denis
92400 Courbevoie
France
Tel. 333 7314

Is a garage specialising in Land-Rover conversion and outfitting; it will do the same for other overland vehicles, especially those with 4WD. Also sells vehicle accessories.

Caravaning
1 place du Théâtre Français
75001 Paris
France

Have camping vehicles and motor homes for sale and hire and will undertake conversions.

Garage Boursault
11 rue Boursault
75017 Paris
France
Tel. 293 65 65

Specialises in preparation and fitting out of Land-Rovers and Range Rovers.

Garage Cardinet
114 rue Cardinet
75017 Paris
France
Tel. 227 06 60

Sometimes have secondhand Land-Rovers for sale.

Jean Rousseau
237 blvd Pereire
75017 Paris
France
Tel. 380 70 31

From time to time has secondhand vehicles for sale. Also undertakes conversions and fittings of overland vehicles.

RVFM
6 route de Flandre
Pont-Yblon
95 Bonneuil
France
Tel. 931 27 27

Sells new and secondhand vehicles suitable for overland use.

Société Marreau et Cie
21 rue de Colombes
Nanterre
France
Tel. 204 24 25

Is a garage specialising in modifying, reconditioning and outfitting overland and off road vehicles.

STRAKIT
20 grande rue
28000 Bonville
France (near Chartres)
Tel. (37) 215482

Prepares vehicles for expeditions and overland travel and sells accessories.

Firms in **France** undertaking conversion of trucks or vans into campervans:

Aquitaine Loisirs
route d'Arcachon
33610 Cestas Gazinet
Tel. (56) 30 63 21

Carrosserie Commerciale
7 ter
rue Duvergier
75019 Paris
Tel. 607 55 72

Carrosserie Villard
100 rue du Vieux Pont
92000 Nanterre
Tel. 204 1248

Holiday Car
nationale 20
91290 Arpajon Sud
Tel. 491 32 31

SODIS
77 route de Senlis
Penchard (near Meaux)
Tel. 434 41 48

STAR Caravanes
51 rue de Gouedic
22000 Saint-Brieux
Tel. (96) 33 21 92

Vehicles: shipment

UK

Austasia Line
Sea Travel Centre (UK) Ltd
10 Haymarket
London SW1
Tel. (01) 839 8844; and

Blue Funnel Line
Elder Dempster Lines Travel Section
India Buildings
Liverpool L2 OR8
Tel. (051) 236 8421 ext 477

Are British agents for steamship lines handling shipping on the Singapore-Fremantle route.

USA

Kuehne & Nagel, Inc
Suite 2451
One World Trade Center
New York
NY 10048

Is one of the largest shipping agents in the USA, with offices in all major ports.

Australasia

George Wills & Co Ltd

140-160 Mary Street
Brisbane
Qld
Tel. (07) 31 1991

45-53 Clarence Street
Sydney
NSW
Tel. (02) 29 8401

66-72 Barry Street
Carlton South
Melbourne
Vic
Tel. (03) 347 9555

12 Todd Street
Port Adelaide
SA
Tel. (08) 47 5366

14 Phillimore Street
Fremantle
WA
Tel. (092) 35 1177

Are the Australian representatives of the Shipping Corporation of India, which handles vehicle shipments on the Madras-Singapore, Australia-Singapore-India and Bombay-Mombasa routes.

Asia

The Shipping Corporation of India

Steel Crate House
4th Floor
Dinshaw Wacha Road
Bombay 20
India

KPV Shaik Mohamed Rowther & Co Pvt Ltd
41 Linghi Chetty Street
Madras 1
India

c/o R.G. Shaw Shipping Ltd
11-13 Southwark Street
London SE1

Handle the shipment of vehicles on the Madras-Singapore, Australia-Singapore-India and Bombay -Mombasa routes.

Visa

USA

Visas International

6331 Hollywood Boulevard
Hollywood
Ca 90028

Provide up-to-date visa information.

Weather information

UK

Meteorological Office (International)

Bracknell
Berkshire
Tel. 0344 20242 ex 2297

USA

National Meteorological Center

5200 Auth Road
Camp Springs
Md 20233

Is the head office. Regional and local offices can also give weather information.

Charts, Lists and Tables

British Government Representatives Overseas

Afghanistan	(E)	Karte Parwan. Kabul	30511/3
Algeria	(E)	Résidence Cassiopée, Bâtiment B, 7 Chemin des Glycines, Algiers. (BP 43, Alger Gare)	605601/605411, 605038, 605831
Argentina	(E)	Luis Agote 2412/52, Buenos Aires. (Casilla de Correo 2050, Buenos Aires).	80-7071/9
Australia	(HC)	Commonwealth Avenue, Canberra, ACT 2600	730422
	(CG)	Gold Fields House, Sydney Cove, Sydney, NSW 2000.	27-7521
	(CG)	The Colonial Mutual Life Building, 330 Collins Street, Melbourne, Victoria 3000	602-1877
	(CG)	BP North House, 193 North Quay, Brisbane, Queensland 4000	(07) 221 4933
	(CG)	Prudential Building, 95 St George's Terrace, Perth, Western Australia 6000	3215611/ 217891
	(CG)	8th Flr, 70 Pirie Street, Adelaide, South Australia 5000	2234572
Bahrain	(E)	PO Box 114, Government Road, North Manama, Bahrain	254002 (7 lines)
Bangladesh	(HC)	DIT Building Annexe, Dillkhusha (PO Box 90), Dacca-2	243251-3, 244216-8, 246867
Benin	c/o	Accra, Ghana	
Bolivia	(E)	Avenida Acre 2732-2754 (Casilla 694), La Paz	51400/29404
Botswana	(HC)	Private Bag 23, Gaborone	2483/5
Brazil	(E)	Setor de Embaixadas Sul, (Caixa Postal 07-0586), Quadra 801, Conjunto K, Brasilia	225 2710, 225 2625, 225 2985, 225 2745
	(CG)	Praia do Flamengo 284-2° andar Rio de Janeiro GB (Caixa Postal 669-ZC-00)	225-7252
	(CG)	Avenida Paulista 1938-17° andar, 01310, São Paulo, (Caixa Postal 846), São Paulo SP	287-7722
Brunei	(HC)	Room 45, 5th Flr, Hong Kong Bank Chambers, Djalan Chevalier, Bandar Seri Begawan, Brunei	26001/26002
Burma	(E)	80 Strand Road (PO Box 638), Rangoon	15700
Burundi	c/o	Kinshasa, Zaïre	
Cameroon	(C)	Boîte Postale 1016, Rue Pau, Douala	42 21 77/42 22 45
Cape Verde	c/o	Dakar, Senegal	
Central African Emp	c/o	Douala, Cameroon	
Chad		*There is no resident mission in Chad: the British ambassador resides in London (the Foreign and Commonwealth office).*	

Chile	(E)	La Concepcion 177 Providencia (Casilla 72-D) Santiago	239166
China	(E)	11 Kuang Hua Lu, Chien Kou Men Wai, Peking	521961/4
Colombia	(E)	Calle 38, 13-35 Pisos 9-11, Bogatá (Apartado Aereo 4508, Bogotá	698100
Congo	c/o	Kinshasa, Zaïre	
Costa Rica	(E)	Paseo Colon, 3202, Apartado 10056, San José	21-55-88
Cuba	(E)	Edificio Bolivar, Capdevila, 101-103, e Morro y Prado, Havana	61-5681
	c/o	Apartado 528, Administracion do Correos No 1 Mexico 1, DF, Mexico	
Djibouti	c/o	Sana'a, Yemen Arab Republic	
Ecuador	(E)	(opposite Hotel Quito), Gonzalez Suarez 111, Quito (Casilla 314 Quito)	230070/3
Equatorial Guinea	c/o	Douala, Cameroon	
Egypt	(E)	Ahmed Raghab Street, Garden City, Cairo	20850/9
El Salvador	(E)	13a Avenida Norte (Bis), No. 611, Colonia Duenas, San Salvador, (Embajada Britanica, Apartado (cc) 2350, San Salvador)	219106, 220590, 223945
Ethiopia	(E)	Papassinos Building, Ras Desta Damtew Avenue. PO Box 858, Addis Ababa	15166, 151252/3
Fiji	(HC)	Civic Centre, Stinson Parade (PO Box 1355) Suva	311033
Gabon	(E)	Bâtiment Sogame, boulevard de l'Indépendence, BP 476 Libreville	72.29.85
Gambia	(HC)	48 Atlantic Road, Fajara, Banjul	Serrekunda 2133, 2134,. 2578, 2672
Ghana	(HC)	Barclays Bank Building, High Street (PO Box 296), Accra The PO Box number must be used for all correspondence	64123, 64134
Guatemala	(C)	Edificio Maya, Via 5, No 4-50, 8° Piso, Zona 4 Guatemala City	61329, 64375
Guinea	c/o	Dakar, Senegal	
Guinea-Bissau	c/o	Dakar, Senegal	
Guyana	(HC)	44 Main Street, (PO Box 625), Georgetown	65881/4
Honduras	(E)	Apartado 290, Tegucigalpa	22-31-91
Hong Kong	(TC)	PO Box 528, 9th Flr, Gammon House, 12 Harcourt Road	5-229541
Iceland	(E)	Laufasvegur 49, Reykjavik	15883/4
India	(HC)	Chanakyapuri, New Delhi, 110021	690371
	(HC)	1 Ho Chi Minh Sarani, PO Box 9073, Calcutta, 700016	44-5171
	(Deputy HC)	PO Box 815, Mercantile Bank Buildings, Mahatma Gandhi Road, Bombay, 400023	274874
	(HC)	PO Box 3710, 150A Anna Salai, Madras, 2-600002	83136
Indonesia	(E)	Djalan Thamrim 75, Djakarta	341091/8
Iran	(E)	Avenue Ferdowsi, (PO Box No 1513), Tehran	375011 (10 lines)
Iraq	(E)	Shaheen Building, Kard Al-Pasha Sa'adoon Street, Baghdad	99001/3

Israel	(E)	192 Hayarkon Street, Tel Aviv 63405	(03) 249171-8
Ivory Coast	(E)	Fifth Flr, Immeubie Shell, Avenue Lamblin, Abidjan. (Boîte Postale 2581, Abidjan)	22-66-15, 32-27-76, 32-49-80
Japan	(E)	1 Ichibancho, Chiyoda-ku, Tokyo 102	(03) 265-5511.
	(CG)	Hong Kong and Shanghai Bank Building, 45 Awajimachi, 4-chome, Higashi-ku, Osaka 541	(06) 231-3355/7
Jordan	(E)	King Hussein Street, PO Box 6062, Amman	37374-5
Kenya	(HC)	PO Box 30133, 13th Flr, Bruce House, Standard Street, Nairobi	335944
Korea, Rep of (South)	(E)	Seoul BFP03 via BFP01 Forces Airmail	75-7341/3
Kuwait	(E)	P.O. Box 300 SaFat, Kuwait	439221/2
Lao People's Dem Rep (Laos)	(E)	Rue Pandit J. Nehru, PO Box 224, Vientiane	2333, 2374
Lebanon	(E)	Avenue de Paris, Ras Beirut, Beirut	36 41 08
Lesotho	(HC)	PO Box 521, Maseru	3961
Liberia	(E)	PO Box 120, Mamba Point, Monrovia	221055, 221107, 221491
Libya	(E)	Sharia Gamal Abdul Nasser, Tripoli	31191/5
Madagascar	(C)	5 Rue Robert Ducrocq, BP167, Antananarivo	251-51
Malawi	(HC)	Lingadzi House (PO Box 300 42), Lilongwe 3	
Malaysia	(HC)	(PO Box 1030), 13th Flr, Wisma Damansara, Djalan Semantan, Kuala Lumpur	941533
Maldive Islands	c/o	Dakar, Senegal	
Mali	c/o	Dakar, Senegal	
Mauritania	c/o	Dakar, Senegal	
Mauritius	(HC)	PO Box 586, Cerne House, Chaussée, Port Louis	20201
Mexico	(E)	Lerma 71, Col Cuauhtemoc, PO Box 96 Bis, Mexico City, 5 DF	5114880/5, 5143327
Mongolia	(E)	30 Enkh Taivny Gudamzh (PO Box 703), Ulan Bator	51033
Morocco	(CG)	60 boulevard d'Anfa (Boîte Postale 762), Casablanca	26 14 40, 26 14 41
Mozambique	(E)	Av Vladimir I Lenine 310, Caixa Postal 55, Maputo	26011/12
Nepal	(E)	Lainchaur (PO Box 106), Kathmandu	11081, 11588/9
New Zealand	(HC)	Reserve Bank of New Zealand Building, 9th Flr, 2 The Terrace, Wellington 1. (PO Box 369)	726-049
	(CG)	9th Flr, Norwich Union Bldg, 179 Queen Street, Auckland (Private bag, Auckland 1)	32-973
	(CR)	PO Box 1762, Christchurch	
Nicaragua	c/o	San José, Costa Rica	
Niger	c/o	Abidjan, Ivory Coast	
Nigeria	(HC)	16 Eleke Crescent, Victoria Island, Lagos (Private Mail bag 12136, Lagos)	51630/1/2
	(HC)	Finance Corporation Building, Lebabon Street, Ibadan. (Private Mail bag 5010 Ibadan)	21551
	(Deputy HC)	United Bank for Africa Building, Hospital Road, Kaduna. (Private Mail bag 2096, Kaduna)	22573/5

Oman	(E)	Muscat (PO Box 300, Muscat)	722411
Pakistan	(E)	Diplomatic Enclave, Ramma 5 (PO Box 1122) Islamabad	22131/5
	(CG)	York Place, Runnymede Lane, Port Trust Estate, Clifton, Karachi-6	53 20 41/46
		Matters concerning the northern areas, including the provinces of Punjab and North West Frontier, are dealt with by the embassy, and for Sind and Baluchistan provinces by the consulate at Karachi	
Panama	(E)	Via España 120, Panama City (Apartado 889, Panama City)	Panama City 23-0451
Papua New Guinea	(HC)	United Church Building, 3rd Flr, Douglas Street, Port Moresby (PO Box 739, Port Moresby)	212500
Paraguay	(E)	25 de Mayo 171, Asunción (Casilla de Correo 404, Asunción)	49146, 44472
Peru	(E)	Edificio Pacifico-Washington, Plaza Washington, Avenida Arequipa (sextra cuadra), Lima (PO Box 854, Lima)	283830
Philippines	(E)	Electra House, 115-117 Esteban Street, Legaspi Village, Makati, Metro Manila (PO Box 1970 MCC)	8910-51/8
Qatar	(E)	PO Box 3, Doha	321991/4
Rwanda	c/o	Kinshasa, Zaïre	
Saudi Arabia	(E)	Jedda Towers, Citibank Building, (PO Box 393), Jedda	27306, 27122
Senegal	(E)	20 rue du Docteur Guillet (BP 6025), Dakar	27051
Seychelles	(HC)	Victoria House, 3rd Flr, PO Box 161, Victoria	23055/6
Sierra Leone	(HC)	Standard Bank of Sierra Leone Ltd Blg, Lightfoot Boston Street, Freetown	23961-5
Singapore	(HC)	Tanglin Circus, Singapore 10 (Tanglin PO Box 19)	639333
Solomon Islands	(HC)	PO Box 676, Honiara	
Somalia	(E)	Waddada Xasan Geeddi Abtoow 7/8, (PO Box 1036) Mogadishu	22088/9, 34072/3
South Africa	(E)	6 Hill Street, Arcadia, Pretoria 0002	74-3121
	(CG)	5th Flr, Nedbank Mall, 145/7 Commissioner Street, Johannesburg 2000 or PO Box 10101	218161
	(CG)	11th Flr, African Eagle Centre, 2 St Georges Street, PO Box 1346, Cape Town 8000	41-1466/8
	(CG)	7th Flr, Barclays Bank Building, Field Street, Durban 4001	313131 (5 lines)
Sri Lanka	(HC)	Galle Road, Kollupitiya, Colombo 3 (PO Box 1433, Colombo)	27611/17
Sudan	(E)	PO Box 801, New Aboulela Building, Barlaman Avenue, Khartoum	70760, 70766-9
Suriname	(HC)	44 Main Street, PO Box 625, Georgetown, Guyana	65881/2/3/4
Swaziland	(HC)	Allister Miller Street, Mbabane (Private Bag, Mbabane)	42581-42586
Tahiti	c/o	(E) 35 rue du Faubourg St. Honoré, 75008 Paris, France	266-91-42

Tanzania	(HC)	Permanent House, corner Azikiwe Street/ Independence Avenue, PO Box 9112, Dar es Salaam	29601
Thailand	(E)	Ploenchit Road, Bangkok	2527161 (Monday to Friday 0800-1630) All other times 2527160
Togo	c/o	Accra, Ghana	
Tonga	(HC)	PO Box 56, Nuku'alofa	
Tunisia	(E)	5 place de la Victoire, Tunis	245100, 245324, 245649, 244805
Turkey	(E)	Sehit Ersan Caddesix, 46a Cankaya, Ankara	274310/5
	(CG)	Tepebasi, Beyoglu, Istanbul	4475459,498874
U.S.S.R.	(E)	Kutuzovsky Prospekt 7/4, Moscow G248	241-1033/4
		Outside office hours	231-8511/2, 231-2331
United Arab Emirates	(E)	PO Box 248, Abu Dhabi	43033/4/5
	(E)	Tariq Ibn Zayed Street, (PO Box 65) Dubai	431070, 431893, 431993
Upper Volta	c/o	Abidjan, Ivory Coast	
Uruguay	(E)	Calle Marco Bruto 1073, Montevideo	791033, 791133, 799173
Venezuela	(E)	12th Flr, Avenida La Estancia No 10, Ciudad Commercial Tamanaco, Chuao, Caracas, (Apartado 1246, Caracas)	91-12-55, 91-14-77
Vietnam	(E)	16 Pho Ly Thuong Kiet, Hanoi	2349, 2510
Yemen Arab Rep (North)	(E)	13 Al Qasr al Jumhuri, (PO Box 1287), Sana'a	5428
Yemen People's Dem Rep (South)	(E)	28 Ho Chi Minh Street, Khormaksar, Aden	24801-4
Zaïre	(E)	9 avenue de l'Equateur, 5th Flr, (BP 8049), Kinshasa	23483/6, 22666
Zambia	(HC)	Independence Avenue, PO Box RW50, Lusaka	51122

Abbreviations:

E	Embassy
HC	High Commission
CG	Consulate General
TC	Trade Commission
CR	Commercial Representative

Embassies, High Commissions and Consulates in London

The Consulates are closed on English public holidays and on the national holidays observed in their own countries. Consulates & visa or consular offices have been listed for preference.

Afghanistan	31 Princes Gate, SW7 IQQ	580 8891
Algeria	54 Holland Park, W11 3RS	221 7800/4
Argentina	53 Hans Place, SW1X 0LB	584 1701/2
Australia	Canberra House, 10-16 Maltravers St, Strand, WC2R 3ER	438 8286
Bahrain	98 Gloucester Road, SW7 4AU	370 5132/3
Bangladesh	28 Queen's Gate, SW7 5JA	584 0081
Benin	125 High St, Edgware, Middx, HA8 7HS	951 1234

Bolivia	106 Eccleston Mews, SW1X 9AD	235 4255
Botswana	162 Buckingham Palace Rd, SW1W 9TJ	730 5216/9
Brazil	6 Deanery Street, W1Y 5LH	499 7441
Brunei	Brunei House, 35 Norfolk Square, W2	402 0045
Burma	19a Charles Street, W1X 8ER	499 8841
Cameroon	84 Holland Park, W11 3SB	727 0771
Chile	12 Devonshire Street, W1N 1FS	580 6392
China	31 Portland Place, W1N 3AG	636 5637
Colombia	Suite 10, 140 Park Lane, W1Y 3DF	493 4565
Costa Rica	1 Culross Street, W1	493 9761
Cuba	57 Kensington Court, W8 5DQ	937 8226
Ecuador	3 Hans Crescent, SW1X 0LN	584 2648
Egypt	19 Kensington Palace Gardens, W8 4QL	229 8810/8/9
El Salvador	Flat 16, Edinburgh House, 9B Portland Place, WIN 3AA	636 9563/4
Ethiopia	17 Princes Gate, SW7 1PZ	589 7212
Fiji	34 Hyde Park Gate, SW7 5BN	584 3661/4
Gabon	48 Kensington Court, W8 5DB	937 5285/9
Gambia	60 Ennismore Gdns, SW7 1NH	584 1242/3
Ghana	38 Queen's Gate, SW7 5HT	584 6311
Guyana	3 Palace Court, Bayswater Rd, W2 4LP	229 7684/8
Honduras	48 George St, W1H 5RF	486 4880
Hong Kong	6 Grafton Street, W1X 3LB	499 9821
Iceland	1 Eaton Terrace, SW1W 8EY	730 5131/2
India	India House, Aldwych, WC2B 4NA	836 8484
Indonesia	38 Grosvenor Square, W1X 9AD	499 7661
Iran	50 Kensington Court, W8 5DD	937 5225/8
Iraq	22 Queen's Gate, SW7 5JG	584 7141
Israel	2 Palace Green, W8 4QB	937 8050
Ivory Coast	1 Upper Belgrave St, SW1X 8BJ	235 6991
Japan	43-46 Grosvenor Street, W1X 0BA	493 6030
Jordan	6 Upper Phillimore Gardens, W8 7HB	937 3685/6/7
Kenya	45 Portland Place, W1N 4AS	636 2371
Korea, South	4 Palace Gate, W8 5NF	581 0247
Kuwait	46 Queen's Gate SW7	580 8471
Lao People's Dem Rep (Laos)	5 Palace Green, W8 4QA	937 9519
Lebanon	15 Palace Gardens Mews, W8 4RB	229 8485
Lesotho	16a St James's St, SW1A 1EU	839 1154
Liberia	21 Princes Gate, SW7 1QB	589 2263
Libya	58 Princes Gate, SW7 2BW	589 5235/7
Madagascar (Malagasy Rep)	33 Thurloe Square, SW7	584 3714
Malawi	33 Grosvenor St, W1X 0HS	491 4172
Malaysia	45 Belgrave Square, SW1X 8QT	245 9221
Mauritius	32 Elvaston Place, SW7	581 0294
Mexico	8 Halkin Street, SW1X 7DW	235 6393

Mongolia	7 Kensington Ct, W8 5DL	937 0150
Morocco	49 Queen's Gate Gardens, SW7 5NE	584 8827/9
Nepal	12a Kensington Palace Gardens, W8 4QU	229 1594
New Zealand	New Zealand House, Haymarket, SW1Y 4TQ	930 8422
Nicaragua	8 Gloucester Rd, SW7 4RB	584 3231
Nigeria	56–57 Fleet Street, EC4	353 3776/7/8/9/0
Oman	64 Ennismore Gardens, SW7 5DN	584 6782
Pakistan	35 Lowndes Square, SW1X 9JN	235 2044
Panama	Wheatsheaf House, 4 Carmelite Street, EC4Y 0BN	353 4792/3
Papua New Guinea	14 Waterloo Place, SW1R 4AR	930 0922/7
Paraguay	Braemar Lodge, Cornwall Gardens, SW7 4AQ	937 6629
Peru	52 Sloane Street, SW1X 9SP	235 6867
Philippines	9a Palace Green, W8 4QE	937 3646/7/8
Qatar	10 Reeves Mews, Grosvenor Square, W1Y 3PB	499 8831
Saudi Arabia	30 Belgrave Square, SW1X 8QB	235 0756
Senegal	11 Phillimore Gardens, W8 7QG	937 0925/6
Seychelles	2 Mill Street, W1	499 9951
Sierra Leone	33 Portland Place, W1N 3AG	636 6483
Singapore	5 Chesham Street, SW1X 8ND	235 9067/9
Somalia	60 Portland Place, W1N 3DG	580 7148/9
South Africa	16 Charles II Street, Haymarket, SW1Y 4QU	839 2313
Sri Lanka	13 Hyde Park Gardens, W2 2LU	262 1841
Sudan	3/5 Cleveland Row, St, James's, SW1A 1DD	839 8080
Swaziland	58 Pont Street, SW1X 0AE	589 5447/8
Syria	5 Eaton Terrace, SW1W 8EX	730 0384
Taiwan	Free Chinese Centre, 100 Baker St, W1	935 5428
Tanzania	43 Hertford Street, W1Y 7TF	499 8951
Thailand	30 Queen's Gate, SW7 5JB	589 2857
Togo	10 New Bond Street, W1Y 9PF	493 6970/9076
Tonga	17th Flr, New Zealand House, Haymarket, SW1Y 4TE	839 3287
Tunisia	29 Princes Gate, SW7 1QG	584 8117
Turkey	Rutland Lodge, Rutland Gdns, Knightsbridge, SW7 1BW	589 0360
U.S.S.R.	5 Kensington Palace Gardens, W8 4QS	229 3215/6
United Arab Emirates	48 Princes Gate, SW7 1PT	589 3434
Upper Volta	104 Park Street, W1	491 7351
Uruguay	48 Lennox Gardens, SW1X 0DL	589 8835
Venezuela	71a Park Mansions, Knightsbridge, SW1X 7QU	589 9916
Vietnam	12-14 Victoria Road, W8	937 1912
Yemen People's Dem Rep (South)	57 Cromwell Road, SW7 2ED	584 6607/9
Yemen Arab Rep (North)	41 South Street, W1Y 5PD	499 4744
Zaïre	26 Chesham Place, SW1X 8HH	235 6137
Zambia	Zambia House, 7/11 Cavendish Place, W1M 0HB	580 0691

The following countries have no official representatives in London but have diplomatic offices as indicated:

Angola	Servicos de Emigracao e Fronteras, Rua Mouzinho de Albuquerque II, Luanda.	
Burundi	Square Marie-Louise, 46 Brussels 1040, Belgium.	(02) 733 55 92, 733 57 15
Central African Rep	29 blvd Montmorency, Paris 16, France.	224 42 56
Chad	65 rue Belles Feuilles, Paris 16, France.	553 36 75/727 39 27, 704 99 95
Congo People's Rep (Brazzaville)	57 bis, rue Scheffer, Paris 16, France.	727 77 09
Djibouti	70 blvd Pereire, Paris 17, France.	
Guatemala	73 rue de Courcelles, Paris 8, France.	227 78 63
Guinea Republic	62 blvd de Courcelles, Paris 17, France.	766 54 98, 766 55 19
Mali	112 rue C. Lemmonier, Brussels 1060, Belgium.	(02) 45 75 89
Mauritania	5 rue de Montevideo, Paris 16, France.	504 88 54
Niger	154 rue de Longchamps, Paris 16, France.	504 80 60
Rwanda	101 blvd St Michel, Brussels 1040, Belgium.	(02) 734 17 63
Uganda	13 ave Raymond Poincaré, Paris 16, France.	727 46 80

Foreign Tourist Offices in London

Australia	Australian Tourist Commission, 49 Old Bond Street, W1	499 2247
Egypt	United Arab Rep Tourist Information, 62a Piccadilly, W1	493 5282
Hong Kong	Hong Kong Tourist Association, 14 Cockspur Street, SW1	930 7955
Iceland	Iceland Tourist Information Bureau, 73 Grosvenor Street, W1	493 9971
India	India Government Tourist Office, 21 New Bond Street, W1	493 0769
Iraq	Iraqi Tourist Office, 4 Lower Regent Street, SW1	930 1155
Israel	Israel Government Tourist Office, 59 St James's Street, SW1	493 2431
Japan	Japan National Tourist Organisation, 167 Regent Street, W1	734 9638
Kenya	Kenya Tourist Office, 318 Grand Buildings, Trafalgar Square, WC2	839 4477
Lebanon	Lebanese Tourist & Information Centre, 90 Piccadilly, W1	409 2031
Malawi	Malawi Department of Tourism, 52 High Holborn, WC1	242 3131
Mauritius	Mauritius Tourist Office, 32 Shaftesbury Avenue, W1	437 6394
Mexico	Mexican National Tourist Council, 52 Grosvenor Gardens, SW1	730 0128
Morocco	Moroccan Tourist Office, 174 Regent Street, W1	437 0073
New Zealand	New Zealand Government Tourist Bureau, New Zealand House, 80 Haymarket, SW1	930 8422

Papua New Guinea	Papua New Guinea Tourist Board, 52 High Holborn, WC1	242 3131
Seychelles	Seychelles Tourist Information Office, 11 Bolt Court, EC4	353 9510
South Africa	South African Tourist Corporation, 13 Lower Regent Street, SW1	839 7462
Sri Lanka	Ceylon Tourist Board, 52 High Holborn, WC1	405 1194
Thailand	Information Service of Thailand, 28 Princes Gate, SW7	584 5421
Tunisia	Tunisian Tourist Office, 7a Stafford Street, W1	499 2234
Turkey	Turkish Tourism Information Office, 49 Conduit Street, W1	734 8681
U.S.S.R.	Intourist Moscow, 292 Regent Street, W1	580 4974
Zambia	Zambia National Tourist Bureau, 129 Finsbury Pavement, EC2	638 8882

US Foreign Service Posts Overseas

Afghanistan	(E)	Kabul, Wazir Akbar Khan Mina	24230-9
Algeria	(E)	Algiers, 4 Chemin Cheikh Bachir Brahimi (ex Beaurepaire)	601425/601255/ 601186/601716/ 601828/603670
	(C)	Oran, 14 Square de Bamako	355502/352665
Argentina	(E)	Buenos Aires, 4300 Colombia, Palermo	774-8811, 774-7611
Australia	(E)	Canberra, Moonah Pl, Canberra, ACT 2600	(062) 73-3711
	(CG)	Melbourne, 24 Albert Rd, South Melbourne, Victoria 3205	699-2244
	(CG)	Sydney, 36th Flr, T & G Tower, Hyde Park Square, Park and Elizabeth Sts, Sydney, NSW 2000	235-7044
	(C)	Brisbane, 141 Queen Street, Brisbane, Qld 4000	(07) 221-1338
(C)	(C)	Perth, 246 St. George's Terrace	22-4466
Bahamas	(E)	Nassau, Mosmar Bldg, Queen St	(809) 322-1700, 322-1181
Bahrain	(E)	Manama, Shaikh Isa Rd, PO Box 431	714151
Bangladesh	(E)	Dacca, Adamjee Court Bldg, (5th Flr), Montijheel Commercial Area, GPO Box 323, Ramna	244220 to 244229
Barbados	(E)	Bridgetown, PO Box 302	63574-7
Belize	(CG)	Belize City, Gabourel Lane and Hutson St	3261
Benin	(E)	Cotonou, Rue Caporal Anani Bernard, Boître Postale 2012	31-26-92/3
Bolivia	(E)	La Paz, Banco Popular Del Peru Bldg, Corner of Calles Mercado y Colon	50251
Botswana	(E)	Gaborone, PO Box 90	2944/7
Brazil	(E)	Brasilia, Lote No 3, Avenida das Nocoes	(0612) 223-0120
	(CG)	Rio de Janeiro, Avenida President Wilson 147	(021) 252-8055/6/7
	(CG)	São Paulo, Edificio Conjunto Nacional, Rua Padre Joao Manuel 20	(011) 289-3155, 289-3355, 289-3560, 289-3722
	(C)	Belem, Avenida Oswaldo Cruz 165	(0912) 223-0800
	(C)	Porto Alegre, Rua Uruguai 155 (11th Flr)	(0512) 31-1888, 32-1046, 32-2046, 32-3046

	(C)	Recife, Rua Goncalves Maia 163	(0812) 221-14-12, 222-66-12, 222-65-77
	(C)	Salvador, Edificio Fundacao Politecnica Bloco A, 4th Flr Avenida Sete de Setembro 73/79	(071) 243-4908, 243-4938
Burma	(E)	Rangoon, 581 Merchant St	18055
	(C)	Mandalay, 71st St & South Moat Rd	555
Burundi	(E)	Bujumbura, Chaussée Prince Louis Rwagasore, Boîte Postale 1720	34-54
Cameroon	(E)	Yaoundé, Rue Nachtigal, Boîte Postale 817	221633/220512
	(C)	Douala, 21 Avenue du Gen de Gaulle, Boîte Postale 4006	423434/425331
Cape Verde	(E)	Praia, Rua Hoji Ya Yenda 81, 3rd Flr	553
Central African Rep	(E)	Bangui, Avenue du 1 er Janvier 1966, BP 924	2050/1
Chad	(E)	N'Djamena, Rue du Lt Col Colonna D'Oranano, BP 413	30-91/2/3/4
Chile	(E)	Santiago, Codina Bldg, 1343 Agustinas	82801-4
China	(E)	Peking, Kuang Hua Lu, Dept of State, Wash, DC 20520	522-033
Colombia	(E)	Bogotá, Calle 37, 8-40	285-1300
	(C)	Cali, Edificio Pielroja, Carrera 3, No 11-55	88-11-36/7
	(C)	Medellin, Edificio Santa Helena, Calle 52 No 49-27, Apartado Aereo 980	313-188
	(C)	Barranquilla, Edificio Seguros Tequendama, Calle 34 No 44 63 (10th Flr), Apartado Aereo 2306	56599
Congo,	(E)	Brazzaville, BP 1015	
Costa Rica	(E)	San José, Avenida 3 and Calle 1	22-55-66
Cuba	(USINT)	Havana, Swiss Embassy, Calcado entre L & M, Vedado Seccion	320551, 329700
Djibouti	(E)	Djibouti, Villa Plateau du Serpent, Blvd Maréchal Joffre, Boîte Postale 185	35-95, 35-49
Dominican Rep	(E)	Santo Domingo, Corner of Calle Cesar Nicolas Pensen & Calle Leopoldo Navarra	682-2171
Ecuador	(E)	Quito, 120 Avenida Patria	548-000
	(CG)	Guayaquil, Casilla X	511570
Egypt	(E)	Cairo, 5 Sharia Latin America, Box 10	28211/9
	(CG)	Alexandria, 110 Ave Horreya	25306, 25607, 28458
	(C)	Port Said, Apt 4, 8 Sharia Aby El Feda Metarch El Baher	8000, 8586, 8622
El Salvador	(E)	San Salvador, 1230, 25 Avenida Norte	25-7100
Equatorial Guinea	(E)	Malabo, Armengol Coll and Asturias Sts	
Ethiopia	(E)	Addis Ababa, Entoto St, PO Box 1014	110666
Fiji	(E)	Suva, 31 Lofters St, PO Box 218	23031
French West Indies	(C)	Martinique, 14 Rue Blenac, Boître Postale 561, Fort de France 97206	71.93.01/3
Gabon	(E)	Libreville, Blvd de la Mer, Boître Postale 4000	72-20-03/4, 72-13-37, 72-03-48
Gambia	(E)	Banjul, 16 Buckle St, PO Box 596	526-7
Ghana	(E)	Accra, Liberia & Kinbu Rds, PO Box 194	66811

Guatemala	(E)	Guatemala City, 7-01 Avenida de la Reforma, Zone 10	31-15-41
Guinea	(E)	Conakry, 2nd Blvd and 9th Ave, Boîte Postale 603	415-20/1/2/3/4
Guinea-Bissau	(E)	Bissau, Avenida Domingos Ramos, CP 297	28-16/7
Guyana	(E)	Georgetown, 31 Main St	62687, Ext 26
Haiti	(E)	Port-au-Prince, Harry Truman Blvd	20200
Honduras	(E)	Tegucigalpa, Avenido La Paz	22-3121/2/3/4/7
Hong Kong	(CG)	Hong Kong, 26 Garden Rd	239011
Iceland	(E)	Reykjavik, Laufasvegur 21	29100
India	(E)	New Delhi, Shanti Path, Chanakyapuri 21	690351
	(CG)	Bombay, Lincoln House, 78 Bhulabhai Desai Rd	363611 to 363618
	(CG)	Calcutta, 5/1 Ho Chi Minh Sarahi, Calcutta 700071	44-3611/6
	(CG)	Madras, Mount Rd 6	83041
Indonesia	(E)	Djakarta, Medan Merdeka Selatan 5	40001-9
	(C)	Medan, Djalan Imam Bondjol 13	22290
	(C)	Surabaya, Djalan Raya Dr Sutomo 33	67545
Iran	(E)	Tehran, 260 Takhte Ave, PO Box 50; or Box 2000	820091/9, 824-001, 820-091, 829-051
	(C)	Isfahan, Trade Ctr Bldg, Rm 201-203, Corner Pahlavi & Chahar Bgh Blvds	32079-9
	(C)	Shiraz, Charkhabi Bldg, Bagh Eram Ave, (Near Eram Garden) PO Box 500	32023/4
	(C)	Tabriz, Shahnaz Ave	2101, 5487
Iraq	(USINT)	Baghdad, 52/5/35 Masbah-Opp For Ministry Club, PO Box 2447 Alwiyah *The Embassy was closed on June 7, 1967. The Government of Belgium serves as protective power for the US in Iraq* Basra: *The consulate at Basra will remain closed until further notice.*	96138/9
Israel	(E)	Tel Aviv, 71 Hayarkon St	54338
Ivory Coast	(E)	Abidjan, 5 Rue Jesse Owens, Boîte Postale 1712	32-46-30
Jamaica	(E)	Kingston, Jamaica Mutual Life Centre, 2 Oxford Rd, 3rd Flr	(809)92-94850
Japan	(E)	Tokyo, 10-1, Akasaka 1-chome, Minato-ku (107)	583-7141
	(CG)	Naha, Okinawa, No 2129, Gusukuma, Urasoe City	0988-77-8142, 0988-77-8627
	(CG)	Osaka-Kobe, Osaka Office, 9th Flr, Sankei Bldg, 4-9, Umeda 2-chome, Kita-ku, Osaka (503)	(06) 341-2754/7
		Kobe Office, 10, Kano-cho 6-chome, Ikuta-ku, Kobe (650)	(078) 331-6865/8
	(C)	Fukuoka, 5-26 Ohori 2 chome, Chuo-ku, Fukuoka-shi 810, or Box 10	(092) 751-9331
	(C)	Sapporo, North 1 West 13	221-5121/3
Jerusalem	(CG)	18 Agron Rd; and Nablus Rd	226312
		Nablus Rd	282231/272681
Jordan	(E)	Amman, Jebel Amman, PO Box 354	44371-6
Kenya	(E)	Nairobi, Cotts House, Wabera St, PO Box 30137	334141
Korea	(E)	Seoul, Sejong-Ro	72-2601 to 72-2619

Kuwait	(E)	Kuwait, PO Box 77 SAFAT	424156/8
Lao People's Dem Rep (Laos)	(E)	Vientiane, rue Bartholonie, Boître Postale 114 Box V.	3126, 3570
Lebanon	(E)	Beirut, Corniche at Rue Ain Mreisseh	361-800
Lesotho	(E)	Maseru, PO Box MS 333	2666/3954
Liberia	(E)	Monrovia, United Nations Dr	22991/2/3/4
Libya	(E)	Tripoli, Shari Mohammad Thabit, PO Box 289	34021/6
Madagascar	(E)	Antananarivo, 14 and 16 rue Rainitovo, Antsohavola, Boîte Postale 620	212-57
Malawi	(E)	Lilongwe, PO Box 30016	30396/30166
Malaysia	(E)	Kuala Lumpur, AIA Bldg, Djalan Ampang, PO Box No 35	26321
Mali	(E)	Bamako, Rue Testard and Rue Mohamed V	246-63/4, 248-34, 248-45
Mauritania	(E)	Nouakchott, Boîte Postale 222	52660/52663
Mauritius	(E)	Port Louis, Anglo-Mauritius House (6th Flr), Intendance St	2-3218/9
Mexico	(E)	Mexico City, Paseo de la Reforma 305, Mexico 5, DF	(905) 553-3333
	(CG)	Guadalajara, Jal; Progreso 175	25-29-98, 25-27-00
	(CG)	Hermosillo, Son; Isssteson Bldg. 3rd Flr, Miguel Hidalgo y Costilla No 15	3-89-22/3/4/5
	(CG)	Monterrey, NL, Avenida Constitucion 411 Poniente	4306 50/59
	(CG)	Tijuana, BC, Topachula 96	386-1001
	(CG)	Ciudad Juarez, Chi, 2286 Ave, 16 de Septiembre	34048
	(C)	Matamoros, Tamps Ave Primera No 232	2-52-50/1/2
	(C)	Mazatlan, Sin, 6 Circunvalacion No 6 (at Venustiana Carrana)	1-26-85, 1-26-87
	(C)	Merida, Yuc, Paseo Montejo 453 Apartado Apartado Postal 1301	2-70-11, 2-70-78
	(C)	Nuevo Laredo, Tamps, Avenida Allende 3330, Col Jardin	2-00-05
Morocco	(E)	Rabat, 2 Ave de Marrakech, Box 99	30361/2
	(CG)	Casablanca, 8 Blvd Moulay Youssef or Box 80	260521/23, 260562, 278457
	(CG)	Tangier, Chemin des Amoureux	(09) 359-04
Mozambique	(E)	Maputo, 35 Rua Da Mesquita, 2nd Flr	26051/2/3
Nepal	(E)	Kathmandu, Pani Pokhari	11199, 12718, 11603, 11604
Netherlands Antilles	(CG)	Curacao, St Anna Blvd 19 PO Box 158 vice John B Gorsirawea 1	13066
New Zealand	(E)	Wellington, 29 Fitzherbert Ter, Thorndon, PO Box 1190	722-068
	(CG)	Auckland, 5th Flr, Old Northern Bldg Society Bldg, Queen & Wellesley St, or PO Box 7140 Wellesley St,	375-102,30-992
Nicaragua	(E)	Managua, Km 4-1/2 Carretera Sur	23061-8, 23881-7
Niger	(E)	Niamey, (No street address) Boîte Postale 201	72-26-61/2/3/4, 72-26-70
Nigeria	(E)	Lagos, 1 King's College Rd, PO Box 554	57320
	(C)	Enugu & Ibadan: *The consulate at Enugu was closed on October 13, 1967 and that at Ibadan on August 1, 1978. Direct inquiries to American Embassy, Lagos.*	

	(C)	Kaduna, 5 Ahmadu Bello Way	23373-7
Oman	(E)	Muscat, PO Box 966	722021
Pakistan	(E)	Islamabad, Diplomatic Enclave Ramna 4	26161-26179
	(CG)	Karachi, 8 Abdullah Haroon Rd	515081
	(CG)	Lahore, 50 Zafar Ali Rd, Gulberg 5	81081-5
	(C)	Peshawar, 11 Hospital Road	73061/73405
Panama	(E)	Panama, Avenida Balboa y Calle 38, Apartado 6959, RP 5	25-3600
Papua New Guinea	(E)	Port Moresby, Armit St, PO Box 3492	211455, 211594, 211654
Paraguay	(E)	Asunción, 1776 Mariscal Lopez Avenue	21041/9
Peru	(E)	Lima, Corner Avenidas Inca Garcilaso de la Vega & España, PO Box 1995	286000
Philippines	(E)	Manila, 1201 Roxas Blvd	598-011
	(C)	Cebu, 3rd Flr, Philippine American Life Insurance Bldg, Jones Ave	7-95-10/24
	(C)	Oporto, Apartado No 88, Rua Julio Dinis 826-30	6-3094
	(C)	Ponta Delgada, Sao Miguel, Azores, Avenida D. Henrique	22216/7
Qatar	(E)	Doha, Farig Bin Omran (opp. TV station), PO Box 2399	87701/2/3
Rwanda	(E)	Kigali, Blvd de la Révolution, BP 28	5601
Saudi Arabia	(E)	Jidda, Palestine Rd, Ruwais	53410, 54110, 52188, 52396, 52589
	(CG)	Dhahran, Between Aramco HQ and Dhahran Int'l Airport, PO Box 81	43200, 43452, 43613
	(LO)	Riyadh	
Senegal	(E)	Dakar, Boîte Postale 49, Avenue Jean XXIII	20206
Seychelles	(E)	Victoria, Box 148	23921/2
Sierra Leone	(E)	Freetown, Corner Walpole and Siaka Stevens St	26481
Singapore	(E)	Singapore, 30 Hill St	30251
Somalia	(E)	Mogadishu, Corso Primo Luglio	2811
South Africa	(E)	Pretoria, Thibault House, 225 Pretorius St	48-4266
	(CG)	Cape Town, Broadway Industries Centre, Heerengracht, Foreshore	021-471280
	(CG)	Durban, Durban Bay House, 29th Flr, 333 Smith St, Durban 4001	324-737
	(CG)	Johannesburg, 11th Flr, Kine Centre, Commissioner and Kruis St, PO Box 2155	(011) 21-1684/7
Sri Lanka	(E)	Colombo, 44 Galle Rd, Colombo 3, PO Box 106	26211
Sudan	(E)	Khartoum, Gamhouria Ave, PO Box 699	74611, 74700
Suriname	(E)	Paramaribo, Dr Sophie Redmondstraat 13, PO Box 1821	73024, 75620
Swaziland	(E)	Mbabane, Embassy House, PO Box 199, Allister Miller St	2272/3/4
Syria	(E)	Damascus, Abu Rumaneh, Al Monsur St No 2 PO Box 29	332315, 332814
	(CG)	Aleppo: *The Consulate General in Aleppo will remain closed until further notice*	
Taiwan	(E)	Taipei: *The American Embassy was closed on 1st January 1979, to be replaced by an American Institute which performs many of the functions of an embassy*	

Tanzania	(E)	Dar es Salaam, National Bank of Commerce Bldg, City Dr, PO Box 9123	22775
	(C)	Zanzibar, 83A Tuzungumzeni Sq, PO Box 4	2118/9
Thailand	(E)	Bangkok, 95 Wireless Rd	252-5040, 252-5171
	(C)	Chiang Mai, Vidhayanond Rd	235566-7
	(C)	Songkhla, 9 Sadao Rd	311-589
	(C)	Udorn, 35/6 Supakitjanya Rd	221548
Togo	(E)	Lomé, Rue Pelletier Caventou & Rue Vouban, Boître Postale 852	29-91
Trinidad & Tobago	(E)	Port-of-Spain, 15, Queen's Park West, PO Box 752	62-26371
Tunisia	(E)	Tunis, 144 Ave de la Liberté	282.566
Turkey	(E)	Ankara, 110 Ataturk Blvd	26 54 70
	(CG)	Istanbul, 104-108 Mesrutiyet Caddesi, Tepebasi	45 32 20
	(CG)	Izmir, 386 Ataturk Caddesi	132135/7
	(C)	Adana, Ataturk Caddesi	14702/3, 14818
Uganda	(E)	Kampala, PO Box 7707 Embassy House, 9-11 Parliament Ave *The Embassy at Kampala was closed on Nov 10th 1973, and all American Diplomatic personnel were withdrawn. The Embassy will remain closed until further notice. The Federal Republic of Germany serves as protective power for the United States in Uganda*	54451
U.S.S.R.	(E)	Moscow, Ulitsa Chaykovskogo 19/21/23	252-00-11 to 252-00-19
	(CG)	Leningrad UL, Petra Lavrova St 15; Box L	274-8235
United Arab Emirates	(E)	Abu Dhabi, Shaikh Khalid Bldg, Corniche Rd, PO Box 4009	61534/35
	(BO)	Dubai, Al Futtaim Bldg, Creek Rd, Deira, PO Box 5343	29003
Upper Volta	(E)	Ouagadougou, Boîte Postale 35	35442/4/6
Uruguay	(E)	Montevideo, Calle Lauro Muller 1776	40-90-51, 40-91-26
Venezuela	(E)	Caracas, Avenida Francisco de Miranda and Avenida Principal de la Floresta	284-7111
	(C)	Maracaibo, Edificio Matema, 1 Piso, Avenida 15 Calle 78	(061) 51-65-06/7
Yemen People's Dem Rep (South)		Aden: *All personnel were withdrawn from Embassy Aden on October 26, 1969. The Embassy will remain closed until further notice. The Government of Great Britain serves as protective power for the United States in the Yemen People's Dem Republic.*	
Yemen Arab Rep (North)	(E)	Sana'a, Box 33	
Zaïre	(E)	Kinshasa, 310 Avenue des Aviateurs	25881/2/3/4/5/6
	(C)	Bukavu, Mobutu Ave, Boîte Postale 3037	2594
	(C)	Lubumbashi, 1028 Blvd de l'Ueac, Boîte Postale 1196	2324/5
Zambia	(E)	Lusaka, PO Box 1617	50222

Abbreviations;

C	Consulate
CG	Consulate General
E	Embassy
LO	Liaison Officer
USINT	U.S. Interests Section
BO	Branch Office

Foreign Embassies and Legations in Washington D.C.

Afghanistan	2001 24th St, NW,DC 20008. Tel. 234-3770
Algeria	2118 Kalorama Road, NW, DC 20008. Tel. 234-7246
Argentina	1600 New Hampshire Ave, NW, DC 20009. Tel. 387-0705 and 265-4557
Australia	1601 Massachusetts Ave, NW, DC 20036. Tel. 797-3000
Bahrain	2600 Virginia Ave, NW, DC 20037. Tel. 965-4930
Bangladesh	3421 Massachusetts Ave, NW, DC 20007. Tel. 337-6644
Benin	2737 Cathedral Ave, DC, 20008. Tel. 232-6656
Bolivia	1625 Massachusettes Ave, NW, DC 20036. Tel. 483-4410
Botswana	4301 Connecticut Ave, NW, DC 20008. Tel. 244-4990
Brazil	3006 Massachusetts Ave, NW, DC 20008. Tel. 797-0100
Burma	2300 S St, NW, DC 20008. Tel. 332-9044
Burundi	2717 Connecticut Ave, NW, DC 20008. Tel. 387-4477
Cameroon	2825 Normanstone Dr. NW, DC 20008. Tel. 265-8790
Cape Verde	1120 Connecticut Ave, NW, Suite 300, DC 20036. Tel. 659-3148
Central African Rep	1618 22nd St, NW, DC 20008. Tel. 265-5637
Chad	2600 Virginia Ave, DC 20037. Tel. 331-7696
Chile	1736 Massachusetts Ave, NW, DC 20036. Tel. 785-1746
China	2300 Connecticut Ave, NW, DC 20008. Tel. 667-9000
Colombia	2118 LeRoy Place, NW, DC 20008. Tel. 387-5828
Congo	1800 New Hampshire Ave, DC 20009
Costa Rica	2112 S St, NW, DC 20008. Tel. 234-2945
Ecuador	2535 15th St, NW, DC 20009. Tel. 234-7200
Egypt	2310 Decatur Place, NW, DC 20008. Tel. 234-0980
El Salvador	2308 California St. NW, DC 20008. Tel. 265-3480
Ethiopia	2134 Kalorama Rd, NW, DC 20008. Tel. 234-2281
Fiji	1629 K St, NW, DC 20006
Gabon	2034 20th St, NW, DC 20009. Tel. 797-1000
Ghana	2460 16th St, DC 20009. Tel. 462-0761
Guatemala	2220 R St, NW, DC 20008. Tel. 332-2865
Guinea	2112 Leroy Place, NW, DC 20008. Tel. 483-9420
Guyana	2490 Tracy Place, NW, DC 20008. Tel. 265-6900
Honduras	4301 Connecticut Ave, NW, DC 20008. Tel. 966-7700
Hong Kong	c/o Embassy of Great Britain, 3100 Massachusetts Ave, NW, DC 20008
Iceland	2022 Connecticut Ave, DC 20008. Tel. 265-6653
India	2107 Massachusetts Ave, NW, DC 20008. Tel. 265-5050
Indonesia	2020 Massachusetts Ave, NW, DC 20036. Tel. 293-1745
Iran	3005 Massachusetts Ave, NW, DC 20008. Tel. 797-6500
Iraq	Iraqi Interest Section, c/o Embassy of India, 1801 P St, NW, DC 20036. Tel. 483-7500
Israel	1621 22nd St, NW, DC 20008. Tel. 483-4100
Ivory Coast	2424 Massachusetts Ave, DC 20008. Tel. 483-2400
Japan	2520 Massachusetts Ave, NW, DC 20008. Tel. 234-2266
Jordan	2319 Wyoming Ave, NW, DC 20008. Tel. 265-1606

Kenya	2249 R St, NW, DC 20008. Tel. 387-6101
Korea, South	2320 Massachusetts Ave, NW, DC 20008. Tel. 483-7383
Kuwait	2940 Tilden St, NW, DC 20008. Tel. 966-0702
Lao People's Den Rep (Laos)	2222 S St, NW, DC 20008. Tel. 332-6416
Lebanon	2560 28th St, NW, DC 20008. Tel. 332-0300
Lesotho	1601 Connecticut Ave, NW, DC 20009. Tel. 462-4190
Liberia	5201 16th St, NW, DC 20011. Tel. 723-0437
Libya	1118 22nd St, NW, DC 20037. Tel. 452-1290
Madagascar	2374 Massachusetts Ave, NW, DC 20008. Tel. 265-5525
Malawi	1400 20th St, NW, DC 20036. Tel. 296-5530
Malaysia	2401 Massachusetts Ave, NW, DC 20008. Tel. 234-7600
Mali	2130 R St, NW, DC 20008. Tel. 332-2249
Mauritania	2129 LeRoy Place, NW, DC 20008. Tel. 232-5700
Mauritius	4301 Connecticut Ave, NW No 134, DC 20008. Tel. 244-1491
Mexico	2829 16th Street, NW, DC 20009. Tel. 234-6000
Morocco	1601 21st St, NW, DC 20009. Tel. 462-7979
Nepal	2131 Leroy Place, NW, DC 20008. Tel. 667-4550
New Zealand	19 Observatory Circle, NW, DC 20008. Tel. 265-1721
Nicaragua	1627 New Hampshire Ave, NW, DC 20009. Tel. 387-4371
Niger	2204 R St, NW, DC 20008. Tel. 483-4224
Nigeria	2201 M St, NW, DC 20037. Tel. 223-9300
Oman	2342 Massachusetts Ave, NW, DC 20008. Tel. 387-2014 and 387-1980
Pakistan	2315 Massachusetts Ave, NW, DC 20008. Tel. 332-8330
Panama	2862 McGill Terr, NW, DC 20008. Tel. 483-1407
Papua New Guinea	1776 Massachusetts Ave, DC 20036
Paraguay	2400 Massachusetts Ave, NW, DC 20008. Tel. 483-6960
Peru	1700 Massachusetts Ave, NW, DC 20036. Tel. 833-9860
Philippines	1617 Massachusetts Ave, NW, DC 20036. Tel. 483-1020
Qatar	600 New Hampshire Ave, NW, DC 20037. Tel. 338-0111
Rwanda	1714 New Hampshire Ave, DC 20009. Tel. 232-2882
Saudi Arabia	1520 18th St, NW, DC 20036. Tel. 483-2100
Senegal	2112 Wyoming Ave, NW, DC 20008. Tel. 234-0540
Sierra Leone	1701 19th St, DC 20009. Tel. 265-7700
Singapore	1824 R St, NW, DC 20009. Tel. 667-7555
Somalia	600 New Hampshire Ave, NW, DC 20037. Tel. 234-3261
South Africa	3051 Massachusetts Ave, NW, DC 20008. Tel. 232-4400
Sri Lanka	2148 Wyoming Ave, NW, DC 20008. Tel. 483-4025
Sudan	600 New Hampshire Ave, NW, DC 20037. Tel. 338-8565
Suriname	2600 Virginia Ave, NW, DC 20037. Tel. 338-6980
Swaziland	4031 Connecticut Ave, Suite 441, NW, DC 20008. Tel. 362-6683
Syria	2215 Wyoming Ave, NW, DC 20008. Tel. 232-6313
Tanzania	2010 Massachusetts Ave, NW, DC 20036
Thailand	2300 Kalorama Rd, NW, DC 20008. Tel. 667-1446
Togo	2208 Massachusetts Ave, NW, DC 20008. Tel. 234-4212
Tunisia	2408 Massachusetts Ave, NW, DC 20008. Tel. 234-6644

Turkey	1606 23rd St, NW, DC 20008. Tel. 667-6400
Uganda	5909 16th St, NW, DC 20011. Tel. 726-7100
United Arab Emirates	600 New Hampshire Ave, NW, DC 20037. Tel. 338-6500
U.S.S.R.	1609 Decatur St, NW, DC 20011. Tel. 882-5829
Upper Volta	5500 16th St, NW, DC 20011. Tel. 726-0992
Uruguay	1918 F St, DC 20003. Tel. 331-1313
Venezuela	2445 Massachusetts Ave, NW, DC 20008. Tel. 265-9600
Yemen	Watergate 600, Suite 8600, New Hampshire Ave, NW, DC 20037
Zaïre	1800 New Hampshire Ave, NW, DC 20009. Tel. 234-7690
Zambia	2419 Massachusetts Ave, DC 20008. Tel. 265-9717

Foreign Consulates and Missions to the UN in New York

Afghanistan	(CG)	122 W. 30th St, NY 10001
	(PMUN)	866 UN Plaza, NY 10017
Algeria	(PMUN)	15 E 47th St, NY 10017
Angola	(MUN)	747 Third Ave, 18th Flr, NY 10017
Argentina	(CG)	12 W. 56th St, NY 10019
	(PMUN)	1 UN Plaza, 25th Flr, NY 10017
Australia	(CG)	636 Fifth Ave, NY 10020
Bahrain	(PMUN)	747 Third Ave, 19th Flr, NY 10017
	(CG)	747 Third Ave, NY 10017
Bangladesh	(CG)	130 E 40th St, NY 10016
Benin	(MUN)	4 E 73rd St, NY 10021
Bhutan	(PMUN)	866 Second Ave, NY 10017
Bolivia	(PMUN)	211 E. 43rd St, NY 10017
Botswana	(PMUN)	866 Second Ave, NY 10017
Brazil	(PMUN)	747 Third Ave, NY 10017
Burma	(CG)	10 E. 77th St, NY 10021
Burundi	(PMUN)	201 E. 42nd St, 28th Flr, NY 10027
Cameroon	(MUN)	22 E. 73rd St, NY 10021
Cape Verde	(PMUN)	211 E. 43rd St, NY 10017
Central African Rep	(PMUN)	386 Park Ave, S, Rm 1614, NY 10016
Chad	(PMUN)	211 E. 43rd St, Suite 1703, NY 10017
Chile	(CG)	866 Second Ave, NY 10017
China	(PMUN)	155 W. 66th St, NY 10023
Colombia	(PMUN)	140 E. 57th St, 5th Flr, NY 10022
Congo	(MUN)	14 E. 65th St, NY 10021
Costa Rica	(CG)	211 E. 43rd Street, NY 10017
Cuba	(PMUN)	6 E. 67th St, NY 10021
Ecuador	(PMUN)	820 Second Ave, 15th Flr, NY 10017
Egypt	(PMUN)	36 E. 67th St, NY 10021

El Salvador	(CG)	211 E. 43rd St, NY 10020
Equatorial Guinea	(PMUN)	440 E. 62nd St, Apt 6-D, NY 10021
Ethiopia	(C)	866 UN Plaza, NY 10017
	(PMUN)	866 UN Plaza, NY 10017
Fiji	(MUN)	1 UN Plaza, NY 10017
Gabon	(PMUN)	820 2nd Ave, Rm 902, NY 10017
Gambia	(CG) c/o	Lee Kolker, PO Box F, Stanfordville, NY 12581
Ghana	(CG)	150 E. 58th St, NY 10022
	(PMUN)	150 E. 58th St, NY 10022
Guatemala	(CG)	1270 Ave of the Americas, NY 10020
Guinea	(MUN)	820 Second Ave, 16th Flr, NY 10017
Guinea-Bissau	(PMUN)	211 E. 43rd St, Suite 604, NY 10017
Guyana	(CG)	622 Third Ave, NY 10017
	(PMUN)	622 Third Ave, NY 10017
Honduras	(C)	290 Madison Ave, NY 10017
	(PMUN)	415 Lexington Ave, NY 10017
India	(CG)	3 E. 64th St, NY 10021
Indonesia	(CG)	5 E. 68th St, NY 10021
	(CG)	630 Fifth Ave, Rm 900, NY 10020
Iran	(PMUN)	622 Third Ave, NY 10017
Iraq	(PMUN)	14 E. 79th St, NY 10021
Israel	(CG)	800 Second Ave, NY 10017
Ivory Coast	(PMUN)	46 E. 74th St, NY 10021
Japan	(CG)	280 Park Ave, NY 10017
Jordan	(MUN)	866 UN Plaza, NY 10017
	(C)	866 UN Plaza, NY 10017
Kenya	(PMUN)	866 UN Plaza, Rm 486, NY 10017
Korea, North	(POUN)	40 E. 80th St, 25th Flr, NY 10021
Korea, South	(CG)	460 Park Ave, NY 10022
	(PMUN)	866 UN Plaza, Suite 300, NY 10017
Kuwait	(C)	235 E. 42nd St, NY 10017
	(PMUN)	801 Second Ave, 5 Flr, NY 10017
Lao People's Dem Rep (Laos)	(PMUN)	321 E. 45th St, Apt 7G, NY 10017
Lebanon	(CG)	9 E. 76th St, NY 10021
Lesotho	(E)	866 UN Plaza, Suite 580, NY 10017
Liberia	(C)	820 Second Ave, NY 10017
	(PMUN)	820 Second Ave, 4th Flr, NY 10017
Libya	(PMUN)	866 UN Plaza, NY 10017
Madagascar	(PMUN)	801 Second Ave, Suite 404, NY 10017
Malawi	(MUN)	777 Third Ave, NY 10017
Maldive Islands	(MUN)	137 E. 36th St, Apt 16-H, NY 10016
Mali	(PMUN)	111 E. 69th St, NY 10021
Mauritania	(PMUN)	600 Third Ave, 37th Flr, NY 10016
Mauritius	(MUN)	301 E. 47th St, NY 10017
Mongolia	(PMUN)	6 E. 77th St, NY 10021
Morocco	(CG)	597 Fifth Ave, NY 10017
	(MUN)	757 Third Ave, NY 10017

Mozambique	(PMUN)	866 UN Plaza, NY 10017
Nepal	(CG) (PMUN)	711 Third Ave, NY 10017 711 Third Ave, Rm 1806, NY 10017
New Zealand	(CG)	630 Fifth Ave, Suite 530, NY 10020
Nicaragua	(CG)	1270 Ave of the Americas, NY 10020
Niger	(MUN)	416 E. 50th St, NY 10022
Nigeria	(CG) (MUN)	575 Lexington Ave, NY 10022 757 Third Ave, NY 10017
Oman	(MUN)	605 Third Ave, NY 10016
Pakistan	(CG) (MUN)	12 E. 65th St, NY 10021 8 E. 65th St, NY 10021
Panama	(MUN)	866 UN Plaza, Rm 544, NY 10017
Papua New Guinea	(PMUN)	801 Second Ave, 12th Flr, NY 10017
Paraguay	(PMUN)	211 E. 43rd St, NY 10017
Peru	(C) (PMUN)	10 Rockefeller Plaza, NY 10020 301 E. 47th St, NY 10017
Philippines	(PMUN)	556 Fifth Ave, NY 10036
Qatar	(MUN)	747 Third Ave, NY 10022
Rwanda	(MUN)	120 E. 56th St, NY 10022
Samoa, Western	(PMUN)	820 Second Ave, NY 10017
Saudi Arabia	(CG) (PMUN)	866 UN Plaza, Rm 484, NY 10017 6 E. 43rd St, NY 10017
Senegal	(PMUN)	51 E. 42nd St, NY 10017
Sierra Leone	(PMUN) (CG)	919 Third Ave, NY 10022 919 Third Ave, NY 10022
Singapore	(PMUN)	711 Third Ave, 11th Flr, NY 10017
Somalia	(MUN)	747 Third Ave, NY 10017
South Africa	(PMUN)	300 E. 42nd St, NY 10017
South West Africa (Namibia)	(POUN)	801 Second Ave, Rm 1401, NY 10017
Sri Lanka	(MUN)	630 Third Ave, NY 10017
Sudan	(CG) (PMUN)	747 Third Ave, NY 10017 747 Third Ave, NY 10017
Suriname	(CG)	1 UN Plaza, NY 10017
Swaziland	(PMUN)	866 UN Plaza, NY 10017
Syria	(PMUN)	964 Third Ave, NY 10022
Tanzania	(MUN)	201 E. 42nd St, 10017
Togo	(MUN)	112 E. 40th St, NY 10016
Tunisia	(MUN)	40 E. 71st St, NY 10021
Turkey	(CG)	821 UN Plaza, NY 10017
Uganda	(PMUN)	801 Second Ave, NY 10017
United Arab Emirates	(PMUN)	747 Third Ave, NY 10017
U.S.S.R.	(PMUN)	136 E. 67th St, NY 10021
Upper Volta	(PMUN)	866 UN Plaza, Suite 265, NY 10017
Uruguay	(CG) (PMUN)	17 Battery Place, Rm 324, NY 10004 301 E. 47th St Rm 16-J, NY 10017
Venezuela	(CG)	7 E. 51st St, NY 10022

Vietnam	(MUN)	20 Waterside Place NY 10010
Yemen Arab Rep (North)	(MUN)	211 E. 43rd St, Rm 2402, NY 10017
Zaire	(PMUN)	866 Second Ave, NY 10017
Zambia	(PMUN)	150 E. 58th St, NY 10022

Abbreviations:	C	Consulate
	CG	Consulate General
	MUN	Mission to the UN
	PMUN	Permanent Missions to the UN
	POUN	Permanent Observer to the UN

Foreign Tourist Offices in New York

Afghanistan	Tourist Organisation: 535 Fifth Ave, NY 10017
Australia	Tourist Commission: 1270 Ave of the Americas, NY 10020
Brazil	Government Trade Bureau: 551 Fifth Ave, NY 10017
Caribbean	Tourism Association: 20 E 46th St, NY 10017. Tel. 682 0435
Chile	Trade Promotion Bureau: One World Trade Center, Suite 5121, NY 10048
China	Information Service: 159 Lexington Ave, NY. Tel. 725 4950
Colombia	Government Tourist Office: 140 E 57th St, NY 10022. Tel. 688 0151
Eastern Caribbean	Tourist Association: 40 E 49th St, NY. Tel. 986 9370
Ecuador	National Tourist Office: 2067 Broadway, NY. Tel. 874 3881
Egypt	Government Tourist Office: 630 Fifth Ave, NY 10020
El Salvador	Tourist Information Office: PO Box 818, Radio City Station, NY 10019; 200 W 58th St, NY 10019
French Guiana	French Government Tourist Office: 610 Fifth Ave, NY 10020
French Polynesia	Tourist Development Board: 200 E 42nd St, NY 10017
Galapagos Islands	Tourist Corporation: 888 Seventh Ave, NY 10019
Ghana	Tourist Office: 445 Park Ave, Suite 903, NY 10022. Tel. 688 8350
Honduras	Information Service: 501 Fifth Ave, NY 10017. Tel. 869 0766
Hong Kong	Tourist Association: 548 Fifth Ave, NY 10036. Tel. 582 6610
Iceland	National Tourist Office: 75 Rockefeller Plaza, NY 10019
India	Government Tourist Office: 30 Rockefeller Plaza, NY 10020
Indonesia	Indonesian Director General of Tourism: 5 E 68th St, NY 10021
Iran	Information and Tourism Centre: 10 W 49th St, NY 10020. Tel. 757 1945
Iraq	Press Information Office: 14 E 79th St, NY 10021
Israel	Government Tourist Office: 488 Madison Ave, NY 10022. Tel. 754 0140
Ivory Coast	Tourist Bureau: c/o Air Afrique, 1350 Ave of the Americas, NY 10019
Japan	National Tourist Organisation: 45 Rockefeller Plaza, NY 10020
Kenya	Tourist Office: 15 E 51st St, NY 10022
Korea, South	National Tourism Corp: 460 Park Ave, Room 628, NY 10022. Tel. 688 7543
Lebanon	Tourist and Information Office: 405 Park Ave, NY 10022. Tel. 421 2201
Mexico	National Tourist Council: 405 Park Ave, Suite 1002, NY 10022
Morocco	National Tourist Office: 521 Fifth Ave, NY 10017. Tel. 421 5771
New Zealand	Government Tourist Office: 630 Fifth Ave, NY 10020
Panama	Government Tourist Bureau: 630 Fifth Ave, NY 10020
Philippines	Tourist Information Office: 556 Fifth Ave, NY 10036. Tel. 575 7915

Senegal	Government Tourist Bureau: 200 Park Ave, NY 10017. Tel. 682 4695
South Africa	Tourist Corp: 610 Fifth Ave, NY 10020. Tel. 245 3720
Sri Lanka	Ceylon Tourist Board: 609 Fifth Ave, NY 10017. Tel. 935 0369
Suriname	Tourist Bureau: One Rockefeller Plaza, Suite 1408, NY 10020
Tanzania	Tourist Corp: 201 E 42nd St, NY 10017
Thailand	Tourist Organisation: 5 World Trade Center, Suite 2449, NY 10048
Tunisia	National Tourist Council: 630 Fifth Ave, Suite 863, NY 10020. Tel. 582 3670
Turkey	Tourism and Information Office: 821 United Nations Plaza, NY 10017. Tel. 687 2194
Uganda	Tourist Office and Mission: 801 Second Ave, NY 10017
U.S.S.R.	Intourist: 630 Fifth Ave, Suite 868, NY 10020
Uruguay	Government Trade Bureau: 301 East 47th St, Apt 21-0, NY 10017
Venezuela	Government Tourist and Information Office: 450 Park Ave, NY 10022
Zambia	National Tourist Bureau: 150 E 58th St, NY 10022. Tel. 758 9450
Zimbabwe-Rhodesia	National Tourist Board: 535 Fifth Ave, NY 10017

Australian Government Representatives Abroad

Afghanistan	c/o	Islamabad, Pakistan	
Algeria	(E)	60 Blvd Colonel Bougara, El-Biar, Algiers	
Argentina	(E)	Santa Fe 846°, Piso 8° (Swissair Bldg), Buenos Aires	
Bahrain	(CG)	Al-Fateh Commercial Bldg, Al-Khalifa Road, Manama, Bahrain (PO Box 252)	55011
Bangladesh	(HC)	Hotel Purbani International, 9th Flr, Dacca	255640/1/2
Bolivia	c/o	Santiago, Chile	
Botswana	c/o	Pretoria, South Africa	
Brazil	(E)	Edificio Venancio, Salas 513-524, Setor Diversoes Sul, Brasilia, DF (Caixa Postal 11-1256)	
	(C)	Rua Barao do Flamengo, 22 Apartment 202, Rio de Janeiro	
	(CG)	2433 Avenida Paulista, 0 1311 São Paulo, SP (Caixa Postal 30-580, São Paulo, SP)	
Burma	(E)	88 Strand Road, Rangoon	15711
Chile	(E)	Moneda 1123 9° Piso (Casilla 1442), Correo 15, Santiago de Chile	
China	(E)	15 Tung Chih Men Wai St, San Li Tun, Peking	
Colombia	c/o	Lima, Peru	
Costa Rica	c/o	Mexico City, Mexico	
Ecuador	c/o	Lima, Peru	
Egypt	(E)	1097 Corniche el Nil, Garden City, Cairo	28190, 28663, 22862
Ethiopia	c/o	Nairobi, Kenya	
Fiji	(HC)	7th & 8th Flrs, Dominion House, Thomson St, Suva	
Ghana	(HC)	Milne Close, Off Dr Amilcar Cabral Road, Airport Residential Area, PO Box 2445, Accra	77972, 75671/2

Guatemala	c/o	Mexico City, Mexico	
Guyana	c/o	Kingston Jamaica	
Hong Kong	(C)	10th Flr, Connaught Centre, Connaught Road	
India	(HC)	Australian Compound, 1/50-G Shanti Path, Chanakyapuri, New Delhi (PO Box 5210)	70337
Indonesia	(E)	Djalan Thamrin 15, Gambir, Djakarta	
Iran	(E)	4th & 5th Flrs, Bank Shahryar Bldg, 248 Ave Soraya, Tehran (PO Box 3409)	
Iraq	(E)	141/377 Masbah, PO Box 661, Baghdad	93256
Israel	(E)	145 Hayarkon St, Tel Aviv	231263/4
Ivory Coast	c/o	Accra, Ghana	
Jamaica	(HC)	4th Flr, National Life Bldg, 64 Knutsford Blvd, Kingston 5	
Japan	(E)	No 1-14 Mita 2-Chome, Minato-ku, Tokyo	
	(CG)	23rd Flr, Osaka International Bldg, Azuchimachi 2-Chome, Higashi-ku, Osaka	
Jordan	c/o	Damascus, Syria	
Kenya	(HC)	AFC/ADC Building Development House, Government Road, PO Box 30360, Nairobi	35666, 34672
Korea, South	(E)	5th Flr, Kudong-Shell Bldg, 58-1, Shinmoonro, IKA, Chongro-Ku, Seoul (KPO Box 562)	
Kuwait	c/o	Jeddah, Saudi Arabia	
Lao Dem People's Rep (Laos)	(E)	Rue Phone Xay, Vientiane (Boîte Postale No 292)	
Lebanon	(E)	Farra Bldg, 463 Bliss St, Ras Beirut	
Lesotho	c/o	Pretoria, South Africa	
Libya	(E)	Room 411, Shati Andalous Hotel, Tripoli (Box 5121)	
Madagascar	c/o	Dar es Salaam, Tanzania	
Malaysia	(HC)	44 Djalan Amphang, Kuala Lumpur	80166/9, 80541
Maldive Islands	c/o	Colombo, Sri Lanka	
Mauritius	c/o	Dar es Salaam, Tanzania	
Mexico	(E)	Paseo de la Reforma 195, 5° Piso, Mexico 5, DF, Mexico City	
Mongolia	c/o	Moscow, USSR	
Morocco	c/o	(E) 4 Rue Jean Rey, 75724 Paris, France	
Nauru	(HC)	Civic Centre, Nauru	
Nepal	c/o	New Delhi, India	
New Caledonia	(C)	8th Flr, 19 Avenue du Maréchal Foch, Nouméa (Boîte Postale 22)	
New Hebrides	(C)	c/o Intercontinental Island Inn,Vila (PO Box 111)	
New Zealand	(HC)	4th Flr, ICI House, Molesworth St, Wellington	
	(CG)	9th Flr, Lorne Towers, Lorne St, Auckland (PO Box 3601)	
	(CG)	Bank of New Zealand House, Cathedral Square, Christchurch (PO Box 2259)	
Nigeria	(HC)	Investment House, 4th Flr, 21/25 Broad St, PO Box 2427, Lagos	25981/2

Pakistan	(E)	Plot 17, Sector G4/4 Diplomatic Enclave No 2, Islamabad	22111-5
	(CG)	14F Clifton, Karachi 6 (PO Box 3939)	
Panama	c/o	Mexico City, Mexico	
Papua New Guinea	(HC)	PO Box 9129, Hohola, Port Moresby	
	(CG)	IPI Bldg, Second St, Lae (PO Box 1180)	
Paraguay	c/o	Buenos Aires, Argentina	
Peru	(E)	6th Flr, Edifico Plaza, Natalio Sanchez, 220 Lima, Peru 2977	
Philippines	(E)	China Banking Corp. Bldg, Paseo de Roxas (cnr Villar St), Makati, Rizal (PO Box 1274)	
Samoa, Western	(HC)	c/o Tusitala Hotel, Apia	
Saudi Arabia	(E)	Villa-Ruwais Quarter, Jeddah (PO Box 4876)	51303
Senegal	c/o	Accra, Ghana	
Seychelles	c/o	Nairobi, Kenya	
Singapore	(HC)	25 Napier Rd, Singapore 10 (Tanglin PO Box 470)	
Solomon Islands	(C)	Lot 35, Mendara Ave, Honiara	
South Africa	(E)	302 Standard Bank Chambers, Church Square, Pretoria	3 7051, 3 4778
	(C)	10th Flr, 1001 Colonial Mutual Bldg, 106 Adderly St, Capetown	
Sri Lanka	(HC)	Cambridge Pl, Colombo (PO Box 742)	96464/5/6
Sudan	c/o	Cairo, Egypt	
Swaziland	c/o	Pretoria, South Africa	
Syria	(E)	Bldg. No 49, Hawaker St, Mouhajreen, West Malki, Damascus.	
Tanzania	(HC)	7th and 8th Flrs, NLC Investment Bldg, Independence Ave, PO Box 2996, Dar es Salaam	20244/5/6
Thailand	(E)	Anglo-Thai Bldg, 64 Silom Road, Bangkok	35970-9
Tonga	c/o	Suva, Fiji	
Tunisia	c/o	Algiers, Algeria	
Turkey	(E)	83 Nenehatun Caddesi, Gaziosmanpasa, Ankara	17 96 18, 18 14 07
Uganda	c/o	Nairobi, Kenya	
U.S.S.R.	(E)	13 Kropotkinsky Pereulok, Moscow	
United Arab Emirates	c/o	Jeddah, Saudi Arabia	
Uruguay	c/o	Buenos Aires, Argentina	
Venezuela	c/o	Lima, Peru.	
Vietnam	(E)	66 Thuong Kiet, Hanoi	
Zambia	c/o	Dar es Salaam, Tanzania	

Abbreviations	E	Embassy
	HC	High Commission
	C	Consulate
	CG	Consulate General

Representatives of Asian Government in Canberra or Sydney, Australia

Country	Address	Telephone
Bangladesh	43 Hampton Circuit, Yarralumla, ACT 2600	(062) 81 1800
Burma	85 Mugga Way, Red Hill, ACT 2603	(062) 95 0045
China	247 Federal Highway, Watson, ACT 2602	(062) 41 2446
India	92 Mugga Way, Red Hill ACT 2603	(062) 95 0034
Indonesia	8 Darwin Avenue, Yarralumla, ACT 2600	(062) 73 3222
Iran	Torres Street, Canberra, ACT 2600	(062) 73 2353
Iraq	Malta House, George Street, Sydney, NSW 2000	(02) 26 1266
Israel	6 Turrana Street, Yarralumla, ACT 2600	(062) 73 1309
Japan	112 Empire Circuit, Yarralumla, ACT 2600	(062) 73 1251
Jordan	20 Roebuck Street, Red Hill, ACT 2603	(062) 95 9951
Kampuchea (Cambodia)	5 Canterbury Crescent, Deakin, ACT 2600	(062) 73 2517
Lebanon	1 Arkana Road, Yarralumla, ACT 2600	(062) 82 1748
Malaysia	71 State Circle, Yarralumla, ACT 2600	(062) 73 1543
Pakistan	59 Franklin Road, Forrest, ACT 2603	(062) 95 0021
Singapore	81 Mugga Way, Red Hill, ACT 2603	(062) 95 1022
Sri Lanka	35 Empire Circuit, Forrest, ACT 2603	(062) 95 0121
Thailand	111 Empire Circuit, Yarralumla, ACT 2600	(062) 73 1149
Turkey	60 Mugga Way, Red Hill, ACT 2603	(062) 95 0227
Vietnam	39 National Circuit, Forrest, ACT 2603	(062) 73 1502

Selected Foreign Tourist Offices in Sydney, Australia

Office	Address	Telephone
India Tourist Information Office	Carlton Centre, Elizabeth Street, Sydney, NSW 2000. (Offices opening shortly in Melbourne and Perth).	(02) 232 1600
Ministry of Tourism and Antiquities, Amman, Jordan	c/o FP Leonard Advertising Pty Ltd, 156 Castlereagh Street Sydney, NSW 2000.	(02) 61 6374
Tourist Development Corporation of Malaysia	11th Flr, R & W House, 92 Pitt Street, Sydney, NSW 2000	(02) 231 1377
Singapore Tourist Promotion Board	8th Flr, Gold Fields House, 1 Alfred Street, Sydney Cove, NSW 2000.	(02) 241 3771
Sri Lanka Tourist Board	c/o FP Leonard Advertising Pty Ltd, 156 Castlereagh Street, Sydney, NSW 2000.	(02) 61 6374
Tourist Organisation of Thailand	12th Flr, Royal Exchange Building, Cnr Bridge and Pitt Streets, Sydney, NSW 2000.	(02) 27 7549

New Zealand Government Representatives Abroad

In countries where there is no New Zealand Government Representative, the British Government Representative protects the interests of New Zealand citizens.

Country		Address	Telephone
Australia	(HC)	Commonwealth Ave, Canberra, ACT 2600	(062) 73 3611
	(CG)	60 Park Street (corner Park & Elizabeth Streets), Sydney, NSW 2000, GPO Box 365, Sydney, NSW 2001	(02) 233 3722
	(CG)	330 Collins St, Melborne, Vic 3000 GPO Box 2136T, Melbourne, Vic 3001	(03) 67 8111
	(C)	Watkins Place Bldg, 288 Edward St, Brisbane, GPO Box 62, Qld 4001	(07) 221 0005
	(C)	5th Flr, St. George's Court, 16 St. George's Terrace, Perth, WA 6000 GPO Box X2227, Perth, WA 6001	(092) 25 7877
Bahrain	(CG)	1st Flr, Manama Centre Bldg, Government Rd, PO Box 5881, Manama	59890
Bangladesh	c/o	New Delhi, India	
Brunei	c/o	Kuala Lumpur, Malaysia	
Burma	c/o	Kuala Lumpur, Malaysia	
Chile	(E)	Avenida Isidora Goyenechea 3516, Las Condes, Casilla 112, Correo Las Condes, Santiago	48 7071
China	(E)	No 1, Street No 2, East Temple of the Sun, Chao Yang District, Peking	522 731/2/3/4
Cook Islands	(Rep)	PO Box 21, Rarotonga	2065 342
Egypt	c/o	Rome, Italy	
Fiji	(HC)	Ratu Sukuna House, corner Victoria Parade and MacArthur Street, PO Box 1378, Suva	311 422
Guyana	c/o	Ottawa, Canada	
Hong Kong	(Comm)	3414 Connaught Centre, Connaught Rd, GPO Box 2790	5-258095
India	(HC)	39 Golf Links, New Delhi 110003	61 8281
Indonesia	(E)	Djalan Diponegoro No 41, Menteng, Djakarta Kotak Pos 2439, DKT	357924/5, 359796/7
Iran	(E)	Avenue Nadershah, Afshin St, No 29, PO Box 128, Tehran	625061, 625083, 623153
Iraq	(E)	2D/19 Zuwiya, Jadriyah, Baghdad (near Baghdad University), PO Box 244-B	7768176/7/8
Jamaica	c/o	Ottawa, Canada	
Japan	(E)	20-40 Kamiyama-cho, Shibuya-ku, Tokyo 150	(03) 460 8711/5, 460 8744
Korea, South	(E)	2nd Flr, Publishers Association Bldg, No 105-2 Sapandong, Chongro-ku, CPO Box 1059, Seoul	75 3707/4645, 70 4255
Lao People's Dem Rep (Laos)	c/o	Bangkok, Thailand	
Malaysia	(HC)	193 Djalan Pekeliling, PO Box 2003, Kuala Lumpur	(03) 201 422/191/262/325
Maldive Islands	c/o	Singapore	
Mexico	c/o	Washington, DC, USA	

Mongolia	c/o	Moscow, USSR	
Nauru	c/o	Suva, Fiji	
Nepal	c/o	New Delhi, India	
New Caledonia	(CG)	4 Blvd Vauban, Boîte Postale 2219, Nouméa	27 25 43 & 27 12 44
Niue	(Rep)	PO Box 78, Niue	
Pakistan	c/o	Tehran, Iran	
Papua New Guinea	(HC)	6th Flr, Australian Government Bldg, Waigani, Port Moresby, PO Box 1144, Boroko, Port Moresby	259444
Peru	(E)	Avenida Salaverry 3006, San Isidoro, Casilla 5587, Lima	62 18 90
Philippines	(E)	10th Flr, Philippine Commercial and Industrial Bank Bldg, 6756 Ayala Ave, PO Box 2208, Makati Commercial Centre, Makati, Rizal	88 37 48, 88 38 89, 88 41 43, 88 44 46, 88 54 71.
Samoa, American		Air New Zealand Ltd., Fagatogo, PO Box 697, Pago Pago	3666/7
Samoa, Western	(HC)	Beach Road, PO Box 208, Apia	380
Singapore	(HC)	13 Nassim Road, Singapore 10	2359 966
Solomon Islands	c/o	Port Moresby, Papua New Guinea	
Sri Lanka	c/o	Singapore	
Tahiti		Air New Zealand Ltd, James Norman Hall Bldg, Rue du Général de Gaulle, PO Box 73, Papeete	20 170
Thailand	(E)	Anglo-Thai Bldg, 64 Silom Rd, PO Box 2719, Bangkok	234 5935/6/7/8/9
Tonga	(HC)	Tungi Arcade, Taufa'ahau Road, Nuku'alofa	120
U.S.S.R.	(E)	44 Ulitsa Vorovskovo, Moscow, 121069	290 34 85, 290 12 77, 290 57 04
Vietnam	c/o	Peking, China	

Abbreviations	HC	High Commission
	CG	Consulate General
	C	Consulate
	E	Embassy
	Rep	Office of the New Zealand Representative
	Comm	Commission

Canadian Government Representatives Abroad

Afghanistan	c/o	Islamabad, Pakistan	
Algeria	(Ch)	27 Bis rue d'Anjou, Hydra, Algiers (PO Box 225, Gare Alger)	60 66 11
Argentina	(Ch)	Brunetta Bldg, Suipacha and Santa Fé, Buenos Aires (Casilla de Correo 1598)	32 90 81/8
Bahrain	c/o	Kuwait City, Kuwait.	
Bangladesh	(Ch)	House No 11, Road No 3, Dhanmondi Residential Area, Dacca	315181/2/3
Belize	c/o	Kingston, Jamaica	
Benin	c/o	Accra, Ghana	

Bolivia	c/o	Lima, Peru	
Botswana	c/o	Pretoria, South Africa	
Brazil	(Ch)	Ave das Nacoes, Number 16, Setor das Embaixadas Sul, Brasilia (Caixa Postal 07-0961, 70000 Brasilia, DF)	(61) 223 7515
	(O)	Edificio Metropole, Ave Presidente Wilson 165, 6 Andar, Rio de Janeiro (Caixa Postal 2164-ZC-00)	242 4140
	(O)	Edificio Top Center, Ave Paulista 854, 5th Flr, São Paulo. (Caixa Postal 22002)	287 2122/3/2234/ 2601, 285 3217/3240
Burma	c/o	Bangkok, Thailand	
Burundi	c/o	Kinshasa, Zaïre	
Cameroon	(Ch)	Immeuble Soppo Priso, Rue Conrad Adenauer, PO Box 572, Yaoundé	22 22 03, 22 29 22, 22 19 36
Cape Verde	c/o	Dakar, Senegal	
Central African Rep	c/o	Yaoundé, Cameroon	
Chad	c/o	Yaoundé, Cameroon	
Chile	(Ch)	Ahumada 11, 10th Flr, Santiago (Casilla 427)	62256/7/8/9
China	(Ch)	10 San Lin Tun Road, Chao Yang District, Peking.	521475, 521571, 521724, 521741, 521684
Colombia	(Ch)	Calle 58, No 10-42, 4th Flr, Bogotá (Apartado Aereo 53531, Bogotá 2)	235 5066
Comores	c/o	Dar es Salaam, Tanzania	
Congo	c/o	Kinshasa, Zaïre	
Costa Rica	(Ch)	6th Flr, Cronos Bldg, Calle 3 y Ave Central, San José (Apartado Postal 10303)	23 0588
Cuba	(Ch)	Calle 30 No 518 Esquina a7a, Miramar, Havana (PO Box 499 (HVA), Ottawa, KIN 8T7, Canada)	26421/2/3
Djibouti	c/o	Addis Ababa, Ethiopia	
Ecuador	c/o	Bogotá, Colombia	
Egypt	(Ch)	6 Sharia Mohamed Fahmi el Sayed, Garden City, Cairo (Post: Kasr el Doubara PO)	2 3110
El Salvador	c/o	San José, Costa Rica	
Ethiopia	(Ch)	African Solidarity Insurance Bldg, Unity Square, Addis Ababa (PO Box 1130)	44 83 35/6
Fiji	c/o	Wellington, New Zealand	
Gabon	c/o	Yaoundé, Cameroon	
Gambia	c/o	Dakar, Senegal	
Ghana	(Ch)	E 115/3 Independence Ave, PO Box 1639, Accra	28555, 28502
Guatemala	(Ch)	Edificio Maya, 5th Flr, Via 5, 4-50, Zona 4, Guatemala City (PO Box 400, Guatemala, CA)	315528, 315547, 319445, 316993, 65497
Guinea	c/o	Dakar, Senegal	
Guinea-Bissau	c/o	Dakar, Senegal	

Guyana	(Ch)	2nd Flr, Bank of Guyana Bldg, Church St and Ave of the Republic, Georgetown (PO Box 660)	72081/5
Honduras	c/o	San José, Costa Rica	
Hong Kong		(Office of the Commission for Canada) 14/15 Flrs, Asian House, 1 Hennessy Road, (PO Box 20264)	5 282222/3/4/5/6/7, 5 282422/3
Iceland	c/o (Ch) (O)	Oscar's Gate 20, Oslo 3, Norway Skulagata 20, Reykjavik	45 69 55 25355, 15337
India	(Ch)	7/8 Shanti Path, Chanakyapuri, New Delhi 110021, PO Box 5207	61 9461
Indonesia	(Ch)	5th Flr, WISMA Metropolitan, Djl Djendral Sudirman, Djakarta (PO Box 52/JKT)	584417, 584566, 584631
Iran	(Ch)	57 Darya-e-Noor Ave, Takht-e-Tavoos, PO Box 1610, Tehran	623177, 623548, 622310, 623549, 623192, 623629, 622975, 623202
Iraq	(Ch)	47/1/7 Al Mansour, Baghdad (PO Box 323, Central Post Office)	552 1459/1932/3
Israel	(Ch)	220 Hayarkon Street, Tel Aviv (PO Box 6410)	22 8122/3/4/5/6
Ivory Coast	(Ch)	Immeuble 'Le Général', 4ème et 5ème étages, Av Botreau-Roussel, BP 21194, Abidjan	32 20 09
Jamaica	(Ch)	Royal Bank Bldg, 30-36 Knutsford Blvd, Kingston 5 (PO Box 1500, Kingston 10)	926 1500/1/2/3/4/5/6/7
Japan	(Ch)	3-38 Akasaka 7-chome, Minato-ku, Tokyo 107	408 2101-8
Jordan	c/o	Beirut, Lebanon	
Kenya	(Ch)	Comcraft House, Hailé Sélassie Ave, Nairobi (PO Box 30481)	334 033/4/5/6
Korea, South	(Ch)	9th Flr, Hankook Ilbo Bldg, 14 Joonghak-dong, Chongro-ku, Seoul (PO Box 6299, Seoul 100)	730116/7
Kuwait	(Ch)	28 Quaraish St, Nuzha District (PO Box 25281 Safat, Kuwait City)	511451, 555754, 555934
Lao People's Dem Rep (Laos)	c/o	Bangkok, Thailand	
Lebanon	(Ch)	Immeuble Sabbagh, Rue Hamra, Beirut, CP2300	350 660/1/2/3/4/5
Lesotho	c/o	Pretoria, South Africa	
Liberia	c/o	Accra, Ghana	
Libya	c/o	Cairo, Egypt	
Macao	c/o	Hong Kong	
Madagascar	c/o	Dar es Salaam, Tanzania	
Malawi	c/o	Lusaka, Zambia	
Malaysia	(Ch)	American International Assurance Bldg, Ampang Road, Kuala Lumpur (PO Box 990)	89722/3/4
Mali	c/o	Abidjan, Ivory Coast	
Mauritania	c/o	Dakar, Senegal	
Mauritius	c/o	Dar es Salaam, Tanzania	

Mexico	(Ch)	Melchor Ocampo 463-7, Mexico 5, DF	533 06 10
	(O)	Hotel el Mirador Plaza, La Guebrada 74, Acapulco, Gro	3 72 91
	(O)	Ave Vallarta No 1373, PO Box 32-6 Guadalajara, Jalisco	(36) 25 9932
Mongolia	c/o	Moscow, USSR	
Morocco	(Ch)	13 Bis Rue Jafaar As-Sadik, CP709, Rabat-Agdal	71375/6/7
Mozambique	c/o	Lusaka, Zambia	
Nepal	c/o	New Delhi, India	
New Zealand	(Ch)	ICI Bldg, Molesworth St, Wellington (PO Box 12-049, Wellington North)	739 577
Nicaragua	c/o	San José, Costa Rica	
Niger	c/o	Abidjan, Ivory Coast	
Nigeria	(Ch)	Niger House, Tinubu Street, PO Box 851, Lagos	653630/1/2/3/4
Oman	c/o	Tehran, Iran	
Pakistan	(Ch)	Diplomatic Enclave, Sector G-5, Islamabad (GPO Box 1042)	21101-4/9, 21302/6, 21318
Panama	c/o	San Jośe, Costa Rica	
Papua New Guinea	c/o(Ch)	Commonwealth Ave, Canberra, ACT 2600, Australia	73 3844
Paraguay	c/o	Buenos Aires, Argentina	
Peru	(Ch)	132 Calle Libertad, Miraflores, Lima (Casilla 1212)	46 38 90
Philippines	(Ch)	4th Flr, PAL Bldg, Ayala Ave, Makati, Rizal, Manila (PO Box 971, Commercial Centre, Makati, Rizal)	87 65 36, 87 78 46
Qatar	c/o	Kuwait City, Kuwait	
Rwanda	c/o	Kinshasa, Zaïre	
Samoa, Western	c/o	Wellington, New Zealand	
Saudi Arabia	(Ch)	6th Flr, Office Tower, Commercial and Residential Centre, King Abdul Aziz St, PO Box 5050, Jeddah	34597/8
Senegal	(Ch)	45 Ave de la République, PO Box 3373, Dakar	20270
Seychelles	c/o	Dar es Salaam, Tanzania	
Sierra Leone	c/o	Lagos, Nigeria	
Singapore	(Ch)	Faber House, 7th-8th Flr, 230 Orchard Road, Singapore 9 (PO Box 845, Singapore 1)	371322
Solomon Is.	c/o	(Ch) Commonwealth Ave, Canberra, ACT 2600, Australia	73 3844
Somalia	c/o	Dar es Salaam, Tanzania	
South Africa	(Ch)	Nedbank Plaza, Cnr Church and Beatrix Streets, Arcadia, Pretoria 0083 (PO Box 26006, Arcadia, Pretoria)	487062/3/4
	(O)	16th Flr, Reserve Bank Bldg, 30 Hout St, Capetown (PO Box 683)	22 5134
Sri Lanka	(Ch)	6 Gregory's Rd, Cinnamon Gdns, Colombo 7 (PO Box 1006)	95841-3
Sudan	c/o	Cairo, Egypt	
Suriname	c/o	Georgetown, Guyana	
Swaziland	c/o	Pretoria, South Africa	

Syria	c/o	Beirut, Lebanon	
Tanzania	(Ch)	Pan Africa Insurance Bldg, Independence Ave, PO Box 1022, Dar es Salaam	20651
Thailand	(Ch)	Boonmitr Bldg, 11th Flr, 138 Silom Road, Bangkok 5 (PO Box 2090)	234 1561-8
Togo	c/o	Accra, Ghana	
Tonga	c/o	Wellington, New Zealand	
Tunisia	(Ch)	2 Place Virgile, Notre-Dame de Tunis, CP 31, Belvédère, Tunis	286 577
Turkey	(Ch)	Nenehatun Caddesi 75, Gaziosmanpasa, Ankara	27 58 03/4/5
Uganda	c/o	Nairobi, Kenya	
U.S.S.R.	(Ch)	23 Starokonyushenny Pereulok, Moscow	241 9034/9155
United Arab Emirates	c/o	Kuwait City, Kuwait	
Upper Volta	c/o	Abidjan, Ivory Coast	
Uruguay	c/o	Buenos Aires, Argentina	
Venezuela	(Ch)	Ave La Estancia No 10, 16 piso Ciudad Commercial Tamanaco, Caracas (Apartado del Este No 62302)	91 30 10, 91 36 10, 91 38 01, 91 34 01
Vietnam	c/o	Peking, China	
Yemen (North) Arab Rep	c/o	Jeddah, Saudi Arabia	
Zaïre	(Ch)	Edifice Shell, coin av Wangata et blvd du 30-juin, PO Box 8341, Kinshasa	227 06, 243 46
Zambia	(Ch)	Barclays Bank, North End Branch, Cairo Road, Lusaka (PO Box 1313)	75187/8

Abbreviations	Ch	Chancery
	O	Office

Motoring Organisations in Off the Beaten Track Countries

Compiled by Colin McElduff, Head of Routes at the Royal Automobile Club, and author of *Trans-Africa Motoring* and *Trans-Asia Motoring*.

Algeria	Automobile Club National d'Algérie, 99, Boulevard Salah-Bouskouir, Algiers, B.P. 67 — Alger Gare. Tel. 64 97 71
Argentina	Automovil Club Argentino, 1850 Avenida del Libertador, Buenos Aires. Tel 836061/830091/8216061
	Touring Club Argentino, Esmeralda 601, Buenos Aires.

Australia

Confederation of Australian Motor Sport (CAMS)
Compétence uniquement sportive,
382 Burke Road,
Camberwell,
Victoria 3124.

Australian Automobile Association (AAA),
88 Northbourne Ave,
Canberra, A.C.T. 2600,
P.O. Box 1555,
Canberra City 2601

R.A.C. of Australia,
89 Macquarie Street,
Sydney,
NSW

The Secretary,
National Roads & Motorists,
N.R.M.A. House Association,
151 Clarence Street,
Sydney, N.S.W. 2000.

Royal Automobile Club of Tasmania,
Cnr. Patrick & Murray Streets,
Hobart,
Tasmania.

R.A.C. of W. Australia,
228 Adelaide Terrace,
Perth,
W. Australia.

R.A.A. of South Australia,
41 Hindmarsh Square,
Adelaide,
South Australia.

R.A.C. of Queensland,
462-470 Queen Street,
Brisbane,
Qld.

R.A.C. of Victoria,
123 Queen Street,
Melbourne,
Victoria.

Bangladesh

Automobile Association of Bangladesh,
3/B Outer Circular Road,
Dacca-17.
Tel 243444/250541.

Bolivia

Automovil Club Boliviano,
2993 Avenida 6 de Agosto,
San Jorge,
La Paz.
Tel 51667/42998

Brazil

Automovel Club do Brasil,
Rua do Passeio 90,
Rio de Janeiro.
Tel 252 4055
52-40-55; 1470 (official)

Touring Club do Brasil,
Praca Maua,
Rio de Janeiro,
Tel. 23-16-60; 43-65-78 (Bureau of Information).

Brunei
Persatuan Automobile Brunei,
(Automobile Association of Brunei-AAB)
Peti Surat — P.O. Box 2150,
Bandar Seri Begawan.

Cameroon
Cameroons Motor Club,
c/o Cameroons Development Corp.,
Tiko,
South Cameroons.

Canada
Canadian Automobile Assoc.,
150 Gloucester Street,
Ottawa,
Ontario.
Tel 237 2105

The British Columbia Automobile Assoc.,
P.O. Box 9900,
Vancouver,
B.C.

Maritime Automobile Assoc.,
Haymarket Shopping Centre,
Saint John,
New Brunswick.

Ontario-Motor League,
Toronto,
Ontario.

Alberta Motor Association,
7th & 8th St. West,
Calgary,
Alberta.

Quebec Automobile Club,
871 Chemin St. Louis,
Quebec,
P.Q.

The Secretary,
Saskatchewan Motor Club Ltd.,
Albert Street & 6th Ave. North,
Box 1427,
Regina,
Saskatchewan.

Touring Club Montreal,
1401 McGill College Avenue,
Montreal 110,
Quebec.

Chile
Automovil Club de Chile,
San Antonio 220,
Santiago.
Tel: 35051/2/3

Automovil Club de Chile,
195 Avenida/Pedro de Valdivia,
Santiago.
Tel: 749516/258040

Colombia
Automovil Club de Colombia,
Carrera 14 No. 46-64,
Bogotá.
Tel: 45-15-34/37, 45-27-00 and 32-08-02 (Secretariat)

Automovil Club de Colombia,
46-61/72 Av. Caracas,
Apartado 46-17,
Apartado Nacional 555,
Bogotá.
Tel: 327580 (office)
452700 (emergency)

Costa Rica
Automobile-Touring Club de Costa Rica,
Apartado 4646,
San José.
Tel: 3570

Ecuador
Automovil Club del Ecuador,
Av. 10 de Agosto y Callejon Negrete,
Quito.
Tel: 37779

Automovil Club del Ecuador (Aneta),
Av. Eloy Alfaro y Berlin,
Apartado 2839,
Quito.

Egypt
Automobile et Touring Club d'Egypte,
10 Rue Kasr-El-Nil,
Cairo,
Tel: 977241/2/3
977340

El Salvador
Automival Club de El Salvador,
Apartado Postal 169,
San Salvador.

Automovil Club de el Salvador (ACES),
Paseo General Escalon 4357,
P.O. Box 1177,
San Salvador.

Fiji
Automobile Association of Fiji Incorporated,
P.O. Box 243,
Suva,
Fiji.

Guatemala
Club de Automovilismo y Turismo de Guatemala (Catgua),
13, Ave. 27-27 Zona 5,
Ciudad de Guatemala.
Tel: 64382/64883.

Federacion Guatemalteca de Motorciclismo y Auto,
Palacio de los Deportes,
Ciudad Olimpica.

Guyana
Georgetown Automobile Association,
Georgetown.

Hong Kong
Hong Kong Automobile Association,
14B Harbour View Mansions,
257 Gloucester Road,
Wanchai,
P.O. Box 20045,
Wanchai.
Tel: 5-767949/5-766198/5-760458

Hong Kong Automobile Assoc.,
Room 58, Caxton House,
1 Duddell Street,
Hong Kong.

Iceland
Icelandic Automobile Association,
Armula 27,
P.O. Box 311,
Reykjavik.

India The Federation of Motor Sports Clubs of India Pvt. Ltd. (FMSCI),
Address for correspondence: c/o Alpha Motor Co.
Liberty Bldg., 2nd Fl,
Marine Lines,
Bombay 400020
Tel: 293899

India	Automobile Association of Eastern India, 13, Promothesh Barua Sarani, Calcutta 700019.
	Automobile Association of Southern India, 38A Mount Road, P.B. 729 Madras 6.
	The Secretary, The A.A. of India, 6, Pratap Building, Connaught Circus, Po.O. Box 269, New Delhi.
	Western India Automobile Association, Dalui Naranji Memorial Bldgs., Churchgate Reclamation, P.O. Box 211, Bombay.
Indonesia	Ikatan Motor Indonesia (IMI) Gedung KONI Pusat Senayan, Djakarta, Kotakpos 609, Djakarta Kota. Tel: 591102
Iran	Touring and Automobile Club of Iran, Khiaban Varzeche, 37 Tehran. Tel: 319040/5
Iraq	Iraq Automobile and Touring Club, Al Mansor, Baghdad. Tel: 35862
Israel	Automobile and Touring Club of Israel, 19 Petah Tikvah Road, P.O. Box 36144, Tel. Tel Aviv 61360
Japan	Japan Automobile Federation, Shiba-Koen, 3-4-8, Minato-ku, Tokyo 105. Tel: 436 2811
Jamaica	Jamaica Motoring Club, B.P. 49, Kingston 10.
	Jamaica Automobile Assoc., 17A Duke Street, Kingston.
Jordan	Royal Automobile Club of Jordan, P.O. Box 920, Amman. Tel: 22467
Kenya	The Automobile Assoc. of East Africa, P.O. Box 87, Nairobi.

Kenya	Automobile Association of Kenya, A.A. House, Westlands, P.O. Box 40087, Nairobi. Tel: 46826
Korea, South	Korea Automobile Association, 1 P.O. Box 2008, Seoul.
Kuwait	Kuwait International Touring and Automobile Club, Siege, Khaldiah, Airport Road, P.O. Box 2100, Kuwait. Tel: 812539/815192/818406
Lebanon	Automobile et Touring Club du Liban, Immeuble Fattal, Rue du Port, B.P. 3545, Beirut. Tel. 221698, 221699, 229222
Libya	Automobile and Touring Club of Libya, Al Fath Boulevard, Maiden al-Ghazala, P.O. Box 3566, Tripoli.
Malaysia	Automobile Association of Malaysia, 3 Djalan Sultan (8/1D) P.O. Box 34, Petaling Jaya, Selangor.
Mauritius	Automobile Assoc. of Mauritius, No. 2, Queen Street, Port Louis.
Mexico	Asociacion Mexicana Automovilistica (A.M.A.) A.C. Orizaba 7, Colonia Roma, Apartado 24-486, Mexico 7, D F Tel: 511-10-84
	Asociacion Mexicana Automovilistica, Av. Chapultepec 276. Mexico 7 D.F.
	Asociacion Nacional Automovilistica, Miguel E. Schultz 140, Mexico 4, D.F.
Morocco	Royal Automobile Club Marocain, 3 Rue Lemercier, B.P. 94, Casablanca. Tel: 250030/253504.
	Touring Club du Maroc, 3 Avenue de l'Armée-Royale, Casablanca.
New Zealand	Motorsport Association New Zealand (M.A.N.Z.) Compétence uniquement sportive, Westbrook House, Upper Willis Street, P.O. Box 27002, Wellington.

New Zealand

Automobile Association Southland, Inc.,
P.O. Box 61,
Invercargill.

Automobile Association (Otago) Inc.,
450 Moray Place,
Dunedin.

Automobile Assoc. Nelson, Inc.,
204 Hardy Street,
Nelson.

Automobile Association of Marlborough, Inc.,
P.O. Box 104,
Blenheim.

Automobile Association Manawatu, Inc.,
P.O. Box 610,
Palmerston North.

Automobile Association (Hawkes Bay), Inc.,
Box 225,
Napier.

The Secretary,
The Automobile Association,
Wairarapa-Inc.,
Chapel Street,
Masterton.

The New Zealand Automobile Association, Inc.,
P.O. Box 1824,
Wellington.

Pioneer Amateur Sports Club,
Club House,
188 Oxford Terrace,
Christchurch.

Automobile Association South Canterbury, Inc.,
37 Sophia Street,
Timaru.

Automobile Association (Taranaki) Inc.,
46 Brougham Street,
New Plymouth.

Automobile Association (Wanganui) Inc.,
P.O. Box 440,
Wanganui.

Automobile Association (Wellington) Inc.,
A.A. House,
166 Willis Street,
Wellington 1.

The Automobile Association (South Taranaki) Inc.,
P.O. Box 118,
Hawera.

Automobile Association (Canterbury) Inc.,
210 Hereford Street,
P.O. Box 994,
Christchurch.

Automobile Association (Auckland) Inc.,
P.O. Box 5,
Auckland.

Paraguay

Touring Automovil Club Paraguayo,
25 de Mayo y Brasil,
Casilla de Correo 1204,
Asunción.
Tel: 1746

Pakistan	Automobile Association of West Pakistan, 8 Multan Road, P.O. Box 76, Lahore.
Panama	Touring Automovil Club de Panama, Av. Tivoli 8, Panama City.
Peru	Touring y Automovil Club del Peru, Cesar Vallejo 699, Lima 14, Casilla 2219, Lima 1. Tel: 403270/225957.
Philippines	Philippines Motor Association, 1011 Hidalgo, P.O. Box 99, Manila 12107 Tel: 609702/609360
Qatar	Automobile Touring Club of Qatar, Al-Bida Street, P.O. Box 18, Doha. Tel: 22734/23415
Senegal	Automobile Club du Sénégal, Immeuble Chambre de Commerce, Place de I'Indépendance, B.P. 295, Dakar. Tel: 266-04/08
Singapore	Automobile Association of Singapore, A.A. House, 336 River Valley Road, Killiney Road, Singapore 9. Tel: 372444
	Automobile Association of Singapore, 90 Orchard Road, P.O. Box 136, Singapore 9.
South Africa	Automobile Association of South Africa, A.A. House, 42 De Villiers Street, P.O. Box 596, Johannesburg 2001. Tel: 28-1400
	The Automobile Association of South Africa, 7 Martin Hammerschlag Way, Foreshaw, P.O. Box 70, Cape Town.
Sri Lanka	Ceylon Motor Sports Club, 4 Hunupitiya Road, P.O. Box 196, Colombo 2. Tel: 26 558
Suriname	Automobile Association of Sri Lanka, Box 338, Colombo.
	Automobilisten Vereniging, Paramaribo.

Syria
Touring Club of Syria,
Rue Baron,
P.B. 28,
Aleppo.
Tel: 10-097

Thailand
Royal Automobile Association of Thailand,
1174 Paholyothin Road,
Bangkok 9.
Tel: 5790430

Trinidad
Trinidad Automobile Assoc.,
Room 2,
94 Frederick Street,
Trinidad.

Tunisia
Touring Club de Tunisie,
15 Rue d'Allemagne,
Tunis.

Turkey
Touring et Automobile Club de Turquie,
364 Sisli Meydani,
Istanbul.
Tel: 46-70-90

United Arab Emirates
Automobile and Touring Club for UAE,
P.O. Box 1183,
Sharjah.

U.S.S.R.
Federacia Automobilnogo Sporta S.S.S.R.
(Fédération Automobile de l'U.R.S.S.)
B.P. 395,
Moscow D-362.
Tel: 491-86-61.

Uruguay
Automovil Club del Uruguay,
Av. Agraciada 1532,
Montevideo.
98 47 10/26/13 (Club),
91 12 64, 91 15 51, 91 12 51/2/3 (Rescue Service)

Centro Automovilista del Uruguay,
Boulevard Artigas 1773,
Montevideo.
Tel: 42-091/2/4-6131/4-5016

Touring Club Uruguayo,
Edificio Propria,
Calle Colonia 1648,
Montevideo.
Tel: 4-48-75

Venezuela
Touring Automovil Club de Venezuela,
Plaza Sur de Altamira,
Edificio Auto Commercial,
Locaux 2 y 4,
Apdo Postal 68102,
Altamira, Caracas
Tel: 32-41-08/09 and 32-72-11/12

Vietnam, South
Automobile Club du Vietnam,
17 Duong Ho Xuan Huong,
Saigon. Tel. 23275

Zaïre
Touring Club du Congo,
Avenue Major Ruwet,
Kinshasa.

Zimbabwe-Rhodesia
Automobile Association of Zimbabwe-Rhodesia,
Fanum House,
57 Jameson Avenue Central,
P.O. Box 585,
Salisbury C.I.
Tel. 61121

Thomas Cook Offices in off the Beaten Track Countries

Thomas Cook and Wagon-Lits together form a vast network of some 900 offices in 145 countries. Thomas Cook offers a complete range of travel services, though not at all branches. It handles ticket and hotel bookings, arranges tours and excursions, is especially handy for obtaining difficult rail bookings overseas, issues travellers' cheques and deals in foreign exchange. The range of Wagon-Lits services is wide, if not quite so comprehensive. In general, Thomas Cook offices are more widespread in English-speaking countries, Wagon-Lits in the French-speaking world.
The following is a list of Thomas Cook and Wagon-Lits offices in off the beaten track countries. Both companies have numerous offices throughout the British Isles, Europe and Asia Minor, Australasia and North America, including Mexico.

Offices of the Thomas Cook Group and Associated Companies

The Thomas Cook Group Ltd., Thorpe Wood, Peterborough PE3 6SB
Telephone 0733 63200 Telex 32581 Telegrams Thomascook Peterborough
Registered Office: 45 Berkeley Street, London W1A 1EB (Telex 266611). Reg. No. 198600 London
(Telegrams for Thomas Cook Bankers Ltd., 45 Berkeley St, London: Cookbanker London W1)

THE MIDDLE EAST

OPERATING COMPANY: THOMAS COOK OVERSEAS LTD. (Reg. No. 433790, London)

	Addresses	**Telegrams/Telex**	**Telephone**
EGYPT	*(Usual hours of opening: 9.00-17.00)*		
CAIRO	4 Champollion Street (P.O. Box 165)	Thomascook Cairo (Telex 92413)	46392/5
„ (Airport)	Bureau de Change, Cairo Airport	—	*(to be advised)*
ALEXANDRIA	15 Midan Saad Zaghloul (P.O. Box 185)	Thomascook Alexandria (Telex 54136)	27830
LUXOR	New Winter Palace Hotel§ *(closes 13.00 Sats.)*.	Thomascook Luxor	2402
	§—Address all communications for Luxor to Cairo Office.		
IRAQ	*(Usual hours of opening: 9.00-13.00 and 16.00-18.00 (17.00-19.00 in summer) Mon. to Thurs. and Sat., 9.00-13.00 Sun. Closed on Fris).*		
BAGHDAD	Thomas Cook Overseas Ltd., Sa'adun Street (P.O. Box 2007), Baghdad.	Thomascook Baghdad (Telex 2464)	88 89721/3
LEBANON	*(Usual hours of opening: 8.30-12.30 and 15.30-18.00 Mon. to Fri., 8.30-12.30 Sat.)*		
BEIRUT	Thomas Cook Overseas Ltd., Al-Moutawakel Building, Monseigneur Messara Street, Achrafié, Beirut (P.O. Box 11-0085).	Thomascook Beirut (Telex 21512)	a.m.: 329569 & 332462 p.m.: 329221 & 346260
BAHRAIN*	*(Usual hours of opening: 8.00-13.00 and 15.00-18.00 Sat. to Thurs.)*		
	Thomas Cook Travco, L.L., Unitag House, Government Road, P.O. Box 20312, Bahrain, Arabian Gulf	Cookbanker Bahrain (Telex 9099GJ)	257444/258000
KUWAIT	*(Usual hours of opening: 8.00-13.00 and 16.00-19.00 Sat. to Thurs.)*		
	Thomas Cook Kuwait Travel and Tourism, Al Sabah Building,	(Telex 3413KT)	424803, 424779/80

	Fahad Al Salem Street, P.O. Box 2156 (Safat), Kuwait		424805, 424820, 424882

—Headquarters for Middle East and India are at Unitag House, Government Road, Manama, Bahrain, Arabian Gulf (Telex 9099; Telephone 258034)

SOUTH AFRICA

OPERATING COMPANY: THOMAS COOK (PROPRIETARY) LTD. (Registered in South Africa)

(Hours of opening vary from office to office, but all offices are open at the following times: 8.30-16.30 Mon. to Fri., 8.30-12.00 Sat.)

BENONI	45 Cranbourne Avenue, (P.O. Box 1962), Benoni 1500	(Telex 84749)	849.4231
CAPE TOWN	African Eagle Centre, 2 St. George's Street (P.O. Box 12), Cape Town 8000	Thomacook Cape Town (Telex 577013)	22-1311
CAPE TOWN (Claremont)	Cavendish Square, Claremont 7700, Cape Province	—	69-1108
DURBAN	57 Gardiner Street (P.O. Box 33), Durban 4000	Thomacook Durban (Telex 62751)	68321
"	Bhoola Centre, 72 Prince Edward St., Durban 4001 (P.O. Box 48227 Qualbert 4078)	—	316878, 317073
DURBAN (La Lucia)	51 La Lucia Mall, La Lucia, Durban 4051 (P.O. Box 22016, Glenashley 4022)	—	52-3245
DURBAN (Montclair)	Old Mutual Centre, Montclair Road, Montclair 4001 (P.O. Box 33093, Montclair 4061)	—	42-8607
JOHANNESBURG	36A Rissik Street, near Commissioner Street (P.O. Box 4569). Johannesburg 2000**	Thomacook Johannesburg (Telex 84115)	23-0363
JOHANNESBURG (Rosebank)	The Firs, Oxford Road, Rosebank (P.O. Box 52261, Saxonwold 2132)	—	47-2379
PORT ELIZABETH	Park Towers, 35 Rink St., Port Elizabeth 6001, (P.O. Box 1456, Port Elizabeth 6000).	Thomacook Port Elizabeth (Telex 74-7908)	2-1708
PRETORIA	Volkskas Centre, 230 Van Den Walt Street (P.O. Box 1550), Pretoria	Thomacook Pretoria (Telex 3686)	48-3020

***—Headquarters for Southern Africa are at N.B.S. Building, 38 Rissik Street, Johannesburg 2000 (P.O. Box 4569) Telephone 23-0363 Telex 84115.*

THE EAST AND FAR EAST

OPERATING COMPANY: THOMAS COOK OVERSEAS LTD.

SINGAPORE	*(Usual hours of opening: 9.00-17.30 Mon. to Fri., 9.00-13.00 Sat.)*		
SINGAPORE	903/4 Orchard Towers, 400 Orchard Rd. Singapore 9	Thomascook Singapore (Telex 23897)	370366
INDIA*	*(Usual hours of opening: 9.30-13.00 and 14.00-17.30 Mon to Fri., 9.30-13.00 Sat.)*		
BOMBAY	Dadabhai Naoroji Road (P.O. Box 46) Bombay, 400.001	Thomascook Bombay (Telex 112096)	258556
BANGALORE	55 Mahatma Gandhi Road, Bangalore 560.001	Thomascook Bangalore One (Telex 0845392)	51066/7
MADRAS	Taj Coromandel Hotel, 5 Nungambakkam High Road, Madras 600.034	Thomascook Madras (Telex 7463)	83092/4
NEW DELHI	Hotel Imperial, Janpath, New Delhi 110.001	Thomascook New Delhi (Telex 313725)	312468, 381481, 383741
HONG KONG	*(Usual hours of opening: 9.00 to 17.00 Mon. to Fri., 9.00-13.00 Sat.)*		
HONG KONG	9th floor, Gammon House, 12 Harcourt Road (P.O. Box 38)	Thomascook Hong Kong (Telex 74858)	235151

Place	Address	Telegraphic address / Telex	Telephone
MALAYSIA	*(Usual hours of opening 8.30-13.00 and 14.00-1700 Mon. to Fri., 9.00-13.00 Sat)*		
KUALA LUMPUR	Block D 3rd Floor, P.O. Box 164, Kompleks Pej, Damansara, Kuala Lumpur 23-04	Cooks Kuala Lumpur (Telex MA 30498)	943177, 943241, 943038
SRI LANKA (CEYLON)	*(Usual hours of opening: 8.30-16.30 Mon. to Fri., closed on Sats.)*		
COLOMBO	Lloyds Building, 15 Sir Baron Jayatilake Mawatha (P.O. Box 36)	Thomascook Colombo (Telex 1278)	22511/3
PHILIPPINES	(OPERATING COMPANY: THOMAS COOK (PHILIPPINES), INC.)		
MANILA (Makati)	Ground Floor, Oledan Building, 131 Ayala Avenue, Makati, (P.O. Box 311) Metro Manila D-708	Thomascook Manila (Telex PN 3942)	857173, 875661, 858573
CEBU	Smith Bell Building, P. Burgos St. (P.O. Box 149)	Thomascook Cebu	7.62.50

Offices of Wagons-Lits Tourisme (Compagnie Internationale des Wagons-Lits et du Tourisme)

Siege Social: 51/53 blvd Clovis, Brussels 1040 Telephone 735.21.70 Telex 26116 Telegrams Wagolits
Direction Generale: 40 rue d'Arcade (BP 729-08), 75381 Paris Cedex 08. Telephone 266.24.00. Telegrams Wagolits
The principal office in each country is that shown first under each heading.

AFRICA

Place	Address	Telegraphic address / Telex	Telephone
CANARY ISLANDS	*(Usual hours of opening: 9.00-13.00 and 16.00-1900 Mon. to Fri., 9.00-13.00 Sat.)*		
LAS PALMAS	248 Calle Leon y Castillo (Apartado 546)	Walits Las Palmas (Telex 95073 WLLPA-E)	23.41.55, 23.43.55
PUERTO DE LA CRUZ	20 Calle de San Telmo (Apartado 53)	Walits Puerto de la Cruz (Telex 92195)	38.14.74
SANTA CRUZ DE TENERIFE	2 Calle del Pilar	Walits Santa Cruz de Tenerife (Telex 92271 SLPTF-E)	24.66.85/6, 24.67.36
EGYPT	*(Usual hours of opening: Daily (except Sundays) 9.00-12.00, 16.00-18.30)*		
CAIRO	Shepheard's Hotel, Sharia Kasr el Ali	Waglits Cairo (Telex 92291 Wagons UN)	31.538
ASWAN***	Grand Hotel, Corniche du Nil (Winter only)	—	2628, 3198
LUXOR***	Winter Palace Hotel	—	2317
MADEIRA	*(Usual hours of opening: 9.00-12.30 and 14.30-18.00 Mon. to Fri., 9.00-12.00 Sat.)*		
FUNCHAL	44 Avenida Arriaga	Walits Funchal (Telex 72179 Wali P)	3.25.18, 3.25.58
MOROCCO	*(Usual hours of opening: Summer: 8.45-12.00 and 16.00-18.00 Mon. to Fri., 9.00-12.00 Sat.; Winter: 9.00-12.00 Mon to Sat., also 14.30-18.00 Mon. to Fri.)*		
AGADIR	26 Avenue des Forces Armées Royales	Walits Agadir (Telex 81040M)	35.28
CASABLANCA (A)	60 Rue de Foucauld	Walits Casablanca (Telex 21605M)	26.12.11-4
FEZ	Imm. du Grand Hotel, Boulevard Mohammed V (Rue No. 5)	Walits Fez (Telex 51022M)	229.58, 244.36
MARRAKECH	122 Avenue Mohamed V	Walits Marrakech (Telex 72063M)	316.87
MEKNES	Imm. Sifiche, 1 Zenkat Ghana	Walits Meknes	219.95/6
OUJDA	Place Mohammed V	Walits Oujda	25.20

RABAT	1 Avenue Al Amir Moulay Abdallah	Walits Rabat (Telex 31671M)	226.45/6
TANGIER	86 Rue de la Liberté	Walits Tangier (Telex 33657M)	316.40
TUNISIA	*(Usual hours of opening: Daily (except Sundays) 8.00-11.45 and 15.00-1800)*		
TUNIS	65 Avenue Habib Bourguiba	Wagoli Tunis (Telex 12033 Wagoli TN)	242.673, 247.320

SOUTH AND CENTRAL AMERICA

ARGENTINA	*(Usual hours of opening: 9.00-12.00 and 14.00-18.30 Mon. to Fri., 9.00-12.00 Sat.)*		
BUENOS AIRES	746 Avenida Cordoba, Buenos Aires 1054	Wagolits Buenos Aires (Telex 121852)	392.5170, 392.1988
ROSARIO	840 Calle Cordoba (local 9), 2000-Rosario	Wagolits Rosario	63778
BOLIVIA	*(Usual hours of opening: 9.00-12.00 and 14.00-1800 Mon. to Fri., 9.00-13.00 Sat.)*		
LA PAZ	Edificio Litoral, Avenida Mariscal Santa Cruz, Esquina Colon. (Casilla 6551)	Wagolits La Paz (Telex 5456 WLITS BX)	58499
BRAZIL	*(Usual hours of opening: 8.30-12.00 and 14.00-17.30 Mon. to Fri., 8.30-12.00 Sat.)*		
RIO DE JANEIRO	156-S-126 Avenida Rio Branco, Edificio Avenida Central	Wagolits Rio de Janeiro (Telex 2122242 Lits BR)	221.7756, 221.0040
,,	Hotel Méridien Copacabana, 1020 Avenida Atlantica	(Telex 2121069 "For Wagons-Lits")	275-8396
PORTO ALEGRE	1.273 Rua dos Andradas, Loja 002-Galeria Edith, 90.000 Porto Alegre	Wagolits Porto Alegre (Telex 051.1411 "For Wagons-Lits")	24-5818
SALVADOR/BAHIA	Hotel Meridien, 216 Rua Fonte do Boi, Rio Vermelho, Loja 8, Salvador 40000	(Telex 71.1029 "For Wagons-Lits")	071.248.4525
SAO PAULO	258 Avenida Sao Luis (Caixa Postal 7091), Loja 15-16	Wagolits Sao Paulo (Telex 011.21090 Lits BR)	256.6491, 256.7437
,, (Jardim)	5719 Avenida 9 de Julho (CP 01407)	Wagojulho Sao Paulo (Telex 011.22640 Lits BR)	852.0484, 852.9703
CHILE	*(Usual hours of opening: 8.45-13.30 and 14.30-18.00 Mon. to Fri.)*		
SANTIAGO	1058 Calle Agustinas	Wagolits Santiago (Telex SGO 260 Wagolits)	65.533, 82.827, 86.254
,, (Providencia)	114 Calle Nueva de Lyon	,, ,, (,, ,, "Para. Providencia")	74.29.47
COLOMBIA	*(Usual hours of opening: 8.00-12.00 and 14.00-18.00 Mon. to Fri., 8.00-12.30 Sat.)*		
BOGOTA (Norte)		Walinor Bogotá (Telex 44523)	257.05.89, 257.03.26
,, (Centro)	Carrera 5 No. 15.89 (Apartado Aéreo 14.495)*, Bogotá 1 DE	Walits Bogotá	241.92.50, 241.95.61
CALI	Edificio Centro Seguros Bolivar, Carrera 4a No. 12-41, Local 204 (Apartado Aéreo 4137)	Walits Cali (Telex 55661 Wali Co)	89.42.51
MEDELLIN	Centro Coltejer, Local 108 (Apartado Aéreo 53034)	Walits Medellin (Telex 06831 Wali Co)	31.83.32; 31.83.12
ECUADOR			
QUITO	698 Colón, Edificio El Dorado, Sucursal No. 1 San Blas Quito	—	235.110

GUATEMALA			
GUATEMALA CITY	Avenida Reforma 12-81, Zona 10 (PO Box 44A)	(Telex 5444 WAGONS-GU)	31.53.64, 32.35.22
PERU	*(Usual hours of opening: 9.00-13.00 and 13.30-17.30 Mon. to Fri., 9.00-13.00 Sat.)*		
LIMA	174 Jirón Ocona (Casilla Postal 3774) Lima 1	Wagolits Lima (Telex 20135PU Wagolit)	27.83.53, 28.76.43
„ (Miraflores)	473 Avenida La Paz, Miraflores Cesars Hotel, Lima-Miraflores	(Telex 25202PU CF MIRAF "Wagons-Lits")	46.50.99
CUZCO	157 Calle Heladeros, (frente al Hotel Turistas) (Casilla Postal 779)	Wagolits Cuzco (Telex 52040PU Wagolít)	2152
URUGUAY	*(Usual hours of opening: 9.00-12.00 and 14.00-19.00 Mon. to Fri., 9.00-12.00 Sat.)*		
MONTEVIDEO	1356 Calle Rio Negro	Wagolits Montevídeo (Telex 398472 Wagolit UY)	91.14.26, 98.60.27
VENEZUELA	*(Usual hours of opening: 8.30-12.00 and 14.00-18.00 Mon to Fri.)*		
CARACAS	Edificio Austerlitz, 33/2 Avenida Urdaneta (Apartado 2404*), Caracas 101	Wagolits Caracas (Telex 22838 Wagolits)	561.24.73, 561.28.79
„	Edificio Deltaven, Local 4, Apartado 61.117, Avenida Andres Bello, Los Palos Grandes, Caracas 106**	Wagonest Caracas (Telex 23337 Wagopaig)	283.64.75, 283.83.46
MARACAIBO	Avenida 20 (Prolongacion 5 de Julio), Centro comercial Las Tejas, Local 212		(061) 51.76.77
VALENCIA	Edificio Hotel Paris, Local 7, Avenida Bolivar	Wagolits Valencia (Telex 41338 Wagoval)	21.39.65

A—Headquarters for Morocco (Agence Générale de Tourisme) are at 62 rue de Foucauld, Casablanca (Telephone 27.35.58/9, Telegrams Walits, Telex 21605M).
—Quote this number in all correspondence.* *—This office handles all incoming traffic.* ****—Address all communications to Cairo office.*

PRINCIPAL THOMAS COOK/WAGONS-LITS AUTHORISED REPRESENTATIVES

(Neither staffed by Thomas Cook or Wagons-Lits, nor under their control)

REPUBLIC OF IRELAND
CORK — JOSEPH BARTER & SONS LTD., 92 St. Patrick Street. Telegrams: Barter 24261. Telephone: (021) 24261. Telex: 6023.

ALGERIA
ALGIERS — SONATOUR, 5 Boulevard Ben-Boulaid (Telex: 52.214).

BERMUDA
HAMILTON — BUTTERFIELD TRAVEL, Harnett & Richardson Building, 51 Front Street, Hamilton 5-24. Telegrams: Buttravel. Telephone: 21510. Telex: 3233 Harnt. *(Open 9.00-17.00 Mondays to Fridays: closed Saturdays).*

FINLAND
HELSINKI — FINLAND TRAVEL BUREAU LTD., 1 Keskuskatu (P.O. Box 319, Helsinki 10). Telegrams: Travelhead. Telephone: 10515. Telex: 12626 (Tours Dept. is at Kaivokatu 10A 6th Floor (P.O. Box 319), SF-00101 Helsinki 10; Telephone: 170515; Telegrams: Travelfit; Telex: 12626). *Branches:* Turku, Tampere, Rovaniemi.

GERMANY
MÜNCHEN (MUNICH) — AMTLICHES BAYERISCHES REISEBÜRO G.m.b.H, im Hauptbahnhof (Postfach 200.125), D8000-München 2. Telegrams: Weltreisen: (089) 590.41. Telex: 524183. *(Representing Thomas Cook only). (Usual hours of opening: 9.00-17.00 Mon. to Fri., 9.00-12.00 Sat.).*

ICELAND
REYKJAVIK — SAMVINN TRAVEL LTD., P.O. Box 1404, Austurstraeti 12 Telephone: 27077. Telex: 2241.

JAPAN

TOKYO — JAPAN TRAVEL BUREAU INC., 1-6-4 Marunouchi, Chiyoda-ku, Tokyo. Telegrams: Tourist; Telex: J24418. Telephone: 23-0521.4141. *(Foreign Tourist Dept. is at 1-13-1 Nihonbashi, Chuo-ku, Tokyo 103. Telephone: (03) 273-2509, Telex: J24418).*

MEDITERRANEAN

GIBRALTAR — J. LUCAS IMOSSI & SONS LTD., 1/5 Irish Town (P.O. Box 167). Telegrams: Pump. Telephone: 3465. Telex: 221.

CYPRUS — HULL, BLYTH, ARAOUZOS, LTD., 116 Archbishop Makarios Avenue (P.O. Box 17), Limassol. Telephone: 62223. Telex: 2253.
(*Branches:* Nicosia, Larnaca, Paphos, Akrotiri. Telegrams: Vaport Cyprus.

NIGERIA

LAGOS — TRANSCAP VOYAGES, Wesley House, 20/21 Marina (P.O. Box 2326). Telegrams: Travoyages. Telex: 21.272. Telephone: 2384012. *Branches:* Apapa, Ikeja, Kaduna, Kano and Port Harcourt.

NORWAY

OSLO — BERG-HANSEN TRAVEL BUREAU LTD., Tollbugaten 27 (P.O. Box 505 Sentrum), Oslo 1. Telegrams: Travel. Telephone: 41.54.10. Telex: 16110.

SENEGAL

DAKAR — TRANSCAP VOYAGES, 20 Boulevard Pinet Laprade (P.O. Box 58). Telex: 618. Telephone: 225.40.

SWEDEN

STOCKHOLM — SJ TRAVEL BUREAU, Vasagatan 22, (P.O. Box 504), 10551 Stockholm. Telephone: (08) 234450. Telegrams: SJ Travel. Telex: 19861. (Business Travel Dept., Telephone: (08) 240450. Telex: 19413).

GÖTEBORG — SJ TOURIST SERVICE, Hotellplatsen 2, 40110 Göteborg 1. Telephone: (031) 17.61.00. Telex: 2390. *Main branches:* Helsingborg, Jönköping, Malmö and Norrköping.

TURKEY

IZMIR — KEY TOURS LTD., Atatürk Caddesi 212/1. Telex: 52659. Telephone: 138150, 139813.

ZAÏRE (CONGO)

KINSHASA — AGENCE MARITIME INTERNATIONALE S.A., (P.O. Box 7597). Telegrams: Agenmarin. Telex: 218. Telephone: 246602. *Branches:* Boma, Goma, Kolwezi, Lubumbashi.

American Express Offices in Off The Beaten Track Countries

Mail Service
American Express offers a free mail service to its clients (ie holders of American Express Cards, Travellers' Cheques, etc) at several American Express offices throughout the world and at many of its representative offices.

It is not always able to say which of its representatives will hold mail for clients. When using the worldwide directory of addresses it is reasonably safe to assume that those offices that list a PO Box number will probably hold clients' letters. In many large cities more than one office is listed. The first office in the list is usually the largest, most centrally located, and will be the address to use for client mail service. If in doubt as to whether or not a particular office holds mail it is best to query that office directly.

American Express states that when an office does offer this service mail will normally be held for 30 days before being returned to the Post Office for return to sender. No telephone queries are allowed; information about the mail and the mail itself will only be released upon presentation of identification, preferably in the form of a passport, and proof of client status. Non-clients are charged a fee of $1.00 (US) or the local equivalent. There is also a fee of $3.00 to register a forwarding address. For receipt of mail only there is no need for prior registration.

(R) = Representative Office

Afghanistan Caravan Travels, Ltd (R) Opposite Indian Embassy, PO Box 3043, Kabul. Tel. 31113 & 31469

Angola Agencias Viagens Expresso, LDA (R) PO Box 977, Rua Joao Belo No 10, Benguela, Tel. 2755 & 3061
Agencias E Viagens Expresso, LDA, (R) Avenida Marechal Carmona 125/27, PO Box 525, Lobito. Tel. 2412
Star Travel Service (R) Rua Pereira Forjaz, Predio Mutamba, Loja G, PO Box 3334, Luanda. Tel. 31901 & 31894/7

Argentina City Service Travel Agency (R), Bartolome Mitre 5, Bariloche. Tel. 2283/4
City Service Travel Agency (R) Florida 890, 4th Flr, Buenos Aires. Tel. 32-8416/18

Australia American Express International, Inc, 24a Gawler Place Adelaide, S. Australia 5000. Tel. 223 5680
American Express International, Inc, 165 Elizabeth Street, Brisbane, Queensland 4000. Tel. 229 2022
Webster's Travel (R) 60 Liverpool Street, GPO Box 333D, Hobart, Tasmania 7000 Tel. 380-200
American Express International, Inc, 327 Collins Street, Melbourne, Vic 3000 (PO Box 5450, Melbourne). Tel. 62 4141
American Express International, Inc, 125 St George's Terrace, PO Box K805, Perth, Western Australia 6000. Tel. 22.1177
American Express International, Inc, 47 York Street, PO Box 7006, Sydney NSW 2000. Tel. 29.5991

Belize Belize Global Travel Service, Ltd, (R) 6 Albert Street, PO Box 244, Belize City. Tel. 2185, 2363/4.

Bolivia Crillon Tours Ltd (R) Avenida Camacho 1223, PO Box 4785, La Paz. Tel. 20222 & 40402

Brazil Kontik-Franstur S.A. (R) Av Almirante Barroso, 91, 7th Flr, PO Box 2952, ZC-00, Rio de Janeiro. Tel 283-3737
Kontik-Franstur SA (R) Praca da Inglaterra, 2, PO Box 973, Salvador. Tel. 20433, 20656
Konik-Franstur SA (R) Rua Marconi, 71, 2nd Flr, São Paulo. Tel. 366301

Cameroon 'Camvoyages' (R) 15, Av de la Liberté, PO Box 4070, Douala. Tel. 42 31 88

Chile Turismo Cocha (R) PO Box 1001, Augustinas 1122, Santiago Tel. 82764-83487, 88002-87426 & 723923

Colombia Tierra Aire Ltda (R) Calle 35, No 44-43, PO Box 21-93, Barranquilla. Tel. 19333 & 171183
Tierra Mar Aire Ltda (R) Edif Bavaria Torre B, Carrera 10, No 27-91, Locale 126, Apartado Aereo 5371, Bogatá. Tel. 83 29 55 Tierra Mar Aire Ltda. (R) Carrera 3, No 8-13, PO Box 44-64, Cali. Tel. 73 13 33 & 74 14 44
Tierra Mar Aire Ltda (R) Calle del Colegio, 34-28, PO Box 27-61, Cartagena. Tel. 43442 & 43646
Tierra Mar Aire Ltda (R) Edif Furatena OF 411, Calle 50 46-36, PO Box 8125, Medellin. Tel. 422288

Costa Rica TAM Travel Agency (R) Avenidas Central-Primera, Calle Primera, PO Box 1864, San José. Tel. 23 51 11

Ecuador Ecuadorian Tours, SA (R) 313 Ballen Street, PO Box 3862, Guayaquil. Tel. 511525, 512980, 512805
Ecuadorian Tours, SA (R) Espejo 935, PO Box 2605, Quito. Tel. 219000
Ecuadorian Tours, SA (R) Amazonas 339, PO Box 2605, Quito. Tel. 528177 & 520777

Egypt American Express of Egypt Limited, 15 Sharia Kasr El Nil, POBox 2160,Cairo. Tel. 970138, 970042
American Express of Egypt Limited, Nile Hilton Hotel, PO Box 2160, Cairo. Tel. 810383
American Express of Egypt Limited, Meridien Hotel, PO Box 2160, Cairo. Tel. 844017
American Express of Egypt Limited, Cairo Airport.

El Salvador El Salvador Travel Service (R) Centro Commercial la Mascota, San Salvador. Tel. 23.0177

Ethiopia ITCO Tourist and Travel Agency (R) Ras Makonnen & Churchill Sts, PO Box 1048, Addis Ababa. Tel. 44 43 34
Hilton Hotel, Addis Ababa, Service Desk. Tel. 44 84 00

Fiji Tapa Tours Limited (R) Nandi International Airport, PO Box 240, Nandi. Tel. 72325
Tapa Tours Limited (R) 189 Victoria Parade, PO Box 654, Suva. Tel. 22345

Ghana Scantravel (Ghana) Limited (R) High St, PO Box 1705, Accra. Tel. 63134 & 64204
Scantravel (Ghana) Ltd (R) Atlantic Hotel, PO Box 693, Takoradi. Tel. 3300/1/2
Scantravel (Ghana) Ltd, (R) Meridian Hotel, PO Box 330, Tema. Tel. 2878/80 Ext. 241

Guatemala Clark Tours (R) Edificio 'El Triangulo', Zona 4-7a Ave y Calle Mariscal Cruz Z-4, PO Box 591, Guatemala. Tel. 60213/16, 67183.

Honduras Mundirama Travel Service (R) Fiallos Soto Bldg, No 131, PO Box 818, Tegucigalpa. Tel. 22 8258 & 22 0385

Hong Kong American Express International, Inc, Yat Fung Building, 2nd Flr, 34-36 D'Aguilar Street, Hong Kong. Tel. 5-241036

Iceland Utsyn Travel Agency (R) 17, Austurstraeti, PO Box 1346, Reykjavik. Tel. 26611

India American Express International Banking Corporation, Majithia Chambers, 3rd Flr, 276 D.N. Naorji Rd, PO Box 10154, Bombay 400 001. Tel. 265615/266078/266120/266594.
American Express International Banking Corporation, 21 Old Court House St, PO Box 2311, Calcutta 1. Tel. 232133/4
Binny Limited (R) No 7 Armenian St, PO Box No 66, Madras 600 001. Tel. 30181 & 26978.
American Express International Banking Corporation, Wenger House, Connaught Place, PO Box537, New Delhi. Tel 344119
Kai Travels Private Limited (R) Oberoi Palace Hotel, Srinagar. Tel. 4366 & 3545.

Indonesia P.T. Pacto Ltd, 88 Djin, Cikini Raya, PO Box 2563, Djakarta. Tel. 46143/49876

Iran Near East Tours (R) Ave Chahar Bagh, Esfahan. Tel. 27697.
Near East Tours (R) Avenue Zand, Shiraz. Tel. 26883, 24843, 28188
Near East Tours (R) 130 Takht Jamshid Ave, Tehran. Tel. 294654.

Israel Meditrad, Ltd. (R) 2 Khayat Square, PO Box 1266, Haifa. Tel. (04) 642266.
Meditrad Ltd (R) 27 King George Street, PO Box 2345, Jerusalem (New City). Tel (02) 222211
Arab Tourist Agency (R) PO Box 19048, Salah-ed-Dine St 25, Jerusalem (Old City). Tel. 283255 & 283266 (Service Desk: Jerusalem Intercontinental Hotel).
Meditrad Ltd (R) 16 Ben Yehuda St, PO Box 4312, Tel Aviv. Tel. 294654 (Service Desk: Sheraton Hotel & Grand Beach Hotel.)

Ivory Coast Socopao-Côte d'Ivoire (R) 14 Blvd de la République, Boîte Postale 1297, Abidjan. Tel. 32 02 11, 22 76 32, 22 83 81

Japan American Express International, Inc, Plaza Shopping Center, 242 Wasoebaru, Aza, Yamazato, PO Box 210, Okinawa City. Tel. 09893-4142/3324/3342
American Express International, Inc, Kita Hankyu Bld (3rd Flr), 55 Shibata-cho, Osaka. Tel. 06 372-8137/8138, 06 371-7792
American Express International, Inc, Halifax Building, 7th Flr, 16-26, Roppongi, 3-chome, Minato-ku, Tokyo. Tel. (03) 586 4321

Jordan International Traders (R) King Hussein Street. PO Box 408, Amman. Tel. 25072, 38213, 42356.

Kenya Express Kenya, Ltd (R) Nkrumah Rd, PO Box 90631, Mombasa. Tel. 24461
Express Kenya, Ltd (R) Consolidated House, Standard Street, PO Box 40433, Nairobi. Tel. 3347277/28.

Korea, South Chunusa Travel Service (R) Sokong Bldg, Room 303, 81 Sokongdong, Chung-ku, CPO Box 1330, Seoul. Tel. 28-9260, 23-3121.
Chosun Hotel, 2nd Flr, Seoul. Tel. 23-0244/6.

Lebanon Amlevco Tours (R) Hotel Phoenicia Intercontinental, PO Box 11-1429, Beirut. Tel. 367 855 & 367 760.

Liberia Morgan Travel Agency (R) Street, PO Box 1260, CDB King Building, Monrovia. Tel. 22149, 26927.

Libya The Libyan Travel Bureau (R) Reufah El Ansari St, Tourist Bldg, PO Box 306, Benghazi. Tel. 3083 & 2565.
North African Maritime and Travel Enterprises (R) 73/75 Baladia St, PO Box 253, Tripoli. Tel. 40565 & 40562.

Malaysia Mayflower Acme Tours Sdn. Bhd. (R) Angkasa Raya Bldg, 123 Djalan Ampang, Kuala Lumpur 01-02. GPO Box 179. Tel. 03-200542, 200550
Mayflower Acme Tours Sdn. Bhd. (R) Hotel Merlin, Leith Street, Penang. Tel. 03-69136

Mauritius The Mauritius Travel & Tourist Bureau Ltd (R) Corner of Sir William Newton & Royal Roads, Port Louis. Tel. 2-2041/42/43

Morocco Voyages Schwartz (R) rue de 'Hôtel Deville Immeuble 'Freres, Agadir.
Voyages Schwartz (R) 112 avenue du Prince Moulay Abdallah, Casablanca. Tel. 222946, 273133, 278054
Voyages Schwartz (R) rue Mauritania, Immeuble 'Mouataouakil', Marrakech. Tel. 333-21
Voyages Schwartz (R) 76 avenue Mohammed 5, Tangier. Tel. 334-59, 334-71

Mozambique Star Travel Service (R) Avenida Manuel Arriaga, 171, PO Box 4271, Maputo (Lourenço Marques). Tel. 23821

Nepal Yeti Travels Pvt Ltd (R) Hotel Mayalu, Ground Flr, Jamal, PO Box 76, Kathmandu. Tel. 12329, 11234

New Zealand Stars Travel United Ltd (R) 87 Queen Street, PO Box 2819, Auckland. Tel. 75-839
Stars Travel United Ltd (R) 236a High Street, PO Box 759, Christchurch. Tel. 40-736, 60-116.
Stars Travel United Ltd (R) 283 Tutanekai St, PO Box 885, Rotorua. Tel. 79091
Stars Travel United Ltd (R) Central House, 26 Brandon Street, PO box 1985, Wellington Tel. 71-029

Nicaragua Agencias Vassalli, SA (R) Lomas de Guadalupe, Apartado 609, Managua. Tel. 27674 & 24209

Nigeria Scantravel Ltd (R) Unity House, 37 Marina, PO Box 1897 Lagos. Tel. 50635, 21469, 24291

Pakistan American Express International Banking Corporation, Standard Insurance House, 1.1 Chundrigar Road, Karachi No 2. Tel. 233686
American Express International Banking Corporation, 112 Rafi Mansion, Shahrah-e-Quaid-e-Azam, PO Box 249, Lahore. Tel. 66078, 312435, 58120.
American Express International Banking Corporation, Cantonment Bldg, Haider Road, PO Box 96, Rawalpindi. Tel. 65766, 65765

Panama Boyd Brothers, Inc (R) PO Box 805, Panama City. Tel. 62-0300

Paraguay Inter-Express, SRL (R) Ntra Sra Asunción, 588, Asunción. Tel. 48-888
Inter-Express SRL (R) Mcal Lopez 393, Encarnación. Tel. 948

Peru Lima Tours, SA (R) PO Box 67, Santa Catalina 120, Arequipa. Tel. 27-6624/33 (10 lines)
Lima Tours SA (R) Av El Sol 954-C, PO Box 531, Cuzco. Tel. 2809
Lima Tours SA (R) Ocona 160, PO Box 4340, Lima. Tel. 27-6624

Philippines American Express International, Inc, 710 Friendship Highway, Riverside Subdivision, Angeles City. Tel. 40-53, 53-91
American Express International, Inc, 7 West Drive Arcade, Makati Commercial Center, Corner Pasay Road/Makati Avenue, Makati. Tel. 88-58-93, 88-58-18, 88-59-61 & 89-75-73
American Express International, Inc, Philamlife Building, United Nations Avenue & Maria Y. Orosa Street, PO Box 1147, Manila. Tel. 50-96-01

Saudi Arabia ACE Arab Commercial Enterprises (R) King Abdul Aziz St, PO Box 6152, Jeddah. Tel. 23731, 31151
ACE Arab Commercial Enterprises (R) Aziziah Bldg, King Faisal St, PO Box 667, Riyadh. Tel. 20021, 20022

Seychelles Travel Services (Seychelles) Ltd (R) Pirates Arms Bldg, PO Box 356, Victoria, Mahé. Tel. 2414

Singapore American Express International, Inc, Holiday Inn Bldg, 3rd Flr, Scotts Road, PO Box 49, Singapore. Tel. 375988

South Africa American Express International, Inc, Shop No 3, Pick 'n Pay Hypermarket, North Reef Road, PO Box 861, Kempton Park, Boksburg North, Transvaal 1620. Tel. 826-2408/9 & 826-1826
American Express International, Inc, Greenmarket Place, Greenmarket Square, PO Box 2337, Cape Town 8000. Tel. 22-8581
American Express International, Inc, Denor House, 1st Flr, Cnr Smith & Field Streets, PO Box 2558, Durban 4000. Tel. 323491
American Express International, Inc, Merbrook, 123 Commissioner Street, PO Box 9395, Johannesburg 2000. Tel. 37-4000
American Express International, Inc, Jan Smuts Holiday Inn Office, PO Box 861, Kempton Park, Transvaal 1620. Tel. 975-5988/89 & 975-1121
American Express International, Inc, SAAU Building, 308 Andries Street, PO Box 3592, Pretoria 0001. Tel. 29182

Suriname Travelbureau C Kersten & Co, NV (R) c/o Hotel Krasnapolsky, Domineestraat 39, Paramaribo. Tel. 74448 & 77148

Syria Amievco Tours (R) rue Fardous Mouradi Bldg, PO Box 507, Damascus. Tel. 111652 & 119553

Tahiti Tahiti Tours (R) rue Jeanne d'Arc, Boîte Postale 627, Papeete. Tel. 2 98 70

Thailand SEA Tours Co Ltd (R) Room No 414 (4th Flr), Siam Center, 965 Rama Road, Bangkok 5, Thailand. Tel. 251-3521/2/3

Taiwan American Express International (Taiwan), Inc, Central Building, 5th Flr, 108 Chung Shan N Rd, PO Box 3977, Taipei. Tel. (02) 581-1176-9

Tunisia Tourafric (R) 52 ave Habib Bourguiba, Tunis. Tel. 245 066

Turkey Turk Ekspres (R) Havacilik ve Turizm Ltd Sti, Sehit Adem Yavuz 12/B, Ankara. Tel. 25 21 82, 17 05 76
Turk Ekspres Havacilik ve Turism Ltd Sti (R) Istanbul Hilton Hotel, PO Box 70-Beyoglu, Istanbul. Tel. 48 39 05 & 40 56 40
Egetur Travel Agency (R) Nato Arkasi Cumhuriyet Cad, Izmir. Tel. 37651 & 33088

USSR American Express Company, Hotel Metropol, Sverdlov Pl 2/4 Moscow. Tel. 225 63 84, 64 89, 64 43

Uruguay Turisport Limitada (R) Bartolome Mitre 1318, Montevideo. Tel. 86300, 94823

Venezuela TMC Consolidado CA (R) Lobby Tamanaco Hotel, Urb Las Mercedes, Caracas 105, PO Box 68459, Caracas. Tel. 914224, 914308 & 923219

Holidays and Business Hours

	Holidays	*Business Hours*
Argentina	May 1, Labour Day; May 25th, Revolution (1810) Day; June 20, Flag Day; July 9, Independence (1816) Day; August 17, Death of General José de San Martin; October 12, Discovery of America (Columbus Day); and December 25, Christmas. On a number of other days, government offices, banks, insurance companies, and courts are closed, but closing is optional for business and commerce. These include: January 1, New Year's Day; January 6, Epiphany; and several days with variable dates — Carnival Monday and Tuesday before Ash Wednesday, Holy Thursday and Good Friday before Easter, and Corpus Christi; August 15, Assumption of the Virgin Mary; November 1, All Saints' Day and December 8, Feast of the Immaculate Conception. In addition, there are local patriotic or religious holidays, which may be observed by part or all of the community in various cities or provinces.	Generally 9am to 7pm from Monday to Friday. Government offices are open 12 noon to 7pm from Monday to Friday, closed on Saturdays and holidays.
Australia	New Year's Day (Jan 1), Australia Day (last Monday in January), Good Friday, Easter Sunday, Easter Monday, Anzac Day (April 25), the Queen's Birthday (except in Western Australia, where it is usually on a Monday in June), Christmas Day and Boxing Day (December 25 and 26).	9am—5.30pm, Monday to Friday. Closed Saturday. Hours vary from one state or business to another.
Brazil	New Year's Day (Jan 1), Carnival (four nights and three days preceding Ash Wednesday), Good Friday, Easter Sunday, Tiradentes Day (April 21), Labour Day (May 1) Independence Day (Sept 7), All Souls Day (Nov 2), Proclamation of the Republic (Nov 15) and Christmas Day.	Generally 8—8.30am to 5.30—6pm.
Hong Kong	The first weekday in January, Lunar New Year's Day, the second and third days of Lunar New Year, Good Friday, the day following Good Friday, Easter Monday, Ching Ming Festival, the Queen's Birthday, Tuen Ng (Dragon Boat Festival), the first weekday in July, the first Monday in August, Liberation Day, the second day of Mid-Autumn Festival, Chung Yeung Festival, Christmas Day, and the first weekday after Christmas.	Mostly 9am to 5pm with lunch between 1 and 2pm. On Saturdays, 9am to 1pm. Some Chinese business houses open at 10am and close around 8pm. Closed Sundays.
India	New Year's Day, Republic Day (January 26), Independence Day (August 15), Mahatma Gandhi's Birthday (October 2) and Christmas Day. Also thirty-two other religious or special occasions which are observed either with national or regional holidays. Government offices are normally closed on these days.	Most Indian government offices are open 10am to 1pm and 2pm until 5pm, Monday to Saturday, except on the second Saturday of each month. Business offices usually open from Monday to Friday 9.30am to 1pm and 2pm to 5pm.

	Holidays	*Business Hours*
Indonesia	New Year's Day, January 1; Good Friday; Hari Maulud Nabi (Muslim festival), March 13; Ascension Day; Waicak Day (celebrating Buddha's birth); Galunggan in Bali (a New Year feast lasting ten days), June 9; Sekaten (birth of Mohammed); Independence Day, August 17; Idul Fitri (Muslim festival), September 25-26; and Christmas Day. Dates for certain holidays change with the lunar calendar. In addition to the holidays listed, business visitors should note the Islamic month of fasting, Ramadan.	During the 28 days following Ramadan, Indonesian government and many other offices work a shorter business day, generally 8am to 12 noon. The month of Ramadan concludes with the holiday of Idul Fitri. Businesses otherwise open generally by 7.30am.
Iran	(Applicable until 1979) March 21 and 22, New Year (Now Ruz); April 2, Thirteenth Day of the New Year (Sizdah); August 5, Constitution Day. In addition there are religious holidays with variable dates, during which most offices are closed.	Government offices are generally open from 8am to 4.30pm from Saturday to Wednesday. Private firms and shops are generally closed on Fridays (the Moslem Sabbath) and open the other six days of the week.
Israel	All business activity ceases on Saturdays and religious holidays, the dates of which vary from one year to another: Passover, first day; Passover, last day; Israel Independence Day; Pentecost; Rosh Hashana (New Year); Yom Kippur (Day of Atonement); First Day of Tabernacles; Last Day of Tabernacles; Hanukkah.	Generally from 8am until 4pm in winter, and from about 7.30am to 2pm during the summer. Most offices close early on Fridays to prepare for Sabbath, which lasts from dusk on Friday until dusk on Saturday. Government offices are generally open to visitors between 9am and 1pm from Sunday to Thursday, and at other times if the matter is pressing.
Japan	January 1, New Year's Day; Jan 2, 3, 4, Bank Holidays (all commercial firms closed). Jan 15, Adults' Day; Feb 11, National Foundation Day; Vernal Equinox Day (variable date); April 29, the Emperor's birthday; May 1 (May Day), most manufacturers closed, service firms open; May 3, Constitution Memorial Day; May 5, Children's Day; September 15, Respect of the Aged Day; Autumnal Equinox Day (variable date); October 10, Physical Culture Day; November 3, Culture Day; November 23, Labour Thanksgiving Day; December 28, New Year's holiday begins (lasts five to ten days). Also 'Golden Week' in late Spring when some firms remain closed. Some manufacturers close for a week during the summer.	9am to 5pm Monday to Friday; 9am to 12 noon, Saturdays. Closed Sundays.
Kenya	January 1 (New Year's Day); Good Friday and Easter Monday (variable); May 1 (Labour Day); June 1 (Mandaraka Day), Id ul-Fitr (variable); October 20 (Kenyatta Day); Id ul-Azhi (variable); December 12 (Independence Day); December 25 (Christmas Day); December 26 (Boxing Day).	Monday to Friday from 8.15am to 12 12.30pm and from 2pm to 4.30pm. Most offices are also open on Saturday from 8.15am to noon.

	Holidays	*Business Hours*
Korea, South	January 1-3, New Year's Celebration; March 1, Independence Movement Day; March 10, Labour Day; April 5, Arbour Day; June 5, Memorial Day; July 17, Constitution Day; August 15, Liberation Day; Korean Thanksgiving Day; October 9, Korean Alphabet Day; October 24, United Nations Day; December 25, Christmas Day.	9am to 5pm Monday to Friday with lunch 1 to 2pm, and 9am to 1pm on Saturday. Service establishments remain open as late as 10pm and are open over weekends and even on public holidays. Except for Cheju Island and Chungchongbukdo province, a curfew is in effect from 11pm to 4am.
Kuwait	Religious holidays vary from one year to another. The only fixed holidays in Kuwait are New Year's Day and Kuwait National Day (February 25). October to May or June is generally considered the best period for foreign business visitors; business slackens off in the summer.	Closed Friday (Muslim Sabbath). Goverment hours generally 7.30am to 1.30pm in winter, 6.30am to 12.30pm in spring and autumn, and 6am to noon in summer. Larger private industries generally open 7.30am to 2.30pm Saturday to Wednesday. Business offices and retail trades generally open from 7am to 1pm and 5—8pm; many, however, do not reopen on Thursday evening.
Malaysia	Many vary in accordance with Muslim or lunar calendars.	Mostly from 8.30 or 9am to 1pm, and from 2.30 to 4.30 or 5pm on Mondays to Fridays and on Saturday from 9am to 1pm. Goverment offices open 9am to 4.30pm on weekdays and 9am to 1pm on Saturdays.
New Zealand	New Year's Day, New Zealand Day (February 6), Good Friday, Easter Monday, Anzac Day (April 25), the Queen's Birthday (generally observed early in June), Labour Day (in October), Christmas Day and Boxing Day (December 25 and 26). Also a holiday for the provincial anniversary in each provincial district i.e. Auckland (January 29), Canterbury (December 16), Hawkes Bay (November 1), Marlborough (November 1), Nelson (February 1), Northland (February 6), Otago (March 23), Southland (March 23), Taranaki (March 31), Wellington (January 22).	Normally 9am to 5pm Mondays to Fridays. Each town has a 'late night,' often Friday, when shops are open until 8pm. Food and ice-cream shops are known as 'dairies' generally open Saturdays 9am to 7pm and sometimes Sundays also.
Peru	January 1, New Year's Day; Holy Thursday (half day only); Good Friday; May 1, Labour Day; June 29, St. Peter and St. Paul; July 28, Independence Day; August 30, Santa Rosa de Lima; October 9, Peruvian National Day; November 1, All Saints' Day; December 8, Immaculate Conception; December 25, Christmas.	Some offices open for a continuous working day with a short lunch break, others for two shifts per day separated by a two-three hour lunch break. Summer hours (January to March) are generally shorter than in winter eg banks ministries, government offices open 8am—1pm in summer for a full 8 hours (from 8.30 or 8.55am in winter). Private offices usually open between 7.45 and 8.15am in summer and between 8 and 9am in winter. Department and retail stores are generally open for trading 10.45am to 7.50pm. Factories open year-

	Holidays	*Business Hours*
Philippines	January 1, New Year's Day; Easter holidays, which include Holy Thursday and Good Friday; May 1, Labour Day; June 12, Independence Day; July 4, Philippine-American Day; November 30, National Heroes' Day; December 25, Christmas; December 30, Rizal Day. Additional holidays such as Bataan Day and General Elections Day may be called by the President of the Republic.	Business and government firms normally from 8 or 8.30am to 5pm with two hours for lunch, the afternoon session beginning between about 2.30 and 3.30pm. Closed Saturdays and Sundays.
Saudi Arabia	During the month of Ramadan (in the Autumn), all Muslims refrain from eating, drinking and smoking from sunrise to sunset. Business hours are shortened. Non-Muslims must also observe the fast while in public.	Government hours Saturday to Wednesday, 8am—4pm during the winter (September 23—May 21), and 7am to 3pm during the summer, with one hour at noon for lunch prayer. During Ramadan, hours are 8am to 2pm. Government offices are open to the public in the morning only. Businesses normally open 8.30am to 1.30pm and 4.30 to 8pm six days a week. Closed Friday, the Muslim Sabbath.
Singapore	New Year's Day, January 1; Labour Day, May 1; Vesak Day, May 17; National Day, August 9; Hari Raya Punsa, October 29; Christmas Day, December 25. Holidays with variable dates are Hari Raya Haji, Chinese New Year, Good Friday and Deepavali. When a holiday falls on a Sunday, the next day is taken as a public holiday.	Office hours generally from 8.30 or 9am to 9pm, Monday to Saturday, and 2.30 to 4.40 or 5pm Monday to Friday. Government offices 9am to 4.30pm, Monday to Friday and 9am to 1pm on Saturday.
South Africa	New Year's Day, Good Friday, Easter Monday, Ascension Day, Republic Day (May 31), Settlers Day (first Monday in September), Kruger Day (October 10), Day of the Convenant (December 16), Christmas Day and Boxing Day.	Generally from 8.30am to 5pm Monday to Friday, with some also open 8.30am to noon on Saturdays. Many businesses close for lunch between 1 and 2pm.
Thailand	New Year's Day, January 1; The Songkran Festival (Buddhist New Year), April 13; Coronation Day Anniversary, May 5; Visakhja Puja (Buddhist Festival), May; Buddhist Lent begins, June/July; Queen's Birthday, August 12; Chulalongkorn Day, October 23; King's Birthday, December 5; New Year's Eve, December 31.	
Turkey	Official holidays: National Sovereignty and Children's Day, April 22 (half day) and April 23; Spring Day, May 1; Youth and Sports Day, May 19; Freedom and Constitution Day, May 26 (half day) and May 27 and 28; Victory Day, August 30; Anniversary of the Declaration of the Republic, October 28 (half day) and October 29 and 30. Religious holidays based on the Arabic calendar: The Feast of Sugar, in autumn; The Feast of Sacrifice, in winter (variable dates).	Offices closed Saturday afternoons and one-two days before and after the holidays. Stores open on Saturday afternoons, closed on the first day only of religious holidays.

Banking Hours

Argentina	12 noon—4 (Mon to Fri)
Australia	10—3(Mon to Th), 10—5(Fri), closed (Sat)
Belize	8.30—noon and 2—4 (Mon to Fri), 8.30—noon (Sat)
Bolivia	9—12 and 2—4.30 (Mon to Fri)
Brazil	10—4 (Mon to Fri)
Burma	9—4 (Mon to Fri), 9—12.30 (Sat)
Chile	9—2 (Mon to Fri)
China	Hours vary drastically. Open Sat, closed Sun.
Colombia	9—3 (Mon to Fri)
Costa Rica	8—11 and 1.30—3 (Mon to Fri), 8—11 (Sat)
Ecuador	9—noon and 2.30—4 (Mon to Fri)
Egypt	8.30—12.30 (Sat and Th) and 10—noon (Sun)
El Salvador	8—4 (Mon to Fri), 8—12 (Sat)
Fiji	10—3 (Mon to Th), 10—4 (Fri) Closed Sat
French Guiana	7.30—1 (Mon to Fri)
Guatemala	8.30—12.30 and 2—4 (Mon to Fri)
Guyana	8—noon (Mon to Fri), 8—11 (Sat)
Honduras	8—12 and 2—4 (Mon to Fri), 8.30—12 (Sat)
Hong Kong	10—3 (Mon to Fri), 9—noon (Sat)
India	10.30—2.30 (Mon to Fri), 10.30—12.30 (Sat)
Indonesia	10—3 (Mon to Fri), 9—noon (Sat)
Israel	8.30—12.30 (Sun to Fri), some 4—5pm
Japan	9—3 (Mon to Fri), 9—noon (Sat)
Kampuchea Cambodia	8.30—3.30 (Mon to Fri), 8.30—noon (Sat), closed on religious holidays
Kenya	9—1 (Mon to Fri),9—11 (Sat)
Korea, South	9.30—4 (Mon to Fri), 9.30—1 (Sat)
Lebanon	8.30—12.30 (Mon to Fri), 8.30—noon (Sat)
Liberia	8—noon (Mon to Fri), 8—2 (Sat), closed Fri
Macao	8—4 (Mon to Fri), 9—1 (Sat)
Malaysia	10—3 (Mon to Fri), 9.30—11.30 (Sat), closed Fri in some states
Mexico	9—1.30 (Mon to Fri)
Micronesia	9.30—2.30 (Mon to Fri)
Morocco	8.15—11.30, 2.15—4.30 (Mon to Fri)
Nepal	10—3 (Sat to Th), 10—noon (Fri), closed Saturday
New Caledonia	7—10.30, 1.30—3.30 (Mon to Fri), 7.30—11am (Sat)
New Zealand	10—4 (Mon to Fri)
Nicaragua	8.30—noon and 2—4 (Mon to Fri), 8.30—11.30 (Sat)
Nigeria	8—3 (Mon to Fri)
Panama	8.30—12.30 and 2.30—3.30 (Mon to Fri)
Papua New Guinea	9—2 (Mon to Fri), 8.30—10 (Sat)
Paraguay	7.30—10.30 (Mon to Fri)
Peru	9.30—12.30 and 4—7 (Mon to Fri)
Philippines	9—6 (Mon to Fri), 9—12.30 (Sat)
Somoa, American	9—2 (Mon to Th), 9—5 (Fri)
Somoa, Western	9.30—3pm (Mon to Fri), 9.30—11.30 (Sat)
Singapore	10—3 (Mon to Fri), 9.30—11.30 (Sat)
South Africa	9—3 (Mon, Tues, Th, Fri), 9—1 (Wed), 9—11 (Sat)
Sri Lanka	9—1 (Mon to Fri), 9—11 (Sat), closed on religious holidays
Suriname	7.30—1 (Mon to Fri), 7.30—11 (Sat)
Tahiti	7.45—3.30 (Mon to Fri)
Taiwan	9—3.30 (Mon to Fri), 9—noon (Sat)
Tanzania	8.30—noon (Mon to Fri), 8.30—11 (Sat)
Thailand	8.30—3.30 (Mon to Fri), Closed Sat
Tonga	No banks. Change money at Treasury Building or hotels.
Turkey	9—noon, 1.30—5 (Mon to Fri), 9—noon (Sat)
USSR	9—1 (Mon to Fri)
Uruguay	1.30—5 (Mon to Fri)
Venezuela	8.30—11.30 and 2.30—5.30 (Mon to Fri)
Vietnam	8—11.30, 2—4 (Mon to Fri), 8—11 (Sat)

Airport Duty-Free Shops

Duty-Free Shops—Available to departing international passengers and at certain airports (marked *) to arriving international passengers

City	Airport
Abu Dhabi	
Abu Dhabi	International
Algeria	
Algiers	Dar-el- Beida
Antigua	
Antigua	Coolidge Field
Argentina	
Buenos Aires	Ezeiza
Australia	
Brisbane	International
Melbourne	Tullamarine
Perth	
Sydney	Kingsford-Smith
Austria	
Graz	Thalerhof
Klagenfurt	Wörthsee
Linz	Hörsching
Salzburg	
Vienna	Schwechat
Bahrain	
Bahrain	Muharraq
Bangladesh	
Dacca	Tezgaon
Barbados	
Barbados	Grantley Adams Int'l
Belgium	
Antwerp	Deurne
Brussels	National
Charleroi	Gosselies
Ostend	
Brazil	
Manaus	Eduardo Gomes
Rio de Janeiro	International
Bulgaria	
Bourgas	International
Sofia	International
Varna	International
Canada	
Calgary	Municipal
Edmonton	International
Gander	Gander 2
Halifax	International
Montreal	Mirabel
	Dorval
Toronto	International
Vancouver	International
Winnipeg	International
Colombia	
Bogotá	El Dorado
San Andres	
Costa Rica	
San José	Santamaria Int'l
Cuba	
Havana	José Marti
Cyprus	
Larnaca	
Czechoslovakia	
Prague	Ruzyné
Denmark	
Copenhagen	Kastrup
Dominican Rep	
Santo Domingo	Las Americas
Ecuador	
Guayaquil	Simon Bolivar
Quito	Mariscal Sucre
Egypt	
Cairo	International*
El Salvador	
San Salvador	Ilopango
Ethiopia	
Addis Ababa	Bole
Asmara	Yohannes IV
Fiji	
Nadi	International*
Finland	
Helsinki	Vantaa
Turku	
France	
Bordeaux	Merignac
Lille	Lesquin
Lyon	Satolas
Marseille	Marignane
Montpellier	Frejorgues
Nantes	Château-Bougon
Nice	Côte d'Azur
Paris	Charles de Gaulle
	Orly
Strasbourg	Entzheim
Tarbes	Ossun
Toulouse	Blagnac
German Democratic Rep	
Leipzig	Schkeuditz
German Federal Rep	
Berlin	Tegel
Bremen	Neuenland
Cologne	
Düsseldorf	
Frankfurt	
Hamburg	Fuhlsbüttel
Hanover	
Munich	Riem
Nuremberg	
Saarbrücken	Ensheim
Stuttgart	Echterdingen
Gibraltar	
Gibraltar	North Front
Greece	
Athens	Hellinikon
Guatemala	
Guatemala City	La Aurora
Guyana	
Georgetown	Timehri Int'l

Haiti	
Port-au-Prince	Duvalier Int'l
Honduras	
San Pedro Sula	Villeda Morales
Tegucigalpa	Toncontin
Hong Kong	
Hong Kong	Kai Tak
Hungary	
Budapest	Ferihegy
Iceland	
Reykjavik	Keflavik*
India	
Bombay	
Calcutta	
Delhi	
Madras	
Indonesia	
Denpasar	Ngurah Rai
Djakarta	Halim Int'l
Iran	
Tehran	Mehrabad
Iraq	
Baghdad	International
Ireland	
Dublin	
Shannon	
Israel	
Tel Aviv	Ben-Gurion Int'l
Italy	
Milan	F. Forlanini-Linate
Pisa	Galileo Galilei
Rome	Leonardo da Vinci(Fiumicino)
Turin	Citta di Torino-Caselle
Venice	Marco Polo-Tessera
Jamaica	
Kingston	Norman Manley Int'l
Montego Bay	Donald Sangster Int'l
Japan	
Fukuoka	Itazuké
Kagoshima	Komaki
Nagoya	International
Okinawa	Naha
Osaka	International
Tokyo	Haneda
	Narita
Jordan	
Amman	
Kenya	
Nairobi	Jomo Kenyatta
Korea, South	
Pusan	Kinkai
Seoul	Kimpo Int'l
Lebanon	
Beirut	International
Luxembourg	
Luxembourg	Findel
Malawi	
Blantyre	Chileka
Malaysia	
Kuala Lumpur	Subang
Penang	Bayan Lepas
Malta	
Malta	Luqa
Mauritius	
Mauritius	Plaisance
Mexico	
Mexico City	Benito Juárez Int'l
Morocco	
Tangier	Boukhalef
Nepal	
Kathmandu	Tribhuyan
Netherlands	
Amsterdam	Schiphol
Rotterdam	
Netherlands Antilles	
Aruba	Prinses Beatrix
Curaçao	Dr A Plesman
Saba	Juancho Yrausquin
St Eustatius	F D Rooseveld
St Maarten	Juliana
New Caledonia	
Nouméa	Tontouta
New Zealand	
Auckland	International
Christchurch	International
Wellington	
Nicaragua	
Managa	Las Mercedes
Norway	
Bergen	Flesland
Oslo	Fornebu
Stavanger	Sola
Oman	
Muscat	Seeb
Pakistan	
Karachi	Civil
Rawalpindi	Chakiala
Panama	
Panama City	Tocumen
Peru	
Lima	Jorge Chavez Int'
Philippines	
Manila	Internationa
Poland	
Warsaw	Okeci
Portugal	
Lisbon	Portel
Puerto Rico	
San Juan	Isla Verde Int'
Romania	
Bucharest	Otopen
St Lucia	
St Lucia	Hewanorr
Senegal	
Dakar	Yof

Seychelles	
Seychelles	Mahé
Singapore	
Singapore	Paya Lebar*
South Africa	
Cape Town	D F Malan
Durban	Louis Botha
Johannesburg	Jan Smuts
Spain	
Alicante	
Barcelona	
Ibiza	
Las Palmas (Canary Is)	
Madrid	Barajas
Mahon (Menorca)	
Malaga	
Palma de Mallorca	
Tenerife (Canary Is)	Los Rodeos
Sri Lanka	
Colombo	Bandaranaike
Sudan	
Khartoum	Civil*
Suriname	
Paramaribo	Zanderij
Sweden	
Gothenburg	Landvetter
Malmö	Sturup
Stockholm	Arlanda
Switzerland	
Basel	Basel/Mulhouse
Geneva	Cointrin
Zurich	Kloten
Syria	
Damascus	International
Taiwan	
Taipei	Sung Shan
Tanzania	
Dares Salaam	
Thailand	
Bangkok	Don Muang
Tobago	
Scarborough	Crown Point
Trinidad	
Port of Spain	Piarco
Turkey	
Istanbul	Yesilköy*
United Arab Emirates	
Dubai	International
Sharjah	International
USSR	
Moscow	Sheremetyevo
United Kingdom	
Birmingham	
East Midlands	
Edinburgh	
Glasgow	Abbotsinch
	Prestwick
Guernsey	
Jersey	
London	Gatwick
	Heathrow
	Stansted
Manchester	
USA	
Anchorage	International
Boston	Logan Int'l
Dallas	Dallas/Ft Worth
Honolulu	International
Houston	Intercontinental
Los Angeles	International
Miami	International
New Orleans	Moisant Int'l
New York	J F Kennedy Int'l
Philadelphia	International
San Diego	Lindberg Int'l
San Francisco	International
Seattle	Seattle/Tacoma Int'l
Tampa	International
Washington	Dulles Int'l
Uruguay	
Montevideo	Carrasco
Venezuela	
Caracas	Simón Bolivar
Yugoslavia	
Belgrade	Surcin
Zagreb	Pleso
Zaïre	
Kinshasa	N'Djili
Zambia	
Lusaka	
Zimbabwe	
Salisbury	

Worldwide Guide to Duty-Free Allowances

This list is intended as a guide to what local Customs authorities will allow into each country free of duty. Every effort has been made to ensure that the allowances shown are correct, but these are subject to alteration without notice.

UNITED KINGDOM

1 If you have come from an EEC country and the goods were not bought in a duty-free shop or on a ship or aircraft.

2 If you have come from a country outside the EEC or if the goods were bought in a duty-free shop or on a ship or aircraft.

	1	2
Tobacco Goods		
Cigarettes	300	●200
OR Cigarillos	150	●100
OR Cigars	75	●50
OR Tobacco	400 grammes	●250 grammes
Alcoholic Drinks		
Over 38.8° proof (22° Gay-Lussac)	1½ litres	1 litre
OR not over 38.8° proof	3 litres	2 litres
OR fortified wine or sparkling wine	3 litres	2 litres
AND still table wine	3 litres	2 litres
Perfume	75 grammes (85.5 ml)	50 grammes (57 ml)
Toilet Water	⅜ litre	¼ litre
Other Goods	£50 worth	£10 worth

The above allowances for tobacco goods and alcoholic drinks are not for people under 17.

		Cigarettes	Cigars	Tobacco	Wine	Spirits	Perfume
EEC COUNTRIES							
Belgium	iEu fEU	300	or 75	or 400 gm	3 ltr	1.5 ltr	75 gm
	iEu K	200	or 50	or 250 gm	2 ltr	1 ltr	50 gm
	oEu fEu	400	or 100	or 500 gm	3 ltr	1.5 ltr	75 gm
	oEu K	400	or 100	or 500 gm	2 ltr	1 ltr	50 gm
Denmark	iE uE	300	or 75	or 400 gm	3 ltr	1.5 ltr	75 gm
	K	200●	or 50●	or 250 gm●	2 ltr	1 ltr	50 gm
	oE uE	400	or 100	or 500 gm	3 ltr	1.5 ltr	75 gm
France	E	300	or 75	or 400 gm	3 ltr	1.5 ltr	⅜ ltr
	N	200	or 50	or 250 gm	2 ltr	1 ltr	¼ ltr
	O	400	or 100	or 500 gm	2 ltr	1 ltr	¼ ltr
Germany	Vo Eu	400	or 100	or 500 gm	3 ltr	1.5 ltr	75 gm
	R&V iEu	200	or 50	or 250 gm	2 ltr	1 ltr	25 gm
Irish Republic	fEu	300	or 75	or 400 gm	3 ltr	1.5 ltr	75 gm
	iEu K	200	or 50	or 200gm	2 ltr	1 ltr	50 gm
	oEu K	1000	or 200	or 1100 gm	1.5 ltr	1.5 ltr	.568 ltr
Italy	VE	300	or 75	or 400 gm	3 ltr	1.5 ltr	75 gm
	VN	200●	or 50●	or 250 gm●	2 ltr	0.75 ltr	50 gm
	R	In accordance with the allowance of the country where the return journey commences.					
Luxembourg	iEu EECnd	300	or 75	or 400 gm	3 ltr	1.5 ltr	75 gm
	iEu EECd	200	or 50	or 250 gm	2 ltr	1 ltr	50 gm
	oEu	400	or 100	or 200 gm	2 ltr	1 ltr	50 gm
Netherlands		200	or 50	or 250 gm	2 ltr	1 ltr	50 gm
OTHER COUNTRIES							
Abu Dhabi		400	or 100	or ½ lb	Nil	Nil	fpu
Antigua		200	or 50	or ½ lb	1 qt	1 qt	nl

		Cigarettes	Cigars	Tobacco	Wine	Spirits	Perfume
Australia		200	or 50	or ½ lb	1 ltr	or 1 ltr	nl*fpu
Austria		200	or 50	or 250 gm	2 ltr	1 ltr	50 gm
Bahamas		200	50	1 lb	1 qt	or 1 qt	fpu
Bahrain		400	or 50	and ½ lb	n 2 btl	alch'c bev	8 oz
Brunei		200	or ½ lb	or ½ lb	1 btl	1 btl	fpu
Bangladesh	V	200	or 50	or ½ lb	⅓ gal	⅓ gal	½ pint
	R (Banga)	200	or 50	or ½ lb	Nil	Nil	½ pint
	R (Foreign)	200	or 50	or ½ lb	1/6 gal	1/6 gal	½ pint
Barbados		200	or 50	or ½ lb	1 qt	or 1 qt	fpu
Bermuda		200	100	1 lb	1 qt	1 qt	fpu (Rnl*)
Bulgaria		250	or 250 gm	or 250 gm	2 ltr	1 ltr	fpu
Canada		200	50	2 lb	1 qt	or 1 qt	value $15 (Rnl*)
Cyprus		200	or 50	or ½ lb	1/6 gal	1/6 gal	2 oz
Czechoslovakia		250	or 50	or ½ lb	2 ltr	1 ltr	value 300 Kcs
Dubai		1000	or 200	1 kg	n 2.5 ltr	n 2.5 ltr	150 gm
Egypt**		200	or ½ lb	or ½ lb	1 btl	1 btl	2 btl
Ethiopia		100	50	½ lb	1 qt	or 1 qt	1 pint fpu**
Finland		200●	or 250 gm●	or 250 gm●	1 ltr●	1 ltr●	fpu
Greece		200	or ½ lb	or ½ lb	1 btl	1 btl	fpu
Guyana		200	or 50	or ½ lb	§1/6 gal	§1/6 gal	fpu
Hong Kong		200	or 50	or ½ lb	1 qt	or 1 qt	fpu
Hungary		250	or 50	or 250 gm	2 ltr	1 ltr	fpu
India		200	or 50	or 250 gm	1 btl	or 1 btl	fpu
Iran		200	or 750 gm	or 750 gm	1 btl	or 1 btl	½ ltr
Iraq		200	50	250 gm	1 ltr	or 1 ltr	or ½ ltr
Israel		250	or 250 gm	or 250 gm	2 ltr	1 ltr	¼ ltr
Jamaica		200	or 50	or ½ lb	1 qt	1 qt	6 oz
Japan	V	400	or 100	or 500 gm	3 btl	or 3 btl	2 oz
	R	200	or 50	or 250 gm	3 btl	or 3 btl	2 oz
Jordan	V	200	or 50	or 200 gm	2 btl	or 1 btl	fpu
	R	200	or 50	or 200 gm	1 btl	or 1 btl	fpu
Kenya		200	or 50	or ½ lb	1 btl	or 1 btl	1 pint
Kuwait		nl	nl	nl	Nil	Nil	nl
Lebanon	V s	400	or 25	or 500 gm	1 btl	or 1 btl	1 oz
	R&V w	200	or 25	or 200 gm	1 btl	or 1 btl	1 oz
Malawi		200	or 225 gm	or 225 gm	1 qt	or 1 qt	Nil
Malaysia		200	or ½ lb	or ½ lb	1 qt	or 1 qt	Value $50
Malta		200	or 225 gm	or 225 gm	1 btl	1 btl	fpu
For duty purposes Malta is taken to be in Europe							
Mauritius	V	200	or 250 gm	or 250 gm	2 ltr	75 cc	fpu
	R	200	or 250 gm	or 250 gm	2 ltr	75 cc	Value MAR.100

		Cigarettes	Cigars	Tobacco	Wine	Spirits	Perfume
Mexico		400	or 1 kg	or 1 kg	2 btl	or 2 btl	3 btl
New Zealand		200	or 50	or ½ lb	1 qt	1 qt	fpu
Norway		200●	or 250 gm●	or 250 gm●	1 ltr	& 0.75 ltr	fpu
Pakistan	fl	50	or 50	or ¼ lb	½ pint	½ pint	2 oz
	fe	200	or 50	or ½ lb	1 btl	1 btl	½ pint
	Tourists	200	or 50	or ½ lb	2 btl	or 2 btl	½ pint
Panama		300	50	3 tins	3 btl	or 3 btl	¼ ltr
Poland		250	or 50	or 250 gm	1 ltr	1 ltr	fpu
Portugal		200	or 50	or 250 gm	1 btl	1 btl	fpu
Qatar	(Vfpu)	nl	nl	nl	Nil	Nil	nl
Romania	(V only)	200	or 50	or 250 gm	2 ltr	1 ltr	fpu
Seychelles		200	or 250 gm	or 250 gm	1 ltr	1 ltr	4 oz
Singapore		200	or 50	or 250 gm	1 btl	1 btl	Nil
South Africa		400	50	250 gm	1 ltr	1 ltr	300 ml
Spain		200	or 50	or 250 gm	1 ltr	1 ltr	250 gm
Sri Lanka	V	200	or 50	or ¾ lb	2 btl	2 btl	fpu
	R3 Mths	200	or 50	or ½ lb	2 btl	2 btl	1 pint
	R2-3 Mths	200	or 50	or ½ lb	2 btl	2 btl	½ pint
	R1-2 Mths	200	or 50	or 125 gm	2 btl	2 btl	8 oz
Sudan		200	or 25	or ½ lb	½ pint	1 pint	2 btl
Sweden		200●	or 50●	or 250 gm●	1 ltr	1 ltr	fpu
Switzerland		200●	or 50●	or 250 gm●	1 ltr	1 ltr	8 oz
Syria		200	25	or 500 gm	1 btl	1 btl	1 oz
Tanzania		200	or 50	or ½ lb	1 pint	or 1 pint	1 pint
Thailand		200	or 200 gm	or 200 gm	1 btl	or 1 ltr	fpu
Trinidad		200	or 50	½ lb	1 qt	or 40 oz	fpu
Turkey		200	or 50	½ lb	1 btl	1 btl	fpu
USA	V g*	nl★	100*f*	nl	ao 1 qt	ao or 1 qt	nl
	V fpu	200	or 50*f*	or 3 lb	ao 1 qt	ao or 1 qt	nl
	R**	nl★	100*f*	nl	× 1 qt	× or 1 qt	nl
USSR		200	or 250 gm	or 250 gm	2 ltr	1 ltr	fpu
Yugoslavia		200	or 50	or 250 gm	1 ltr	1 ltr	.75 ltr

KEY

V	visitors
R	residents
ao	adults only
oEu	who live outside Europe
iEu	who live in Europe
fEu	from Europe
fl	from India
fe	from elsewhere
n	non-Moslems only
s	who arrive 1 Jun-31 Oct
w	who arrive 1 Nov-31 May
g	for gifts
§	vodka, champagne and beer prohibited
fpu	for personal use
3 mths	away for at least 3 months
2-3 mths	away for 2-3 months
1-2 mths	away for 1-2 months
*	cost deducted from gift or personal allowance
**	costs for *residents* deducted from personal allowance
★	maximum two cartons if for use in New York City
×	returning residents must be 21 years old to bring in one quart of wine or spirits
f	cigars of Cuban origin not admitted

●	double quantity if resident outside Europe	K	entering from countries outside EEC
EECnd	from EEC *not* bought duty-free	O	entering from countries outside Europe
EECd	from EEC bought duty-free	N	entering from non-EEC European countries
E	entering from EEC	nl	no limit
C	entering from Europe		

APPROXIMATE CONVERSION TABLE Metric & Imperial Measures

1 oz	1 litre	0.7 litres
28 grams	1¾ pints	1 btl
½ lb	1.2 litres	4.6 litres
220 grams	1 quart	1 gallon

SOURCES OF MAPS:
Worldwide Guide to National Civilian Mapping Agencies

Afghanistan Ministry of Mines and Industry, Mines and Geological Survey Department, Kabul.

Algeria Direction des Mines et de la Géologie, Service Géologique, Immeuble Maurania —Agha, Algiers.

Angola Servicio de Geologia e Minas, Caixa Postal No 1260-C, Luanda.

Argentina Instituto Nacional de Geología y Minería, Avenida Julio A, Roca 651, Buenos Aires.

Australia Bureau of Mineral Resources, Geology and Geophysics, Box 378, PO, Canberra City, 2601 ACT.

Benin Service of Mines and Geology, Department of Public Works, Transport and Telecommunications, Box 249, Cotonou.

Bolivia Ministerio de Minas y Petróleo, Servicio Geológico de Bolivia, Avenida 16 de Julio No 1769, Casilla Correo 2729, La Paz.

Botswana Geological Survey, PO box 94, Lobatsi.

Brazil Department Nacional da Produção Mineral, Avenida Pasteur 404, Rio de Janeiro.

Brunei The Government Geologist, Brunei Town.

Burma Geological Department, 226 Mahabandoola St, Box 843, Rangoon.

Burundi Ministère des Affaires Economiques et Financières, Département de Géologie et Mines, Boîte Postale 745, Bujumbura.

Cameroon Ministère des Transports, des Mines, des Postes et des Telécommunications, Direction des Mines et de la Géologie, Boîte Postale 70, Yaoundé.

Chad La Service des Mines de la République de Tchad, Service des Travaux Publiques, Fort Lamy.

Chile Instituto de Investigaciones Geológicas, Augustinas 785, 5° Piso, Casilla, 10465.

China Chinese Ministry of Geology, Peking.

Colombia Instituto Nacional de Investigaciones Mineras, Carrera 30, No 51-59, Bogotá.

Congo (Brazzaville) Service des Mines et de la Géologie, Boîte Postale 12, Brazzaville.

Costa Rica Geological Survey of Costa Rica, Geologic Dept, University of Costa Rica, San Pedro de Montes de Oca, San José.

Djibouti	Service des Travaux Publiques, Djibouti.
Ecuador	Servicio Nacional de Geología, y Minería, Casilla 23-A, Quito.
Egypt	Geological Survey and Mineral Department, Dawawin Post Office, Cairo.
El Salvador	Centro de Estudios e Investigaciones Geotecnicas, Apartado Postal 109, San Salvador.
Ethiopia	Imperial Ethiopian Government, Ministry of Mines, PO Box 486, Addis Ababa.
French Guiana	Bureau des Recherches Géologiques et Minières, Bo'te Postale 42, Cayenne.
Gabon	Direction des Mines, Bo'te Postale 576, Libreville.
Ghana	Ghana Geological Survey, PO Box M 80, Accra.
Guatemala	Sección de Geologia, Dirección General de Cartografia, Avenida los Amerécos 5 - 76 Zona 13, Guatemala City.
Guinea	Service des Mines et de la Géologie, Conakry
Guyana	Geological Survey Department, PO Box 789, Brickdam, Georgetown.
Honduras	Departament Minas e Hidrocarburos, Direccion General de Recursos Naturales, Tegucigalpa.
Iceland	Department of Geology and Geography, Museum of Natural History, PO Box 532, Reykjavik.
India	Geological Survey of India, 29 Jawaharal Nehru Road, Calcutta 16.
Indonesia	Geological Survey of Indonesia, Djalan Diponegoro 57, Bandung.
Iran	Iranian Geological Survey, PO Box 1964, Tehran.
Iraq	Geological Survey Department, Directorate of Minerals, Ministry of Oil, Baghdad
Israel	Geological Survey of Israel, 30 Malkhei Israel St, Jerusalem.
Ivory Coast	Service Géologique, Boîte Postale 1368, Abidjan.
Japan	Geological Survey of Japan, 135 Hisamoto-cho, Kawaasaki, Kanagawa.
Jordan	Natural Resources Authority, Mines and Geological Survey Department, Box 2220, Amman.
Kampuchea (Cambodia)	Service des Mines, Phnom Penh.
Kenya	Ministry of Natural Resources, Mines and Geological Department, PO Box 30009, Nairobi.
Korea, South	Geological Survey of Korea, 125 Namyoung-dong, Seoul.
Kuwait	Kuwait Oil Affairs Department, Ahmadi 94, Kuwait.
Lao People's Dem Rep (Laos)	Département des Mines, Ministère du Plan, Boîte Postale 46, Vientiane.
Lebanon	Direction Générale des Travaux Publiques, Beirut.
Liberia	Liberia Geological Survey, Ministry of Lands and Mines, PO Box 9024, Monrovia
Libya	Geologic Division, Ministry of Industry, Tripoli.
Madagascar	Minitère de l'Industrie et des Mines, Direction des Mines et de l'Energie, Service Géologique, Boîte Postale 280, Tananarivo.
Malawi	Geological Survey Department, PO Box 27, Zomba.
Malaysia	Geological Survey, Scrivener Road, PO Box 1015, Ipoh.
Mali	Ministère des Travaux Publiques, des Télécommunications, de l'Habitat et des Ressources Energétiques, Bamako.
Mauritania	Direction des Mines et de la Géologie, Nouakchott.
Mexico	Instituto de Geologia, Universidad Nacional Autínoma de México, Ciudad Universitaria, Mexico 20, DF
Morocco	Division de la Géologie, Direction des Mines et de la Géologie, Rabat.
Mozambique	Servicio de Geologia e Minas, Caixa Postal 217, Lourenço Marques.
Nepal	Ministry of Industry and Commerce, Nepal Geological Survey, Lainchaur, Kathmandu.

New Zealand	Department of Scientific and Industrial Research, New Zealand Geological Survey, PO Box 30368, Andrews Avenue, Lower Hutt, Wellington.
Nicaragua	Servicio Geológico Nacional, Ministerio de Economia, Apartado Postal No 1347, Managua, DN.
Niger	Service des Mines et de la Géologie, Ministère des Travaux Publiques, Niamey.
Nigeria	Ministry of Mines and Power, Geological Survey Division, PMB 2007, Kaduna South.
Pakistan	Geological Survey of Pakistan, PO Box 15, Quetta.
Panama	Departamento de Recursos Minerales, Apartado Postal 1631, Panama City.
Paraguay	Dirección de la Producción Mineral, Tucari 271, Asunción.
Peru	Servicio de Geología y Minería, Apartado 889, Lima.
Philippines	Department of Agriculture and Natural Resources, Bureau of Mines, PO Box 1595, Manila.
Rwanda	Service Géologique, Boîte Postale 15, Ruhengeri.
Senegal	Direction des Mines et de la Géologie, Boîte Postale 1238, Dakar.
Sierra Leone	Geological Survey Department, Freetown.
Somalia	Geological Survey Department, Box 41, Hargeisa.
South Africa	Geological Survey, PB 112, Pretoria.
	Geological Survey, PO Box 2168, Windhoek.
Sri Lanka	Geological Survey Department, 48 Sri Jinaratana Road, Colombo 2.
Sudan	Geological Survey Department, Ministry of Mining and Industry, PO Box 410, Khartoum.
Suriname	Geologisch Mijnbouwkundige Dienst, Dept van Opbouw, Klien Wasserstraat 1, Paramaribo.
Swaziland	Geological Survey and Mines, PO Box 9, Mbabane.
Syria	Directorate of Geological Research and Mineral Resources, Ministry of Petroleum, Fardos Street, Damascus.
Taiwan	Geological Survey of Taiwan, PO Box 31, Taipei.
Tanzania	Ministry of Commerce, Mineral Resource Division, PO Box 903, Dodoma.
Thailand	Royal Department of Mineral Resources, Rama VI Road, Bangkok.
Togo	Ministère des Travaux Publiques, Mines, Transports et Télécommunications, Direction des Mines et de la Géologie, BP 356, Lomé.
Tunisia	Geological Survey of Tunisia, 95 Avenue Mohamed V, Tunis.
Turkey	Maden Tetkik ve Arama Enstutusu, Posta Kutusu 116, Eskisehir Road, Ankara.
Uganda	Geological Survey and Mines Department, PO Box 9, Entebbe.
Upper Volta	Direction de la Géologie et des Mines, Ouagadougou.
Uruguay	Instituto Geológico del Uruguay, Calle J Herrer y Obes 1239, Montevideo.
Venezuela	Ministerio de Minas e Hidrocarburos, Dirección de Geologia, Torre Norte, Piso 19, Caracas.
Vietnam, South	Direction Générale des Mines, de l'Industrie et de l'Artisanat, Service Géologique, 59 Rue Gia Long, Saigon.
Yemen Arab Rep (North)	Office of Mineral Resources, Ministry of Public Works, Sana'a.
Zaïre	Service Géologique Nationale de la République du Zaïre, Boîte Postale 898, Kinshasa.
Zambia	Ministry of Labour and Mines, Geological Survey Department, PO Box RW 135, Ridgeway, Lusaka.
Zimbabwe-Rhodesia	Ministry of Mines and Lands, Geological Survey Office, Causeway PO Box 8039, Salisbury.

Steamship Lines in USA and Canada

Alaska Marine Highway System	Dept of Pub Works, Div of Marine Transportation, Pouch R, Juneau, Ak 99811	
American Export Lines, Inc	see Farrell Lines	
American President Lines	1950 Franklin Street, Oakland, Ca 94612	(415) 271 8148
	PO Box C-81411, Seattle, Wa 98108	(206) 292 4646
Armement Deppe, SA	Hansen & Tideman, Inc, 310 Sanlin Bldg, 442 Canal Street, New Orleans, La 70130	(504) 586 8755
Bergen Line	505 Fifth Ave, New York, NY 10017	(212) 986 2711
Black Star Line Ltd	FW Hartmann and Co, Inc, 17 Battery Place, 5th Flr, New York, NY 10004	(212) 425 6100
Blue Star Line	Three Embarcadero Centre, Suite 2260, San Francisco, Ca 94111	(415) 434 3780
Delta Line, Inc	One Market Plaza, Suart St Tower, San Francisco, Ca 94106	(415) 777 8200
	444 W Ocean Blvd, Long Beach, Ca 90802	(213) 436 4275
	One World Trade Centre, New York, NY 10048	(212) 466 0061
	Pacific Bldg, 720 Third Ave, Seattle, Wash 98104	(206) 682 0090
	Board of Trade Bldg, 310 SW Fourth St, Portland, Ore 97204	(503) 228 4126
	C Gardner Johnson Ltd, 505 Burrard St, Vancouver 1, BC, Canada	(604) 684 4221
	Sailings from the west coast of Canada and the USA around South America	
Egon Oldendorff	Biehl & Co, Inc, 416 Common St, New Orleans, La 70130	(504) 581 7788
Farrell Lines, Inc	One Whitehall St, New York, NY 10004	(212) 425 6300
	Biehl & Co, 416 Common St, New Orleans, La 70130	(504) 529 4211
Hansa Line	FW Hartmann & Co, Inc, 17 Battery Place, 5th Flr, New York, NY 10004	(212) 425 6100
Hellenic Lines Ltd	39 Broadway, New York, NY 10006	(212) 482 5694
	1133 International Trade Mart, 2 Canal St, New Orleans, La 70130	(504) 581 2825
	303 Petroleum Bldg, 1314 Texas, Houston, Tx 77002	(713) 224 8607
Ivaran Lines	United States Navigation, Inc, 17 Battery Place, New York, NY 10004	(212) 269 6000
Johnson Line	General Steamship Corp, 400 California St, San Francisco, Ca 94104	(415) 392 4100
	Sailings from San Francisco to Europe through the Panama Canal	
Knutsen Line	Bakke Steamship Corp, 650 California St, San Francisco, Ca 94108	(415) 433 4200
	600 S Commonwealth Ave, Los Angeles, Ca 90005	(213) 487 2640
Knutsen Line	1411 Fourth Ave Bldg, Seattle, WA 98104	(206) MU 2-9080
	520 SW Sixth Ave, Portland, Ore 97204	(503) 228 7671
	1075 W Georgia St, Suite 1460, Vancouver, BC, Canada	(604) 687 2721
Lykes Bros Steamship Co, Inc	Lykes Centre, 300 Poydras St, New Orleans, La 7013	(504) 523 6611
	PO Box 1339, Houston, Tx 77001	(713) 227 7211
	Sailings from New Orleans to the west coast of South America	
Moore-McCormack Lines	Two Broadway, New York, NY 10004	(212) 363 2600

Nauru Pacific Line	North American Maritime Agencies, 100 California St, San Francisco, Ca 94111	(415) 981 0343
	North American Maritime Agencies, 510 W Sixth St, Los Angeles, Ca 90014	(213) 624 4094
Polish Ocean Lines	McLean Kennedy Ltd, PO Box 1086, Place d'Armes, Coristine Bldg, 410 St Nicholas St, Montreal, Que H2Y 2P5, Canada	(514) 849 6111
	Gdynia-America, One World Trade Centre, Suite 3557 New York, NY 10048	(212) 938 1900
	Gdynia-America Line Inc, 1121 Walker St, Suite 210, Houston, Tx 77002	
Prudential Lines, Inc	see Delta Line, Inc	
Robin Line	see Moore-McCormack Lines	
Royal Netherlands Steamship Co	Lislind International, 5 World Trade Centre, New York, NY 10048	
	Sailings from New York and New Orleans to Central, and South America and Europe	
Schlüssel Reederei	Sea and Land Shipping Inc, 305 N Morgan St, Tampa, Fla. 33602	(813) 229 7284
South African Marine Corp	One Bankers Trust Plaza, New York, NY 10006	(212) 775 0400
Splosna Plovba	Kerr Steamship Company, One Market Plaza, Spear St Tower, Suite 2400, San Francisco, Ca 94105	(415) 764 0200
States Line (States Steamship Co)	320 California Street, San Francisco, Ca 94104	(415) 982 6221
	612 S Flower St, Los Angeles, Ca 90017	(213) 628 7151
	900 SW Fifth, Portland, Ore 97204	(503) 226 2611
	327 S La Salle St, Chicago, Ill 60604	(312) 939 2245
	Two Broadway, New York, NY 10004	(212) 363 6544
	618 Second Ave, Seattle, Wash 98104	(206) MA 2-7501
United Brands Company	1271 Ave of the Americas, New York, NY 10020	(212) 397 4039
United Yugoslav Line	see Splosna Plovba	
Westfal-Larsen Line	General Steamship Corp, 400 California Street, San Francisco, Ca 94104	(415) 392 4100
Yugoslav Line	NEWS Shipping Inc, 87 Washington St, New York NY 10006	(212) 248 4500

Other Freighter/Shipping Lines

Bruns Line	Oberhafenstrasse 5, 2000 Hamburg 1, West Germany
	General Passenger Agent, 20 Moorfields High Walk, London EC2Y 9DN, England
	Sailings from Hamburg to Florida, Central America and Ecuador
Fyffes Line	1 Queen's Way, Southampton, England
	Sailings from Southampton to Jamaica and Belize
Geest Industries Ltd	London Freight Exchange, Brushfield St, London E1, Tel. (01) 247 8581 or 377 8909
	Sailings on banana boats between Britain and the Caribbean
Jamaica Banana Producers Shipping Line	60 Oxford Road, Kingston 5, Jamaica
	Sailings on banana boats between Britain and Jamaica

Lauro Lines 99-119 Roseberry Ave, London EC1R 4RE, England
Tel. (01) 837 2157 or 278 6358

Sailings between the USA and Europe via the Caribbean

Singapore/ Australia Shipping Co British agent: Far East Travel Centre Ltd, 32 Shaftesbury Ave, London W1
Tel. (01) 734 6822

Passengers only, no vehicles. Vessels ply the Singapore-Fremantle route.

Ports Normally Served By Passenger—Carrying Freighters

Algeria	Algiers/Annaba/Oran
Angola	Cabinda/Landana/Lobito/Luanda/Mossamedes/Nova Redondo/Porto Alexandre/Porto Amboim
Argentina	Buenos Aires
Australia	Adelaide/Brisbane/Burnie/Cairns/Devonport/Fremantle/Hobart/Melbourne/Newcastle/Port Pirie/Sydney
Australian Trust Terr	Allardyce/Lunga/West Bay/Yandina
Azores	Ponta Delgada
Bahrain	Bahrain
Bangladesh	Chalna
Belize	Belize
Benin	Cotonou
Brazil	Angra Des Reis/Bahia: see São Salvador/Belem/Cabedello/Fortaleza/Maceio/Manaus/Natal/Paranagua/Parnaiba/Pernambuco: see Recife/Porto Alegre/Recife/Rio de Janeiro/Rio Grande/Salvador/Santos/Sao Luis/São Salvador/Vitoria
Burma	Rangoon
Cameroon	Douala
Cape Verde	Preia/São Vicente
Chile	Antofagasta/Arica/Chanaral/Iquique/San Antonio/Talcuhuano/Tocopilla/Valparaiso
China	Hsinkang/Shanghai
Colombia	Barranquilla/Buenaventura/Cartagena/Santa Marta
Congo	Boma
Costa Rica	Golfito/Puntarenas
Cuba	Guantanamo/Santiago
Djibouti	Djibouti
Dominican Rep	Santo Domingo
Ecuador	Guayaquil
Equatorial Guinea	Rio Benito
Egypt	Alexandria/Port Said/Suez
El Salvador	La Libertad/La Union
Ethiopia	Assab (Eritrea)/Massawa
Fiji	Lautoka/Suva
Gabon	Libreville/Port Gentil
Ghana	Accra/Cape Coast/Takoradi/Tema
Guadalcanal Is	Honiara
Guatemala	Puerto Barrios/San José
Guinea	Conakry
Gulf of Guinea	São Tomé/Principe Is
Guyana	Georgetown/Mackenzie
Haiti	Cap Haitien/Port-au-Prince
Honduras	Puerto Cortes/Tela
Iceland	Reykjavik
India	Alleppey/Bombay/Calcutta/Cochin/Madras/Mangalore/Marmagao
Indonesia	Balikpapan (Borneo)/Belawan Deli (Sumatra)/Djakarta (Java)/Macassar (Celebes)/Tjeribon/Semarang (Java)/Surabaya (Java)
Iran	Abadan/Bandar Shapur/Bushire/Khorramshahr
Iraq	Basra
Israel	Ashdod/Haifa/Tel Aviv
Ivory Coast	Abidjan
Japan	Kobe/Moji/Nagoya/Okinawa Ryukyu Islands/Osaka/Otaru/Shimizu/Yawata/Yokkaichi/Yokohama
Jordan	Aqaba
Kampuchea	Sihanoukville
Kenya	Mombasa

Korea	Busan (Pusan)/Incheon (Inchon)
Kuwait	Al Kuwait/Kuwait
Lebanon	Beirut
Liberia	Cape Palmas/Grand Bassa/ Monrovia
Libya	Tripoli
Madagascar	Majunga/Tamatave
Malaysia	Jesselton (Sabah)/Kuching (Sarawak)/Kudat (Sabah)/ Labuan (Sabah)/Malacca/Miri (Sarawak)/Penang/Port Kelang/Sandakan (Sabah)/ Sibu (Sarawak)/Tawau (Sabah)
Mariana Is	Agana (Guam)
Mexico	Acapulco/Tampico/Vera Cruz
Morocco	Agadir/Casablanca/ Ceuta (Spanish)/Melilla (Spanish)/ Safi/Tangiers
Mozambique	Beira/Lourenco Marques/ Mozambique/Nacala/Porto Amelia
New Caledonia	Noumea
New Hebrides	Espiritu Santo/Port Vila
New Zealand	Auckland/Bluff/Christchurch/ Dunedin/Invercargill/ Lyttleton/Napier/New Plymouth/Opua/Wellington
Nicaragua	Corinto
Nigeria	Apapa/Calabar/Lagos/ Port Harcourt
Pakistan	Chittagong/Karachi
Panama	Balboa/Cristobal
Papua New Guinea	Kavieng (New Ireland Is)/Lae/ Lombrum/Lorengau/Madang/ Manus Island(Admiralty Is)/ Port Moresby/Rabaul (New Britain Is.)/Samarai (Samarai Is.)/Wewak
Paraguay	Asunción
Peru	Callao/Iquitos/Mollendo/ Salaverry
Philippines	Cebu/Davao/Iloilo/Manila
Puerto Rico	Mayaguez/Ponce/San Juan
Qatar	Umm Said
Samoa American	Pago Pago
Samao Western	Apia
Saudi Arabia	Demmam/Jeddah
Senegal	Dakar
Sierra Leone	Freetown
Solomon Is	Kieta
Somalia	Mogadishu
South Africa	Cape Town/Durban/East London/Port Elizabeth
(Namibia)	Walvis Bay
Sri Lanka	Colombo/Trincomalee
Sudan	Port Sudan
Suriname	Moengo/Paramaribo/Paranam
Syria	Lattakia
Taiwan	Kaohsiung/Keelung/Takao
Tahiti	Papeete
Tanzania	Dar es Salaam/Mtwara/Tanga/ Zanzibar
Thailand	Bangkok/Songkhia
Togo	Lomé
Tonga	Ha'apia (Vavau Is)/Nukualofa
Tunisia	Sfax/Tunis
Turkey	Derince/Iskenderun/ Istanbul/Izmir/Mersin
United Arab Emirates	Dubai
Uruguay	Montevideo
Venezuela	Caracas: see LaGuaira/Ciudad Bolivar/Guanta/La Guaira/ Maracaibo/Puerto Cabello/ Puerto Cardon/Puerto Sucre
Vietnam,	Haiphong Saigon
Yemen People's Dem Rep (South)	Aden
Zaïre	Matadi

Worldwide Average Temperatures and Humidities

The information given below details temperature and humidity at important cities throughout the world. As given at ACCRA the first and second lines in each case indicate Minimum and Maximum Temperature respectively in degrees Centigrade and the third and fourth lines Humidity percentage a.m. and p.m. respectively. For conversion from Centigrade to Fahrenheit, see foot of table.

		Jan.	Feb.	Mar.	Apr.	May	June	July	Aug.	Sept.	Oct.	Nov.	Dec.
Accra													
Temperature °C	Min.	23	24	24	24	24	23	23	22	23	23	24	24
	Max.	31	31	31	31	31	29	27	27	27	29	31	31
Humidity %	a.m.	95	96	95	96	96	97	97	97	89	97	97	97
	p.m.	61	61	63	65	68	74	76	77	68	71	66	64
Amsterdam		1	1	3	6	10	13	15	15	13	9	8	2
		4	5	8	11	16	18	21	20	18	13	5	5
		90	88	85	80	75	74	76	78	82	86	89	90
		84	79	71	66	63	63	65	66	69	75	82	86
Athens		6	6	8	11	16	19	22	22	19	16	11	8
		12	13	16	19	25	29	32	32	28	23	18	14
		77	77	74	70	66	61	53	53	60	73	77	79
		62	61	54	47	44	40	32	33	38	52	60	63
Auckland		16	16	15	13	11	9	8	8	9	11	12	14
		23	23	22	19	17	14	13	14	16	17	18	21
		71	72	74	78	80	83	84	80	76	74	71	70
		62	61	65	69	70	73	74	70	68	66	64	64
Bahrain		14	15	17	21	26	28	29	29	27	24	21	16
		20	21	24	29	33	36	37	38	36	32	28	19
		85	83	80	75	71	69	69	74	75	80	80	85
		71	70	70	66	63	64	67	65	64	56	70	77
Bangkok		20	22	24	25	25	24	24	24	24	24	22	26
		32	33	34	35	34	33	32	32	32	31	31	31
		91	91	92	90	91	90	91	92	94	93	92	91
		53	55	56	58	64	67	66	66	70	70	65	56
Beirut		11	11	12	14	18	21	23	23	23	21	16	13
		17	17	19	22	26	28	31	32	30	27	23	18
		72	72	72	72	69	67	66	65	64	65	67	70
		70	70	69	67	64	61	58	57	57	62	61	69
Berlin		−1	−3	0	3	8	11	13	12	9	5	1	2
		2	3	8	13	18	21	23	22	19	13	6	3
		89	87	87	83	79	78	79	86	90	92	91	90
		81	73	63	56	50	53	55	58	60	68	79	84
Bermuda		14	14	14	15	18	21	23	23	22	21	17	16
		20	20	20	22	24	27	29	30	29	26	23	21
		78	76	77	78	81	83	81	79	81	79	76	77
		70	69	69	70	75	74	73	69	73	72	70	70
Bombay		19	19	22	24	27	26	35	24	24	24	23	21
		28	28	30	32	33	32	29	29	29	32	32	31
		70	71	73	75	74	79	83	83	85	81	73	70
		61	62	65	67	68	77	83	81	78	71	64	62
Brussels		−1	−1	2	4	8	10	12	12	10	7	2	1
		6	6	9	13	18	21	23	22	19	14	8	6
		94	95	93	92	93	94	95	94	96	95	96	96
		82	79	71	62	62	62	63	63	67	76	82	87
Buenos Aires		17	17	16	12	8	5	6	6	8	10	13	16
		29	28	26	22	18	14	14	16	18	21	24	28
		81	83	87	88	90	91	92	90	86	83	79	79
		61	63	69	71	74	78	79	74	68	65	60	62
Cairo		8	9	11	14	17	20	21	33	20	18	14	10
		18	21	24	28	33	35	36	35	32	30	26	20

	Jan.	Feb.	Mar.	Apr.	May	June	July	Aug.	Sept.	Oct.	Nov.	Dec.
Cairo	69	64	63	55	50	55	65	69	68	67	68	70
	40	33	27	21	18	20	24	28	31	31	38	41
Calcutta	13	15	21	24	25	26	26	26	26	23	18	13
	27	29	34	36	38	33	32	32	32	32	29	26
	85	82	79	76	77	82	86	88	86	85	79	80
	52	45	46	56	62	75	80	82	81	72	63	55
Christchurch	12	12	10	7	4	2	2	2	4	7	8	11
	21	21	19	17	13	11	10	11	14	17	19	21
	65	71	75	82	85	87	87	81	72	63	64	67
	59	60	69	71	69	72	76	66	69	60	64	60
Colombo	22	22	23	24	26	25	25	25	25	24	23	22
	30	31	31	31	31	29	29	29	29	29	29	29
	73	71	71	74	78	80	79	78	76	77	77	74
	67	66	66	70	76	78	77	76	75	76	75	69
Copenhagen	−2	−2	−1	3	7	11	13	12	9	6	2	0
	2	2	5	10	16	19	22	21	17	12	6	3
	89	92	87	80	73	73	77	82	87	90	89	89
	86	86	77	66	60	61	63	67	70	77	83	87
Delhi	7	9	14	20	26	28	27	26	24	18	11	8
	21	24	31	36	41	39	36	34	34	34	29	23
	72	67	49	35	35	53	75	80	72	56	51	69
	41	35	23	19	20	36	59	64	51	32	31	42
Djakarta	23	23	23	24	24	23	23	23	23	23	23	23
	29	29	30	31	31	31	31	31	31	31	30	29
	95	95	94	94	94	93	92	90	90	90	92	92
	75	75	73	71	69	67	64	61	62	64	68	71
Frankfurt	−2	−1	2	5	9	12	13	13	11	6	2	−1
	3	6	9	14	19	22	24	23	19	13	7	4
	89	88	86	82	80	80	82	87	90	92	91	90
	79	70	60	52	50	51	53	54	60	69	77	82
Hong Kong	13	13	16	19	23	26	26	26	25	23	18	15
	18	17	19	24	28	29	31	31	29	27	23	20
	77	82	82	87	87	86	87	87	83	75	73	74
	66	73	74	77	78	77	77	77	72	63	60	63
Istanbul	2	3	4	7	12	16	18	19	16	12	9	5
	7	8	11	16	20	25	27	27	25	19	15	11
	82	82	80	81	82	78	78	79	80	82	82	82
	74	71	65	62	62	57	55	55	59	64	71	74
Johannesburg	14	14	13	10	6	4	4	6	9	12	13	14
	26	25	24	22	19	17	17	20	23	25	25	26
	75	78	79	74	70	70	69	64	59	64	67	70
	50	53	50	44	36	33	32	29	30	37	45	47
Kuala Lumpur	22	22	23	23	23	23	22	23	23	23	23	22
	32	33	33	33	33	33	32	32	32	32	32	32
	97	97	97	97	97	96	95	95	96	96	97	97
	60	60	58	63	66	63	63	62	64	65	66	61
Lima	19	19	19	17	16	14	14	13	14	14	16	17
	28	28	28	27	23	20	19	19	20	22	23	26
	93	92	92	93	95	95	94	95	94	94	93	93
	69	66	64	66	76	80	77	78	76	72	71	70
Lisbon	8	8	9	11	13	16	17	18	17	14	11	8
	13	14	16	18	21	24	26	27	24	21	17	14
	83	80	76	69	67	64	61	61	67	72	80	83
	72	66	63	58	57	53	48	46	53	59	68	72
London	2	2	3	4	7	11	13	12	11	7	4	2
	7	7	11	13	17	21	23	22	19	14	9	7
	87	84	79	72	69	68	68	73	78	83	87	87
	80	72	63	58	57	57	55	58	63	70	79	81

	Jan.	Feb.	Mar.	Apr.	May	June	July	Aug.	Sept.	Oct.	Nov.	Dec.
Madrid	1	2	4	7	10	14	17	17	13	9	4	1
	8	11	14	18	22	27	31	30	25	19	12	9
	89	87	84	78	75	69	63	63	73	83	89	90
	71	64	58	52	51	43	37	36	47	57	68	73
Manila	21	21	22	23	24	24	24	24	24	23	22	21
	30	31	33	34	34	33	31	31	31	31	31	30
	89	88	85	85	88	91	91	92	93	92	91	90
	63	59	55	55	61	68	74	73	73	71	69	67
Melbourne	14	14	13	11	8	7	6	6	8	9	11	12
	26	26	24	20	17	14	13	15	17	19	22	24
	58	62	64	72	79	83	82	76	68	61	60	59
	48	50	51	56	62	67	65	60	55	52	52	51
Mexico City	6	6	8	11	12	13	12	12	12	10	8	6
	19	21	24	25	26	24	23	23	23	21	20	19
	79	72	68	66	69	82	84	85	86	83	82	81
	35	28	26	29	29	48	50	50	54	47	41	37
Miami	16	16	18	19	22	23	24	24	24	22	19	17
	23	24	26	27	29	30	31	31	31	28	26	24
	81	82	77	73	75	75	75	76	79	80	77	82
	66	63	62	64	67	69	68	68	70	69	64	65
Montreal	−14	−13	−7	1	8	14	16	15	11	4	−3	−11
	−6	−5	1	10	18	23	26	24	19	12	4	−3
	78	76	75	72	72	75	75	79	82	80	79	78
	73	69	62	56	54	55	52	54	57	60	70	75
Moscow	−13	−12	−8	−1	7	11	13	11	6	1	−5	−11
	−6	−5	0	8	18	23	24	22	16	8	−1	−5
	89	88	88	83	72	74	79	83	89	89	89	88
	85	80	70	60	50	54	56	56	62	70	81	86
Nairobi	12	13	14	14	13	12	11	11	11	13	13	13
	25	26	25	24	22	21	21	21	24	24	23	23
	74	74	81	88	88	89	86	86	82	82	86	81
	44	40	45	56	62	60	58	56	45	43	53	53
Nassau	18	18	19	21	22	23	24	24	24	23	21	19
	25	25	26	27	29	31	31	32	31	29	27	26
	84	82	81	79	79	81	80	82	84	83	83	84
	64	62	64	65	65	68	69	70	73	71	68	66
New York	−4	−4	−1	6	12	16	19	19	16	9	3	−2
	3	3	7	14	20	25	28	27	26	21	11	5
	72	70	70	68	70	74	77	79	79	76	75	73
	60	58	55	55	54	58	58	60	61	57	60	61
Oslo	−7	−7	−4	1	6	11	13	12	7	3	−2	−4
	−1	0	4	10	17	21	23	21	16	9	3	−1
	85	84	81	74	67	68	73	79	84	86	87	87
	82	75	65	56	51	54	57	61	65	72	82	85
Papeete	22	22	22	22	21	21	20	20	21	21	22	22
	32	32	32	32	31	30	30	30	30	31	31	31
	82	82	84	85	84	85	83	83	81	79	80	81
	77	77	78	78	78	79	77	78	76	76	77	78
Paris	0	1	2	5	8	11	13	12	10	7	3	1
	6	7	11	16	19	23	24	24	21	15	9	6
	90	90	88	80	79	79	81	84	91	94	93	92
	77	69	59	50	52	55	55	54	59	68	76	80
Port of Spain	21	20	20	21	22	22	22	22	22	22	22	21
	31	31	32	32	32	32	31	31	32	32	32	31
	89	87	85	83	84	87	88	87	87	87	89	89
	68	65	63	61	63	69	71	73	73	74	76	71

	Jan.	Feb.	Mar.	Apr.	May	June	July	Aug.	Sept.	Oct.	Nov.	Dec.
Prague	−4	−2	1	4	9	13	14	14	11	7	2	−2
	1	3	7	13	18	22	26	23	18	12	5	1
	85	83	83	78	76	74	77	81	85	87	87	87
	75	66	56	49	47	48	47	49	51	61	73	79
Rio de Janeiro	23	23	22	21	19	18	17	18	18	19	20	22
	29	29	28	27	25	24	24	24	24	25	26	28
	82	82	87	87	87	87	86	84	84	83	82	82
	70	71	74	73	70	69	68	66	72	72	72	72
Rome	4	4	6	8	13	16	18	18	16	12	8	5
	12	13	17	20	23	28	31	31	28	23	17	13
	85	86	83	83	77	74	70	73	82	86	87	86
	68	64	56	54	54	48	42	43	50	59	66	70
San Francisco	7	8	9	9	11	11	12	12	13	12	11	8
	13	15	16	17	17	19	18	18	21	20	17	14
	85	84	83	83	85	88	91	92	88	85	83	83
	69	66	61	61	62	64	69	70	63	58	60	68
Singapore	23	23	24	24	24	24	24	24	24	23	23	23
	30	31	31	31	32	31	31	31	31	31	31	31
	82	77	76	77	79	79	79	78	79	78	79	82
	78	71	70	74	73	73	72	72	72	72	75	78
Stockholm	−5	−6	−3	0	5	9	13	12	8	4	−1	−3
	−1	−1	3	7	14	18	21	19	14	9	3	1
	85	82	80	77	67	66	71	79	85	88	88	87
	82	75	67	62	54	55	59	64	68	76	84	86
Sydney	18	18	17	14	11	9	8	9	11	13	16	17
	26	26	24	22	19	16	16	17	19	22	23	25
	68	71	73	76	77	77	76	72	67	65	65	66
	64	65	65	64	63	62	60	56	55	57	60	62
Tehran	−3	0	4	9	14	19	22	22	18	12	6	1
	7	10	15	21	28	34	37	36	32	24	17	11
	77	73	61	54	55	40	51	47	49	53	63	76
	75	59	39	40	47	49	41	46	49	54	66	75
Tokyo	−2	−2	2	8	12	17	21	22	19	13	6	1
	8	9	12	17	21	24	28	30	26	21	16	11
	73	71	75	81	85	89	91	92	91	88	83	77
	48	48	53	59	62	68	69	66	68	64	58	51
Vancouver	0	1	3	4	8	11	12	12	9	7	4	2
	5	7	10	14	18	21	23	23	18	14	9	6
	93	91	91	89	88	87	89	90	92	92	91	91
	85	78	70	67	63	65	62	62	72	80	84	88
Vienna	−3	−2	1	5	10	13	15	14	11	7	2	−1
	1	3	8	14	19	22	24	23	19	13	7	3
	82	81	79	78	82	80	81	82	84	86	85	84
	74	68	57	51	53	54	54	55	58	67	75	78
Warsaw	−6	−5	−2	3	9	12	13	13	9	5	0	−4
	−1	0	5	12	19	22	24	23	18	12	4	0
	92	90	89	81	78	77	80	85	89	91	92	92
	87	83	70	58	55	57	58	62	66	70	83	88
Zurich	−3	−2	1	4	8	12	13	13	10	6	2	−2
	3	6	10	14	20	23	25	24	21	14	7	3
	90	89	85	82	81	79	81	84	90	92	92	90
	76	66	58	56	58	57	58	56	62	70	78	79

Fahrenheit = C° × 9/5 + 32

Altitudes of Selected Principal Cities (in feet)

City	Feet	City	Feet
Asunción, Paraquay	253	Kingston, Jamaica	25
Bangkok, Thailand	40	La Paz, Bolivia	12,200
Beirut, Lebanon	25	Lima, Peru	501
Bogotá, Colombia	8,500	Manila, Philippines	25
Buenos Aires, Argentina	45	Mexico City, Mexico	7,349
Calcutta, India	85	Montevideo, Uruguay	30
Caracas, Venezuela	3,164	Panama City, Panama	40
Cayenne, French Guiana	25	Quito, Ecuador	9,248
Damascus, Syria	700	Rangoon, Burma	55
Hong Kong	25	Rio de Janeiro, Brazil	30
Istanbul, Turkey	30	Santiago, Chile	1,800
Jerusalem, Israel	2,500	Singapore	25
Karachi, Pakistan	50	Tokyo, Japan	30

Worldwide Currencies

Country	*Unit*	*1 Unit = 100 (unless otherwise stated)*
Afghanistan	Afghani	Puls
Albania	Lek	Quintar
Algeria	Dinar	Centimes
Angola	Kwanza	Centavos
Antigua	East Caribbean Dollar	Cents
Argentina	Peso	Centavos
Australia	Dollar	Cents
Austria	Schilling	Groschen
Bahamas	Dollar	Cents
Bahrain	Dinar	1000 Fils
Bangladesh	Taka	Paise
Barbados	Dollar	Cents
Belgium	Franc	Centimes
Belize	Dollar	Cents
Benin	Franc	Centimes
Bermuda	Dollar	Cents
Bolivia	Peso	Centavos
Botswana	Pula	Thebe
Brazil	Cruzeiro	Centavos
British Virgin Islands	Dollar US and BVI	Cents
Brunei	Dollar	Cents
Bulgaria	Lev	Stotinki
Burma	Kyat	Pyas
Burundi	Franc	Centimes
Cameroon	Franc	Centimes
Canada	Dollar	Cents
Cape Verde	Escudo	Centavos
Cayman Islands	Cayman Islands Dollar	Cents
Central African Rep	Franc	Centimes
Chad	Franc	Cents

Country	*Unit*	*1 Unit = 100 (unless otherwise stated)*
Chile	Peso	10 Centesimos
China	Ren Min Bi	Fen
Colombia	Peso	Centavos
Congo	Franc	Centimes
Cook Islands	New Zealand Dollar	Cents
Costa Rica	Colon	Centavos
Cuba	Peso	Centavos
Cyprus	Pound	1000 Mills
Czechoslovakia	Koruna	Halers
Denmark	Krone	Ore
Djibouti	Franc	Centimes
Dominica	East Caribbean Dollar	Cents
Dominican Republic	Peso	Centavos
Ecuador	Sucre	Centavos
Egypt	Pound	Piastres
El Salvador	Colon	Centavos
Equatorial Guinea	Ukuele	Centimos
Ethiopia	Bir	Cents
Fiji	Dollar	Cents
Finland	Markka	Penni
France	Franc	Centimes
French Guiana	Franc	Centimes
French Polynesia	Franc	Centimes
French West Indies (Guadeloupe and Martinique)	Franc	Centimes
Gabon	Franc	Centimes
Gambia	Dalasi	Batut
German Democratic Republic	Mark	Pfennig
German Federal Republic	Mark	Pfennig
Ghana	New Cedi	Pesawas
Gibraltar	Pound	Pence
Greece	Drachma	Lepta
Grenada incl Carriacou and Petit Martinique	East Caribbean Dollar	Cents
Guatemala	Quetzal	Centavos
Guinea — Bissau	Escudo	Centavos
Guinea	Sily	Couris
Guyana	Dollar	Cents
Haiti	Gourde	Centimes
Honduras	Lempira	Centavos
Hong Kong	Dollar	Cents
Hungary	Florint	Fillers
Iceland	Krona	Aur

Country	*Unit*	*1 Unit = 100 (unless otherwise stated)*
India	Rupee	Naya Paise
Indonesia	Rupiah	Sen
Iran	Rial	Dinars
Iraq	Dinar	Fils
Ireland	Pound	Pence
Israel	Pound	Agorot
Italy	Lira	—
Ivory Coast	Franc	Centimes
Jamaica	Dollar	Cents
Japan	Yen	—
Jordan	Dinar	1000 Fils
Kampuchea	Riel	Centimes
Kenya	Shilling	Cents
Korea, North	Won	Jun
Korea, South	Won	—
Kuwait	Dinar	1000 Fils
Lao People's Democratic Republic (Laos)	Kip	Centimes
Lebanon	Pound	Piastres
Lesotho	Rand	Cents
Liberia	Dollar	Cents
Libya	Dinar	1000 Dirham
Liechtenstein	*as Switzerland*	
Luxembourg	Franc	Centimes
Macao	Pataca	Avos
Madagascar	Franc	Centimes
Malawi	Kwacha	Tambala
Malaysia	Ringgit	Cents
Maldive Islands	Maldivian Rupee	Laree
Mali	Franc	Centimes
Malta	Pound	Cents
Mauritania	Ouguiya	—
Mauritius	Rupee	Cents
Mexico	Peso	Centavos
Micronesia	*as USA*	
Montserrat	East Caribbean Dollar	Cents
Morocco	Dirham	Centimes
Mozambique	Escudo	Centavos
Nauru	*as Australia*	
Nepal	Rupee	Pice
Netherlands	Guilder or Florin	Cents
Netherlands Antilles	Guilder or Florin	Cents
New Caledonia	Franc	Centimes
New Hebrides	Franc	Centimes

Country	*Unit*	*1 Unit = 100 (unless otherwise stated)*
New Zealand	Dollar	Cents
Nicaragua	Cordoba	Centavos
Niger	Franc	Centimes
Nigeria	Naira	Kobos
Norway	Krone	Ore
Oman	Rials Omani	—
Pakistan	Rupee	Paisa
Panama	Balboa	Cents
Papua New Guinea	Kina	Cents
Paraguay	Guarani	Centimos
Peru	Sol	Centavos
Philippines	Peso	Centavos
Poland	Zloty	Groszy
Portugal	Escudo	Centavos
Puerto Rico & US Virgin Islands	*as USA*	
Qatar	Ryal	Dirhams
Réunion	Franc	Centimes
Romania	Lei	Bani
Rwanda	Rwandese Franc	Centimes
St Kitts-Nevis-Anguilla	East Caribbean Dollar	Cents
St Lucia	East Caribbean Dollar	Cents
St Vincent & the Grenadines	East Caribbean Dollar	Cents
Samoa American	*as USA*	
Samoa Western	Tala	Sene
São Tome & Principe	Escudo	Centavos
Saudi Arabia	Ryal	Hallalah
Senegal	Franc	Centimes
Seychelles	Rupee	Cents
Sierra Leone	Leone	Cents
Singapore	Dollar	Cents
Solomon Islands	*as Australia*	
Somalia	Shilling	Cents
South Africa	Rand	Cents
Spain	Peseta	Centimos
Sri Lanka	Rupee	Cents
Sudan	Pound	Piastres (1Piastre = 100 Milliemes)
Suriname	Guilder or Florin	Cents
Swaziland	Lilangeni	Cents
Sweden	Krona	Ore
Switzerland	Franc	Centimes
Syria	Pound	Piastres
Taiwan	New Taiwan Dollar	Cents

Country	*Unit*	*1 Unit = 100 (unless otherwise stated)*
Tanzania	Shilling	Cents
Thailand	Baht	Satang
Togo	Franc	Cents
Tonga	Pa'anga	Seniti
Trinidad & Tobago	Dollar	Cents
Tunisia	Dinar	Millimes
Turkey	Lira	Kurus
Turks & Caicos Islands	*as USA*	
Tuvalu Island	*as Australia*	
Uganda	Shilling	Cents
USSR	Rouble	Kopeks
United Arab Emirates	UAE Dirham	Fils
United Kingdom	Pound	Pence
USA	Dollar	Cents
Upper Volta	Franc	Centimes
Uruguay	Peso	Centimos
Venezuela	Bolivar	Centimos
Vietnam Socialist Republic	Dông	10 Hào
Yemen Arab Republic	Riyal	Fils
Yemen People's Democratic Republic (South)	Dinar	1000 Fils
Yugoslavia	Dinar	Paras
Zaïre	Zaïre	Makutas
Zambia	Kwacha	Ngwee
Zimbabwe-Rhodesia	Dollar	Cents

Note: The *ABC Guide to International Travel,* published quarterly by ABC Travel Guides Ltd, Oldhill, London Road, Dunstable, Bedfordshire LU6 3EB, England, is an excellent source of up-to-date information on regulations relating to the import and export of currency for each country.

Worldwide Voltage Guide

In general, all references to 110V apply to the range from 100V to 160V. References to 220V apply to the range from 200V to 260V. Where 110/220V is indicated, voltage varies within country, depending on location.

Aden 220V
Afghanistan 220V
Algeria 110/220V
Angola 220V
Anguilla 220V
Antigua 110/220V
§**Argentina** 220V
Aruba 110V
§**Australia** 220V
Austria 220V
Azores 110/220V
Bahamas 110/220V
Bahrain 220V
Balboa Heights Panama Canal 110/220V
Bangladesh 220V
Barbados 110/220V
Belgium 110/220V
Belize 110/220V
Benin 220V
Bermuda 110/220V
Bhutan 220V
Bolivia 110/220V

Bonaire	110/220V
Botswana	220V
§**Brazil**	110/220V
Brit. Virgin I.	110/220V
Bulgaria	110/220V
Burma	220V
Burundi	220V
Cambodia	110/220V
Cameroon	110/220V
Canada	110/220V
Canary I.	110/220V
Cayman I.	110V
Cen. African Rep.	220V
Chad	220V
Channel I. (Brit)	220V
§**Chile**	220V
China	220V
Colombia	110V
Costa Rica	110/220V
Curaçao	110V
Cuba	110V
***Cyprus**	220V
Czechoslovakia	110/220V
Denmark	220V
Dominica	220V
Dominican Rep.	110/220V
Ecuador	110/220V
Egypt	110/220V
El Salvador	110V
Ethiopia	110/220V
Fiji	220V
Finland	220V
France	110/220V
French Guiana	110/220V
Gabon	220V
Gambia	220V
§**Germany**	110/220V
Ghana	220V
Gibraltar	220V
***Great Britain**	220V
§**Greece**	110/220V
Greenland	220V
Grenada	220V
Grenadines	220V
Guadeloupe	110/220V
Guatemala	110/220V
Guinea	220V
Guyana	110/220V
Haiti	110/220V
Honduras	110/220V
***Hong Kong**	220V
Hungary	220V
Iceland	220V
§**India**	220V
Indonesia	110/220V
Iran	220V
Iraq	220V
Ireland	220V
Isle of Man	220V
Israel	220V
Italy	110/220V
Ivory Coast	220V
Jamaica	110/220V
Japan	110V
Jordan	220V
Kenya	220V
Korea, South	220V
Kuwait	220V
People's Dem Rep (Laos)	110/220V
Lebanon	110/220V
Lesotho	220V
Liberia	110/220V
Libya	110/220V
Liechtenstein	220V
Luxembourg	110/220V
Macao	110/220V
Madagascar	220V
§**Madeira**	220V
Majorca	110V
Malawi	220V
Malaysia	220V
Mali	110/220V
Malta	220V
Martinique	110/220V
Mauritania	220V
Mexico	110/220V
Monaco	110/220V
Montserrat	220V
Morocco	110/220V
Mozambique	220V
Nepal	220V
Netherlands	110/220V
Neth. Antilles	110/220V
Nevis	220V
New Caledonia	220V
New Guinea	220V
New Hebrides	220V
New Zealand	220V
Nicaragua	110/220V
Niger	220V
***Nigeria**	220V
Northern Ireland	220V
Norway	220V
Okinawa	110V
Oman	220V
Pakistan	220V
Panama	110V
§**Paraguay**	220V
Peru	220V
Philippines	110/220V
Poland	110/220V
Portugal	110/220V
Puerto Rico	110V
Qatar	220V
Romania	110/220V
Rwanda	220V
Saba	110/220V
St Barthélemy	220V
St Eustatius	110/220V
St Kitts	220V
St Lucia	220V
St Maarten	110/220V
St Vincent	220V
Saudi Arabia	110/220V
Senegal	110V
Seychelles	220V
Sierra Leone	220V
***Singapore**	110/220V
Somalia	110/220V
South Africa	220V

Spain	110/220V
Sri Lanka	220V
Sudan	220V
Suriname	110/220V
Swaziland	220V
§**Sweden**	110/220V
Switzerland	110/220V
Syria	110/220V
Tahiti	110/220V
Taiwan	110/220V
Tanzania	220V
Tobago	110/220V
Togo	110/220V
Tonga	220V
Trinidad	110/220V
Tunisia	110/220V
Turkey	110/220V
Turks & Caicos I.	110V
Uganda	220V
Upper Volta	220V
Uruguay	220V
United Arab Emirates	220V
USA	110V
USSR	110/220V
US Virgin I.	110V
Venezuela	110/220V
Vietnam	110/220V
Yemen	220V
Yugoslavia	220V
Zaîre	220V
Zambia	220V
***Zimbabwe-Rhodesia**	220V

**Denotes countries in which plugs with 3 square pins are used (in whole or part).*
§Countries using DC in certain areas.

Metric Data Conversion Table

Imperial units to Metric (SI) units

Quantity	*Imperial unit*	*SI equivalent* Accurate	*SI equivalent* Rough approximation
Length	inch	0.0254 m	4 in = 10 cm
	foot	0.3048 m	10 ft = 3 m
	yard	0.9144 m	10 yds = 9 m
	mile	1.6093 km	10 miles = 16 km
Area	sq. inch	645.16 mm^2 (sq. millimetres)	
	sq. foot	0.0929 m^2	100 sq. ft = 9.3 m^2
	sq. yard	0.836 m^2	100 sq. yd = 84 m^2
	sq. mile	2.58999 km^2	10 sq. miles = 26 km^2
	acre	0.40469 ha (hectare)	10 acres = 4 ha
		4046.9 m^2	1 acre = 4000 m^2
Volume	cu. inch	1.6387×10^{-5} m^3	1 cu. in = 16 cm^3
	cu. foot	0.0283 m^3	1 cu. ft = 28 litres 1000 cu. ft = 28 m^3
	UK gallon	0.004546 m^3	1 UK gal = 4.5 litres 1000 UK gal = 4.5 m^3
Mass	ounce	0.0284 kg	1 oz = 28 gm
	pound	0.4536 kg	10 lb = 4.5 kg
Pressure	pound per sq. inch	6894.8 N/m^2 (newtons per sq. metre)	1 lb/sq. in = 7000 N/m^2 10 lb/sq. in = 7 N/cm^2
Density	pound per cu. foot	16.018 kg/m^3	1 lb/cu. ft = 16 kg/m^3
Temperature	degree Fahrenheit (°F)	0.555 degree Celsius (°C)	
Temperature scale	t°F	5/9(t°F − 32)°C (e.g. 70°F = 21.1°C 80°F = 26.7°C 90°F = 32.2°C)	
Power	horsepower	0.7457 kw	10 hp = 7.5 kw
Force	pound force	4.4482N	10 lbf = 44.5N

Metric (SI) units to Imperial units

Quantity	*SI unit*	*Imperial equivalent* *Accurate*	*Rough approximation*
Length	centimetre (cm)	0.394 in	10 cm = 4 in
	metre (m)	3.281 ft 1.094 yds	10 m = 33 ft
	kilometre (km)	0.621 miles	100 km = 62 miles
Area	square metre (m^2)	10.764 sq. ft	10 m^2 = 12 sq. yds
	hectare (10000 m^2)	2.471 acres	1 ha = 2.5 acres = 12,000 sq. yds
	square kilometre (km^2)	0.386 sq. miles	100 km^2 = 40 sq. miles
Volume	cu. metre (m^3)	35.315 cu. ft	1 m^3 = 35 cu. ft 1 m^3 = 1.3 cu. yds
	litre (1)	0.22 UK gal	100 litres = 3.5 cu. ft = 22 UK gal
Mass	kilogramme (kg)	2.205 lb	10 kg = 22 lb
	gramme (g)	0.035 oz	100 g = 3.5 oz
Pressure	newtons per sq. metre (N/m^2)	1.4504×10^{-4} lb/sq. in	1 N/m^2 = 1.5×10^{-4} lb/in^2 or 16/sq. in
Density	kilogramme per cu. metre (kg/m^3)	0.062 lb/cu. ft	100 kg/m^3 = 6 lb/cu. ft
	gramme per cu. cm (g/cm^3)	0.036 lb/cu. in	30 g/cm^3 = 6 lb/cu. ft
Temperature	degree Celsius (°C)	1.8° Fahrenheit (°F)	
Temperature scale	t°C	= (9/5t°C + 32)°F (e.g. 20°C = 68°F 30°C = 86°F 40°C = 104°F)	
Power	kilowatts	1.341 hp	1 kw = 1.3 hp
Force	newton	0.22481 lbf	IN = 0.22 lbf

Other Conversion Data

	unit	*equivalent*
Velocity	knot	0.8684 miles per hour
	mile per hour	1.1515 knots
Volume	UK gallon	0.8333 US gallons
	US gallon	1.2 UK gallons

Metric Tyre Pressure Conversion Chart

Pounds per sq in	*Kilograms per sq cm*	*Atmospheres*
14	0.98	0.95
16	1.12	1.08
18	1.26	1.22
20	1.40	1.36
22	1.54	1.49
24	1.68	1.63
26	1.83	1.76
28	1.96	1.90
30	2.10	2.04
32	2.24	2.16
36	2.52	2.44
40	2.80	2.72
50	3.50	3.40
55	3.85	3.74
60	4.20	4.08
65	4.55	4.42

Diseases

Name of Disease	*Type of Disease*	*How acquired*	*Characteristics*	*Where found*	*Vaccine*
Smallpox (Variola)	Virus	By contact with infected person, articles handled by infected person, flies or by inhalation of polluted air.	Acute infectious fever with headache, vomiting, backache, rigor, temperature of 103°F and spots which turn to blisters, then pustules, then crusts. High mortality rate.		yes
Diphtheria	Bacterium	By close contact with infected person, articles handled by infected person.	Toxins absorbed by body, causing systemic poisoning, with sore throat, fever and extreme difficulty in breathing. Sometimes fatal.		yes. Immunity conferred by attack is short-lived.
Polio	Virus	From food, droplets, from an infected person or a carrier. Enters body through intestinal tract.	Affects nerve roots, with headache, vomiting, muscular pains, fever, sometimes convulsions. Severe paralysis and death are not uncommon.	Endemic in Himalaya.	Salk and Sabin vaccines.
Whooping cough (Pertussis)	Bacterium	By droplet infection.	Fever, running nose, explosive cough followed by long drawn out crowing inspiration, perhaps vomiting.		yes—lessens severity of attack.
Mumps	Virus	From infected person.	Affects salivary glands and in males may lead to orchitis, inflammation of the testes.		yes—gives protection for up to 2 years.
Tetanus	Bacterium	Through wound or open sore.	Intense and extremely painful muscle spasms.		yes
Measles	Virus	Through respiratory tract, from infected person.	Rather like acute cold at first, then sensitive eyes, cough, sore throat, rash, maybe headache and temperature of 102°F or more.		yes
German measles (Rubella)	Virus	From infected person.	Rash, mild fever and cold-like symptoms.		yes—may be recommended for women of child-bearing age who have not had the disease.

Name of Disease	*Type of Disease*	*How acquired*	*Characteristics*	*Where found*	*Vaccine*
Chickenpox (Varicella)	Virus	From infected person—child with chickenpox or adult with shingles.	Rash beginning on trunk and spreading outwards, spots turning to blisters, then scabs. Complications rare.		no. One attack confers immunity for life.
Malaria	Plasmodium	Bite of Anopheles mosquito.	Alternating hot, cold and sweating stages with high fever and aches. Disease recurs.	Low-lying marshy land, hot and humid, tropics and sub-tropics, esp. India and parts of America and Africa.	no—but prophylactics are available, e.g. Chloroquine, Paludrine.
Plague	Bacterium	Flies that live on rats; and droplet infection from another person.	Fever, swollen glands. High mortality rate.	Endemic in southern Asia and Africa.	yes—gives partial protection only. Modern treatment is to give sulpha drugs combined with streptomycin where outbreaks occur.
Yellow Fever	Virus	Bite of Aëdes aegypti mosquito.	Attacks liver and other organs, with fever, rigor, vomiting, jaundice and partial or total suppression of urine. May be fatal.	Moist tropical areas of East and West Africa, Central and South America.	yes—given by clinics only, as vaccine is easily damaged or contaminated.
Cholera	Bacterium	From the human intestine, spread by faecally contaminated food or water.	Intestinal infection, with fever, cramps, vomiting, extremely severe diarrhoea, loss of body water and salts.	Endemic in southern Asia, outbreaks occur elsewhere, esp. in war or famine.	yes—effective 6 months only; lessens severity.
Typhoid Fever	Bacterium of salmonella family	Carriers or infected persons, contaminated food or water.	Generalised fever, headache, often acute diarrhoea, nose-bleeds, pink rash on abdomen and chest; delirium and temporary loss of memory common.	Endemic in under-developed areas, both temperate and tropical, esp. India.	yes—gives protection for a year or more, lessens severity.
Typhus Fever	Rickettsial organism (intermediate between bacterium and virus)	Bite of skin parasites, e.g. lice, fleas, ticks, mites.	High fever, headache, rash, clots, pains in back and limbs, rigors; great prostration and delirium.	Endemic in Third World, esp. in uplands of East Africa, Latin America and Asia.	yes

Name of Disease	*Type of Disease*	*How acquired*	*Characteristics*	*Where found*	*Vaccine*
Infective Epidemic Hepatitis	Virus	Droplets or close contact with infected person; contaminated food and water.	Acute inflammation of liver; similar to intestinal influenza, followed by swollen, very tender liver, jaundice.		no*
Serum Hepatitis	Virus	Contaminated hypodermic needles or other medical instruments.	Virus extremely tough, even survives boiling water for a time. Disease similar to but more severe than Infective Hepatitis.		no**
Rabies (Hydrophobia)	Virus	Bite of infected animal (mammals only).	Painful muscle spasms, esp. those brought on by breathing and swallowing; mania, paralysis and death from heart failure. Death almost inevitable after symptoms have manifested themselves.		yes—can be given if appropriate, on doctor's advice.

*Gamma globulin, an extract of blood serum, gives temporary passive immunity, for 4-6 months, to this and various other viral diseases, including German Measles. It is given by intramuscular injection. It gives complete protection against some diseases and attenuates others. Against Serum Hepatitis there is no evidence that Gamma globulin is effective, except perhaps in very large doses.

**As a preventive measure, disposable syringes or other instruments may be used where there is a risk of infection from instruments inadequately sterilised. Travellers may wish to carry a supply of disposable syringes in their medical kit.

Guide to International Road Signs

▲ Warning signs ● Regulative signs ■ Informative signs

Colors may vary from country to country, but are usually red and black on a white background.

WARNING SIGNS

Right bend · Double bend · Dangerous bend · Danger! Train

Cross roads · Intersection w/minor road · Merging traffic · Road narrows · Uneven road · Slippery road · Other dangers

Round-about · Give way · Dangerous descent · Road work · Tunnel · Opening bridge · Animals

Level crossing with barrier · Level crossing without barrier · Pedestrians · Children · Two-way traffic · Falling rocks · Traffic signals ahead

INFORMATIVE SIGNS

Motorway exit · Priority road · End of priority road · One-way traffic · Hospital · First-aid station · Mechanical help

REGULATIVE SIGNS

Road closed · No entry · No right turn · Direction obligatory

No U-turns · No entry for motorcars · No entry for motor-cycles · No entry for all motor vehicles · No entry for bicycles · No entry for pedestrians · Priority to oncoming vehicles

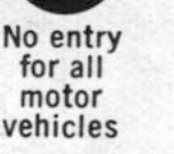
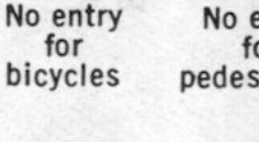
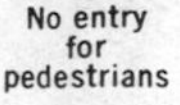

No overtaking · End of no overtaking · Maximum load · Axle weight limit · Width limit · Height limit · No parking

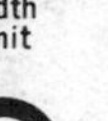
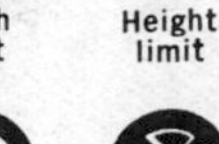
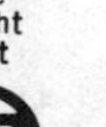

Maximum speed limit · End of speed limit · End of all restrictions · Halt sign · Customs · No stopping · Use of horns prohibited

The International Search and Rescue Signs

I	Require medical attention
II	Require medical supplies
F	Require food & water
L	Require fuel & oil
W	Require an engineer
⩔	Require firearms
LL	All is well
N	No
Y	Yes
△	Safe to land
↑	Proceeding in this direction

Addressing Mail in most Third World Countries

Delivery of mail off the beaten track is usually slow —up to two weeks for mail to remote places—and may be unreliable. Have mail addressed to you as follows:

A. Citizen
Poste Restante
Main Post Office (or Grande Poste in French-speaking countries)
Town
Country

Many Moslem countries use a Christian name predominantly in place of a surname. If mail is addressed to you as 'Any Citizen', the post office that receives it may file it under 'A' instead of 'C' and have difficulty in finding it when you come to collect. It is best to have your correspondents address your letters with one initial followed by your surname.

Alternative mailing addresses include

i) your consulate or embassy. Check, however, that there is a consulate in the place concerned and that it accepts mail.
ii) offices of a shipping line if you are a passenger on one of its vessels.

VOCABULARIES

English

1. Do you speak English?
2. I do not understand
3. Yes/No
4. Your name?
5. I, you, he, she
6. Good morning
7. Good evening
8. Good night
9. Good-bye
10. How are you?
11. How far?
12. How much is it?
13. Too much
14. Very well
15. Thank you
16. You are welcome
17. Excuse me
18. I am sorry
19. Please
20. I want
21. Airport
22. Automobile/car
23. Bank
24. Barber
25. Beauty salon/parlour
26. Breakfast
27. Bus
28. Change (money)
29. Check (bill)
30. Church
31. Dentist
32. Dinner
33. Doctor
34. Flat tyre
35. Gasoline/petrol
36. Hospital
37. Information
38. Lavatory (toilet)
39. Lunch
40. Men (gentlemen)
41. Occupied (sign)
42. Pharmacy
43. Post Office
44. Registered (letter)
45. Room (hotel)
46. Shop (store)
47. Sick
48. Soap
49. Stamp (postage)
50. Station (railroad)
51. Suitcase
52. Telephone
53. Ticket (travel)
54. Time (of day)
55. Today
56. Tomorrow
57. Towel
58. Train
59. Waiter
60. Water (drinking)
61. Women/ladies

French

1. Parlez-vous anglais?
2. Je ne comprends pas
3. Oui/Non
4. Votre nom?
5. Je, vous, il, elle
6. Bonjour
7. Bonsoir
8. Bonne nuit
9. Au revoir
10. Comment allez-vous?
11. A quelle distance?
12. Combien est-ce?
13. Trop cher
14. Très bien
15. Merci
16. Je vous en prie
17. Excusez-moi
18. Je regrette
19. S'il vous plaît
20. Je veux
21. Aéroport
22. Auto/voiture
23. Banque
24. Coiffeur
25. Salon de beauté
26. Petit déjeuner
27. Autobus
28. Change
29. Addition
30. Eglise
31. Dentiste
32. Dîner
33. Docteur
34. Pneu crevé
35. Essence
36. Hôpital
37. Renseignements
38. Toilettes (WC)
39. Déjeuner
40. Messieurs
41. Occupé
42. Pharmacie
43. Bureau de Poste
44. Recommandé
45. Chambre
46. Boutique
47. Malade
48. Savon
49. Timbre-poste
50. Gare
51. Valise
52. Téléphone
53. Billet
54. Heure
55. Aujourd'hui
56. Demain
57. Serviette
58. Train
59. Garçon
60. Eau potable
61. Dames

Spanish

1. ¿Habla usted ingles?
2. No entiendo
3. Si/No
4. Su nombre?
5. Yo, usted, él, ella
6. Buenos dias
7. Buenas noches
8. Buenas noches
9. Adiós
10. ¿Cómo está usted?
11. ¿Hasta dónde?
12. ¿Cuánto vale?
13. Demasiado
14. Muy bien
15. Gracias
16. De nada
17. Dispénseme
18. Lo mucho
19. Por favor
20. Yo quiero
21. Aeropuerto
22. Automóvil/coche
23. Banco
24. Barbero
25. Salón de belleza
26. Desayuno
27. Autobús
28. Cambio
29. Cuenta
30. Iglesia
31. Dentista
32. Cena
33. Médico
34. Neumático reventado
35. Gasolina
36. Hospital
37. Información
38. Retrete/excusado
39. Almuerzo
40. Señores/cabelleros
41. Ocupado
42. Farmacia
43. Oficina de correos
44. Certificado
45. Habitación
46. Tienda
47. Enfermo
48. Jabón
49. Sello/estampilla
50. Estación
51. Maleta
52. Teléfono
53. Billete
54. Hora
55. Hoy
56. Mañana
57. Toalla
58. Tren
59. Camarero
60. Aqua potable
61. Señoras/damas

Portuguese

1. Você fala inglês?
2. Nâo compreendo
3. Sim/Nâo
4. Seu nome?
5. Eu, você, êle, ela
6. Bom dia
7. Boa noite
8. Boa noite
9. Adeus
10. Como vai você?
11. A que distância?
12. Quanto custa?
13. Demasiado
14. Muito bem
15. Obrigado
16. De nada
17. Desculpe
18. Perdâo
19. Faz favor
20. Eu quero
21. Aeroporto
22. Automóvel/carro
23. Banco
24. Barbeiro
25. Salâo de beleza
26. Café de manhâ
27. Onibus
28. Trôco
29. Conta
30. Igreja
31. Dentista
32. Jantar
33. Doutor
34. Pneu furado
35. Gasolina
36. Hospital
37. Informaçâo
38. Banheiro
39. Almôço
40. Homens
41. Ocupado
42. Farmácia
43. Correio
44. Registrado
45. Quarto
46. Loja
47. Doente
48. Sabâo/sabonete
49. Sêlo
50. Estaçâo
51. Valise/mala
52. Telefone
53. Bilhete
54. Hora
55. Hoje
56. Amanhâ
57. Toalha
58. Trem
59. Garçom
60. Agua
61. Mulheres/senhoras

VOCABULARIES

English

1. Good day
2. Good evening
3. Goodbye
4. What is your name?
5. How are you?
6. How much?
7. Too expensive
8. How far?
9. Excuse me
10. Please
11. Thank you
12. I am sorry
13. What is this?
14. Where?
15. When?
16. Which way?
17. Do you have?
18. I want
19. Good
20. Bad
21. Broken
22. Do you understand?
23. I do not understand
24. Yes
25. No
26. I
27. You
28. Hospital
29. Doctor
30. Sick
31. Today
32. Tomorrow
33. Time
34. Maybe
35. Room
36. Bed
37. Sleep
38. Store/shop
39. Station
40. Exchange
41. Money
42. Automobile
43. Food
44. Milk
45. Cheese
46. Yogurt
47. Eggs
48. Meat
49. Tomato
50. Sugar
51. Water
52. Do you speak English?

Arabic

1. Illah Bilkhayr
2. Masaa al khayr
3. Fii amaan illah
4. Shismak?
5. Shlonak?
6. Cham floos?
7. Waajid floos
8. Cham masaafah?
9. Saamah
10. Minfudluk
11. Ashkurak
12. Anna mitassif
13. Shinoo haatha?
14. Wayn?
15. Mata?
16. Ay tariig?
17. Andak?
18. Ariid
19. Zahn
20. Mu zayn
21. Kharaab
22. Tuf ham?
23. Ma ta arif
24. Naam
25. Laa
26. Ana
27. Inta
28. Mistashfa
29. Tabiib
30. Mariidh
31. Al yom
32. Baachir
33. Wagt
34. Yimkin
35. Daar
36. Margad
37. Naam
38. Makhzan
39. Mahattah
40. Baddal
41. Floos
42. Sayyaara
43. Akil
44. Haliib
45. Jibin
46.
47. Bayl
48. Laham
49. Tamaat
50. Shakarr
51. Maay
52. Takallam Ingleesi?

Farsi

1. Rooz beghair
2. Shab beghair
3. Houda hafez
4. Esmeh shomah chic?
5. Chetori?
6. Chande?
7. Khely geraan
8. Cheraad rohay?
9. Bebak shiid
10. Me tavaoni (= will you please?)
11. Merci
12. Mutasefam
13. In chir?
14. Kajao?
15. Kai?
16. Kodaam raah?
17. Shomar dariid?
18. Me kham
19. Hoob
20. Bad
21. Sheycastere
22. Me fah me?
23. Neme fah mam
24. Balleh
25. Na
26. Man
27. Shomar
28. Bemar estan
29. Doktor
30. Mareez
31. Emrooz
32. Farrdoh
33. Vakht
34. Shoyad
35. Otagk
36. Takht
37. Khab
38. Furooshka
39. Iisgah
40. Tab dill
41. Pool
42. Auto
43. Chazoh
44. Shiir
45. Paneer
46. Masst
47. Togh morghe
48. Goosht
49. Gorghe farangi
50. Shikarr
51. Ub
52. Metu needh Ingleezi so- paht Koneedh?

Important Moslem religious greetings/sayings, applicable throughout the Middle East and southern Asia:
Assalaam alaykum - Peace be on you; the response: Wa alaykum - And on you peace. Allah kariim - God is generous. Al hamdalillah - Praise be to God. Inshallah - If God wills it.

Hindi ⋆

1. Namestey
2. Namestey
3. Phir theykengey/ Phir thik aenge
4. Tumhara naam kiya hai?
5. Tum khaiseh ho?
6. Kitna?/Yih to kitna dãm lagta? (how much does this cost?)
7. Bohuut mungha/ Ziyada
8. Kitna dour hai?
9. Kom maaf korna/ Mãp karna/khama ũarna
10. Putchiyeh
11. Dan ya waad/ Shukriya
12. Maaf kiijayeh
13. Yek kiya hai/ Yih kiya hai?
14. Kahan?
15. Kãb?
16. Kiya rasta hoga?/ Rasta kahan hai?
17. Tumhara pas hai?/ Ap ke pas hai?
18. Mujeh chahiyeh/ Mujh ko chahiyeh
19. Atcha
20. Karaab
21. Tut gaya
22. Tum-ko samajdj gaya?/ Tum-ko malum hai?
23. Mujav mulum nahe hua
24. Haan
25. Nahee
26. Main (pron. mai)
27. Tum
28. Hospitaal
29. Doktur
30. Bimarr
31. Ãj
32. Kal
33. Wãqt (pron waqat)
34. Shayad
35. Camrah
36. Palang
37. Sona (also = gold!)
38. Dukaan
39. Station
40. Hudle buddal
41. Paisa
42. Garhi
43. Khana
44. Dũdh
45. Panĩr
46. Dahi/Lassi
47. Andar
48. Goasht
49. Tomaato
50. Cheeni (white sugar)
51. Pãni
52. Ap Angrighe boltai hai?

Urdu⋆

1. Salam
2. Salam
3. Khuda Hafiz
4. Apka naam kiya hai?
5. Ap khaiseh ho?
6. Kitneh?
7. Bohuut mungha
8. Kitneh dour?
9. Maaf korna
10. Mahirbani
11. Shukryia
12. Maaf kiijayeh
13. Ye kiya hai?
14. Kahan?
15. Kub?
16. Consa rasta?
17. Ab ka pas hai?
18. Mujeh chahiyeh
19. Atcha
20. Bura
21. Toot gayar
22. Apko samajh lagti hai?
23. Mujay samajh nahee ati
24. Haan
25. Nahee
26. Mein
27. Aap
28. Huspital
29. Daktar
30. Bimar
31. Aaje
32. Kull
33. Waqt
34. Hu Sukta hai
35. Camrah
36. Chaar pai
37. Sona (also = gold)
38. Dukaan
39. Station
40. Bidulna
41. Pesai
42. Car
43. Khana
44. Dood
45. Paneer
46. Dahee
47. Andai
48. Goasht
49. Tamatar
50. Khand
51. Paani
52. Ap Angrighe boltai hai?

⋆ There is an infinite gradation between Hindi and Urdu, with variations influenced by other regional languages. For 'you' (no.27), the form used to anyone of moderately good 'class' is 'ap', pronounced *aap* or *arp*; while the word 'tum' is (or was, in the days before independence) used by a superior to address an inferior.

Turkish

1. Iyi günler
2. Iyi geceler
3. Allahaismarladik
4. Isminiz nedir?
5. Nasilsiniz?
6. Ne kadar?
7. Çok pahali
8. Nerede?
9. Pardon
10. Lütfen
11. Tesekkür ederim
12. Özür dilerim
13. Bu ne dir?
14. Neresi?
15. Ne zaman?
16. Nereye?
17. Sizde varmidir?
18. Ben isterim
19. Iya
20. Fena
21. Kirildi
22. Anliyormusun?
23. Anlamiyorum
24. Evet
25. Hayir
26. Ben
27. Sen
28. Hastahane
29. Doktor
30. Hasta
31. Bugün
32. Yarin
33. Saat
34. Belki
35. Oda
36. Yatak
37. Uyku
38. Dükkân
39. Istasyon
40. Exchange
41. Para
42. Araba
43. Yemek
44. Süt
45. Peynir
46. Yoğurt
47. Yumurta
48. Et
49. Domates
50. Şeker
51. Su
52. Ingilizce konuşurmusunuz?

Swahili

1. Jambo (all purpose greeting)
2. Jambo
3. Kwa heri (all purpose farewell)
4. Jina lako nani?
5. Habari yako?/ Habari gani? (= what news?)
6. Gapi?/ Bei gani?
7. Khari sana
8. Ni bali gagi?/ Maili ngapi? (= how many miles?)
9. Aabu
10. Please/Tafadhali
11. Ahsante sana
12. Sorry
13. He yko nini?
14. Wapi?
15. Saa gani?
16. Yia gani?
17. Wawa yiko?
18. (Meme) nataka
19. Mzuri/Mzuri sana (= very good)
20. Baya
21. Bhungika
22. Wawa na sikia?
23. Meme apana sikia/ Sifahamu
24. Ndio/Dhio
25. Sio/Hapana
26. Meme (used objectively)/Ni (with verbs)
27. Wawa (used objectively)/ U (with verbs)
28. Hospitali
29. Daktari
30. Ugonjwa/ Makanjo
31. Leo
32. Kesho
33. Saa/Saa ngapi (= what is the time?)
34. Labda
35. Chumba (cha hoteli = hotel room)
36. Kitanda
37. Singishi
38. Duka
39. Stesheni (ya gari la moshe) (= railway station)
40. Baddalawisha
41. Pesa
42. Gari/ Motokaa
43. Chakula
44. Mazewa
45. Cheese
46. Mazewa lala
47. Marie
48. Yaama
49. Yanya
50. Sukhari
51. Maji (ya kunywa) (= drinking water)
52. Unajua kusema Kiingereza?

Gujrati

1. Namastey
2. Namastey
3. Awejaw
4. Tamaru naam soo cha?
5. Tame kam chu?
6. Katla?
7. Bohu monghu
8. Katlu dour?
9. Maf kar sho
10. Mahirbani
11. Aabhar
12. Hoondil gir cho
13. Ah soo cha?
14. Kia?
15. Kiara?
16. Qua rastea?
17. Tamre pasa cha?
18. Mane yoya cha
19. Saru
20. Kaarab
21. Tuttalu
22. Tame samji shoko cho?
23. Hoon nathe samji shukto
24. Haan
25. Na
26. Hoon
27. Tame
28. Easpital
29. Daaktar
30. Bimarr
31. Arge
32. Cull
33. Samai
34. Cadache
35. Ordo
36. Khathallo
37. Oongh
38. Dukan
39. Stationa
40. Baddaluu
41. Paisa
42. Motor gardi
43. Anaj
44. Dood
45. Paneer
46. Dahee
47. Eanda
48. Mas
49. Tamato
50. Khan
51. Paani
52. Tamei Angrighe bolo cho?

Malay/Indonesian °

1. Selamat hari
2. Selamat petang
3. Selamat jalan (MS)/ Selamat tinggal (MN/I)
4. Apakah nama anda?/ Apakah nama awak? °°
5. Apa kabar?
6. Berapa harganya?
7. Mahal sangat
8. Berapa jauh? (MS)/ Berapa jauhnya? (MN/I)
9. Minta maaf
10. Silakan
11. Terimah kaseh
12. Ma'afkan saja
13. Apa-kah ini?
14. (Di) mana?
15. Bila?
16. Jalan mana?
17. Adakah anda ada?
18. Saya mahu
19. Baik
20. Jahat (people) Busuk (fruit)
21. Pechah (glasses) (MS)/ Pecah (glasses) (MN/I)/ Patah (legs)
22. Adakah anda faham?
23. Saya tidak mengerti/ saya tidak faham
24. Ya (MS/MN)/ Ja (I)
25. Tidak
26. Saya
27. Anda (written or spoken)/ Awak (spoken only)
28. Ospital (MS/MN)/ Rumah sakit (I)
29. Doktor (MS/MN)/ Tukang ubat (I)
30. Sakit
31. Hari ini
32. Esok
33. Masa
34. Barangkali
35. Bilik
36. Katil
37. Tidur
38. Kedai
39. Stesyen
40. Tukar
41. Wang/Duit
42. Kereta
43. Makanan
44. Susu
45. Keju
46. Yogat
47. Telur
48. Daging
49. Buah tomato/ Labu (old Kelantan village word)
50. Gula
51. Air (minum)
52. Anda bercakap Bahasa Inggeris, kah? (MS/MN)/ Ja bercakap Bahasa Inggeris, kah? (I)

° (MS) Malay, South; (MN) Malay, North; (I) Indonesian. Indonesian is very similar to Malay, especially to that spoken in northern Malaysia.
°° Anda is the polite form for 'you'; Awak is less formal and is used in village Malay.

ARABIC NUMERALS

	is written as	
1		١
2		٢
3		٣
4		٤
5		٥
6		٦
7		٧
8		٨
9		٩
10		١٠
11		١١
12		١٢
13		١٣
14		١٤
15		١٥
16		١٦
17		١٧
18		١٨
19		١٩
20		٢٠
21		٢١
22		٢٢
30		٣٠
40		٤٠
50		٥٠
60		٦٠
70		٧٠
80		٨٠
90		٩٠
100		١٠٠
200		٢٠٠
1000		١٠٠٠

International Time Comparison

Hours ahead (+) or behind (−) Greenwich Mean Time (GMT)

Argentina	− 3
Australia	
New South Wales	+10
Queensland	+10
South Australia	+ 9.30
Tasmania	+10
Victoria	+10
Western Australia	+ 8
Austria	+ 1
Bahrain	+ 3
Belgium	+ 1
Brazil	− 3
Canada	
Eastern Time	− 5
Central Time	− 6
Mountain Time	− 7
Pacific Time	− 8
Chile	− 4
mid-Oct — mid Mar	− 3
Cook Islands	− 9.30
Denmark	+ 1
Fiji	+12
Finland	+ 2
France	+ 1
Germany	+ 1
Gilbert Islands	+12
Greece	+ 2
Hong Kong	+ 8
India	+ 5.30
Indonesia	+ 7
Iran	+ 3.30
Ireland	GMT
Israel	+ 2
Italy	+ 1
Japan	+ 9
Korea, South	+ 9
Kuwait	+ 3
Malaysia	+ 7.30
Mexico	− 6
Nepal	+ 5.40
Netherlands	+ 1
New Caledonia	+11
New Hebrides	+11
New Zealand	+12
Norfolk Island	+11.30
Norway	+ 1
Pakistan	+ 5
Papua New Guinea	+10
Peru	− 5
Philippines	+ 8
Samoa (American and Western)	−11
Singapore	+ 7.30
Solomon Is.	+11
South Africa	+ 2
Spain	+ 1
Sri Lanka	+ 5.30
Sweden	+ 1
Switzerland	+ 1
Tahiti	−10
Taiwan	+ 8
Thailand	+ 7
Tonga	+13
United Arab Emirates	+ 4
United Kingdom	GMT
mid-Mar — late Oct	+ 1
USA	
Eastern Time	− 5
Central Time	− 6
Mountain Time	− 7
Pacific Time	− 8
Hawaiian Islands	−10
USSR	
Moscow	+ 3
Yugoslavia	+ 1

Off the Beaten Track Guide to Rainy Seasons

Within each region, the places mentioned in the table are arranged in order of decreasing latitude north of the equator, increasing latitude south of the equator. This is a reminder that at any given time of the year opposite seasons are to be found north and south of the equator. December to February, for example, bring winter to the northern hemisphere, summer to the southern hemisphere. In the belt stretching about 5°-10° north and south of the equator, the equatorial climate tends to prevail: the seasons are almost indistinguishable from each other and rain, broadly speaking, more evenly spread throughout the year than elsewhere. But a lot depends on altitude and other features of geographical location; proximity to the sea or to mountains, and the nature of prevailing winds and currents.

The places listed are representative of different types of climate: this means that they are not necessarily typical of other places within the same region or country. And these places represent only a minute sample globally. Total annual rainfall should always be taken into account, since the rainy season in one place may be less wet than the dry season in another. At best, this table can be a very rough guide only.

+ represents a month having more than 1/12 of the annual total rainfall.
– represents a month having less than 1/12 of the annual total rainfall.
● indicates the month(s) with the highest average rainfall of the year.

	Total annual rainfall, cm.	J	F	M	A	M	J	J	A	S	O	N	D	*Latitude*
Asia														
Tokyo, Japan	156.5	–	–	–	+	+	+	+	+	●	+	–	–	35° 45'N
Nagasaki, Japan	191.8	–	–	–	+	+	●	+	+	+	–	–	–	32° 47'N
Hankow, China	125.7	–	–	–	+	+	●	+	–	–	–	–	–	30° 32'N
Delhi, India	64.0	–	–	–	–	–	+	●	●	+	–	–	–	28° 38'N
Cherrapunji, India	1079.8	–	–	–	–	+	●	●	+	+	–	–	–	25° 17'N
Calcutta, India	160.0	–	–	–	–	+	+	●	●	+	–	–	–	22° 36'N
Hong Kong	216.1	–	–	–	–	+	●	●	●	+	–	–	–	22° 11'N
Mandalay, Burma	82.8	–	–	–	–	●	●	+	+	+	+	–	–	22° 0'N
Hyderabad, India	75.2	–	–	–	–	–	+	+	+	●	+	–	–	17° 10'N
Rangoon, Burma	261.6	–	–	–	–	+	+	●	+	+	–	–	–	16° 45'N
Bangkok, Thailand	139.7	–	–	–	–	+	+	+	+	●	+	–	–	13° 45'N
Mangalore, India	329.2	–	–	–	–	–	●	●	+	–	–	–	–	12° 55'N
Colombo, Sri Lanka	236.5	–	–	–	+	●	+	–	–	–	●	●	–	6° 56'N
Sandakan, Malaysia	314.2	●	+	–	–	–	–	–	–	–	–	+	+	5° 53'N
Singapore	241.3	●	–	–	–	–	–	–	–	–	+	●	●	1° 17'N
Djakarta, Indonesia	179.8	●	●	+	–	–	–	–	–	–	–	–	+	6° 9'S
Africa														
Zungeru, Nigeria	115.3	–	–	–	–	+	+	+	+	●	–	–	–	9° 45'N
Harar, Ethiopia	89.7	–	–	+	+	+	+	+	●	+	–	–	–	9° 20'N
Freetown, Sierra Leone	343.4	–	–	–	–	–	+	●	●	+	+	–	–	8° 30'N
Lagos, Nigeria	183.6	–	–	–	–	+	●	+	–	–	+	–	–	6° 25'N
Accra, Ghana	72.4	–	–	+	+	+	●	–	–	–	+	–	–	5° 35'N
Mongalla, Sudan	94.5	–	–	–	+	+	+	●	+	+	+	–	–	5° 8'N
Libreville, Gabon	251.0	+	+	+	+	+	–	–	–	–	+	●	+	0° 25'N
Entebbe, Uganda	150.6	–	–	+	●	+	–	–	–	–	–	+	–	0° 3'N
Nairobi, Kenya	95.8	–	–	+	●	+	–	–	–	–	–	+	+	1° 20'S
Mombasa, Kenya	120.1	–	–	–	+	●	+	–	–	–	–	–	–	4° 0'S
Luluabourg, Zaïre	158.2	+	+	+	+	–	–	–	–	–	+	●	●	5° 55'S
Salisbury, Zimbabwe	82.8	●	+	+	–	–	–	–	–	–	–	+	+	17° 50'S
Tamatave, Madagascar	325.6	+	+	●	+	–	+	+	–	–	–	–	–	18° 2'S
Beira, Mozambique	152.2	●	+	+	–	–	–	–	–	–	+	+	+	19° 50'S
Australasia and South Pacific														
Tulagi, Solomon Islands	313.4	+	●	+	–	–	–	–	–	–	–	–	+	8° 0'S
Port Moresby, Papua New Guinea	101.1	+	●	+	+	–	–	–	–	–	–	–	+	9° 24'S
Thursday Is., Australia	171.5	●	+	+	+	–	–	–	–	–	–	–	+	10° 30'S
Darwin, Australia	149.1	●	+	+	–	–	–	–	–	–	–	+	+	12° 20'S
Cairns, Australia	225.3	+	+	●	+	–	–	–	–	–	–	–	+	16° 55'S
Suva, Fiji	297.4	+	+	●	–	–	–	–	–	–	–	–	+	18° 0'S

	Total annual rainfall, cm.	J	F	M	A	M	J	J	A	S	O	N	D	*Latitude*
C. America, W. Indies, S. America														
Mazatlan, Mexico	84.8	–	–	–	–	–	–	+	+	●	–	–	–	23° 10′N
Havana, Cuba	122.4	–	–	–	–	+	+	+	+	+	●	–	–	28° 8′N
Kingston, Jamaica	80.0	–	–	–	–	+	+	–	+	+	●	+	–	18° 0N
Salina Cruz, Mexico	102.6	–	–	–	–	–	–	+	+	●	–	–	–	16° 10′N
Dominica, Leeward Islands	197.9	–	–	–	–	–	+	●	+	+	+	+	–	15° 20′N
Tegucigalpa, Honduras	162.1	–	–	–	–	–	●	+	–	+	+	–	–	14° 10′N
Caracas, Venezuela	83.3	–	–	–	–	+	●	●	●	●	●	+	–	10° 30′N
Balboa Heights, Panama	177.0	–	–	–	–	+	+	+	+	+	●	●	–	9° 0′N
Ciudad Bolivar, Venezuela	101.6	–	–	–	–	+	+	●	+	+	–	+	–	8° 5′N
Georgetown, Guyana	225.3	+	–	–	–	+	●	+	–	–	–	–	+	6° 50′N
Bogotá, Colombia	105.9	–	–	+	+	+	–	–	–	–	●	+	–	4° 34′N
Quito, Ecuador	112.3	+	+	+	●	+	–	–	–	–	–	–	–	0° 15′S
Belém, Brazil	243.8	+	●	●	+	+	–	–	–	–	–	–	–	1° 20′S
Guayaquil, Ecuador	98.5	+	+	●	+	–	–	–	–	–	–	–	–	2° 15′S
Manaus, Brazil	181.1	+	+	●	–	–	–	–	–	–	–	+	+	3° 0′S
Recife, Brazil	161.0	–	–	+	+	+	●	+	+	–	–	–	–	8° 0′S
Bahia, Brazil	190.0	–	–	–	●	+	+	+	–	–	–	–	–	13° 0′S
Cuiabá, Brazil	139.5	+	+	●	–	–	–	–	–	–	+	+	+	15° 30′S
La Paz, Bolivia	57.4	●	+	+	–	–	–	–	–	–	–	+	+	16° 20′S
Rio de Janeiro, Brazil	108.2	+	+	+	+	–	–	–	–	–	–	+	●	23° 0′S
Sao Paulo, Brazil	142.8	+	+	+	–	–	–	–	–	–	+	+	●	23° 40′S
Tucumán, Argentina	97.0	●	+	+	–	–	–	–	–	–	–	+	+	26° 50′S
Valdivia, Chile	260.1	–	–	–	+	+	●	+	+	–	–	–	–	39° 50′S

Pollo al Jerez (for 4)

1 chicken, jointed.
2 chopped onions
3 cloves garlic, mashed
2 bay leaves
1 tsp. oregano
½ tsp chilli powder
24 stuffed olives
2 glasses sherry
cornflower
(1 can tomatoes)

Simmer chicken pieces in water. Add spices. Simmer 30 mins more. Add sherry & olives. Serve with rice

Picadillo (for 6)

2 lb mince
1 large green pepper, chopped
1 clove garlic
1 large onion
2 large tomatoes
½ tsp cumin
½ tsp chilli
½ tbs jamaica pepper
2 oz raisins
1 tbsp capers
2 tbsp stuffed olives
oil.

Fry onion + Meat. Add rest - Simmer. Serve w. rice, black beans + plantains (fried).

Meat Stew & Aubergines. (for 4)

2 medium aubergines. Chopped.
1 large onion
$1\frac{1}{2}$ lb beef, veal or lamb, cubed.
2-3 tomatoes, cut.
1 tbsp tomato concentrate.
juice of 1 lemon.
$\frac{1}{2}$ tsp. cumin.
$\frac{1}{2}$ tsp allspice (jamaica pepper)
salt or beef cube., oil.

Simmer $1\frac{1}{2}$ hrs.

Goulash. (for 4)

1 lb shin.
2 large onions
2 cloves garlic
1 tin tomatoes
paprika 1-2 tsp.
1 tsp caraway seeds
$\frac{1}{4}$ pt Hot water
$1\frac{1}{2}$ lbs tatties
oil, salt, pepper
Sour cream

Simmer meat 2 hrs. Add sour cream.
Boil potatoes.

(Could use mince for faster results)

Salade niçoise (for 4)

1 iceberg lettuce
1 thinly chopped onion
2 sliced tomatoes
¼ sliced cucumber
1 sliced green pepper
12 black olives, pitted
4 anchovy fillets
1 7oz can tuna
1 tbs fresh basil
1 clove garlic, crushed.
4 boiled eggs
1 can haricot beans.

Vinaigrette:

¼ tsp. salt, pepper, mustard.
3 tbs good olive oil
1 tbsp vinegar.
parsley, basil

Serve with hot French bread & garlic butter.

Ragù (for 4)

8 oz mince
4 oz chicken livers
3 oz bacon, diced.
1 carrot
1 onion
1 stick celery
1 can tomato puree
1 glass white wine
2 glasses stock
Salt, pepper, nutmegs, oregano.

Simmer 30 minutes.
Serve with spaghetti.